World Regional Geography Book Series

Series Editor
Eduardo F. J. De Mulder

What does Finland mean to a Finn, Sichuan to a Sichuanian, and California to a Californian? How are physical and human geographical factors reflected in their present-day inhabitants? And how are these factors interrelated? How does history, culture, socio-economy, language and demography impact and characterize and identify an average person in such regions today? How does that determine her or his well-being, behaviour, ambitions and perspectives for the future? These are the type of questions that are central to The World Regional Geography Book Series, where physically and socially coherent regions are being characterized by their roots and future perspectives described through a wide variety of scientific disciplines. The Book Series presents a dynamic overall and in-depth picture of specific regions and their people. In times of globalization renewed interest emerges for the region as an entity, its people, its landscapes and their roots. Books in this Series will also provide insight in how people from different regions in the world will anticipate on and adapt to global challenges as climate change and to supra-regional mitigation measures. This, in turn, will contribute to the ambitions of the International Year of Global Understanding to link the local with the global, to be proclaimed by the United Nations as a UN-Year for 2016, as initiated by the International Geographical Union. Submissions to the Book Series are also invited on the theme 'The Geography of…', with a relevant subtitle of the authors/editors choice. Proposals for the series will be considered by the Series Editor and International Editorial Board.

This book series is published in cooperation with the International Geographical Union (IGU). The IGU is an international, non-governmental, professional organization devoted to the development of the discipline of Geography. The purposes of the IGU are primarily to promote Geography through initiating and coordinating geographical research and teaching in all countries of the world.

More information about this series at http://www.springer.com/series/13179

Eduardo F. J. De Mulder • Ben C. De Pater
Joos C. Droogleever Fortuijn

The Netherlands and the Dutch

A Physical and Human Geography

In cooperation with Len A. De Klerk
and Jerry Van Dijk

Springer

Eduardo F. J. De Mulder
Delft Technical University (Emeritus)
Haarlem, The Netherlands

Joos C. Droogleever Fortuijn
Department of Geography, Planning
and International Development Studies
University of Amsterdam
Amsterdam, The Netherlands

Ben C. De Pater
Department of Human Geography and
Spatial Planning
Utrecht University
Utrecht, The Netherlands

ISSN 2363-9083 ISSN 2363-9091 (electronic)
World Regional Geography Book Series
ISBN 978-3-030-09141-5 ISBN 978-3-319-75073-6 (eBook)
https://doi.org/10.1007/978-3-319-75073-6

Printed on acid-free paper

This Springer imprint is published by the registered company Springer International Publishing AG part of Springer Nature.
The registered company address is: Gewerbestrasse 11, 6330 Cham, Switzerland

Acknowledgments

Apart from the essential contributions provided by Len A. De Klerk and Jerry Van Dijk who contributed a full chapter to this book, we wish to acknowledge contributions to the contents provided by the following persons: Aat Barendregt, Meindert Van De Berg, Mart Van Bracht, Frans Claessen, Maurice Hanegraaf, Jan Jellema, and Peter Vos.

A special word of thanks is expressed to Ton Markus and Margot Stoete (Faculty of Geosciences, Utrecht University) who prepared all drawings and most of the figures in this book.

We also wish to thank Michiel Van Der Meulen, Nikki Trabucho, and Mart Zijp (Geological Survey of the Netherlands [TNO]) for granting their permissions to publish figures.

The authors, February 2018

Contents

About the Authors

Eduardo F. J. De Mulder is a geologist, and throughout his professional career, he has been employed by the Geological Survey of the Netherlands (TNO) in various positions, from field geologist to manager of the Sustainable Management of the Subsurface business unit. He was a professor at Delft Technical University. Internationally, Eduardo F. J. De Mulder has been president of the International Union of Geological Sciences, initiator and executive director of the UN-proclaimed International Year of Planet Earth, and the director of the Earth Science Matters Foundation. He has published 135 scientific publications and books. He is Springer's editor of the World Regional Geography series.

Ben C. De Pater is a human geographer and associate professor at the Department of Human Geography and Spatial Planning, Faculty of Geosciences, Utrecht University. He was editor in chief of the journal *Geografie*, published by the Royal Dutch Geographical Society, in 1992–1999. He is the (co)author and (co)editor of (text)books in Dutch about the history and theory of geographical sciences, Europe, North America, beach resorts, Randstad Holland, the geographical history of the Netherlands, and more. He was coauthor of the *Comprehensive Atlas of the Netherlands 1930–1950* (2005) and scientific advisor for the Bosatlas van Nederland (2007).

Joos C. Droogleever Fortuijn is a human geographer and was associate professor and chair of the Department of Geography, Planning and International Development Studies of the University of Amsterdam. She is the first vice president of the International Geographical Union. She published on urban geography, rural geography, gender, ageing, and geography in higher education. The main topics in her publications relate to the gendered aspects of activity involvement and networks of families with children and older women and men from an urban-rural comparative perspective.

Len A. De Klerk worked in various cities as a planning officer, and he retired as a professor of urban and regional planning at the University of Amsterdam. He wrote articles on practical planning issues and several books and book chapters on the history of Dutch city planning in the nineteenth century. He is coauthor of The Self-made Land: Culture and Evolution of Urban and Regional Planning in The Netherlands (2012).

Jerry Van Dijk is assistant professor restoration ecology and biodiversity conservation at the Copernicus Institute of Sustainable Development at Utrecht University, the Netherlands. He graduated as an environmental scientist and holds a PhD in wetland restoration ecology. He was editor-in-chief of Landschap, the scientific journal of the Dutch Society for Landscape Research, from 2009 to 2012. His research and teaching focus on the impact of environmental change on ecosystems, community dynamics, and biodiversity in the context of nature conservation.

1 Introduction

Eduardo F. J. De Mulder

Abstract

The Physical and Human Geography of the Netherlands has been written as a model for Springer's World Regional Geography Book Series. This Book Series aims to address physically and socially coherent regions that are characterised by common roots and future perspectives as described through a wide variety of scientific disciplines according to the best traditions of modern Geography. Geography is a unique scientific discipline indeed, encompassing about the entire spectrum of sciences that search for understanding and description of the complexity of Planet Earth, its natural phenomena, their development over time and its people. The wide range of scientific disciplines involved in this ambition include two main fields, i.e. those that might best be labelled as Physical Geography and those that best fit under the umbrella of Human Geography. Both fields are well developed in the International Geographical Union (IGU) that was founded in 1922 in Brussels, Belgium.

This Springer Book Series does not just has the ambition to properly describe the lands (and waters) and people of Planet Earth, it also aims to identify where and how physical and human geographical issues interact. The Netherlands might perhaps serve as an example par excellence as the landscape is almost entirely shaped by humans and some features of Dutch society and its people may be (at least partly) explained by its roots in the delta landscapes.

Keywords

Physical and human geography · Evolution of the netherlands · Interaction nature and society · Landscaping

E. F. J. De Mulder (✉)
Delft Technical University (emeritus), Haarlem, Noord-Holland, The Netherlands
e-mail: e.demulder@planet.nl

1.1 The Past

Most of the Netherlands came into existence only when the ice masses and glaciers of the last Glaciation had retreated by the onset of the Holocene about 10,000 C^{14} years Before Present (BP) which is roughly 11,500 calendar years ago. By that time, the Noordzee/North Sea, dividing the UK from the Netherlands, did not yet exist and early Dutchmen and Brits were hunting for deer at the floor of the current Noordzee, while River Thames was a branch of River Rijn/Rhine. Deglaciation generated sea level rise and the North Sea was filled with seawater from its deeper parts in the south (The Channel) and the north. It took several millennia to fill the Noordzee entirely which was completed about 7000 years BP. By that time agriculture had arrived in this part of Europe and the first farmers settled in the high sand areas. The Noordzee expanded eastwards and flooded large areas in the current west, centre and north of the country until this marine area reached its maximum extension some 5800 years BP. Next, sea level rise slowed down somewhat and sets of westward migrating sandy beach barriers developed eventually closing the Dutch coastline some 1000 years later while leaving a few tidal inlets. Protected by these beach barriers swampy conditions developed east of them and peat began to grow successively, initially in the lowest places as fen peat, later also as high moor peat bogs fed by rainwater only.

Most of the Netherlands turned into one big swamp for millennia. Human settlements there could only develop along river banks, around river mouths, tidal inlets and on dwelling mounds at northern shores. People lived from sea food, some agricultural products and from trade via rivers or overseas. Peat bogs continued to grow in most of the Netherlands well into Roman Times (2000 BP) and even later. When the Romans arrived in the south and centre of the delta, flood hills in the north already were well populated and relatively prosperous. Roman occupation caused intensive interaction with Mediterranean and other cultures and introduction of

E. F. J. De Mulder et al., *The Netherlands and the Dutch*, World Regional Geography Book Series,
https://doi.org/10.1007/978-3-319-75073-6_1

new techniques in the delta. Population density increased, agriculture intensified, larger areas were cultivated and many forests were cut. Romans dug the first canals and deferred water courses in the Dutch delta while for the first time really interfering in the still more or less natural delta landscapes. Meanwhile, the (almost) uninterrupted natural coastal defence system began to breach, first in the southwest (Zeeland), and new tidal inlets found their ways eastward. People moved away from the delta when the Romans left, about 350 AD.

Some 600 years later, around 1000 AD, coastal erosion intensified and breached the chain of natural beach barriers further northward, in Holland, Groningen and Friesland where new tidal inlets developed. Simultaneously, the early Dutch began developing, draining and excavating the vast volumes of accumulated peat behind the beach barriers which caused these peat lands to subside considerably (Chap. 3). In turn, that increased vulnerability for marine flooding into the hearth of the Dutch delta. By then, the the first river dikes and sluices to manage water supply and discharge in the upcoming towns in the west and centre of the delta were built. Such interventions in the water system of the delta caused further drainage and settlement of the peat lands which further reduced resistance against flooding. Expanding cities from the fifteenth century onwards required more fuel that was provided by more systematic and large-scale excavation of the peat lands. The former peat area behind the coastal dunes eventually turned into a landscape of shallow lakes between remnants of peat lands.

People in the Dutch delta had always been quite independent and preferred remaining united in smaller entities (cities, provinces) rather than in a centralised state. In Medieval Times, the area of the current countries of the Netherlands and Belgium were a decentralised society of farmers and autonomous towns, cloisters and nobility. After the Dutch Revolt (1568–1648), the independence war against King Philips II of Spain who was the sovereign of the Dutch provinces, the northern part of the low lands became an independent federal state of seven provinces with some major, rather autonomous cities. Provinces and cities only united against joint enemies in times of war or when they were threatened by natural hazards as floods. Major dike constructions as early as the twelfth and thirteenth centuries were feasible only by the joint efforts of landowners, e.g. cloisters, and individual local farmers. Together, they also reclaimed the numerous lakes in the peat areas, supported by new power generating technologies as windmills and their improved successors. Such costly operations were possible when the Netherlands, and in particular the west (Holland), had become an economic powerhouse in the seventeenth century, the Dutch Golden Age. Former lake floors were turned into new agricultural lands. Also the higher sands in the east, south and north and the river belt were further developed for food production. Agriculture and cattle breeding were intensified and their products were transported all over the delta and beyond.

After Napoleonic occupation the northern and southern parts of the Netherlands (including current Belgium) were finally united in 1815 in the centralised Kingdom of the Netherlands. The current nation state of the Netherlands dates from secession of Belgium in 1830. Infrastructure rapidly developed in the nineteenth century by digging networks of canals, followed by railways, tramways and, in the twentieth century, a dense network of roads (Chap. 9).

Flood protection and food production remained the prime targets for the Dutch also during the past century. Afsluitdijk (1932) was built to protect the Zuyderzee area against more marine floods and the catastrophic 1953 flood in Zeeland triggered construction of the Delta Works in the southwest (Chap. 2). Large parts of IJsselmeer (former Zuyderzee) were reclaimed and to further increase agricultural efficiency, country-wide, vast land areas were reshuffled. However, in the last 50 years food production grew faster than population and the Netherlands became one of the world's leading exporting nations for agricultural products. Less land was then needed for food production and with growing interest in nature conservation, increasing areas of land were turned into nature. Growing populations and increasing needs for space for residential use, industry and infrastructure resulted in an increase of the built environment in the Netherlands from 2% in 1900 to 15% in 2017 (Chap. 8).

1.2 Interactions Between Nature and Society

From this brief history it may appear that the Netherlands is perhaps somewhat atypical for pointing at interactions between Physical and Human Geographical phenomena in general. Nevertheless, some of the observed interrelations are given below.

As the Dutch delta developed in a vast and inaccessible, swampy area some five millennia ago, early Dutch people settle only along their fringes, at river mouths, along rivers, at tidal inlets and on flood hills along the north coast. Such sites proved to be fine trading locations from where goods would rather easily be transported overseas and via rivers to the delta's hinterlands. That made such sites relatively prosperous and attracted more inhabitants. As such towns and regions were faced by tidal floods on the one hand and by emerging groundwater on the other due to land subsidence through drainage and reclamation, dikes were built for safety reasons and sluices for water discharge. Such prosperous and busy sites attracted more traders and various, often navigation-related industries and, in addition, societal meeting places as churches and a town hall. The maritime position of the Netherlands provided the basis for a trade-centred, globalised economy (Chap. 6).

Dutch cities as Amsterdam and Rotterdam (Fig. 1.1) thank their origin and (former) prosperity to their geographic position at the interface between land and sea (Chap. 2). Amsterdam developed as a major trade and financial centre in the Hanseatic and East Asian trades and emerged as one of the most prominent cities in the world, in the seventeenth century. Thanks to industrialisation of the German Ruhr area in the hinterland of the Dutch delta, Rotterdam became the largest harbour city of Europe in the twentieth century and, during several decades after WWII, the largest harbour in the world (Chap. 7). Both cities are part of the larger Randstad Holland, a ring of cities and towns in the maritime west of the nation with currently some 7 million people, 40% of the Dutch population (Chap. 8).

The Dutch have been fighting water threats since they settled in the delta. Initially they built small dikes to protect individual family houses from flooding but they soon learned that this would only be effective if done jointly to save their houses, cattle and lands as a community. These people were independent, free farmers who joined their forces to fight common enemies, either in conflicts of war or against hostile natural events, as floods. Already by 1232, such cooperation for flood protection had been institutionalised through Water Boards: supra-regional, public bodies responsible for flood prevention. Water Boards are the oldest Dutch democratic institutions ran by elected individuals and they still exist. Natural water threats evoked a Dutch tradition of cooperation and a spirit of finding solutions together for common problems (Chaps. 2 and 10).

Dutch swamps may have been inaccessible for millennia, these peat lands also provided great opportunities for exploitation, for both salt production as in Zeeland and for providing fuel to cities in Holland. Digging peat for fuel developed into major and systematic operations from the tenth century onwards, peaking in the sixteenth to eighteenth century when in particular Holland was economically powerful and attracted many immigrants. Abundant availability of relatively cheap energy considerably contributed to Dutch prosperity. Peat has been sold for fuel as late as the 1960s although most people had then switched to coal.

A comparable situation where physical conditions (availability of natural resources) have had major impact on society has been the coal resources in south Limburg and the societal impact both of opening (about 1900–1915) and of closing (1963–1974) of these mines. A similar case has been the exploitation of giant natural gas resources in the deep subsurface of Groningen. The Dutch have benefited greatly from these energy resources and from its additional income for the State but Groningers paid a serious price as they faced the negative impacts from earthquake risks (Chap. 4).

Water played a decisive role in Dutch freedom fights against Spanish troops in the seventeenth century, on several occasions. For example, the Westerschelde, the vital access route from the North Sea to Antwerpen, was controlled by Dutch ships since re-occupation of the city by the Spanish army in 1585. Control of the Westerschelde, in combination with other historical factors, may eventually have been the impetus for the northward shift of major overseas trade along the delta and the emergence of political and economic power of Amsterdam (Chap. 7).

Interaction between society and the physical conditions of the land has also been obvious when first local towns and farmers, later groups of rich merchants and, from the nineteenth century onwards also the public sector, reclaimed the shallow peat lakes in the west and centre of the Dutch delta. These lakes were drained by windmills first and former lake floors then turned into new lands (polders) while adding to the nation's security against frequent marine floods. Such new lands were aimed to be used for agriculture, a potentially profitable target for investors by then.

The Dutch river belt comprises the east-west running downstream parts of Rivers Rijn and Maas. This belt is some 10 km wide and has split the country for long until railway and later road bridges were built. Before that, the river belt could only be crossed by boat. This physical divide coincides with a major cultural distinction between people living south, in the provinces of Noord Brabant and Limburg, and north of this divide. People in the south generally adhered to the Roman Catholic religion while those north of the river belt were often Protestants (Reformed or orthodox). Apart from religious distinctions people in both regions often differ in other cultural aspects as demography, education level and voting behaviour (Chaps. 8, 11, 12, and 13).

In their attempts to improve and rationalise delta lands for agricultural purposes the Dutch have interfered quite significantly in the landscape and soils. Since the early twentieth century, parcel re-allotment has been a legal and formal way to exchange pieces of land among landowners and farmers to arrive at larger, coherent land areas that could be elaborated more efficiently. Simultaneously, winding waterways and roads were stretched, differences in height were equalised and the lands were deeply ploughed. Production certainly increased but many original (geomorphological) features got lost. Today, this type of landscaping is no longer practised as appreciation for past landscapes and biodiversity has significantly grown since and historical landscape preservation has become a serious policy issue now (see Chaps. 3 and 5).

Fig. 1.1 Main topographical features of the Netherlands. For details and modifications over time (between 1815 and 2016) one is referred to http://www.topotijdreis.nl/

A final example where physical and human geography meets in the Netherlands is described in Chap. 3 and concerns de-poldering, the opposite of land reclamation. To maintain or increase nature values, nature compensation for infrastructural or for other development works has become a regular policy item in the Netherlands. To compensate for deepening of the Westerschelde estuary in the southwest to accommodate for larger ships to enter Antwerpen harbour, new nature had to be developed in the region. But not many options remained. Eventually, the *Hertogin Hedwigepolder* was assigned to be flooded which generated significant debate. The dikes surrounding that polder will be cut and flooding is anticipated by 2019. In fact, this is an example of double human impact on nature, first that polder has been reclaimed from the sea in 1907, a century later this piece of agricultural land is given back to the sea (nature) due to strongly modified views on human values about the physical environment.

Several more such interactions are described in the 14 Chapters of this book. Chapters 2, 3, 4, and 5 mainly deal with the physical geographical aspects of the Netherlands while Chaps. 6, 7, 8, 9, 10, 11, 12, and 13 mainly deal with the human geography of the Dutch. Chapter 14 depicts the possible future of the Netherlands based on both long-term physical processes and trends in Dutch society. In the future, as in the past, water will be a major player in the interaction between nature and society.

Water

2

Eduardo F. J. De Mulder

Abstract

Water is an issue of prime relevance in the Netherlands. The Dutch have been combatting river and sea floods since they settled in the Noordzee delta of Rivers Rijn, Maas and Schelde. Flood protection has been and still is the main tool for their survival. Ancient and modern types of dike construction will be discussed in this chapter. In addition to flood protection against aggressive rivers and the sea, the Dutch also reclaimed significant parts of their territory from the sea and from shallow lakes. This resulted in a low-lying country which surface level is for 25% below sea-level.

The Dutch did not only combat but also heavily depend on water, both on surface water and on groundwater. Fresh groundwater is their main source of drinking water, their industries and for their agriculture. Water is also essential to maintain vital ecosystems. Environmental protection, combatting contamination and (wet) nature conservation are important issues in Dutch policy. This multiple attitude concerning water requires cautious management. Control over water issues at both national, regional and local level has always been and still is a matter of significant attention in Dutch management and water management is well advanced in the Netherlands. Anticipating future developments, including impact of climatic change, is integral part of Dutch policy making and European legislation.

Keywords

Flood protection · Land reclamation · Drainage · Delta Works · Land subsidence · Groundwater · Water management · Sea level rise

E. F. J. De Mulder
Delft Technical University (emeritus), Haarlem, Noord-Holland, The Netherlands
e-mail: e.demulder@planet.nl

The Netherlands lays on a delta of three main rivers in the Noordzee. ‘Water’ is probably the first issue that would come up in one’s mind when thinking about the Netherlands. Indeed, the Netherlands are rooted in water. Figure 2.1 shows the part of the nation that is below sea-level. The lowest point in the Netherlands is east of Rotterdam: 6.7 m below Mean Sea Level (MSL). Its highest point is in south Limburg: 322 m above MSL. About 65% of the land would be flooded if all natural and artificial water defence and pumping systems would fail (NHV 1998). The Dutch are in a permanent battle to protect themselves against flooding and their government spends considerable sums annually to maintain and improve such systems. Large parts of these low lying areas were reclaimed from the sea. This low topographic situation is well reflected in the name of the country: the Netherlands, which means: Low Lands.

Such battles and numerous measures taken over the past millennium and longer resulted in a country able to address water issues quite properly and in a balanced way. But water may still jeopardise the Dutch delta and from three directions: horizontally by the sea and the rivers, vertically from below by excessive or shortage of groundwater, and from above by (excessive or scarcity of) precipitation. However, water is not just a threat, the Dutch also benefit from abundance of water in their country. Water refreshes the population by providing drinking water. (Fresh) water is essential for many of the Dutch industries, not only for beer breweries (Box 2.6). Coping with water has been a leading issue throughout Dutch history. The Dutch historian Huizinga stated that their permanent battle against water made the Dutch united and shaped their national identity. Dutch fascination for water is also reflected in its culture, for example in the Dutch ambitions in water sports and through numerous expressions and sayings in the Dutch language (Chap. 12). Here, we discuss two main realms of water: Surface water and Groundwater. In addition, information about precipitation, water quality, contamination and water management & policy in the Netherlands will be provided by the end of this chapter.

E. F. J. De Mulder et al., *The Netherlands and the Dutch*, World Regional Geography Book Series,
https://doi.org/10.1007/978-3-319-75073-6_2

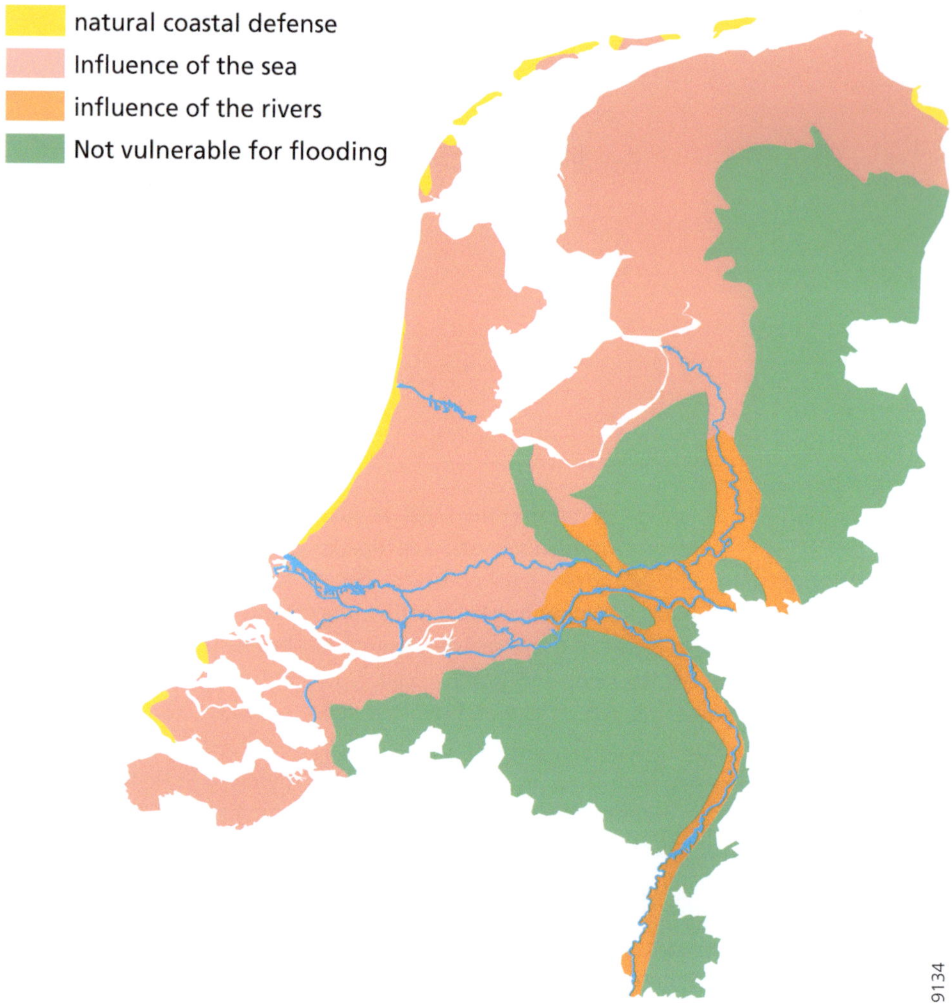

Fig. 2.1 Vulnerability for flooding of the Netherlands; without pumping, dikes and other measures these lands would be drowned. (Source: After NHV 1998)

2.1 Surface Water

A considerable part of the Netherlands' territory is covered by water: over 17% (7160 km^2; Donkers and Monincx 2007). This does not only include all fresh water bodies as rivers, lakes, ditches, canals, but brackish and fully marine water bodies too, as Waddenzee and sea arms as Dollard, Oosterschelde and Westerschelde. Most of these water bodies occur in the low southwest, west and north of the nation. The European Water Framework Directive states that most Dutch water bodies are highly modified and/or man-made water, very few are more or less natural water bodies, as the estuarine waters, upstream River Maas and some streams. Current and former estuarine water bodies constitute most of Dutch surface water bodies. Fresh water ditches occur in dense networks almost throughout the entire Dutch territory.

Water flows. Water volumes entering and leaving the Dutch delta by rivers and canals fluctuate and are controlled by precipitation, at least to some extent (Sect. 2.3). In a particularly dry year (e.g. 1976), the Netherlands received more than 67 km^3 of water, precipitation contributed 31% and 69% of the water entered via rivers and canals from Germany and Belgium. Under average weather conditions the Netherlands annually receives almost 110 km^3 water (precipitation + supplied by rivers and canals; NHV 1998). Most of the delta is also drained by its rivers. River Rijn is accountable for some 62% of all Dutch river water discharges (Donkers and Monincx 2007). Rivers do not only carry water to the Noordzee, significant volumes of fresh water from the rivers are also abstracted for consumption (after treatment), for industry, for irrigation and to prevent salt water intrusion in the polders. Before describing individual Dutch water bodies in more detail some insight is provided in the evolution of the Dutch delta first.

2.1.1 Evolution of the Coast, Rivers and Lakes

The Dutch coast and most of the major rivers (except for river Maas) and lakes in the Dutch delta are relatively young, some developed less than a half a century ago. The coastline is dynamic and would still be quite dynamic if the Dutch would not have (flexibly) fixed it since 1990 (Water 2016).

In short, the coastline moved eastward driven by rapid sea level rise in the earlier part of the Holocene. Maximum eastward marine extension by the Noordzee was reached about 5800 C^{14} years Before Present (BP) ago. Next, it shifted westward and about 4000 BP the Dutch coast was almost entirely closed by chains of dune-topped beach barriers, except for a few river mouths. That hardly interrupted coastline hampered normal (gravity driven) discharge of fresh delta waters into a still rising Noordzee. Together with an associated rise in groundwater levels swampy conditions and peat growth began prevailing in the west of the Dutch delta. This situation lasted some three millennia until about 1000 years ago when peat drainage and exploitation began.

Land subsidence resulting from drainage of peat for top soil improvement has been the driving force for making the low lying peat lands more vulnerable to flooding and more and larger inland lakes developed (Fig. 2.2). Storms further eroded peat lands along the lake borders. As the surface of the entire west and centre of the peaty delta gradually went down, more and more peat lands turned into surface water. Small peat-draining rivers found connection with the coastline that was breached during high tides and major storms. Such storms eventually demolished remaining peat islands and tidal channels came in contact with growing fresh water bodies. That happened in Zeeland as early as the third century AD (Vos 2015), and much later in the north (tenth to twelfth centuries AD). Deposition of marine clays and sands from such ingressions further aggravated the situation as these deposits further compressed underlying peat beds while accelerating land subsidence.

In the central Dutch river belt comparable processes took place. Numerous river floods occurred as sea level rose hampering rivers to properly discharge their waters and sediment loads into the Noordzee. Simultaneously, peat areas surrounding the rivers subsided also due to impacts of local water management and land developing activities and locally submerged, especially when flooding rivers laid down river clays on top of the highly compactable peat beds. As a consequence of such surface level subsiding processes and activities in the coastal areas and in the river belt more and more room capacity emerged to accommodate new, marine sediments through tidal floods.

Actions to counter both marine and river floods are discussed after a brief description of the main water bodies in the Netherlands, below.

IJsselmeer and Markermeer jointly constitute the largest Dutch fresh water lake. Both water bodies are separated by a dam with locks and sluices. It is a shallow lake (on average 5.5 m deep) in the heart of the country. *IJsselmee*r (Lake IJssel) has an area of 1100 km^2 and adjacent *Markermeer* (Lake Marken) 700 km^2. Formerly, they were in open connection with the Noordzee and were then called *Zuyderzee* (Southern Sea). In 1932, an almost 33 km long and over 7 m high sea wall: *Afsluitdijk* (Closing dike) was built separating Zuyderzee from Waddenzee. Due to inflow of a River Rijn branch (IJssel) this marine water body turned into a fully fresh water lake in just 5 year time. From the 1930's onwards major parts of former Zuyderzee have been reclaimed and were turned mostly into agricultural land. Not only IJsselmeer and Markermeer are quite recent, former marine Zuyderzee did not exist before Roman Times (Vos 2015). Before, a smaller fresh water lake occurred there: *Almere*.

Waddenzee is a wide tidal flat area in the north which continues into Germany and Denmark. The northern Waddenzee border consists of a chain of low islands, *Waddeneilanden,* (West Frisian Islands), topped by coastal dunes. Its southern border is formed by the landmasses of the provinces of Noord-Holland, Friesland and Groningen. Waddenzee covers about 8000 km^2. Currently it is a protected nature reserve and this wetland consists of shallow shoals that surface during ebb tides, intersected by deep gullies linking Waddenzee with Noordzee (Fig. 2.3). Due to its tidal character Waddenzee morphology has been and still is changing. Its origin is related to Holocene drowning of the Dutch coastal landscape and successive development of moving chains of coastal barriers protecting these northern tidal flats against further erosion by the SW-NE directed Noordzee currents.

Dollard is part of Waddenzee and a sea arm of the Noordzee in the extreme northeast of the Netherlands. Dollard originated from oxidised, excavated and eroded peat beds and, eventually by the disastrous 1362 Marcellus flood (Jongmans et al. 2015). Today, Dollard hosts the border with Germany, in the centre of the sea arm. Its size is about 100 km^2. German river Ems discharges in Dollard and is responsible for the brackish water quality in this estuary. Recently, Dollard was proclaimed a nature reserve in both countries.

Westerschelde (Western Scheldt) is an estuary of river Schelde in the extreme SW (Zeeland). Westerschelde evolved from a tiny peat river (*Honte*) into a sea arm by the end of the twelfth century. Today, Westerschelde is a 2–8 km wide and 113 km long and maximally 67 m deep tidal channel, connecting Antwerpen harbour with the Noordzee. It is one of the most intensively navigated waters on earth. To

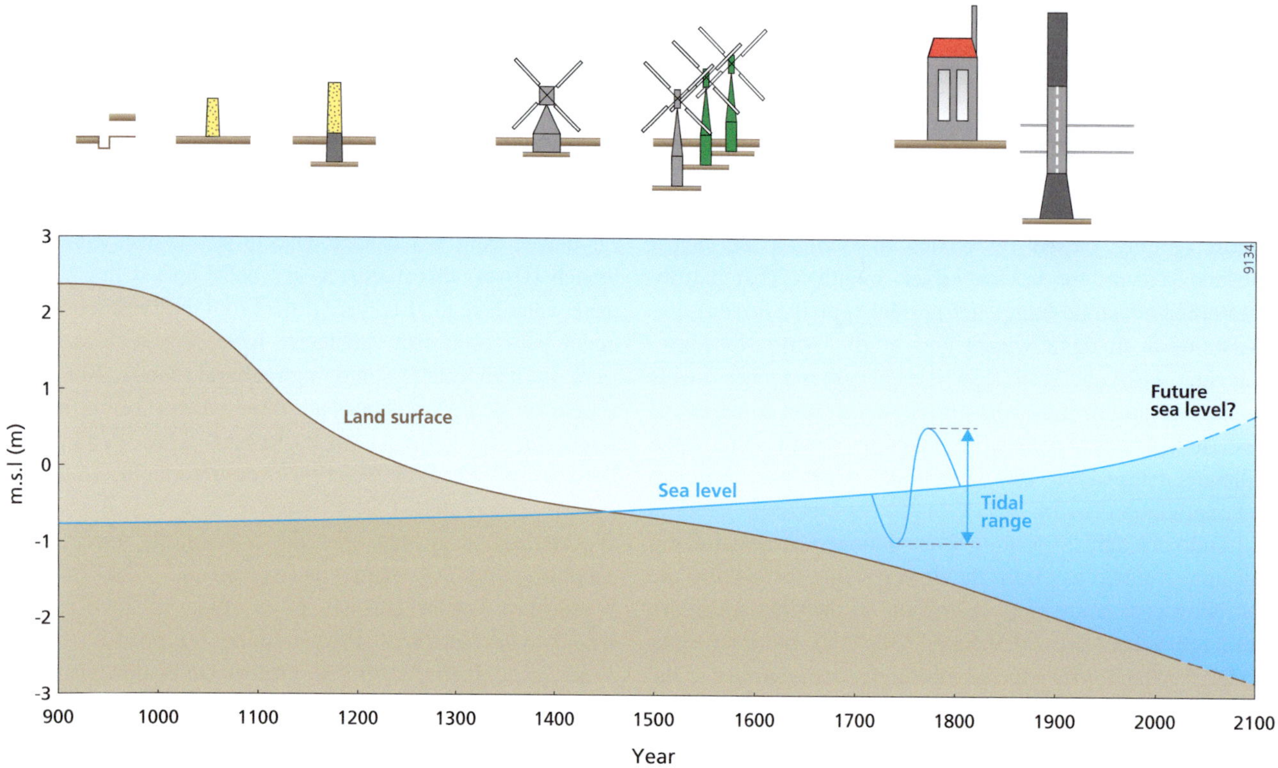

Fig. 2.2 Stepwise response to increasing land subsidence and sea level rise from about 900 AD. (Source: NHV 1998). Step one refers to the first ditches dug to drain the topmost peat beds in the tenth century; the second step corresponds with the creation of dwelling mounds the tenth and eleventh centuries; ring dikes and sea walls were built in the twelfth and thirteenth centuries while the land surface further subsided and approached twelfth century's sea level. Around 1400 AD, the first polders drained by windmills (step 4) giving a boost to further subsidence of the peatlands while long term sea level continued to rise. From the sixteenth century onwards, deeper polders were drained by series of successive windmills while the surface of the land dropped under the sixteenth century's sea level (step 5). When the big polders were drained by means of steam power in the nineteenth century mean surface level already was about 1.50 m below sea level (step 6); this process went on with the construction of the twentieth century polders which were drained by electricity driven pumping stations (step 7) with average land surface levels deep below mean sea level

Fig. 2.3 Air photograph of Waddenzee, a tidal flat area and wetland with its shoals and gullies

maintain its navigation ambitions for deep vessels dredging is required to remove the sand daily brought in by tidal currents.

Oosterschelde (Eastern Scheldt) is an (former) estuary in Zeeland and, in former times, served as the connection between river Schelde and the Noordzee until this role was taken over by Westerschelde. As all Dutch sea arms and estuaries its size changed over time and Oosterschelde covers an area of about 350 km^2 today. Since 1986, this estuary has been closed from the Noordzee by a storm surge barrier (see below). For environmental and commercial fishing reasons a connection with the Noordzee is built in this structure to maintain tidal influence. The barrier, however, is fully closed in times of high tides and severe storms.

River Rijn (Rhine) is one of Europe's longest rivers and originates in SE Switzerland (Graubunden). River Rijn discharges its waters in the Noordzee, along with fine-grained sediments. Together with rivers Maas and Schelde, river Rijn constitutes the Dutch delta. Close to the German border, in Spijk, river Rijn splits into two branches among which River Waal, carrying about 2/3 of its water volume. Further downstream more branches are added, one of these (river IJssel) discharges in IJsselmeer (see above). In times of low discharge, most of this river water can be diverted for irrigation and water level control to polders and lakes in the north of the Netherlands. River Rijn has a meandering character and frequently changed its bedding until Medieval Dutchmen began to build dams to avoid flooding of the hearth of the Dutch delta (Stouthamer et al. 2015). In the 1990s, a barrier gate built in the river mouth prevents flooding of Rotterdam harbour (Fig. 2.4; Delta Works). Numerous measures have been taken to properly channel water discharge from the hinterland and to improve river navigation abilities, also in its downstream branches. River Rijn and its connecting canals is one of the world's most intensively navigated rivers. In Roman Times, river Rijn discharged west from the present city of Leiden and formed the north boundary of the Roman Empire (*Limes*). The Rijn and Maas river belt divides the Netherlands into a northern and a southern part. For many centuries this river belt has been a geographical barrier coinciding with significant cultural differences between the people living at both sides (Chap. 12).

River Maas (Meuse) is the second largest river in the Netherlands and originates in NE France. Water levels may fluctuate considerably as its supply entirely depends on hinterland precipitation and not from melting ice and snow. Where river Maas enters the country, in Limburg, it flows north and forms the border between Belgium and the Netherlands. There, the currently meandering river Maas deposited thick gravel beds (Chap. 4). Downstream and north of Venlo, river Maas proceeds westward. Measures have been taken to improve navigation options upstream including building six weirs and sluices and the construction of parallel canals. River Maas is much older than river Rijn and its oldest (coarse gravel) deposits occur in Limburg and date from Late Miocene times (about 6 million years ago) when the Ardennes and Eiffel began to emerge. Pleistocene Maas river clays have been used for coarse ceramics in Roman Times already and are now still applied to produce roofing tiles and bricks in northern Limburg.

Friesland Lakes Friesland is the Dutch province most abundantly covered with fresh water lakes. These lakes are mainly in Friesland's SW and centre, an area once covered by thickest peat bogs. Drainage of these bogs for agricultural use made these to subside and submerge from Medieval times followed by further abrasion of the remaining peat islands in stormy seasons. Friesland's lakes attract many tourists for water sports as sailing.

Limburg Lakes Limburg also hosts many fresh water lakes but being by-products of gravel and coarse sand extraction, Limburg lakes have entirely different origins as Frisian lakes. Since the 1960's large-scale extraction of construction materials took place by dredging from several tens of metres depth. As in Friesland, these lakes are popular destinations for tourist industries.

2.1.2 Counter Measures against Floods

The early Dutch tried to prevent flooding by raising their homes and farms by dwelling mounds (Chap. 3). Initially, this was done on an individual/family basis, but with increasing risks for marine ingressions flood protection became a more communal issue. The first, primitive dikes had been erected earlier but next dike generations were stronger and more effective from the tenth century onwards (Hacquebord 2010). But sea level rose and peat lands subsided and eroded further and more rapidly. More and more drastic measures had to be taken to cope with this challenge while further aggravating the problem (Fig. 2.2).

One of the first measures to regulate delta waters and to prevent rivers and tidal channels flooding the land has been the construction of dams locking these waterways, or by building sluices to drain the land at low tides and to prevent floods at high tides or by high river water levels. The lower courses of river Rijn belonged to the first regions where such water regulation measures were taken to drain peat lands at both sides of the river. Apart from draining such measures

Fig. 2.4 Flood prevention gate in the mouth of river Rijn

also contributed to improve water traffic, transport of goods and trade (Meyer 2017). Continued sea level rise also caused river mouths to silt-up and to shift their locations. River Rijn eventually began to discharge via more southerly branches. In the course of the twelfth and thirteenth century a major tidal channel system in open connection with the Noordzee began to develop in the northwest and found connection with a fresh water lake, *Almere*. Eventually, that region developed into the tidal flat area of the *Zuyderzee*.

2.1.3 Urban Settlement and Trade Connected Flood Protection

Small rivers draining peat lands were dammed at their discharge points and trade and water transport settlements developed around such dams. Some of these later evolved into major cities, as Amsterdam and Rotterdam. Success and growth of many of such settlements depended on their very location with access to open seas, enabling international trade, and at the mouths of peat rivers providing access to more upstream delta settlements. In the meantime, such dike-related settlements could operate relatively free from flooding. Economy boomed and these settlements attracted many immigrants. That explains why such towns became targets of fights with earls and counts in the thirteenth to sixteenth centuries. That may, at least partly, have fuelled spirits of independency of citizens as well. Such spirits have been an essential drive in the Dutch Revolt (1568–1648) against the Spanish empire and eventually resulted in the establishment of the independent Republic of the Seven United Provinces. Water-related geography thus became a controlling factor in local and regional prosperity and a relevant political dimension as well.

Flood protection has been a major and direct issue for all inhabitants of the Dutch delta until today. In their fight against the water all people were united regardless their political or religious backgrounds. However, there were regional differences as how to approach flood protection. Cities in Holland (Box 2.3) mainly developed from building dams or sluices at river mouth or tidal channel entry points by the sea. Ships for international traffic landed at the sea side of the dam while vessels for domestic inland traffic were landed landwards. River mouths were widened into a major, central canal along which harbours, shipyards and repair shops emerged. Near the dam, a town hall and one or more churches were built (Meyer 2017).

Sluices were built where small peat streams discharged drainage water from the vast peat lands into the central canal. Water discharge served several purposes. One of these was to discharge the stagnant, contaminated waters that accumulated at its land side when high external water levels prevented any discharge. That also generated temporary higher (ground) water tables in the peat lands which became even more wet while preventing cattle to graze and elaborating the lands. A settlement or town was further protected by a dike surrounding this town and the central canal. Flood protection and controlling water discharge from the wet hinterlands belonged to the first priorities in towns in Holland. In contrast, flood protecting in the many islands in Zeeland was mainly geared at building dikes around entire islands, thus safeguarding agricultural revenues, rather than protecting individual settlements (Meyer 2017).

Their geographic position and spirits of their inhabitants more and more determined the prominent economic and political positions of water-facing provinces as Holland and Zeeland in the former Republic of the Seven Provinces (1588–1795). That was well reflected in their respective capitals: Amsterdam and Middelburg. Internationally, these provinces and cities were united and jointly combated hostile nations and other entities, e.g. against the Spanish armies in the sixteenth and seventeenth centuries when the Netherlands formally belonged to the Spanish House of Habsburg. Internally, however, both provinces often were in competition. They both ruled different parts of the delta, Holland the northern and Zeeland the southern parts, separated by a sea arm, *De Grevelingen*. Moreover, significant rivalry and conflicts occurred within these provinces, among cities, among water boards (public water management bodies, see Sect. 2.3) and between islands who all felt and behaved as independent entities. In addition, open sea facing cities as Vlissingen and Amsterdam rivalled with upstream cities and towns having access to the hinterlands, as Dordrecht and Rotterdam (Meyer 2017).

The mostly clayey soils in the southwest of the delta were more fertile than the wet, peaty soils in the centre and west of the nation and are superb for high quality corn and crop production. From the fifteenth century onwards agriculture became the dominant economic factor in Zeeland, nourishing Holland's inhabitants as well. Transport and trade of agricultural products was booming in the sixteenth century in Zeeland. Vlissingen, with its deep Noordzee facing harbour played a key role in Zeeland's international trade ambitions, in particular with England and the Mediterranean countries.

Holland gained its prosperity in these times mainly by maritime transport and trade with *Hanseac* cities as Lübeck and Danzig along the East Sea coast. In the early seventeenth century Holland's harbours became more relevant and economically more powerful than those in Zeeland. The access route of the Flemish cities of Gent, Brugge and Antwerpen to the Noordzee, the Westerschelde, had been vital for their economies but controlled by Holland's ships when these cities were re-occupied by Spanish troops (Zijlmans 2017). Although united in their battle against Spanish armies, Zeeland and Holland were driven by different motives. In contrast to Holland, Zeeland's people more strongly adhered the orthodox Dutch Reformed religion which fiercely opposed the Roman catholic religion imposed by Spain.

2.1.4 Floods and Old Clogged Waterways

But again, unity prevailed in times of common hazards, both political, military, and natural. Apart from Late Medieval epidemic plagues, flooding has been the most threatening of all natural enemies since about 1000 AD. Two types of floods may be distinguished, marine and fluvial floods.

Marine floods were triggered by both Holocene sea level rise and subsidence of peat lands. In turn, land subsidence was caused by a variety of human-induced factors, including artificial drainage by digging ditches and later by wind mills, oxidation and cultivation of peat lands, compression by younger sediments, storm erosion, and most seriously by large-scale excavation of peat for salt or fuel production. Originally (by the year 1000 AD), peat land surface was about 2–3 m above MSL in the west and centre of the Dutch delta. Since, sea level rose roughly 50 cm (NHV 1998), 23 cm of it in the past 125 years (Zeespiegel 2016). Subsiding peat land surfaces increased areas that were susceptible to flooding and thus in accommodating new tidal deposits. Tidal inlets providing such deposits widened during high tides and storms allowing tidal channels to penetrate deeper landwards.

In Holland marine floods occurred frequently from the ninth century onwards and their impacts on lives and properties were quite significant. Major historic marine floods took place in Zeeland and south of Rotterdam. From the thirteenth to the eighteenth century such often devastating floods happened almost every winter and autumn (Stormtides 2016). The worst took thousands of lives, major losses of land while entire towns were swallowed by the sea. One of the best described, historic marine floods was in November 1421: St. Elizabeth flood (Gottschalk 1975). Significant marine floods still occurred in the twentieth century, for example in 1916 causing major losses of land along the shores of Zuyderzee near Amsterdam. That flood eventually pushed the Dutch government to approve plans for complete closure of the Zuyderzee by the Afsluitdijk which was realised in 1932. Another and quite recent marine flood occurred in 1953 (Noordzee flood) and drowned parts of the Province of Zeeland and Zuid-Holland, Belgium, England and Scottish coasts while causing 2533 casualties (Stormtides 2016), 1836 of whom in the Netherlands. That flood prompted politicians to approve the Delta Plan aimed to secure Zeeland and Zuid-Holland against future flooding (see below).

However, floods have not always been an enemy only to the Netherlands, floods might be applied deliberately in a military sense, as a weapon, too, to combat hostile troops. (Box 2.1).

Box 2.1 Water, a Weapon

Combatting water has been a key issue throughout Dutch history and been conducted in all imaginable ways. Where water has been seen as an enemy, it may, reversely, also be seen as a friend and be applied as a weapon to defend Dutch territory against hostile human armies. That tool has been applied on various occasions in history. That strategy may have implemented from opening sluices to prevent hostile armies approaching their cities to constructing major water defence systems. Often that weapon proved to be quite successful but not in winter and particularly not during the Little Ice Age (fifteenth to nineteenth centuries) when water was frozen and ice could barely halt hostile armies.

Since the seventeenth century significant efforts have been made to build military water defence systems. The (Oude) Hollandse Waterlinie (the Old Dutch Water defence line) had been operational in the seventeenth and eighteenth centuries and was meant to prevent approaching French armies occupying Holland's cities. It was composed of fortresses and other defence works and was situated east and south of Amsterdam, excluding the city of Utrecht. This water defence line had been only partly successful until Napoleonic times when a new water defence line was built.

The *Nieuwe Hollandse Waterlinie* (New Dutch Water defence line) was the most important Dutch military defence structure ever and was operational from 1871. This new defence line was 85 km long and 3–5 km wide, had 46 fortresses, 5 fortified cities and included the city of Utrecht. Inundation was regulated via complex sluice systems and fortresses were built in places where defensive flooding was no option. But this Nieuwe Hollandse Waterlinie has never been put in practise. It was proclaimed a National Landscape in 2005 and proposed for UNESCO World Heritage site proclamation in 2020. The *Stelling van Amsterdam* (Defence system of Amsterdam) is the latest of the Dutch Water defence systems, was built in a ring around the city of Amsterdam and consisted of 45 fortifications. It was built from 1880–1920 but has never used. This structure was proclaimed UNESCO World Heritage Site in 1996.

Box 2.1 (continued)

Water was used for military purposes in WWII, but not by the Dutch but rather by the German occupation army to block progress of the Allied (Brits and Canadians) forces. In turn, the Brits bombed the sea walls of the Island of Walcheren to force the Germans to withdraw. By the end of WWII, in May 1945, some 9% of the Dutch land territory had been submerged.

The last time that water has been used as a weapon in the Netherlands was during the Cold War, to protect Randstad Holland against potentially advancing Soviet armies following a possible nuclear attack. The IJssel water defence system was believed to halt Soviet troops for maximally 1 week time. However, in the 1970's this water defence system was finally abolished (De Bos atlas van Nederland 2007).

A comparable pattern of interaction between humans and natural forces has been responsible for river flooding. Clogged rivers and accumulation of ice-floes by the end of winter enhanced such flooding. Moreover, river dikes were often undermined by muskrats. River floods were even more frequent than marine floods and may also have been more killing, at least until the twentieth century (Fig. 2.5). This was because authorities always gave less priority to river flood protection systems than to such marine systems. That changed only recently when flood protection in general became a more coherent issue in Dutch politics.

In Roman Times, clogged rivers were combatted by digging connecting canals between downstream river branches and by closing some river mouths (Box 2.2). That not only improved river water discharge and navigation options, but also prevented river beds to shallow, silt-up and clog. More recently, river beds were narrowed by constructing groynes forcing river beds to deepen and thus allowing for larger and deeper transport vessels. Dutch river networks are 3500 km long and transport some 250 million tonnes of goods annually, most of which is transboundary and serves Germany, Belgium and France. Transboundary transport by ship comprised 49% of all transboundary transports in the Netherlands in 1992 (NHV 1998).

Fig. 2.5 Dike burst in the Bommelerwaard, 1861. (Source: Van de Ven 1993)

Box 2.2 A Roman Canal

A remarkable canal connecting the mouths of the rivers Rijn and Maas was dug in the Dutch delta in Roman Times, around 50 AD, as described by Roman historian Tacitus. This canal was named after the leading Roman general Corbulo who ordered this work. Parts of the canal have been operational until about 270 AD. By then, river Rijn was the northern boundary of the Roman Empire (Limes) on the European continent. Roman troops, however, were transported via river Maas. To accelerate military transports between both rivers, a 34 km long connecting canal was dug while making use of existing, small regional waterways. Normally, such transports would have taken place via an offshore route but most of the small Roman transport vessels were unable to sustain the turbulent Noordzee waters.

To cope with different water levels in both rivers, a dam was constructed near the present town of Leidschendam, as early as 49 AD. Dendrochronological evidence proves that this canal was renovated in 121 AD at the occasion of a site visit of Roman Emperor Hadrianus. Its course has been modified several times as parts of the canal became too shallow. Parallel to the Corbulo Canal, a Roman Road was built. The Canal lost its relevance when most people left the region around 270 AD. Parts of it were uncovered and

Box 2.2 (continued)

identified during twentieth century excavations. Construction of the Corbulo Canal proved availability of advanced water management knowledge as early as Roman Times.

2.1.5 Dike Construction

In Dutch flood protection works several successive stages may be distinguished. Already prior to about 500 years Before Christ (=2500 BP, Iron Age) low, discontinuous dikes were made to protect agricultural lands from floods. The first dwelling mounds were raised by individual families living in salt marshes in the north to protect their homes against floods (Vos 2015; Chap. 3). When marine ingressions became a more serious threat, dwelling mounds were heightened and connected from the tenth century onwards (Hacquebord 2010). Next, families and villagers joined their efforts and began building ring dikes to protect their settlements. Finally, individual ring dikes were incorporated in larger and taller dike units surrounding entire islands or regions. Over time, more sophisticated dams and dikes were built, (smaller) tidal inlets were closed and river water discharge was regulated by sluices. That stage in dike construction occurred during the twelfth and thirteenth centuries and could develop only by close cooperation among seriously threatened people (Box 2.3).

Box 2.3 Westfriese Omringdijk

The *Westfriese Omringdijk* (Dike around Westfriesland) is a fine example of Dutch dike construction. This dike is situated in the northeast of Noord-Holland and surrounds a rather large area. Geologically, Westfriesland is a Late Holocene tidal flat area nourished with marine sediments by a major tidal channel in open connection with the Noordzee, via a major tidal inlet near the present town of Bergen, north of Alkmaar. That tidal inlet has been active since about 6500 BP and deposited sand in the channels and silt and clay in the tidal flats until around 3500 BP (Vos 2015; Chap. 3). Upon closure, swampy conditions prevailed and peat began to grow (again), as elsewhere behind the Dutch coastline until cultivation of the area commenced in the tenth century AD.

This explains why Westfriesland is topographically relatively high as compared to most of the west and centre of the Dutch Delta. The body of Late Holocene clastic tidal flat deposits may have provided more physical resistance against later, aggressive marine ingressions than other parts of Noord and Zuid Holland which mostly consisted of peat alone.

During the first cultivation stages of the peat lands some settlements developed in Westfriesland. But this area has been beyond any central authority for quite some time. Westfrisian villages remained independent until they were seized by the counts of Holland in 1288.

When surrounding peat regions were heavily eroded by storm floods, simple, low dikes were constructed from 1000 AD onwards in Westfriesland. These were typically built with a rather wide foreland to protect dike toes. Upon increased flood risks and some major dike bursts in the thirteenth century AD, Westfrisians began combating such risks as against a joint enemy. Although any written evidence is missing, there must have been some kind of coordination to build the 126 km long Westfriese Omringdijk, in terms of location, common height and width, maintenance and construction materials. As the Westfrisian society was not clearly governed by a single ruler, decision making on such a complex project could only have been realised by consent of all parties involved. Probably some leading religious leaders in the influential Egmond Abbey have played a major role in this process. Most sources agree that the dike construction project was completed by 1250 AD.

In the thirteenth century, average tidal range has been between 1.60 + and 0.80 m minus MSL (Schilstra 1974). Accordingly, dikes should have been at least

Box 2.3 (continued)

2–3 m higher to withstand periodic storm floods. Recent sections through the Westfriese omringdijk (Fig. 2.6) provided insight in construction methods and in materials applied to build the dike which was constructed in phases over centuries. The oldest dikes here were mainly composed of dug-out heaps of peat put directly on the remaining peat layer, later (fourteenth century) covered by clay. Peat and clay were put in as sods. In the fifteenth and sixteenth century walls of wave-absorbing sea weed were placed in front of the (sea-facing side of) the Westfriese omringdijk, kept in position by rows of wooden poles. But these poles were successively demolished by pole worms which arrived in the Netherlands not before 1731. Finally, this dike was further protected by stones and boulders derived from nearby glacial boulder clay deposits (Schilstra 1974; Kwaad 2016; Omringdijk 2016). Upon construction of the Afsluitdijk in the 1930s (see below), the Westfriese omringdijk lost its function as a sea-facing dike. Parts of this dike have been declared a Provincial Monument today.

Normally, dikes were built from materials in their direct vicinity. Initially, such dikes mainly consisted of peat, clay and seaweed, later more clay was applied to raise these further. More recently, dikes and sea walls are generally composed of an impermeable clay core near the water front and a permeable sand core to absorb water fluctuations towards the land interior. Their open water facing sides are often covered by naturals stone (blue Palaeozoic limestones as derived from mining activities in Limburg, from the Ardennes, imported basalts, or Scandinavian boulders from boulder clays) and/or by asphalt, today. Dike profiles normally show a rather steep water-facing side and gentle slopes towards the land. Most of the ring dikes enclosing polders are built on peat and must be monitored to check peat quality, subject to periodic maintenance. Some of the major sea walls and larger Dutch dikes are discussed below.

Hondsbossche Zeewering This sea wall is a prominent element in the Dutch coastal defence system and may serve as an example of its seemingly never ending story of dike maintenance. In the fifteenth century the Dutch dune coast extended about 1 km further west than today. Heavy storm surges severely eroded the coastline in 1421 (St. Elisabeth flood). Although dunes restored afterwards and some lower dikes were built westward from the current dike, the Noordzee could not be halted to further erode, nor could the sixteenth century rows of wooden poles and groynes. This

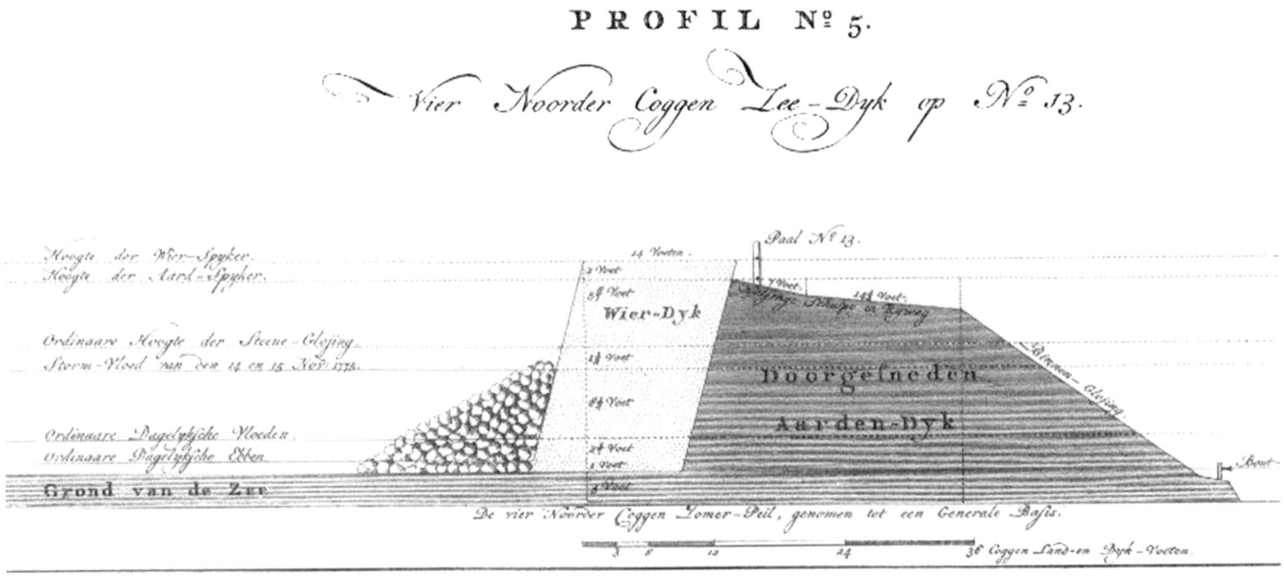

Fig. 2.6 Schematised cross section Westfriese Omringdijk. (Source: Schilstra 1974; Kwaad 2016). Sea-facing side is to the left. From left to right, first the outer protection of stones are visible. Further to the right, a zone of wave-absorbing sea weed is visible. Right from the sea weed, horizontal layers of clay and peat have been placed, being the core of the Westfriese Omringdijk

rather weak link in the Dutch coastal defence system further deteriorated until a new dike was built there with groynes in 1792. That new dike is at the same position as the present dam and was made of sand and clay, excavated nearby the dam where proper clay beds occur 3–5 m in the subsurface (Schilstra 1981).

However, that structure could prevent further erosion of the coastline neither. About one century later, in 1880, the sand dike was significantly improved and its seaward facing toe was strengthened by blocks of basalt. The Hondsbossche Zeewering served its purpose well and withstood many major storms. Meanwhile, the natural dune ridges north and south of the Hondsbossche Zeewering were eroded further while positioning this sea wall more prominently as a bulge in the Noordzee. Upon the disastrous 1953 storm surge, heights and strengths of all Dutch sea walls were re-checked and upgraded to 'Delta height standards', i.e. 11.50 m above MSL. Reconstruction of the Hondsbossche Zeewering occurred in 1981.

A next re-appraisal of the Dutch coast, mainly inspired by the IPCC forecasts for future sea level rise, was in 2004. The Hondsbossche Zeewering did not fully comply with these new standards and was again considered a weak link in the Dutch coastal defence system as flood risk would exceed 1:10,000 years. A new approach was chosen by creating a wide foreland by inserting some 20 million cubic metres of Noordzee floor sands at the seaward toe of the sea wall. A 'soft', dynamic coastline would thus replace the armoured sea wall and natural coastal dunes would be able to develop while further protecting the hinterland against marine flooding (Fig. 2.7). These works were completed in 2015.

Afsluitdijk This 32.5 km long sea wall separates the (current) fresh water basins of IJsselmeer and Markermeer from the tidal marine Waddenzee. At both abutments sluices were constructed between both water bodies which serve navigation and water discharge purposes. The prime purpose of this sea wall was to protect the former Zuyderzee banks against further marine erosion and ingression. The first plans for constructing this dam, submitted by Cornelis Lely, date from 1891. Decisions to actually build this sea wall were taken by the same person but now in his capacity as a Minister and shortly after a significant flood in 1916. That flood eroded vast areas along the borders of the Zuyderzee, drowning more than 50 people. Construction works started in 1920 and were completed by 1932. The dam is 7.25 m high and has a core of resistant Pleistocene boulder clays excavated from the sea floor close by the dike. The Afsluitdijk is an important regional and national infrastructural element and connects Friesland and Noord-Holland. It encompasses a highway and (originally) a railway track. After 80 years of service the sea wall will soon be significantly renovated and raised further.

Houtribdijk This dam is 33 km long and partitions former IJsselmeer in two fresh water components: Markermeer in the west and (small) IJsselmeer in the east. Navigation between both lakes is maintained via two sluice complexes. Originally, this dam was meant to serve as the northern border of

Fig. 2.7 The sea wall of the Hondsbossche Zeewering. Top: coastal protection consisted of sand and wooden poles in the sixteenth century; middle: situation before and after (below) sand suppletion works and 'sand motor' (in 2015)

the Markermeer polder, which has never been realised, however. Construction took place between 1963 and 1976 and decisions to postpone and later abandon construction of a Markermeer polder were taken in 2003. Increasing public pressure to maintain large surfaces of open fresh water in the hearth of this densely populated country for nature and recreation purposes and a reduced need for more agricultural land inspired these decisions. Permanent pumping of water to keep the anticipated Markerwaard polder dry would also have major impact on the groundwater balance in the bordering land areas and would affect the foundations of old towns as Hoorn and Edam (Claessen et al. 1987). Instead, new plans creating small islands for nature development (Markerwadden) in Markermeer are being implemented in 2018 (Houtribdijk 2016).

Deltawerken (Delta Works) One of the most impressive and internationally best known coastal defence works in the Netherlands are Delta Works. These consist of a series of separate coastal defence constructions that together would protect Zeeland and Zuid-Holland against flooding. This part of the country had been seriously affected by a catastrophic storm surge in 1953 when dikes breached at about 100 places, 1836 Dutch people died, tens of thousands cattle drowned and 200,000 hectares of fertile lands were inundated. By then, some dikes in the central parts of the Netherlands could barely be prevented from breaching. If so, major cities and even more land in the Dutch delta would have been flooded. All this urged the Dutch government to immediately start implementing earlier developed plans to safeguard the southwest of the nation.

Broadly, Delta Works comprised two types of flood protection: (1) permanent or semi-permanent closure of all tidal inlets, except for Westerschelde, the Netherlands' southernmost sea arm which is of vital navigation interest to the port of Antwerpen (Belgium), and (2) raising and reinforcing the locally low and weak dikes around Wester- and Oosterschelde sea arms. Upon implementation, the Dutch coastline between Flanders and Hoek van Holland would drastically be shortened (Fig. 2.8). That would not only protect the (is)lands behind such new sea walls against future flooding and significantly reduce future dike maintenance works, these dams would also create another series of fresh water lakes. Such fresh water lakes would considerably reduce salt water intrusion problems and thus ameliorate agricultural capacity in Zeeland. Moreover, Delta Works would resolve many of the infrastructural challenges and (cultural) isolation problems in Zeeland. In addition, Delta Works would prompt socio-economic development in this region which lagged behind in relation to the booming harbour area of Rotterdam. Finally, Delta Works would also boast Dutch knowledge infrastructure in sophisticated water constructions as solutions had to be generated for many major technical challenges.

Delta Works began by constructing a storm surge barrier in the mouth of *Hollandse IJssel* to prevent future flooding of the city of Rotterdam and neighbouring towns and cities (Fig. 2.8). That work began 1 year after the storm surge, in 1954 and was completed 4 years later, in 1958. This barrier is normally open but is closed about five to six times per year during extreme high tides or storms. Closing the major and very active tidal inlets in the southwest of the Dutch delta has been a matter of 'doing by learning' and the most complex closures in the deepest channels (Oosterschelde) would therefore be done last. Coastal zone studies showed that closing tidal inlets would generate significant erosion elsewhere along the coast and their negative side effects had to be mitigated.

As Delta Works would have major impacts on many aspects of the entire Dutch society, a separate law (Delta Law) was issued for these works in 1961. But by then, the first works were already completed. Among others, the Delta Law ensured that overall management was done by one single, public entity: Rijkswaterstaat.

Although there was broad political support for protecting the southwest of the Netherlands from future flooding, some Delta Works implementation issues faced significant opposition. These dealt with anticipated lack of local and regional (Zeeland) participation in decision-making and with drawbacks foreseen for the fishery sector. Mussel and oyster cultures require marine tidal waters to flourish, conditions that would disappear if tidal inlets would be fully closed. Both issues were addressed only during implementation.

What should be the minimum level of protection to the people and industries in the southwest of the Dutch delta? To resolve such questions probabilistic risk analyses and other scientific studies were conducted. It was agreed that dikes surrounding the islands should not fail more often than once every 4000 years while failure risk for strongly urbanised Randstad Holland (Chap. 8) was further reduced to 1:10,000 years. To check validity of such calculations, models of new structures were tested in a laboratory specially developed for this purpose.

Originally, all main tidal channels except Westerschelde were supposed to be fully locked by massive dams. But as public awareness about values of nature and environment to society increased by the end of the 1960s and in the 1970s, human interferences in land and water ecosystems were more and more mistrusted. Eventually, this urged Rijkswaterstaat and Dutch parliament to modify some of the original Delta Works plans. Closure of some sea arms was reconsidered and 'open dams' or storm surge barriers with gates that could be shut down entirely in cases of high flood risks would be constructed instead.

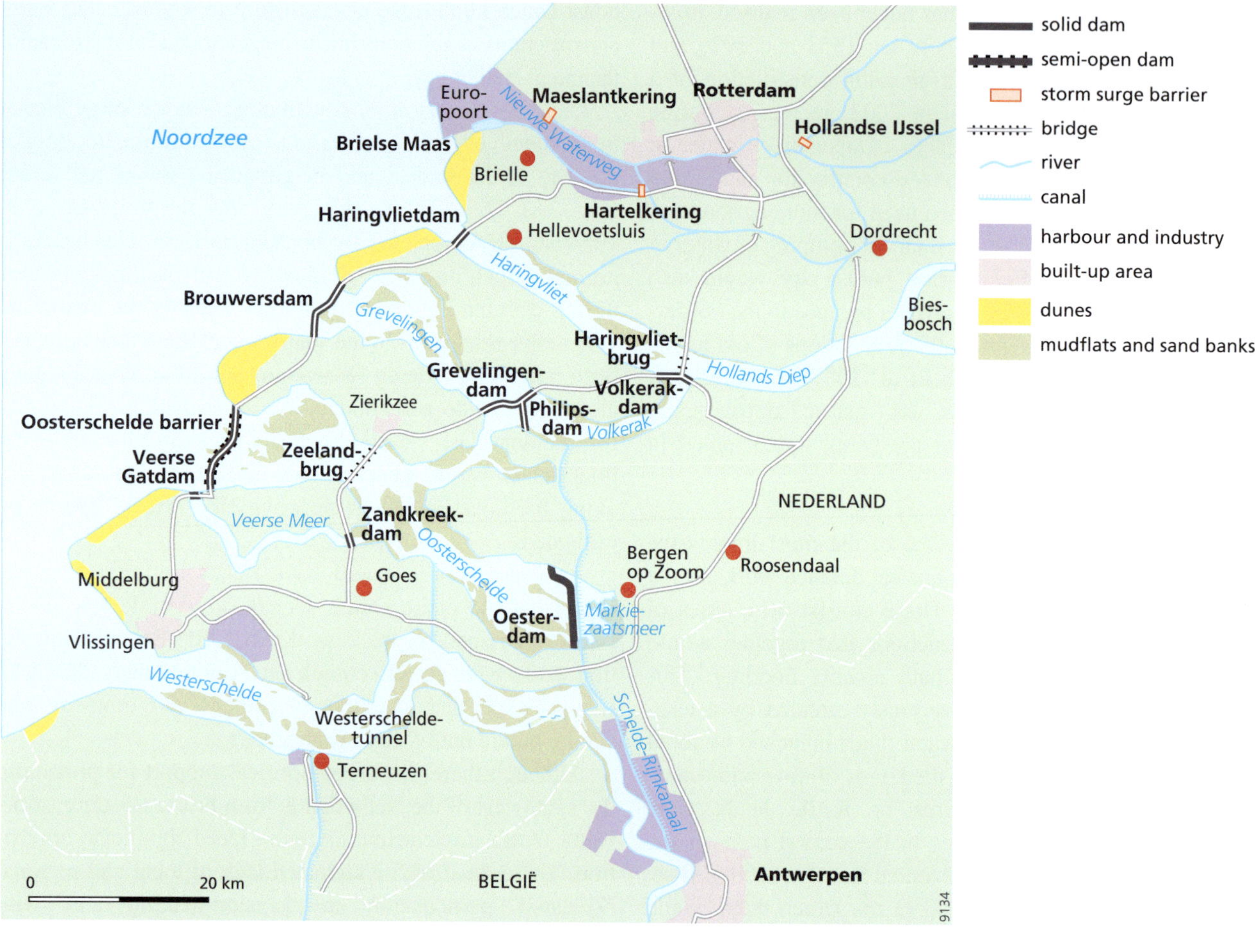

Fig. 2.8 Overview of Delta Works (in blue)

The best and most innovative example of such a storm surge barrier was built in Oosterschelde. This barrier can fully close three up to 30 m deep tidal channels during storm surges but would let tidal currents flow in normal conditions. The Oosterschelde barrier was a revolutionary design that exists of 66 concrete piers, built in dry docks (Fig. 2.9) and brought in position by specially designed vessels. Prior to this, the rather weak sea floor in these tidal channels was reinforced, densified and rock containing matrasses were laid down on the sea floor to carry these heavy structures. Finally, the piers were connected by beams and steel gates were hung in between the piers (Van de Ven 1993). The Oosterschelde storm surge barrier was officially opened in 1986, terminating Delta Works as a project.

Upon approval of the Delta Law in 1961, implementation of Delta Works was planned to last 25 years and to cost 700–900 million €. Upon completion (25 years later indeed), however, total costs were 5 billion € (Deltawerken 2016). Meanwhile, the Dutch coastline was shorted by 700 km and water management in the southwest of the Netherlands was considerably improved. That also held for the battle against salt water intrusion, socio-economic aspects in Zeeland and beyond and the nature and ecological conditions of the region. Moreover, Delta Works generated a knowledge spike in water construction and coastal zone management sciences, successfully applied in other major international projects around the world since. In 1994, the American Society of Civil Engineers declared Delta Works one of the seven wonders of the modern world (7Wonders 2016).

2.1.6 Land Reclamation

Dike construction is one way to prevent flooding, another, potentially more profitable way, is land reclamation. Four main types of land reclamation have been realised in the Netherlands. The earliest such type closely followed the natural process of land accretion in tidal flat areas where tidal currents eroded certain parts and deposited sand, silt or clay in others. Salt marshes may develop in such tidal deposition areas. Their highest parts hosted the earliest settlers in the Dutch delta, both in the north and in its southwest.

Fig. 2.9 Oosterschelde piers in construction in one of the dry docks. (Source: Van de Ven 1993)

In the north, salt marshes were drained, cultivated and inhabited since the Early Iron Age, i.e. around 500 Before Christ (BC) (Vos 2015). By building dwelling mounds (*terpen*) on top of salt marsh ridges the cultivated zone progressively moved further towards the sea shore as the sea retreated and salt marsh areas raised and extended further. Ever larger areas of their highest and central parts were reclaimed (Vos 2015; Van de Ven 1993; Jongmans et al. 2015). Successively, tidal channels silted-up and did no longer threaten the newly acquired lands.

Land reclamation in the north has been a process of natural silting-up of tidal flats and a receding sea, stimulated by human cultivation actions. Precipitation naturally desalinised such areas which could thus be added to the adjacent agricultural land realms. In the north, this process progressed relatively fast, e.g. between 500 BC and 500 AD some 5–10 km of new land was acquired, an average land extension of 5–10 m/year. Even earlier and during Roman Times low, discontinuous dikes were made to protect the new agricultural lands that developed in the lower salt marshes (Vos 2015; (Fig. 2.10).

In the southwest, salt marshes were inhabited later as tidal flat deposition did not occur there before the fourth century AD. The earliest dwelling mounds in Zeeland have been reported from the tenth and eleventh centuries AD. There, the first, major dikes for salt marsh reclamation were made in the eleventh century. In Zeeland, systematic dike construction to safeguard humanly occupied salt marshes from flooding took place during the twelfth and thirteenth centuries (Vos 2015). However, dike construction had an adverse effect here. Dikes hampered the natural storing capacity of salt marshes for flooding. That, in turn, increased storm surge amplitudes which required construction of even higher dikes to keep dry feet, and so on.

A second type of land reclamation dealt with turning virgin peat lands into land suitable for agricultural purposes. This was done by draining wet peat lands. Such reclamation activities are extensively described in Chap. 3.

Where land accretion basically was a natural process beneficial to the early Dutch, land reclamation by turning former lakes into new land has been an entirely human operation. This third type of land reclamation had the dual advantage of turning numerous worthless and potentially dangerous, storm prone surface waters into new and often profitable agricultural lands. Typically, such lakes were quite shallow (less than 5 m) caused either by human peat excavation or by storm and flood erosion, as described above. Initially, water was removed from such shallow lakes by means of man- or horse powered paddling wheels to adjacent open waters, discharging to the sea during low tides. Since the thirteenth century AD water was removed by wind power (wind mills) (NHV 1998).

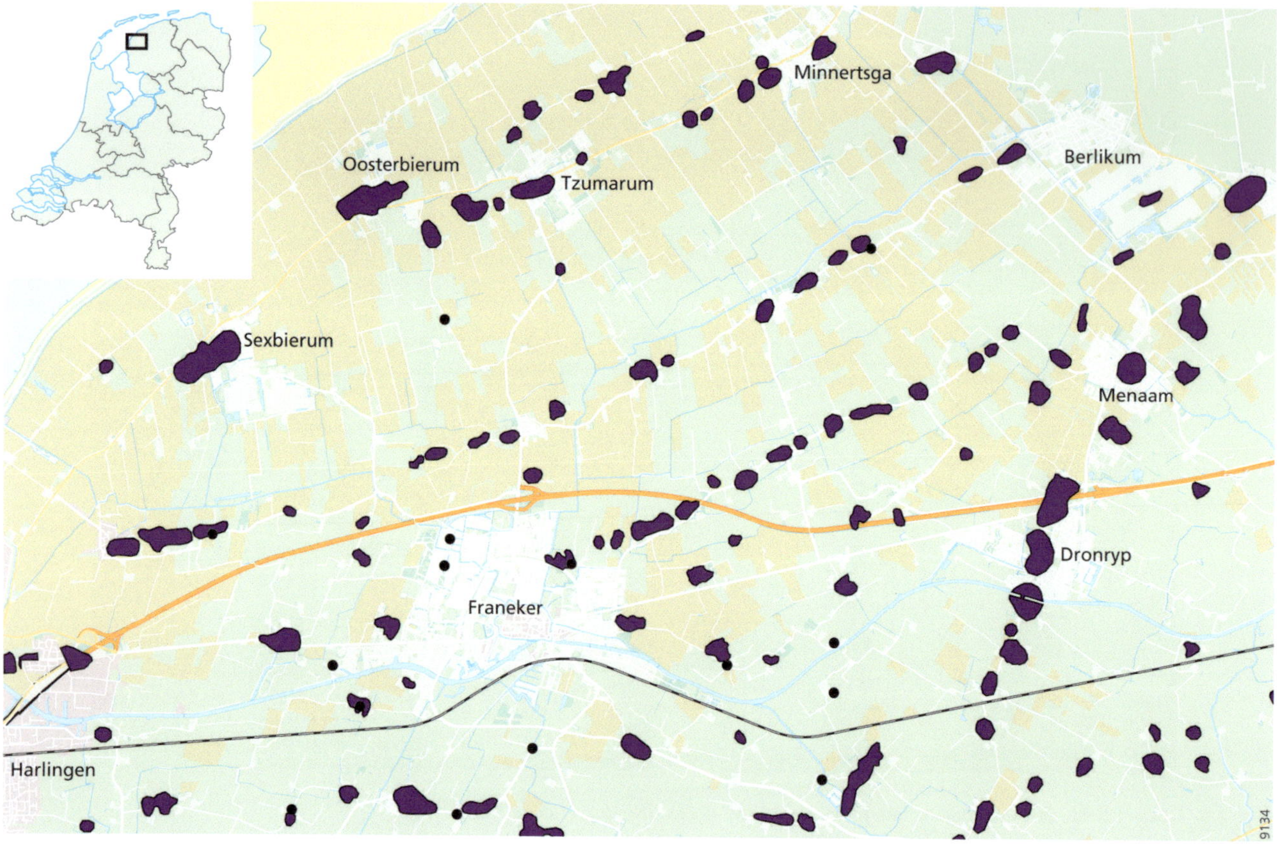

Fig. 2.10 Dwelling mounds (Terpen) in Westergo. (Source: De Bosatlas van de geschiedenis van Nederland 2011)

This process brought the lake floor to surface. Often unexcavated peat remains saved from erosion were uncovered at lake floors. More frequently however, former lake floors were composed of clays deposited before peat growth started. Once exposed to the atmosphere and drained freshly exposed peat was subject to oxidising processes, resulting in decomposition and compaction. Freshy exposed clays also underwent significant compaction. Upon continuation of the artificial draining process a reclaimed lake or wetland surface subsided while approaching groundwater levels. This made these newly reclaimed areas wet again and maintaining an unsaturated zone required deeper drainage, resulting in even more subsidence. As soon as this process would stop the polder would successively submerge shortly. Such knowledge has been used to the advantage of the Dutch in times of war while creating extensive submerged areas as part of a military defence system Box 2.1).

Solutions for excess water problems were found in surrounding polders by ring dikes and ring canals. Windmills were built along the dikes and the wheels paddled water from the (former) lakes or from ditches excavated in the polder floors into ring canals from where it flowed into main storage basins waiting to be released to open waters or to the sea during low tide. (Box 2.4; Fig. 2.11).

Box 2.4 Windmills

Polder reclamation by artificial drainage was made possible by a series of successive innovations. Water had to levelled-up from deeper water levels in the polder to a drainage outlet, a pool or ring canal (at '*boezem*' level). Application of windmills in the early fifteenth century in the west, turning wind power into vertical water levelling power through paddle wheels has been a first step. Earlier, water paddling was done by horse or man power and at a much smaller scale. Windmills could drain a polder in a relatively short period of time. Another invention was the mill head that could be turned in the wind direction to maximally benefit from the wind power. But paddles on a wheel had a limited depth range and for deeper polders a second or a third mill was needed to level water up to a next higher level. Such rows of windmills to jointly level the water up to the boezem were applied in the seventeenth century and allowed deeper (and often larger) lakes to be drained. Some polders, e.g. De Beemster (see below) required even a fourth

(continued)

Fig. 2.11 Typical set of three windmills to drain a deep polder (left) and to lift water from the polder into the ring canal. (Source: NHV 1998)

Box 2.4 (continued)
windmill in a row to drain water from the deep polder floor, as strong settlement occurred.

Application of the Archimeadian screw (*vijzel*) in the seventeenth century reduced the number of windmills in a row as these screws could be constructed for longer depth ranges and gradually but eventually replaced paddle wheels. Although wind energy was for free, one major disadvantage of windmills was that sufficient wind was not always available, in about 25% of the time only, regardless ability to turn the mill head in the best position. In winter days polder floors were often flooded making the land unsuited for agriculture.

Draining polder floors made also swampy lands applicable for cattle. As a consequence, more and more people became involved in dairy (cheese and butter) production. Such products could be kept fresh longer and were sold well in the cities. That, in turn, prompted farmers to improve their dairy production and cattle breeding methods. In the sixteenth century, on average Dutch cows produced twice as much milk than elsewhere in Europe. That reduced prices of dairy products which made Dutch milk, butter and cheese attractive, commercial products.

Box 2.4 (continued)
The often wet polders in winter accelerated introduction of the steam engine for polder drainage. The first steam powered water mill was applied for this purpose in 1787, in addition to windmills. Most nineteenth century polders were drained by means of steam engines although fuel costs were rather high. By then many older windmills were demolished. Almost all 20th polders were drained by electrical pumps. However, windmills are still widely spread over the Dutch land in the west and many of them may still be functional, if required.

To further illustrate Dutch polders, ring dikes and canal construction some examples are given below.

Polder De Beemster This polder has an area op 72 km^2 and is situated in Noord Holland, some 20 km northwest of the city of Amsterdam. The former Beemster Lake developed in the twelfth to thirteenth century partly by peat excavations and partly by erosion. The polder was built in the early seventeenth century by Amsterdam-based investors who earned their capital from overseas trade on the East Indies. With this investment they aimed to protect adjacent cities and lands in Amsterdam's direct vicinity and to generate new profitable, fertile lands. First, a 42 km long ring dike was built around the former Beemster Lake with an adjacent ring canal. Fifty windmills pumped the lake dry which was completed in 1612. The former lake floor was drained further by a unusual, rectangular pattern of ditches and canals. Such drainage further attributed to land subsidence in the new polder by about one extra meter. That, in turn, required an extra, fourth stage of windmills in lifting water out of the polder (Van de Ven 1993). Quality of the new, agricultural land was high and brought back earn-back times to one single year only (Beemster 2016). This polder acquired a World Heritage status by UNESCO in 1999.

Haarlemmermeer Polder Construction of this polder has been the last and largest land reclamation project in the nineteenth century. In the thirteenth century, the Haarlemmermeer consisted of three fresh water lakes. Peat excavation to fuel the cities of Amsterdam, Haarlem and Leiden caused expansion of the lakes and they became united into one large freshwater lake with an area of 17,000 hectares in the sixteenth century. More and more adjacent peatland eroded since due to storms and connection with the marine IJ sea arm could barely be halted. Serious plans to reclaim the Haarlemmermeer date back as far as the seventeenth century but this was technically not feasible with the limited water displacing capacities of windmills in those times. Halfway nineteenth century reclamation of this large lake was considered a State rather than an endeavour by private parties as had been the case for almost all earlier polders (Van der Woud 1987). But, opposition by the cities of Leiden and Haarlem emerged as reclamation would jeopardise commercial and fishery interests of their citizens. Only after significant damage was done by storms to surrounding cities in 1836 and 1839, reclamation began in 1840 by constructing a 60 km long ring dike and ring canal. Construction works were completed 8 years later in 1848. By then pumping began, not by the 200 windmills as anticipated in the seventeenth century, but by just three steam-powered pumping engines. Four years later (in 1852) the Haarlemmermeer Polder was dry (Haarlemmermeer 2016). Today this polder hosts Schiphol Airport, the largest Dutch airport, at some 4 m below sea level. A major new technological and spatial accomplishment in Haarlemmermeer Polder construction had been significant reduction of water storage basins to accommodate pumped polder water (Van der Woud 1987).

IJsselmeerpolders The largest Dutch polders, however, are in IJsselmeer, the former Zuyderzee, and all have been constructed in the twentieth century. The tidal marine Zuyderzee developed when one of the largest inland freshwater lakes (Almere) became connected with the Noordzee in Medieval Time by a tidal channel north of the present town of Den Helder. The often peaty borders of Zuyderzee were increasingly eroded during stormy weather, eventually jeopardising major cities in the interior and western parts of the nation. Plans for reclaiming (parts of) Zuyderzee date back to the seventeenth century. But since reclamation of Haarlemmermeer Polder (1852), no more technical limitations for implementation remained. From then onwards, a variety of reclamation plans were made. Two motives for polder construction prevailed by then: to protect adjacent cities, including Amsterdam, from marine flooding and to gain new, productive, agricultural land. A marine flood in 1916 triggered decisions for immediate measures. Despite their high implementation costs and protests from both nature preservationists and the fishing industry, the anticipated significant gains in national safety and positive prospects for agriculture prevailed. New governmental bodies were established to realise this enormous project (De Pater 2011; Van de Grift 2013; Meyer 2017).

First, Afsluitdijk was built to prevent new marine floods to threaten the Zuyderzee/IJsselmeer area. Soon after its completion this area turned into a large freshwater lake. Concurrently, ring dikes were built for construction of the first of the IJsselmeerpolders: the *Wieringermeer Polder* (309 km^2). Construction of this polder served as a test for successive large polders in the IJsselmeer. However, this polder was already completed by 1930, i.e. 2 years before the Afsluitdijk was done and was thus built in the marine Zuyderzee. Two pumping stations, one powered by electricity and one by diesel, milled the water out of the polder. The first parcels of land were distributed in 1934. Design, distribution of land and polder urban settlements was done according to by then prevailing views by governmental agencies driven by socio−/plan economic motives, e.g. to spread and mix people of different religions in the villages.

The next, but real IJsselmeer polder was the 480 km^2 *Noordoostpolder* (North-East Polder), which became dry during WWII, in 1942. Here, new techniques for land cultivation were introduced. As Wieringermeer Polder, Noordoostpolder was directly connected with the hinterland which had major impacts on the peatlands in the east. These peat lands were rapidly drained by the water pumping activities for the new polder and thus subsided. The latest views on land-use planning were introduced and implemented in Noordoostpolder, e.g. one central town (Emmeloord) surrounded by a ring of smaller villages. This pattern was, perhaps, inspired by the central place-theory of (Christaller 1993; Van der Wal 1997). This polder includes two small former islands: Urk and Schokland (Figs. 2.12 and 3.4).

Oostelijk Flevoland has been the next IJsselmeer Polder in line for reclamation. In contrast to Noordoostpolder, the

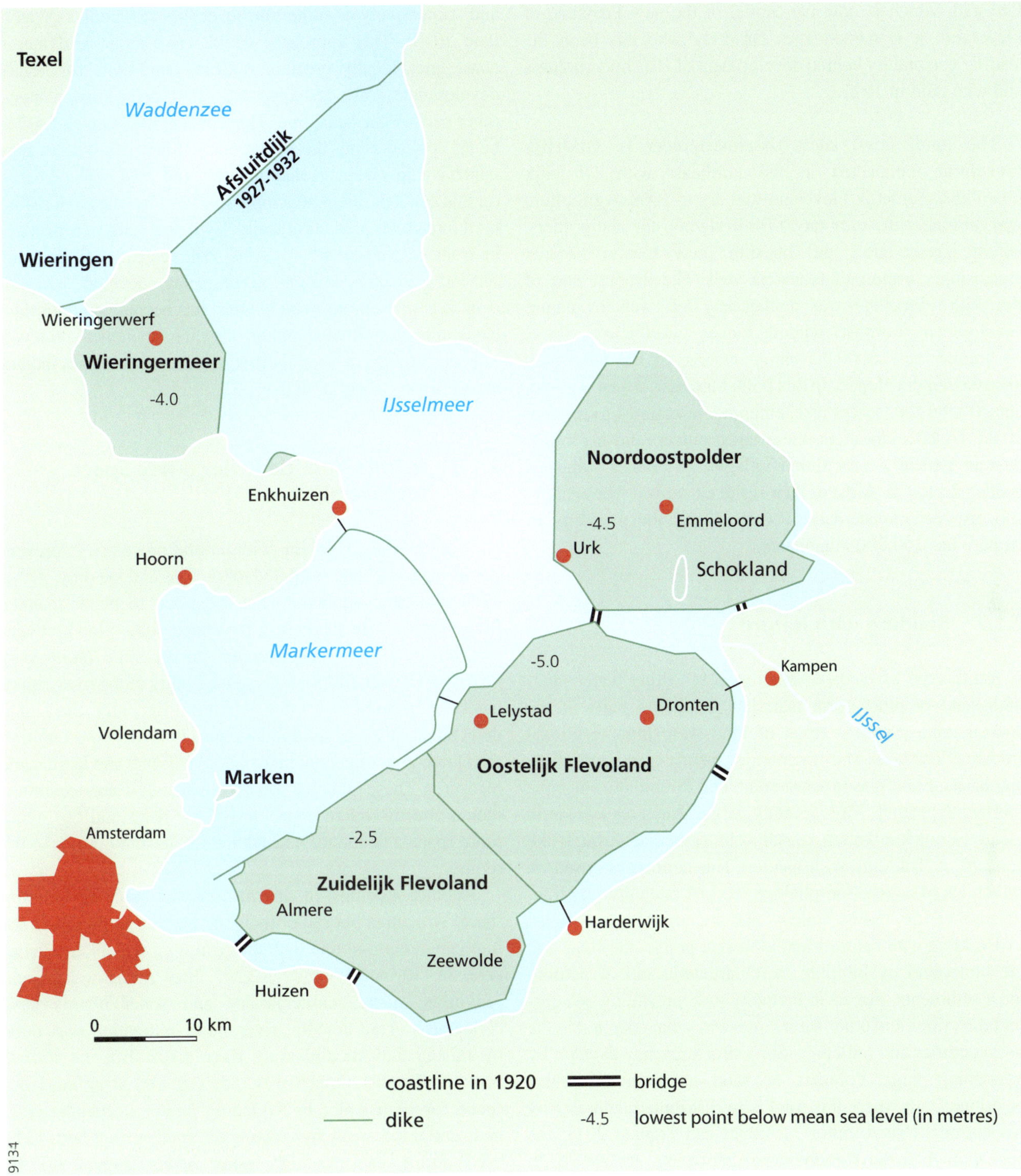

Fig. 2.12 Situation IJsselmeerpolders

540 km^2 Oostelijk Flevoland Polder is entirely surrounded by water, except for its boundary with Zuidelijk Flevoland in the SW. The purpose of this surrounding water body was in the first place to prevent surrounding lands to suffer from permanent pumping in the polder. Construction began in 1950 and was completed in 1957. The main city in this polder is

Lelystad, which is now the capital of the new Province of Flevoland; it is named after Dr. Lely who has been the leading personality behind development of IJsselmeerpolders and who died in 1927.

The (until now) latest IJsselmeerpolder is **Zuidelijk Flevoland**, connected in the northeast with Oostelijk Flevoland. Zuidelijk Flevoland was dry in 1968. Agriculture was no longer the only target for designing the last polders, nature conservation and housing (new towns) became increasingly important issues as well. The deepest part of Zuidelijk Flevoland was deliberately left wet for nature development, (almost) without human interference. Since, an unique, new 6000 ha nature reserve of the *Oostvaardersplassen* developed. In this polder the new city of Almere, named after the former lake, emerged. In contrast to the first of the IJsselmeerpolders, these latest polders did no longer contain typically agricultural villages, but rather suburban residential towns. Most of their residents moved from nearby and crowded Amsterdam. Today (2017), the new city of Almere has 200,000 inhabitants.

2.1.7 Building with Nature

A fourth type of land reclamation is by filling water areas with sand or other earth materials, including parts of the coastal zone. Smaller strips of land were thus reclaimed from shallow lakes and water ways, mainly to expand urban and harbour areas, as in Amsterdam and Rotterdam, since the fourteenth century. More recently, larger structures were built locally along the Dutch coastline in particular through the concept of 'Building with nature'. This concept is based on principles of coastal morphology, coastal evolution, hydraulic engineering, sedimentology, and sea currents.

Building with nature's ambition is to protect the coast and the hinterland by building 'soft' structures out of natural, local sediments, placed in harmony with prevailing sea current directions and other natural processes that operated in the past centuries and millennia. Such structures may develop by nourishing large volumes of sand at offshore coastal positions from where this sand is re-distributed naturally by sea currents ('sand motor') (Mulder and Tonnon 2011), as e.g. applied in the Hondbossche Zeewering, see above. In turn, natural beaches and beach barriers will emerge protecting the hinterland.

Once sufficient sand would have accumulated at such beaches, sand dunes may develop by natural wind action. Such dunes give additional protection to the coast while simultaneously serving as catchment areas for precipitation and successively as subterranean fresh water bodies (Waterman 2010). This approach would result in more dynamic coast lines locally creating optimal conditions for nature development while reducing risk for coastal erosion along other parts of the coast line. This concept has been and still is being applied along the Dutch coast. Some 1050 ha of new land have now been created in Noord and Zuid Holland based on this concept, often attaching soft sand beaches to existing hard (concrete) protection walls. By developing new beaches in front of coastal urbanisations e.g. recreational elements serving citizen's leisure needs, may develop as well (Fig. 2.13). 'Building with Nature' has become an important element in new Dutch dynamic coastal management policy. It has been applied in several other, sea-facing countries including the Palm Island in Dubai.

2.1.8 Ruimte Voor de Rivier (More Space for Rivers)

New approaches in coping with natural processes to prevent floods have not been restricted to the coastal zone only. These have also been applied in the river belt to better manage future risks for river floods in the Dutch delta. Flood prevention, however, has not been the only target of the *Ruimte voor de Rivier* Project. Improving spatial quality of the river region has been an important other ambition. Recently, the Dutch delta faced serious river flood threats (1993, 1995) prompting the Dutch government to take unconventional and large-scale measures. These included increasing overall water accommodation and discharge capacities of the delta and by giving more space and attention to nature and leisure options to the region.

Another ambition of this project was avoiding building 'hard' structures but rather focusing on natural processes for implementation, in line with the 'Building with Nature' concept. In 'Ruimte voor de Rivier' three ministries, several provinces, many municipalities and water boards have cooperated. Traditionally, river flood mitigation was done by raising and strengthening river dikes. But the *Ruimte voor de Rivier* approach was rather geared at giving more space for river water by widening and/or deepening river beds and flood areas, by creating alternative river beds to be used during extreme high water levels (Fig. 2.14), by replacing dikes and by cutting some dikes to allow some specific polders to be temporarily submerged. In addition, some weak river dikes were raised and strengthened. The project is in its final implementation phase now (2017), several parts have been realised already. The Project is scheduled to be fully completed ultimately in 2019.

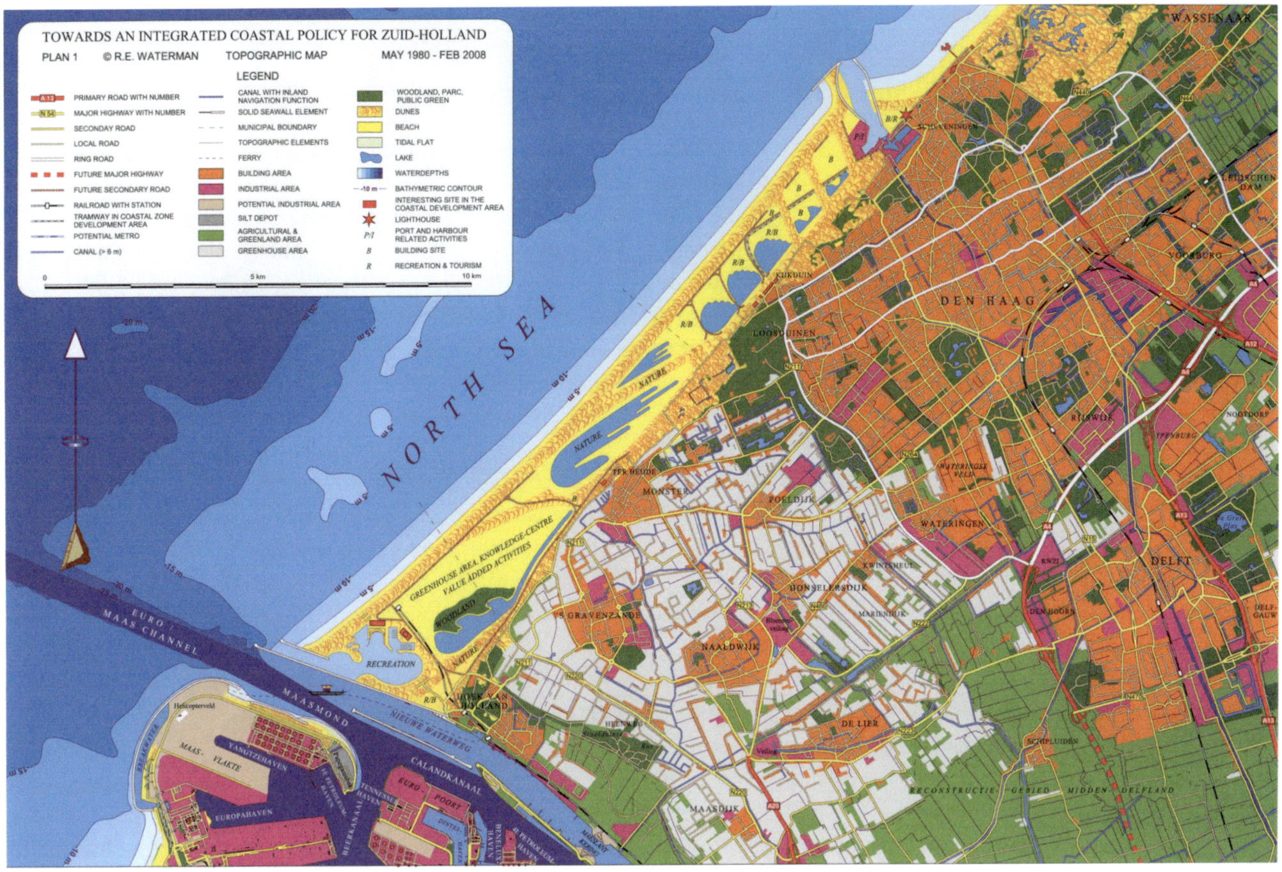

Fig. 2.13 Coastal extension by 'Building with Nature' as envisaged near the city of Den Haag. (Source: Waterman 2010)

Fig. 2.14 Ruimte voor de rivier; Hondsbroeksche Pleij, completed in January 2012

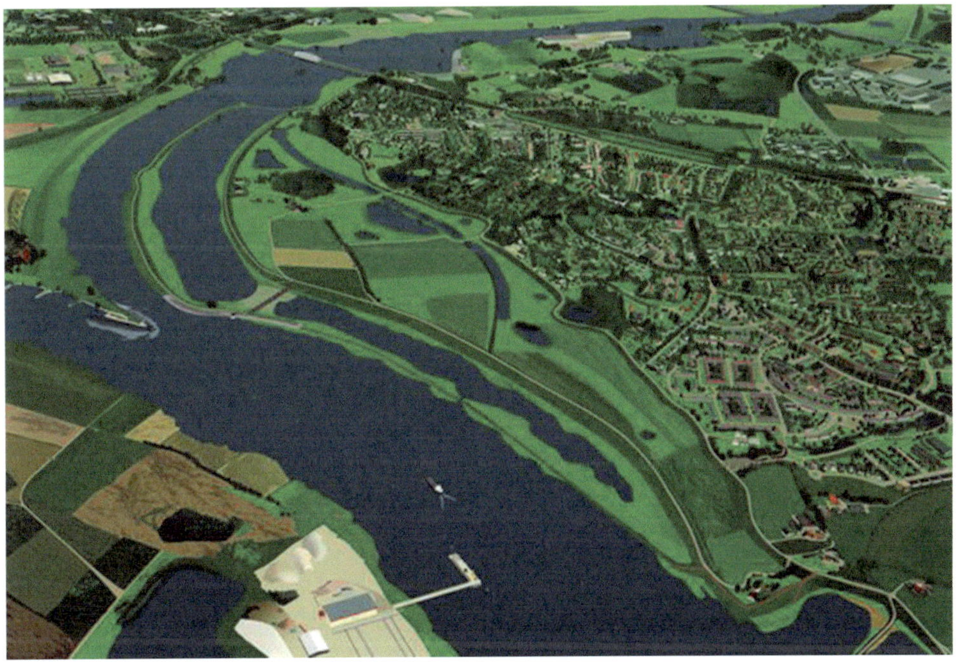

2.1.9 Surface Water Quality

Water quality is controlled by both natural and by human or human-induced factors. Chemical quality of surface water has fluctuated significantly in course of time. Originally, water quality in Dutch rivers was mainly controlled by composition of rocks and sediments in the hinterland of the river basis, i.e. the geological conditions in nine countries, including Switzerland, Germany, France, Belgium, and upstream Netherlands. In the Dutch delta another factor played a major role in chemical water quality. That is the impact of salt water encroachment by the Noordzee and Waddenzee. In coastal regions marine water may percolate through the dunes and Holocene sediments, or via groundwater into water courses and lakes. That impact significantly increased following large-scale reclamation of the inland lakes since the fifteenth century while water was pumped or milled permanently attracting lateral and deeper salt waters from the wide surroundings. Rivers in peat areas take up decayed plant remains and deposit these downstream in lakes or and seas.

Human-induced surface water pollution, in particular in river Rijn, repeatedly arose fierce debates among nations in the hinterland. Waste water from potash and phosphate mines in the French Alsace region are being discharged in river Rijn. Downstream, the Dutch receive salt-polluted waters. To counter this and to take measures in halting this process the Dutch government made some first diplomatic moves already by 1932 (NHV 1998). In 1950, a special International Commission for the Protection of the Rhine (ICPR) was installed to evaluate effectiveness of measures for water quality improvement and flood prevention.

Pollutants from other industries along the main rivers in the Dutch delta and in their hinterlands also deteriorated water quality. For centuries domestic wastes have been transported seawards without significantly compromising delta water quality at large. Retention capacity of the river waters was sufficient to maintain rather sustainable water qualities. Before the 1970s, industries discharged their often heavily polluted waste waters together with agricultural and domestic wastes directly or indirectly into the rivers or inland lakes. But both population and industry increased dramatically since. River and lake water qualities became so poor that swimming in such waters was no longer feasible. In 1970, the Pollution of Surface Waters Act became effective and polluting agents as oxygen-consuming substances from industries dropped 300% since, while heavy metal pollution dropped 20-fold from 1970 to 1995 and river water oxygen contents increased simultaneously.

However, pollution by diffuse sources, e.g. nitrogen and phosphate by agriculture, did not reduce comparably in the same time span. New depollution targets were set but only significant reduction of phosphate was achieved since. Pollution was not only limited to river waters, river floor sediments in industrial zones may be strongly contaminated as well. To keep a minimum river depth for navigation some rivers and canals, including the access route to Rotterdam harbour, *Nieuwe Waterweg,* are being dredged on a regular basis. Contaminated river floor sediments were normally discharged in the Noordzee. To avoid further, uncontrolled Noordzee pollution, such dredged materials are being stored today in special, large basins situated along the coastline in the mouth of *Nieuwe Waterweg*. Since, the 1990s water quality of Dutch surface waters gradually improved and swimming in rivers and lakes is allowed again today. River fauna improved considerably too and salmons re-appeared in Dutch rivers since 1998 (CLO 2010).

European rules determine maximum amounts for many tens of chemical components permitted to occur in European surface waters. Surface water quality also has an ecological dimension determining how aquatic life (plants and animals) best flourish and thus (partly) controls biotope conditions. More about this topic may be found in Sect. 5.2.

2.2 Groundwater

The Netherlands may have abundant surface water, even more water occurs in the subsurface. As groundwater is underground the general public normally pays less or no attention to it. But groundwater is quite relevant as it is the main source of Dutch drinking water. In addition, many industries, including agriculture, rely on groundwater for their production processes. Moreover, almost all bio-ecological systems depend on groundwater. Proper groundwater management is therefore vital for sanitation, industry and for nature conservation. Groundwater and surface water are closely related. Relatively high groundwater tables in the lower parts of the delta prevent small rivers and lakes to dry-up in summer and keep Dutch wetlands wet. In this section both groundwater quantity and quality will be discussed and their relations to human interference.

2.2.1 Groundwater Presence in the Subsurface

The Dutch delta consists of many hundreds metres of poorly to unconsolidated sands, clays and peat beds. This succession of Quaternary and Tertiary sediments is thickest in the northwest and thinnest in the southeast, as related to their position on the edge of the subsiding Noordzee Basin. All these relatively young sediments and also the older Tertiary, Mesozoic and Palaeozoic rocks are saturated with groundwater. In the Netherlands, groundwater tables are generally high and often near to the surface. Unsaturated zones normally are a few decimetres thick but may be over 10 metres in sandy, ice-pushed ridges in the centre and east of the Netherlands which also form the nation's highest spots, together with the older, southern parts of Limburg. Dutch groundwater resources are mostly recharged by precipitation excess which, in turn, is controlled by the temperate maritime climate in the Netherlands (Box 2.5).

Box 2.5 Precipitation and Aquifer Recharge
Mean precipitation in the Netherlands is 792 mm/year and mean annual evaporation is 501 mm. Thus, mean annual precipitation excess is 291 mm: almost 11 billion m^3/year. This volume varies considerably and ranged between −100 and +800 mm from 1930 to 1999 in De Bilt, reference site in the centre of the Netherlands and seat of Royal Dutch Meteorological Institute (KNMI) (Dufour 2000).

Most of the precipitation excess leaves the nation by surface waters (river, lakes), straight or indirectly into the Noordzee. Significant volumes of ground and surface water (together almost 2 billion m^3/year) is consumed for industrial, agricultural, domestic uses and for water regulation measures. Part of the precipitation excess remains in the soil and is used by plants and other biota which are part of the ecosystem. Much rainwater previously absorbed by the soil has become part of run-off today as more and more land surface is covered by concrete, asphalt or by stone now. They do no longer replenish Dutch aquifers constituting the nation's underground freshwater resources.

About 10% of annual precipitation volume (equivalent to some 2.6 billion m^3/year) only is estimated to eventually reach and recharge Dutch aquifers (Dufour 2000).

Hydrogeological units often coincide with geological units in the Dutch subsurface. In the west and centre of this country Early Pleistocene marine clay beds form the basis of the first and main aquifer which are more than 500 m deep in the northwest. But this basis is much shallower in the east where they consist of Tertiary clays.

Groundwater levels ('hydraulic heads') may fluctuate by the season or by the year. Normally such levels relate to precipitation values. Hydraulic heads and water quality data are monitored through a narrow grid of observation wells equipped with piezometers. These and other relevant groundwater data are registered often electronically and remotely every 2 weeks. With some 31,400 observation wells (Dufour 2000), the Netherlands has one of the world's finest meshed groundwater monitoring networks (slightly less than one observation point per km^2). These data are mainly used for (strategic) water management purposes and are all freely accessible online through DINOloket (DINO 2016) and through BRO (Basis Registratie Ondergrond) in the near future.

Groundwater levels in the topographically low west must be regulated artificially, if not, polders would submerge almost immediately. This is realised by continuous and long-term pumping which significantly contributes to land subsidence. Not only polders are artificially drained, some industries and cities withdraw groundwater at smaller or larger scales as well. Small-scale groundwater extraction may occur e.g. if a building pit is pumped to keep the construction site dry. Large-scale industrial groundwater abstraction happens in some industries, e.g. the former (about 1830–1965) textile industry in Twente, or for cooling purposes in the chemical industry creating e.g. a cone-shaped depression in hydraulic heads in Delft (De Mulder et al. 2003). However, if such often long-term extractions would cease the consequently rising groundwater levels may cause significant problems as citizen and their houses and infrastructures were adapted to the former deep abstraction. Cellars of houses and tunnels might then drown. Other geotechnical problems may arise if groundwater levels would sink while exposing older and wooden foundation heads to the air. These would successively rot and eventually fail.

Polder construction might dramatically change groundwater flow regimes in their wide surrounding areas too as demonstrated in studies concerned with construction of Oostelijk and Zuidelijk Flevoland Polders (Dufour 2000) and with the formerly planned Markerwaard Polder, where anticipated groundwater flow directions would reverse upon reclamation of the new polder. Without drastic counter measures such effects would seriously impact the often fragile foundation poles of the old cities along IJsselmeer (Claessen et al. 1987). A comparable situation occurred in Limburg where groundwater flow conditions fundamentally changed by construction of over 400 m deep open pits for large-scale lignite mining at just 30 km from the Dutch border, in Germany. Associated massive pumping drains the east of Limburg and threatens the very existence of nature reserves in the Limburg *Meinweg* area.

2.2.2 Groundwater Flow and Interaction with Subsurface Components

Groundwater flows due to varying heights of their levels. Such gravitationally driven flows run towards the lowest levels, i.e. often from topographical highs in the (south)east to the (north)west. Groundwater flow velocity is not only controlled by gravitation but by porosities and permeabilities of groundwater-hosting strata as well. Well-consolidated, Early Tertiary clays are saturated with groundwater but their low porosities and permeabilities hamper rapid flow. The opposite is true for poorly consolidated, Holocene sands where velocities may be reached of 10 m/year or more. Even higher velocities (30–50 m/year) may be achieved in shallow, gravel bearing sand beds. Horizontal groundwater flow in deep-seated aquifers normally is 10 cm/year (Dufour 2000). Horizontal flow directions are much more prominent than vertical flow directions which, in clays are often barely measurable. Groundwater is in direct and intense chemical, physical and biological contact with its surrounding sediments and rocks.

2.2.3 Groundwater Quality

Before environmental awareness became a public and political concern in the mid 1970's, groundwater quality was almost exclusively expressed in terms of salinity, or chloride concentration. According to hydrogeochemical standards fresh, unpolluted groundwater contains less than 150 mg chloride/l (Dufour 2000). Salty or brackish groundwater has been a major threat to drinking water exploitation and to agriculture as (in general) salty or brackish groundwater hampers crop production.

Seepage of saline waters is a common feature in sea-facing polders where groundwater is withdrawn from the wide, also marine surroundings. As salinity normally increases with depth (due to early Holocene marine ingressions) in the Netherlands, long-term pumping withdraws water from ever deeper and more saline sources. Hydrogeological conditions are relevant, in particular when poorly permeable aquicludes would have been cut by natural (e.g. by former tidal channels) or artificial causes (e.g. by human digging or drilling activities). Saline (ground)water is a major problem in the Netherlands.

But fresh water seepage might occur as well, mainly in the higher central parts of the Dutch delta where sandy ice-pushed ridges or sandy coastal dunes containing large bodies of fresh water resources are nearby low-laying areas, e.g. lakes. Interfaces between fresh and salt water are generally quite deep in these sand accumulations due to long-term infiltration of (fresh water) precipitation. These may be as deep as 100 metres below mean sea level (MSL) in the coastal dune region and 150 to 200 m – MSL below the higher, Pleistocene, ice-pushed ridges of Veluwe and Utrechtse Heuvelrug.

Apart from salinity, groundwater quality is determined by its calcium and magnesium contents as well. In turn, these are controlled by geochemical composition of subsurface beds and groundwater residence times. Calcium and magnesium control groundwater 'hardness': calcium and magnesium contents of marine sediments are higher ('harder') than fluviatile sediments (Dufour 2000). The best (lowest chloride, calcium and magnesium concentrations) Dutch groundwater bodies are situated under the higher ice-pushed ridges in the centre of the country. Natural methane and iron components may also impact groundwater quality. Natural methane often occurs in aquifers nearby major rivers.

Groundwater pollution has been a matter of major concern since the mid 1970's. Two types of groundwater pollution sources may be distinguished. Diffuse pollution which is often associated with agriculture by extensive use of fertilizers (nitrogen, phosphorus and potassium) and pesticides. Manure spreading significantly contributes to diffuse groundwater pollution, particularly in sandy soils. Since the 1990s, more and more detailed standards for nitrate and phosphate contents have been set by the Dutch government to combat this type of pollution in agricultural areas. For 2017 specific limits vary for soil type (clay or sand), for type of land use (grassland or arable land), and for crop type. Phosphate supply is confined to 50–100 kg P/ha/year and nitrate from 140 to 260 kg N/ha/year for arable land and from 250 to 385 kg N/ha/year for grasslands (CLO 2014). These strict regulations have been rather effective as demonstrated by levels of e.g. nitrate in groundwater under arable lands which dropped from over 180 mg/l in 1992 to 51 mg/l in 2015.

More locally, groundwater pollution may also have occurred under (chemical) waste deposit sites, leaking transport lines or cargo trucks with chemical fluids including oil, etc. Investigation of local pollution sites in 2011 determined some 250,000 possibly or seriously polluted sites in the Netherlands. More recent (2014) investigations pointed out that 1518 of these would urgently require some kind of decontamination action.

Dutch land-use and (drinking) water managers and authorities are mainly concerned in preventing pollutions to penetrate into the hydrological system and their successive spreading by groundwater flows. To anticipate and combat such threats requires proper knowledge of groundwater flow systems. Measures may include hydrological engineering techniques. Many such techniques have been tested and applied in the Netherlands, often by modification of groundwater flow regimes through pumping activities.

2.2.4 Groundwater Production and Consumption; Groundwater Resources

Most of the water (approximately 60%) used for Dutch consumption (industry, agriculture and households) is produced from groundwater resources, primarily because of its rather high and constant quality. The remaining part is derived from surface water, mainly from large basins situated alongside downstream rivers Rijn and Maas.

Fresh groundwater for consumption is mainly abstracted from aquifers below Holocene coastal dunes in the west and under buried Pleistocene river deposits in the centre, the east and the south. Due to high consumption in the west, fresh groundwater aquifers under coastal dunes were depleting. That challenge was overcome by inserting surface waters into the dunes where these percolate down into 50–100 m deep sandy aquifers, via infiltration wells. During this process surface water is purified naturally by chemical reactions between surface water and the sand grains in the subsurface. Such large-scale surface water infiltrations began in 1956. Today, large quantities of surface water from IJsselmeer, Haringvliet and rivers, are transported to infiltration basins in the dunes by pipelines. Risk for excessive nutrification of the dune area was countered by pre-purification of surface waters before infiltration. Groundwater extraction for drinking water consumption continued without major interruption since (Fig. 2.15).

Industrial companies and the agricultural sector extract major quantities of groundwater too. These account for about half of total annual groundwater production in the Netherlands. The remaining 50% is extracted by (semi-public) water extraction companies. The first of these served the population of Amsterdam by extracting groundwater from the dunes in the west since 1853. Their numbers increased rapidly to 229 such water companies in 1938. Since, many such companies merged and only 23 of them remained by 2001. Semi-public water extraction companies are being held responsible for production of sufficient, high-quality water for consumption. Quality control is a key issue in these companies. Groundwater and soil contamination belongs to their prime concerns.

Water extraction companies securely monitor their water infiltration areas for groundwater production, in particular developments in (suspected) contaminated sites. Intake of surface water may (temporarily) be halted to cope with water quality issues. Hydrological engineering methods may be applied to deviate contaminated groundwater flows from the abstraction wells. Purification of produced groundwater is a rather complex process and may consist of intensive exposure to the air, filtration, adsorption over activated carbon, biological reduction and/or ion-exchange, decalcification, pH-correction, and UV-disinfection (Dufour 1998).

Fresh groundwater reserves in the Dutch subsurface are estimated at some 800 billion m^3 (Van Montfrans et al. 1988). Annual recharge is about 2.6 billion m^3. Technically, potentially extractable volumes of fresh Dutch groundwater are some 1.9 billion m^3/year. Annual consumption of fresh water by industry and for domestic use has been 1.1 billion m^3 in 2007, 0.8 billion m^3 of which is groundwater (Box 2.6). This is why groundwater is a renewable resource in the Netherlands while current and future Dutch demands seem to be guaranteed rather easily.

Since the past few decades average water consumption per person is decreasing: from 131 litre/person/day in 2001 to 119 litre in 2013 (Waterverbruik 2016; Box 2.4). This reduction is mainly due to more efficient use of water facilities in Dutch households (toilets, washing machines etc.). Water prices are determined, among others, by their production costs (extraction + purification) which may vary considerably among water extraction companies. Other cost components include transport, water management, pollution control, and taxes (17%). In 2015, the price for 1 m^3 (1000 litres) water from the tap ranged between 1.12 and 1.83 € in the Netherlands.

Box 2.6 Beer and Groundwater

Many Dutch like beer and 678 Dutch breweries produced beer in 1819. Their numbers decreased since together with declining water quality caused by increasing industrialisation and high taxes. 160 years later, in 1980, only 16 beer breweries of some significance survived in the Netherlands. But since, their numbers are going up again and numerous small, home breweries emerge. Today, all breweries together produce 2.5 million m^3 beer, most is exported. The Dutch are rather modest beer drinkers: 80 litres per year. Germans drink 131 litres and Czechs even 161 litres/year.

All beer breweries use groundwater for their production. Before significant surface water quality deterioration, in the fifteenth and sixteenth centuries, breweries collected their waters straight from canals and ditches. Alternative sources were sought and Dutch breweries found clean water springs at the bases of coastal dune areas in the west. They dug shallow ditches at the often wet dune fringes. Water type, taste and quality of beer are related. Water poor in calcium withdraws more attributes from the malt than calcium-rich waters. For brewing dark beers CO_2 rich waters are required; Ca ions add more taste to beer than Na and Mg ions. Breweries modify their beer tastes by filtration, sterilisation and iron distraction from the water. Pilsner beer, e.g. is brewed from relatively Ca-poor water with few dissolved components. But *Dortmunder* beer requires Ca-rich water with large quantities of sulphate, Cl and Na ions. By the end of the nineteenth century Dutch breweries began to extract water from groundwater wells. Today, almost all large beer breweries control their own groundwater wells. Smaller breweries often use local tap water. Heineken brewery in Den Bosch uses ancient water (30,000 years old) from the Roerdal Graben (De Mulder et al. 2003).

Fig. 2.15 Infiltration basins in the dunes of Noord Holland. Pre-purified river and IJsselmeer water is infiltrated here to supply the city of Amsterdam with clean drinking water. (Source: Rijkswaterstaat 2011)

2.3 Water Management: Past, Present and Future

2.3.1 Public Water Management Organisations

The origin of water management in the Netherlands goes back to ancient times and exists as long as people jointly combatted water threats. Essentially, water management implies finding proper ways to cope with threats and opportunities provided by marine and/or fresh water, both surplus (e.g. floods) and shortages (droughts). Prime aim of water management in the Netherlands was to keep the land dry and to move water to where this is most needed. Water management originates from a common threat which forced people to cooperate. Water management also finds ways to drain low and wet regions, including polders. Moreover, it aims to prevent or at least reduce salt water encroachment, a permanent challenge in delta areas.

Water management deals with both surface and groundwater and their interactions. In case of soil contamination, water management provides measures to prevent (further) spreading of pollutants via groundwater flows. Water management also predicts impacts on land subjected to any development plan. Land subsidence is primarily calculated by geotechnical engineers but depends on groundwater characteristics. It may impact groundwater tables in the wide surroundings and may implicitly impact ecosystems, a matter of direct concern to water managers as well. Throughout the centuries the Dutch have gained significant reputation in water management and their expertise is often demanded for resolving complex water related issues abroad. The Dutch King Willem Alexander has been specialised in and is an international water management promotor. Water management issues have been described earlier in this chapter. Here some information about formal organisation of water management in the Netherlands is given. Water management is organised at local/regional levels (Water Boards) and at a national level (*Rijkswaterstaat*).

Water Boards Since the eleventh century landowners began realising that reclamation and drainage of peat lands and dam and dike construction requires a more than local approach. In the thirteenth century, concerned people formed co-operatives, resulting in the first water boards. Their co-operation was not only limited to jointly working on dikes, such cooperation also implied participation in governance, making water boards the oldest form of democratic government in the Netherlands. Water board *Hoogheemraadschap van Rijnland*, established in 1232, is the oldest water authority still operational. Most of the originally hundreds of water boards merged into larger units. Today (2017), only 23 water boards are operational in the Netherlands.

Water boards are democratic public bodies with their own parliaments, their own public elections (every 4 years) and their own public tax systems. Water Board' parliaments consist of representatives elected by inhabitants in their region, by land owners (farmers), by industries and by nature organisations. Water Boards occur in parallel with the normal Dutch governmental system. They operate on a supra-regional scale and govern water issues only; they are responsible for flood prevention and for sufficient, high quality surface water.

Rijkswaterstaat By the end of the eighteenth century, parliament decided to establish a powerful central organisation to prevent the country from future marine and river floods. By then, coastal and river defence systems were in poor condition. Measures were taken at local scales only while challenging safety of areas beyond. Coping with water problems would evidently require a national approach. In 1798, an 'Agency for Public Works and Water Management', consisting of just three persons, was established. Today, Rijkswaterstaat is the implementing agency for the Ministry of Infrastructure and Environment. It has 8700 employees, it governs the Dutch road network (3260 km), the national waterways network (1686 km) and the water system (65,250 km^2), including the Dutch part of the Noordzee and the Waddenzee (Rijkswaterstaat 2011).

Anticipated future developments Based on natural, long-term geological processes the land surface in the west will further subside. This process will be strongly enhanced by drainage of polders and peat lands and by further extraction of natural resources, as e.g. gas in the northeast. Sea level rise has a natural and a man-made component too. Moreover, sea level rise and deep drainage will generate more salt water encroachment and further challenge Dutch agricultural activities, particularly in the coastal provinces. These factors together will put the Dutch coastal defence system under further pressure. Rivers in the Dutch delta will receive more water from their hinterlands threatening safety of the lands and the people in the hearth of the delta.

Climate change is expected to generate more droughts in summer times and periods of heavier downpour may occur throughout the year while adding to increased flood hazards. In its Water Vision in 2007, the Dutch government determined five key targets for the future. Resilience to climate change is the first of these targets. Water management should become 'climate proof' (Rijkswaterstaat 2011). To realise this ambition an Adaptation Strategy was developed aiming to reduce vulnerability by taking 'no-regret' measures (PBL 2011). Such measures are primarily geared at reducing flood risks (hazards and/or impacts) in terms of casualties and capital losses to society. These may include construction of flood-proof dikes and dams around low laying cities and assigning (rural) areas as potential flood lands, areas that would be allowed to be submerged during extreme high floods while releasing river dike pressures and increasing water absorption capacities in the Netherlands.

This Strategy challenges communication between governmental authorities and local people as allowing fertile lands to be flooded would be in sharp conflict with traditional Dutch mindsets to keep their lands dry. On the other hand this Strategy would provide new options in combining land and water functions, e.g. for agricultural and nature development. In 2016, the Dutch coastal defence system was declared climate- and flood-proof for storms that may occur once in every 10,000 years in Randstad Holland, or once in every 4000 years in the sparsely populated north.

Moreover, land use will be changing as well. Economic activities and urbanisation will require more physical space in this small country. Multiple use of space and utilisation of underground space will most probably increase. Harbours and cargo transport over water are key issues in Dutch economy. Expansion of water ways and better access to harbours are priorities of the Dutch government. However, such measures would bear significant risk for further salinisation of the delta. All of these developments will have implications for water issues and demand for innovative approaches by water managers. The Future of the Netherlands will be discussed further in Chap. 14.

References

Beemster (2016) https://nl.wikipedia.org/wiki/Beemster. Accessed 28 June 2016

Christaller W (1933) Die Zentralen Orte in Süddeutschland. Jena (English translation: Central Places in Southern Germany. Englewood Cliffs 1966: Prentice-Hall)

Claessen FAM, Van Bruchem AJ, Hannink E, Hulsbergen JG, De Mulder EFJ (1987) Secondary effect of the reclamation of the Markerwaard Polder. Geologie & Mijnbouw 69

De Bosatlas van de Geschiedenis van Nederland (2011) Wolters-Noordhoff Atlasproducties, Groningen

De Bosatlas van Nederland (2007) Wolters-Noordhoff Atlasproducties, Groningen

De Mulder EFJ, Geluk MC, Ritsema IL, Westerhoff WE, Wong TE (2003) De ondergrond van Nederland. Wolters Noordhof, Groningen

De Pater B (2011) Conflicting images of the Zuider zee around 1900: nation-building and the struggle against water. J Hist Geogr 37 (1):82–94

Deltawerken (2016) http://www.deltawerken.com/De-Deltawerken/1548.html. Accessed 27 June 2016

DINO (2016) www.dinoloket.nl. Accessed 21 July 2016

Donkers HWHA, Monincx JF (2007) Water. In: De Bosatlas van Nederland. Noordhof Atlas Productions, Groningen, pp 244–259

Dufour FC (1998) Grondwater in Nederland. Geologie van Nederland nr. 3. Nederlands Instituut voor Toegepaste geowetenschappen – TNO, Delft

Dufour FC (2000) Groundwater in the Netherlands. Netherlands Institute of Applied Geoscience TNO – National Geological Survey, Delft

Gottschalk E (1975) Stormvloeden en rivieroverstromingen in Nederland. Deel II: de periode 1400–1600. Sociaal-geografische studies, vol 13, Van Gorcum, Assen

Haarlemmermeer (2016) https://nl.wikipedia.org/wiki/Haarlemmermeer. Accessed 28 June 2016

Hacquebord L (2010) Noordelijk zeekleilandschap. In: Barends S et al (eds) Het Nederlandse landschap, een historisch-geografische benadering. Matrijs, Utrecht

Houtribdijk (2016) https://nl.wikipedia.org/wiki/Houtribdijk. Accessed 23 June 2016

Jongmans AG, Van den Berg MW, Sonneveld MPW, Peek GJWC, Van den Berg van Saparoea RM (2015) Landschappen van Nederland. Geologie, bodem en landgebruik. Academic Publishers, Wageningen

Kwaad (2016) Het Ontstaan van West Friesland. http://www.kwaad.net.West-Friesland.htm. Accessed 5 Nov 2016

Meyer H (2017) The state of the delta. Hydraulic engineering, urban development and nation building in the Netherlands. Vantilt, Nijmegen

Mulder JPM, Tonnon PK (2011) 'Sand Engine': background and design of a mega-nourishment pilot in The Netherlands (unpublished)

NHV (1998) Water in the Netherlands. NHV Special 3. Netherlands Hydrological Society, Delft

Omringdijk (2016) http://www.omringdijk.nl/node/415. Accessed 6 Nov 2016

PBL (2011) Een delta in beweging. Bouwstenen voor een klimaatbestendige ontwikkeling van Nederland. Planbureau voor de Leefomgeving, Den Haag

Rijkswaterstaat (2011) Water management in the Netherlands. Den Haag, WDO111VV007B

Schilstra JJ (1974) In de ban van de dijk. West-Friesland, Hoorn

Schilstra JJ (1981) De Hondsbossche. Hoogheemraadschap Noordhollands Noorderkwartier. Meijer, Wormerveer

Stormtides (2016) https://en.wikipedia.org/wiki/Storm_tides_of_the_North_Sea#Major_storm_tides. Accessed 22 June 2016

Stouthamer E, Cohen KM, Hoek WZ (2015) De vorming van het land, Geologie en morfologie. Perspectief uitgevers, Utrecht

Van de Grift L (2013) On new land a new society: internal colonisation in the Netherlands, 1918–1940. Contemp Eur Hist 22(4):609–626

Van de Ven GP (1993) Man-made lowlands. History of water management and land reclamation in the Netherlands. Matrijs, Utrecht

Van der Wal C (1997) In praise of common sense. Planning the ordinary. A physical planning history of the new towns in the IJsselmeerpolders. 010 Publishers, Rotterdam

Van der Woud A (1987) Het lege land. De ruimtelijke orde van Nederland 1798–1848. Meulenhoff, Amsterdam

Van Montfrans HM, De Graaff LWS, Van Mourik JM, Zagwijn WH (1988) Delfstoffen en samenleving. Geologie van Nederland nr. 2. Rijks Geologische Dienst, Haarlem

Vos P (2015) Origin of the Dutch coastal landscape. Long-term landscape evolution of the Netherlands during the Holocene, described and visualized in national, regional and local palaeogeographical map series. Dissertation, Deltares. Barkhuis, Groningen

Water (2016) http://www.helpdeskwater.nl/onderwerpen/waterveiligheid/programma'-projecten/beheer-onderhoud. Accessed 4 Nov 2016

Waterman RE (2010) Integrated coastal policy via building with nature. Dissertation, TU Delft

Waterverbruik (2016) https://nl.wikipedia.org/wiki/Drinkwater#Waterverbruik. Accessed 21 July 2016

7Wonders (2016) http://www.7wonders.org/images/europe/netherlands/delta-works/delta-works-big.jpg. Accessed 24 June 2016

Zeespiegel (2016) http://www.clo.nl/indicatoren/nl0229-zeespiegelstand-nederland-en-mondiaal?i=9-54. Accessed 22 June 2016

Zijlmans R (2017) Troebele betrekkingen. Grens, scheepvaart- en waterstaatskwesties in de Nederlanden tot 1800. Verloren, Hilversum

3 Landscapes

Eduardo F. J. De Mulder

Abstract

Dutch landscapes are rather unique as they all have been more or less heavily influenced and modified by humans, one of these (polder landscape) would not even exist without people. Here, we distinguish seven different landscapes in the Dutch delta. Laterally and chronologically these are often interrelated in rather complex patterns. Original and undisturbed landscapes are determined by their geological origin, their lithological composition and their vegetation. They range from limestones covered by loess in the southeast to periglacial cover sands and ice pushed ridges in the east and centre of the country. The Dutch delta is subsiding in the west and north and their associated landscapes consist of Holocene marine sand and clays. River sands and clays are found in a broad zone in the centre of the country. Polder landscapes occur in the west, north and in the centre of the Netherlands, whereas coastal dunes can be found all along the present coastal zone.

Although humans lived here before, they began to markedly influence Dutch landscapes from the Iron Age onwards and more intensively since the tenth century AD. From the start of the twentieth century the Dutch more and more detached from their substrate. Major parts of the thick peat beds have disappeared by excavation, by oxidation after drainage, and by erosion. Peat has literally fuelled the Dutch economy for several centuries. Dutch soils host flourishing agriculture and dairy productions. Many Dutch landscapes have almost entirely been remodelled to address marine and river flood risks and to comply with growing demands for more productive agricultural use. Urbanisation and infrastructure have further fragmentised the original landscapes. Nature development currently has priority for policy makers, often at the expense of agricultural land. Some lands once recovered from the sea are now being flooded deliberately to comply with Dutch and European nature ambitions.

Keywords

Higher sands area · Loess and limestone area · Peatlands · River belt · Marine area · Polder area · Coastal dune landscape · Land use · National landscapes · National parks

Landscape remodelling, engineering and land reclamation are typical Dutch features. Part of the Dutch landscapes are fully artificial. If the Dutch would not have reclaimed major parts of their territory, the Netherlands would have only half of its present size.

The Netherlands is part of the delta by rivers Rijn (Rhine), Maas (Meuse) and Schelde (Scheldt) and by the former major Eridanos river system that drained most of Scandinavia and the Baltic region in the Pleistocene. Sixty percent of the delta is situated at or below sea level. The areas below Mean Sea Level (MSL) mainly occur in the west while the higher areas are in the east and south. The SW-NE trending tectonic hinge line separating higher and lower regions roughly connects the cities of Breda in the south and Groningen in the north (Fig. 3.1). Long-term and large-scale tectonic processes as the evolution of the North Sea Basin will cause further subsidence in the northwest and uplift in the southeast (Wong et al. 2007). Land subsidence in the Dutch delta is not only determined by geological processes but also by human activities, as oxidation of peat through drainage, by excavation, settlement and by the extraction of oil, gas, coal and salt from the deeper subsurface (Chap. 4).

The Netherlands is a wet country. Long-term tectonic forces, water and the Dutch have been responsible for development of the delta (Chap. 2). As described in Chaps. 7 and

E. F. J. De Mulder
Delft Technical University (emeritus), Haarlem, Noord-Holland, The Netherlands
e-mail: e.demulder@planet.nl

E. F. J. De Mulder et al., *The Netherlands and the Dutch*, World Regional Geography Book Series,
https://doi.org/10.1007/978-3-319-75073-6_3

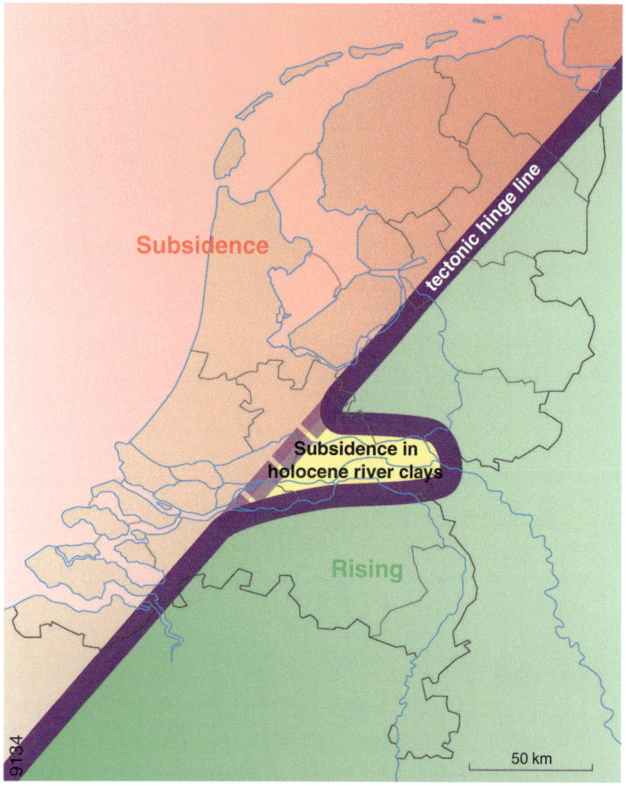

Fig. 3.1 Division of the Netherlands in a subsiding part in the Northwest and a rising part in the Southeast, separated by a tectonic hinge line; note the subsiding river belt in the centre and east which are subsiding caused by drainage and to their high clay and peat contents. (Sources: Rijkswaterstaat 1997; De Mulder et al. 2003)

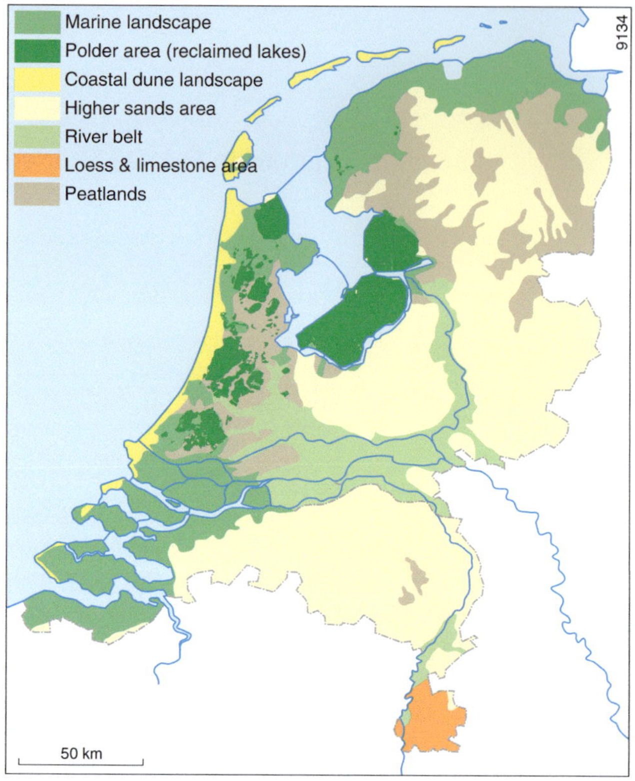

Fig. 3.2 Distribution of Dutch landscapes. (Source: modified after Jongmans et al. 2015)

12 cultural and economic distinctions between their inhabitants appear to roughly coincide with physical geographical characteristics of the higher and lower parts of the Dutch delta.

Natural landscape development in the lower parts of the country has been and still is driven by Holocene marine-tidal and fluvial (or river) processes in the delta where both realms meet. Natural landscape genesis and evolution in the higher regions was determined by Late Pleistocene glaciations and by Late Pleistocene to Late Holocene aeolian (wind) processes. Landscape distinction is primarily based on geologic genesis and successive anthropogenic-driven evolution. For the Netherlands this may broadly result in a division of seven main landscapes: the higher sands area, the loess and limestone area, the river belt, the peatlands area, the marine area, the polder area, and the coastal dune area (De Bosatlas van Nederland 2007). Some of these may be further divided into regional subtypes. In landscape development three successive stages may be distinguished: the natural landscape, human cultivation and present land use.

3.1 The Higher Sands Area

This landscape dominates east of the SW-NE hinge-line (Fig. 3.2). The higher sands areas mainly consist of glacial, fluviatile and periglacial deposits of Pleistocene age. North of the line connecting the cities of Haarlem and Nijmegen these sediments may locally have been pushed-up by Middle Pleistocene glaciations, the last time (Saalien: 200,000–130,000 years ago) that the Netherlands has (partly) been covered by ice caps. There, glaciation may have disturbed the original successions of fluviatile strata locally to significant depths and formed ridges and hills that clearly stand out in the otherwise rather flat Dutch landscape, attaining (present) heights of more than 100 m (Veluwe), Fig. 3.3. Depending on their geographic position such ice-pushed ridges may consist of sands of the Pleistocene Eridanos or of the Rijn/Maas river systems. In the higher sands areas glacial deposits may also include boulder clays. Boulder clay accumulations, marking ultimate extension lines of glaciations, may outcrop as rather isolated hills as in the island of Texel, the former islands of Schokland, Urk and Wieringen, and in Friesland (Gaasterland). In the north and east boulder clay deposits are widespread but mostly covered by younger wind-blown sands. There, such clays may locally occur in ice-pushed ridges together with e.g. Tertiary marine clays.

Fig. 3.3 Excavation in an ice-pushed ridge composed of fluviatile Pleistocene sands. (Source: De Mulder et al. 2003)

Except on top of such ridges most of these sands and clays are normally covered by periglacial medium to fine grained, mainly wind-blown sands deposited during the last glaciation (Weichselian: 115,000–11,500 years ago) when northern ice caps advanced close to the NE Dutch borders. During these times tundra, polar desert and periodic permafrost conditions prevailed. Wind-blown cover sands form slightly undulating landscapes often hampering natural drainage conditions in such areas hampering options for later agricultural use. Not all sand areas are high and dry, some of them are in rather wet topographical depressions.

The higher sands belong to the first landscapes inhabited by humans in the Dutch delta (Box 3.2). Early Neolithic people moved into these areas some 6300 years ago. Initially, they preferred the forested areas where they could practise primitive types of agriculture in high-quality humus ('Brown-Forest') soils. Upon exhaustion of the soils, people moved to other parts (shifting-cultivation). Exhausting soils forced people and settlements to migrate while urging them to invent or adopt new cultivation methods. Around 5000 BP mixed types of agriculture and successively heaths appear in the higher sand areas. In course of time, cultures and civilisations changed as shown e.g. in burial types. Earlier, burials in collective graves as the large *hunnebedden,* made of large Scandinavian boulders in Drenthe. Later, people were buried individually. Cattle was held in the open air until farms with connected stables were built from the Bronze Age (around 3800 BP) onwards. By then cattle's manure was used to fertilise lands, first as Celtic fields, small, square and walled parcels, dating from Iron and Roman Times (from 2800 BP). Such parcels were temporarily used as crop land and, upon exhaustion, left to recover. Cultivation was always controlled by soil and subsurface conditions. Presence of boulder clay in the shallow subsurface made soils more fertile but harder to elaborate, e.g. boulder clay plateaus as in Drenthe remained unexploited until Medieval people discovered tools to elaborate such tough soils and built their settlements.

Around 1000 BP agriculture in the higher sands increased. Abandoned heath fields were re- cultivated. The Dutch population (mostly beyond the higher sand areas) grew demanding for more grain and grain types including hop and barley for beer, and fax for linen to be produced. Population density was still low, on average 3 persons per km^2 (Vervloet 2010). Wooden farm houses were arranged around a central, common field (*brink*). While food demands further increased open field cultivation spread over all higher sand areas. This resulted in a strongly diversified, rather open and varied landscape. Common open fields were used by all, they were surrounded by walls made of sand or shrubs to prevent cattle to enter or to shelter from periodic sand storms. Cattle was supposed to roam in the barren areas outside the common open crop fields. Manure was collected to fertilise these fields, mostly combined with organic fertiliser from soils in the woods. Once supply of organic soils was exhausted, grass sods were dug from the surrounding sand areas and deposited on the common open fields, layer after layer. This is how pillow-shaped *es-* type complexes with thicknesses of up to 1.20 m, characteristic for the high sands areas developed. As a consequence, surrounding grasslands disappeared and turned into heathlands and over-exploited areas turned into bare-grounds. Meanwhile, the open fields raised as ever more sod layers were attributed.

Initially, Medieval farmers settled on the higher, sandy ice-pushed ridges. Since the twelfth century, however, they moved downward to the river valleys to benefit from better drinking water conditions. Farm lands remained at topographically higher positions but pasture lands migrated to the valleys while settlements laid in between. That lasted for several centuries and upon Medieval Times a new and quite successful type of manure collection was introduced in the higher sand areas. Cattle deposited manure in depressions made in the stable floors (*potstal*), and successively heath sods were added and mixed by cattle feet. Such mixtures were then used to fertilise the farm lands (Vervloet 2010).

Since Medieval Times the Noord-Brabant higher sand areas were more intensively populated (about 27 people/km^2) than Drenthe (less than 8). Drenthe mainly produced sheep and horses while more intensive cattle breeding occurred in Noord-Brabant. The higher sands in the centre of the Netherlands were exploited more intensively from the sixteenth century onwards. *Es*-grounds were kept barren to recover but were more permanently used for agriculture since the seventeenth century. That was done by digging more sods while heath fields deteriorated. Two centuries later common ownership of heath fields was abandoned and given out for different destinations, including reforestation for wood production in the higher and dryer parts.

Further population growth and emergence of the wool industry resulted in breeding more and more sheep which in turn led to wider heathlands and more open landscapes in the nineteenth century. Open landscapes, deforestation, falling groundwater tables and systematic removal of vegetation in the higher sands areas triggered sand drift during storms from the sixth century AD onwards with abundant Late Pleistocene cover sand resources at surface available. This process strongly increased in the mid-nineteenth century when human cultivation and the effects of climatic change (Little Ice-age) reshaped the higher sands landscape.

In the late nineteenth and early twentieth centuries demands for animal (sheep) manure dramatically declined as more and more artificial fertilisers became available. Moreover, domestic wool production became redundant by the wide supply of cheap, imported wool from Australia. Both factors dramatically changed land use in the higher sands areas. Century-long traditions of sod supply to common open fields and self-supportive farming practices changed into external supplies of nutrients and cattle fodder. The same held for traditional interdependence and coherence of agricultural and pasture lands. Education, infrastructure, general economic growth and an increasing role of the (local) government as a public landowner, spurred larger scale cultivation. Large-scale reclamation of un- and poorly developed higher sands areas began in the early twentieth century by new organisations as *Heidemaatschappij* (Box 3.1).

Vast regions were reshaped, also by parcel re-allotment. Sand dunes were planted with *Pinus* trees and heathlands were deeply ploughed and drained. Sand areas were levelled and turned into agricultural lands. By 1900, an area of 800 km^2 of active drift sands occurred in the Netherlands, today this is 1.5 km^2 only. For example, 20% of the Province of Noord-Brabant was covered by heathlands in 1900, 2% only of it is left today. Simultaneously, urbanised areas increased from 1% surface area in 1900 in Noord-Brabant

Box 3.1: Parcel Re-allotment

Parcel re-allotment has had devastating impacts on geomorphology and character of Dutch landscapes. This method to exchange land amongst land owners to their mutual agreement to arrive at more coherent parcels enabling more efficient farming practices, originated in the USA and took off in 1916 in the Netherlands. Organisations and companies as the *Nederlandse Heidemaatschappij*, (today's ARCADIS) and *Grontmij* (SWECO today), formerly involved in developing un- or poorly exploited regions in the Netherlands, served as implementation agencies.

During the 1930s economic crisis the Dutch government ordered major projects to rationalise and reconstruct rural areas, and simultaneous development of remaining uncultivated areas. Many thousands of people were employed in such projects and national food production was improved. In addition to mutual exchange of lands, local and regional soil & water management was improved and transport infrastructure in rural areas were updated as well. Small, winding often unpaved tracks were replaced by new, broad and straight asphalt roads, deep drainage pipes were installed and numerous deep and wide ditches were dug in strictly rectangular patterns. Mutual exchange of land was done on a voluntary basis in principle. However, if three or more farmers agreed, the government facilitated such agreement by handling the paper work. As not all farmers agreed, a Re-allotment Law was put in place in 1924 to force unwilling farmers to cooperate.

After WWII, parcel re-allotment was considered a great success and the Dutch government decided to continue and to significantly expand such activities in the 1950s (Fig. 3.4). Large areas became subject to re-allotment which by then also included levelling of the land. A disadvantage of such often drastic landscape interventions was the almost complete destruction of all smaller geomorphological structures and features in Dutch landscapes. That, in turn, also had a negative impact on biodiversity. For understanding landscape genesis and evolution today's geomorphologists often rely on historic maps and old air-photographs. In the past few decades other land use targets were introduced including designating larger and more coherent regions for nature preservation and development. Although farming practice improved and became more economical, re-allotment of parcels has made Dutch landscapes less diversified and distinctly more uniform.

to 15% in 2004 (Jongmans et al. 2015). Eventually, large-scale land reclamation halted in 1961 when nature conservation became a policy issue.

Where cattle breeding has been a minor activity in traditional farming here, since the agricultural crisis in the 1880s this type of farming gradually and initially slowly became dominant in the higher sands areas. Cropland became grassland and maize production replaced traditional crops. Irrigation has become an important tool for adequate land use in the higher sands areas to combat impacts of droughts. Phosphate and nitrate concentrations in the higher sandy soils in the Netherlands often exceed EU limits and these belong to the most polluted soils and related groundwaters in Europe. Abolition of the EU-policy on milk quotas in 2015 resulted in increasing cattle stocks and more manure-linked phosphate pollution.

Nine out of the 20 National Parks in the Netherlands are situated in the glacial and periglacial higher sands and constitute the backbone of the Ecological Network (EHS) in the Dutch delta. Between 1850 and 1950 more than 95% of all heathlands and other types of uncultivated lands in the Netherlands were turned into agricultural lands. Current Dutch policy is geared to transform many of these into heathlands and other former land types again. One of the main challenges is to de-fertilise the sandy soils. To that end humus-rich top soils are being excavated today. Moreover, the higher sands areas are often too dry to meet EU biodiversity standards. Therefore, groundwater levels are being raised. However, this may mobilise chemically-bound phosphates contributing to further eutrophication of these areas. Landscape engineering for nature restauration may thus generate adverse effects. Other measures to increase biodiversity include diversification of mono-cultural forests on the higher sands by additional hydrological measures, by replanting new, endemic trees, and by restoring inland drift sand areas in some relatively poorly populated regions.

3.2 The Loess and Limestone Area

This area shows distinct topographic diversities and is quite untypical for the Netherlands which is otherwise a dominantly flat country, their distribution is smallest of all Dutch landscapes and exclusively occur in the southeast (province of Limburg; Fig. 3.2). They are part of the tectonically uplifting eastern and southern rims of the North Sea Basin. Late Cretaceous flint-containing limestones are exposed at the surface in the southern part of Limburg only and display escarpments, hills and rather steep river valleys. River Maas valleys often consist of gravel-bearing terraces reflecting various Late Tertiary and Pleistocene incision stages of this river and tectonic uplift in Limburg.

Limestones and river terraces are almost always covered by wind-blown, fine sandy to silty loess beds of Middle to Late Pleistocene age giving this landscape its undulating character (Fig. 3.5). Loess deposits may reach thicknesses of up to 35 m (in Roerdal Graben); these may locally extend beyond Limburg and may occur in the centre and east of the Dutch delta as well. In general, loess soils are rather deeply (2.5–3 m) decalcified. Very locally, however, at some of the highest spots in this landscape (e.g. the *Ubachsberg*) loess is lacking. In general, loess deposits are fertile, easy to elaborate, permeable but retaining water rather well in the soil. In short, loess belongs among the best soils for agriculture.

Not surprisingly, agriculture in the Netherlands first appeared in the loess area, about 7300 years ago when early Neolithic people (re-)occupied the loess region (Box 3.2). Pollen contents of loess demonstrate that these people had almost no impact on landscape and vegetation. That changed around 6000 BP when successive human cultures began to exploit and elaborate the widespread Cretaceous flint occurrences while simultaneously expanding cattle breeding. During the Late Bronze Age (3100–2800 BP) larger loess areas were used for crop production. Forests had been cleared causing soil erosion along hill slopes and successive deposition of anthropogenic colluvial deposits along the base of such slopes. Erosion processes intensified when population grew in the Iron Age and even more so in Roman Times (500 BC to 250 AD). The Romans built several major terrestrial transport routes connecting their garrisons and fortifications in Trier and Xanten and between Cologne and Boulogne in the west. Where these transport axes crossed, the settlement of Coriovallum (present day *Heerlen*) developed. Along these roads large grain-producing farms and villas were situated and most of the large plateaus were cultivated.

After Roman Times population drastically declined but the best positioned settlements along River Maas and the valleys remained occupied. These sites had access to (surface) water, a major asset as groundwater is deep in Limburg. Forestation recovered on the now abandoned plateaus until population grew again from the eleventh century onwards. The old villages along the valleys created daughter settlements higher-up on the plateaus from where these were re-cultivated (Renes 2010). Around 1300, almost the whole region had been developed and (drinking) water wells had been dug. Forests survived in the steepest and most infertile parts (near *Vaals*) of this landscape only.

In the nineteenth century most of south Limburg was covered by crop lands, even the erosion-prone slopes. Grassland and orchards were placed close by settlements. Most a

b

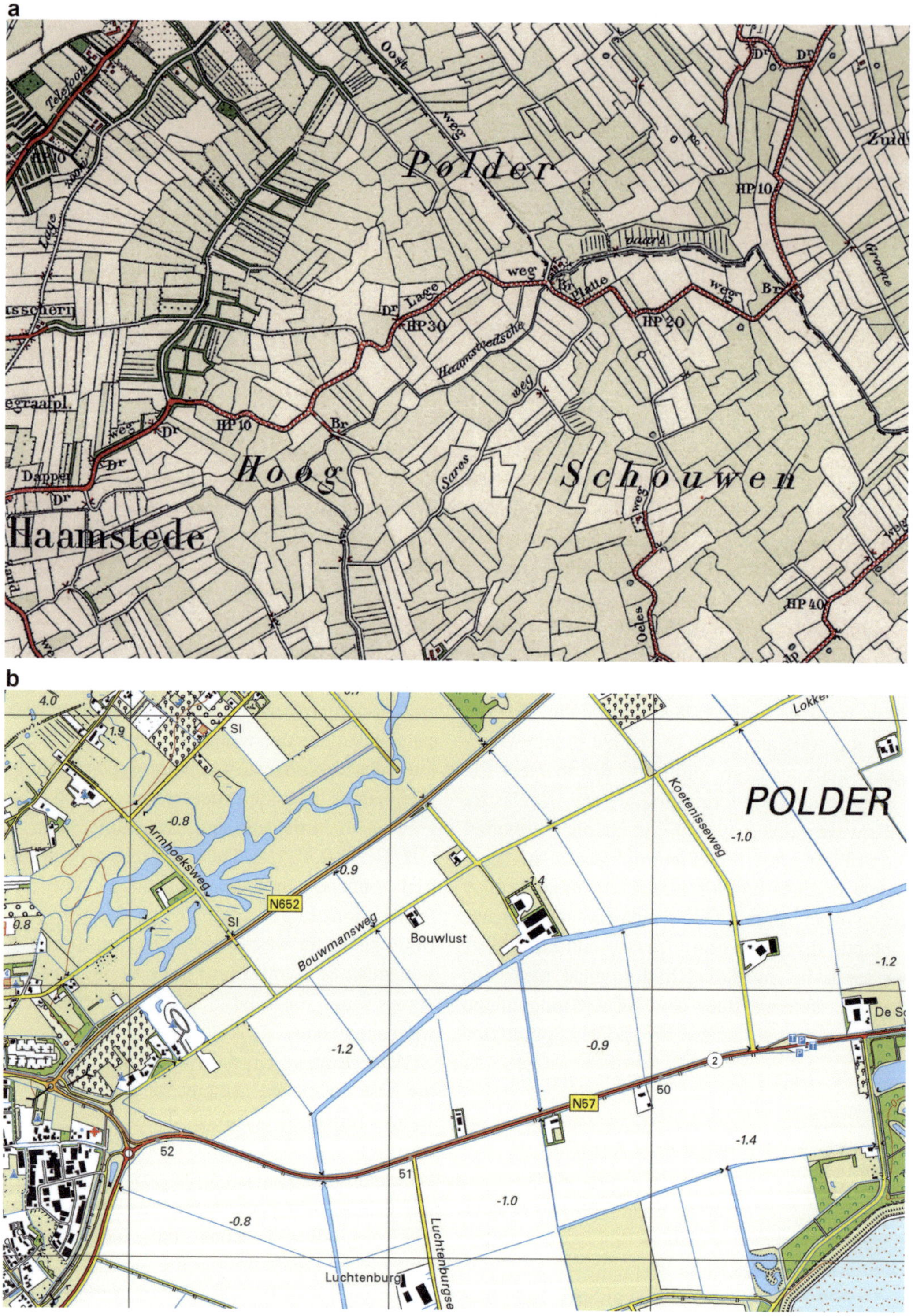

Fig. 3.4 An example of parcel re-allotment in Walcheren, Zeeland. Top: before parcel re-allotment, bottom: after re-allotment. (Source: Barends 1989)

Fig. 3.5 The loess and limestone landscape in south Limburg. Limestone beds covered by thick loess deposits resulting in undulating landscapes with relatively small grass parcels bordered by scrubs. Steeper slopes are often covered by forests. (Photograph by Martin van Lokven)

farms were mixed types: they produced crops but held cattle as well. Initially, grasslands were public property. New agricultural techniques including fertilisers were applied to feed increasing populations. Among these techniques supplying limestone dust was used particularly here. Mining contributed significantly to population growth in Limburg but considerably changed landscapes as well, mainly since 1900. Mining led to overall prosperity for the miners, many of them former farmers. Thus, agriculture became less intensive and more and more farmlands turned into grasslands, often associated with fruit trees. In the 1960s, coal mine tailings occupied major areas in the loess and limestone areas. After closure of the coal mines around 1970, they were partly removed and dumped in abandoned mining pits. Other mine waste hills have since been used for leisure and touristic attractions.

Over the past century significant parts of the landscape in south Limburg have been removed by limestone excavation at a rate of 1.5 million tonnes average per year, in the period 1980–2000 (De Mulder et al. 2003). Gravel excavation from the terraces and deep in the valleys further downstream river Maas have made the province of Limburg the second largest in water surface (after Friesland). Despite of its attractive character, no National Parks have been designated in the loess and limestone landscape. Numerous projects geared at turning crop lands into grassland and simultaneous reduction of fertilisation by cattle grazing have been initiated in this landscape. However, except the cities, the entire region of south Limburg has been declared a National Landscape (*Heuvelland*) and is part

Box 3.2: The First Dutch

About half a million years ago humans crossed the Pyrenees and Alps from the Mediterranean realm to settle in northern Europe. They were hunter-gatherers and lived in small, isolated groups, well-equipped to sustain deteriorating climatic conditions. Well-preserved remains of 350,000 years old hunting spears in nearby Schöningen (Germany) witness of their hunting habits. As in the Dutch delta almost all simultaneous sediments have been covered under thick layers of younger deposits since, direct or indirect remains of these earliest humans have not been found (yet). Pre-Saalien flintstone tools have been identified in fluviatile sands in some ice-pushed ridges. In an excavation near the city of Maastricht (province of Limburg), well-preserved remains of a camp site and tools, dating back 245,000 years ago were unearthed. Such tools were made of flintstones which occur abundantly in the limestone landscape in south Limburg. Neanderthal people must have lived in the Dutch Delta too, as witnessed by discovery of (a part of) a Neanderthal-skull in sediments presently in the bottom of the North Sea, off Zeeland. About 33,000 years ago, during the Weichselian glaciation when the Netherlands suffered from severe periglacial conditions, *Homo neanderthalensis* made place for *homo sapiens* in this region (NEA 2016).

of a larger, cross-border Landscape covering adjacent parts of Germany and Belgium (*Drielandenpark*).

The twenty Dutch National Landscapes (Fig. 3.15) combine (small-scale) agriculture, nature and cultural history of a region, while the National Parks are solely meant for nature preservation, nature-linked recreation, nature education and research.

3.3 Peatlands

Peatlands are assigned to (mostly former) occurrences of peat at or near the surface. Peat growth has been quite common in the Dutch delta since the onset of the Holocene and more than half of the Dutch territory was once covered by peat. Most of the peatlands developed since the closing process of the coastline that began around 5000 BP. That hampered river discharge in the North Sea while groundwater levels rose simultaneously as a consequence of the Holocene sea level rise. In combination with increased all-season precipitation and cooler weather conditions in the course of the Holocene, this created rather wet and swampy conditions in expanding areas behind and east of the Dutch coast. When and where plant production exceed their decay, peat may begin to accumulate. Such conditions prevailed in wide realms in the Dutch delta eventually resulting in massive peat growth which peaked about 2500 BP (= 500 BC).

Timewise, peat growth normally starts with water plants growing under high and often stagnant and nutritious groundwater conditions as eutrophic fen peat. If sufficient plant remains have accumulated, peat may grow further, while being fed by rainfall only, as oligotrophic high moor or bog peat which developed as relative topographic highs in the peatland-scape. Such successions occur regularly in vertical peat sections.

Also geographically such distinctions occur. Fenlands were dominant in the lower west of the Netherlands while high moor bogs prevailed in the higher parts of the country, in the south, the centre and the north. Their position in association with (former) groundwater tables have also been responsible for the excavation techniques applied. Peat excavation in lower fenlands was normally done while creating fresh water pools but peat in bogs could be dug under rather dry conditions.

Box 3.3: Holland

Often the term 'Holland' is used as an equivalent for the Netherlands. However, that is incorrect. The Netherlands is part of the Kingdom of the Netherlands, which is a nation composed of four countries: the Netherlands, Aruba, Curacao and Sint Maarten. These four countries are equal parts of the Kingdom but their political position is not the same. Although the Caribbean islands are autonomous, the Netherlands represents them, both internally and externally.

Holland is an informal term that refers to a region in the west of the Netherlands only, the area roughly covered by the provinces of Noord- en Zuid-Holland. 'Holland' is derived from the term 'Holt' land. 'Holt' stands for the (present) Dutch word 'Hout' which means wood. So, Holland seems to refer to Wood Land. But no woods occur in this region today. However, abundant forests must have existed, in particular along the borders of the main peat areas in the tenth and eleventh centuries. Historic archives show that the term 'Holtland' was first used in the tenth century for places in the centre and west of the delta. Archives also show that Count's Floris II name was associated with 'Hollant' for the first time in 1101. In the eleventh century inhabitants of his county were supposed to associate themselves with Holland rather than with 'Frisia' as this county was called earlier.

An alternative explanation for 'Holland' might be found in Dutch speaking Northern France (Flemish-France) where 'Holland' might refer to a waterway, the 'Holland-becque', situated inland from the town of Duinkerke, as described in written sources dating from the ninth century.

Not everybody in the Netherlands is happy with the term 'Holland'. People in the southwestern province of Zeeland ('Zeeuwen') did not always share the same economic interests as the rulers of Holland and fiercely opposed the 'Hollanders' occasionally (Chap. 2). People in the southern provinces of Noord-Brabant ('Brabanders') and Limburg ('Limburgers') usually do not wish to be associated with Holland at all. They tend to use 'Holland' as a nickname.

3.3.1 Fenlands

On a European scale, peatlands have been the latest of all landscapes occupied by humans. Human occupation of the Dutch peatlands began in the west in Iron Times, about 2800 BP (= 800 BC). It took another 600 years before they settled on peat lands in the north. But humans lived here well before peat lands began to develop. Archaeological evidence from

Flevoland (*Swifterbant*) proves human occupation there about 6000 BP. They lived on the levees of tidal channels. During peat formation (i.e. roughly between 5000 and 2000 BP) they must have abandoned these swampy and hostile areas and moved to higher places from where they began exploring and exploiting surrounding peatlands areas. Farmers near the estuaries and river mouths began entering the fringes of the peat areas, primarily for cattle breeding (De Bosatlas van de geschiedenis van Nederland 2011). First indications (Early Iron Age) for fenland peat exploitation were found from a farm house built on peat, 15 km west of Amsterdam. In Zeeland fenland peat exploitation began a few centuries later, in the 2nd century BC (Late Iron Age). There, they began draining the fringes of periodically flooded peatlands to collect salt by burning peat. Such peat contained significant amounts of sulphur and was less suited for fuel but good enough for fertilisation (Borger 2010). Salt was a valuable commodity and was primarily used for food conservation and widely sold to neighbouring regions.

In Roman Times, settlements developed close to Roman fortresses along river Rijn (*Limes*), the border of their empire. Agricultural products were traded there, driving growing populations deeper into the hearts of the peatlands which were drained and prepared for cattle. That generated major subsidence and eventually caused the natural coastal defence system to fragmentise and collapse. Marine floods eroded the peatlands turning these into a tidal flat area with channels and sandy beach barriers. In the second and third century AD, Romanised, native people erected temples dedicated to a local goddess *Nehalenia* on some of these beach barriers (Lendering 2000). However, soon hereafter marine floods destroyed these temples many of which have been found in recent times. Unwillingly, flood conditions were facilitated by draining the peatlands. Tidal channels penetrated deep into the hearts of the peatlands while eroding large volumes of the poorly resistible fenlands. In Zeeland such marine invasions occurred about 270 AD (Vos 2015). To the north, in Holland, peat growth continued behind an unbroken chain of beach barriers.

Initially, reclamation of the fenlands was on a very local basis only in the direct vicinity of farms to satisfy their agricultural needs, mainly for cattle breeding or to collect salt. This resulted in small-scale and irregular drainage patterns. Such block patterns frequently occur in Noord-Holland and may point to a lack of cooperation between groups of colonists. Natural drainage was provided by the dome-shaped surfaces of the peat bogs where rainwater collected in small rivers or streams. Peat exploitation often began where such streams discharged into major rivers. Peat streams normally served as the bases from where ditches were dug further into the peatlands. Natural drainage allowed cattle to graze and farmers to grow crops. But, both drainage and agriculture caused rapid oxidation of the top parts of the peat. Peat's surface level went down and peat became wet again. That prompted peat developers to exploit and drain higher and more inland peat bogs. Rising Holocene sea levels made groundwater levels to rise too, requiring artificial drainage, in turn accelerating further land subsidence. Overall and over the centuries, peat surfaces in fenlands have dropped some 4 m, locally more, bringing peatlands only a few decimetres above sea level while making these lands extremely vulnerable for flooding.

Written statements witness of peat development as early as 800 AD, near Texel. Early large scale peat development was reported from the region near the current Frisian town of Sneek, in 975 AD and around 1000 AD peat colonisers developed an area near Amsterdam and dug calcareous clay under the peat layer as fertiliser. Today such clay pits may be seen in winter landscapes (*daliegaten*). Real systematic peat exploitation commenced in the tenth and eleventh centuries. Prime aims of this and other types of reclamation were agriculture and cattle breeding. De-population of the agricultural peatlands stimulated land owners to sell their unexcavated peatlands for fuel production (Borger 2010). By then, nobility and clergy possessed most of the fenlands wilderness. They ordered for large-scale peat reclamation. In course of time, however, free farmers became the owners of such peatlands. Reclamation was done by systematic digging; most frequently applied was the so-called Cope development. 'Cope' stands for a contract between owner and exploiter regulating mutual rights and duties. This type of development is characteristic for the Holland-Utrecht peat landscapes: over 1 km long straight ditches were dug into the hearth of the peatlands at fixed locations. These regular strips of land were arranged perpendicular to the access river used as development basis. If no access stream or river was available, new canals were dug serving as development basis to reclaim the target area. Parallel, other ditches were dug at fixed distances of 100 m or more. Natural fenlands were fully remodelled into lands full of parallel ditches (Figs. 3.6 and 3.7). Many new settlements developed in this region, between 950 and 1300 AD. Most peat exploiters or colonisers came from Friesland and from the dune regions. This systematic methodology has been exported abroad, to Germany where such expertise and their workers were hired by the Bisschop of Bremen in the early twelfth century (Borger 2010). In the north of the country reclamation begun from the higher Pleistocene sand areas.

In contrast to Holland, most of the peat in Zeeland was excavated for salt collection, at a small scale since the Iron Age and Roman Times but on a much larger scale in the eleventh century. It boomed around 1300 AD and went on until the seventeenth century (Jongmans et al. 2015). Salt production from peat has been a prosperous business for Zeeland while attributing greatly to their significance and rather independent position in the Netherlands. The downside

Fig. 3.6 Characteristic reclamation pattern of peatlands: cope reclamation in Tienhoven (Utrecht)

of this type of peat excavation has been its significant contribution to the loss of land due to marine erosion.

As in Zeeland, excavation fragmentised the extensive fenlands and peat bogs in Holland and erosion further contributed to this fragmentation. More and more shallow pools expanded into lakes emerged which rapidly widened further by wind action and by marine and river floods. Peat exploitation in the fenlands often continued below the water level by means of a spoon-like device (*baggerbeugel*), invented in Brabant in the fifteenth century and shortly later introduced in fenlands in Holland and Utrecht. Reclamation of some of the smaller lakes began in the sixteenth century and the construction of larger polders developed in the seventeenth century (Chap. 2).

By today, very little peat of the former fen lands is left in Zeeland and in the west of Noord- Brabant. In the north and centre of the country some remains of the vast peat landscape may still be present along the borders of the Pleistocene sands. A century ago, land use in the remaining fenlands was limited to grass land and cattle breeding only. Apart from these traditional applications most of the current fenlands are used for horticulture and for maize production, introduced only recently here. Fertilisers and improved water management were used to intensify crop production on the remaining fenlands in the twentieth century.

Urbanisation and infrastructure has been expanding, also in peat areas. Land subsidence is still a significant feature in fenlands, active and accurate water management is crucial there. Higher water levels may slow down land subsidence.

Another, promising method is the application of underwater drainage by means of which water may also be infiltrated in the peat during dry weather conditions. Construction works in peatlands is another challenge due to its low bearing capacity. Apart from pile foundations to be applied even for all roads and sewerage systems in peat areas, sophisticated methods have been developed to overcome such problems, including application of imported, light-weight volcanic ash and polystyrene foam elements providing significant bearing capacity.

To human interference also belongs the conservation and protection of the few remaining peatlands. Since the mid-1970s nature conservation and nature protection and, since the 1990s, nature development and management belonged to ruling policies in the Netherlands. Such policies

Fig. 3.7 Cope reclamation in the west of province of Utrecht (Source: Poelman 1966)

emerged from the national government and were based on agreements among EU member countries to actively promote and implement landscape diversification and biodiversity. To that end scattered pieces of EU legislation were brought together and the ambition to physically connect all European nature landscapes into an EU Ecological Network had been leading. Implementation was delegated to the provincial governments (Chap. 10). Following this ambition three main types of policy measures were taken: transforming cultural landscapes into nature landscapes, agricultural management and transforming existing nature reserves.

This policy also affected fenlands. With public funds nature organisations acquired peatlands and successively transformed these into nature reserves. Biodiversity was often stimulated by hydrological measures and by artificial modification of the landscape (landscaping). On the other hand peatlands also were to be de-fertilised, i.e. N and P contents in the peatlands were to be reduced and water tables were to be raised, returning these into swamps. Water quantity and water quality was strictly controlled in these new nature reserves and costly measures including hydrological isolation techniques have been taken in some places.

A second step towards establishing an ecological network was discouraging farmers conducting their conventional farming practices. They were now subsidised to surrender their lands to nature, to stop cutting their grasslands early in the season to protect ground-breeding birds, to exclusively use natural fertilisers and to let their lands to become wetter. Finally, existing nature reserves in the fenlands were to be turned into 'original' peatlands. Landscape diversity in these reserves was increased by dredging patchy patterns with open water to regenerate vegetation successions ultimately leading to peat formation. On the other hand, unmanaged peat landscapes will turn into brook-forests.

As cultural historic core characteristics of Dutch fenlands are under threat, the government assigned five National Landscapes in the Dutch fenlands to protect such core qualities, including their wide and open character, their typical historic reclamation patterns (e.g. the cope-type of ditch digging), their dominant land use as grassland and their typical high water tables.

Box 3.4: Landscaping

Dutch authorities never showed much concern for natural heritage issues and the cultural historical values of landscapes as e.g. may be deduced from the devastating effects of parcel re-allotment on the original geomorphology (Box 3.1). However, that changed in 1999 when the Dutch government issued a policy vision (Nota Belvedere 1999) concerning the relation between cultural history and spatial planning. This policy vision acknowledges the relevance of cultural historic issues as a source of inspiration in landscape architecture and spatial design and aims to prioritise cultural historic values in spatial planning. A new balance between these issues should be found as Dutch cultural heritage is at threat. Unique landscape elements and in the built environment should not only be preserved but revitalised as well, also for educational purposes. Moreover, that would improve ecology, aesthetics, biodiversity, economy (tourism) and would counter loss of landscape relief. As a first step, cultural historic values of landscapes and built environment have been mapped resulting in assigning 70 localities with a 'Belvedere' status which eventually became part of the 20 National Landscapes. These localities receive special protection and provinces and municipalities have been charged with implementation of this policy.

3.3.2 High Moorlands/Raised Peat Bogs

Once most forests had been cleared alternative types of fuel had to satisfy demands of growing urban populations and of their emerging industrial activities (Fig. 3.8). That alternative fuel was peat, in particular bog peat. Fen peat was less suited for fuel than bog peat as the first often contained sand and clay particles as this peat was fed mainly by groundwater and by streams instead of the bogs that were rather fed by rainwater only. Peat for fuel had been applied earlier but only on smaller scales.

Extensive peat bogs occurred both in the coastal plain and in the higher sands areas. Higher positioned peat bogs (high moorlands) typically and widely occurred in the north, the east and in the south of the Netherlands during several thousands of years. Peat growth in the higher sands areas began about 6500 BP on top of Pleistocene cover sands when groundwater levels began to rise there as well. Peat accumulation intensified about 4000 BP when climate conditions became wetter and peat growth continued well into Medieval Times. In the north of the Netherlands raised peat bogs extended almost continuously over Friesland, Groningen, Drenthe and Overijssel and further eastwards into adjacent Germany. The largest of these occurrences (*Boertangerveen*: 1600 km^2) once covered the entire east of Groningen, parts of Drenthe and Germany.

Larger-scale peat cultivation and exploitation in the raised peat bogs did not start before the fifteenth century (Jongmans et al. 2015). Excavation of peat has been relatively easy here as these peat occurrences were all positioned well above the (ground)water table. Large scale peat bog exploitation for

Fig. 3.8 Peat landscape, partly excavated in the province of Zuid-Holland

fuel commenced from their fringes along the Pleistocene cover sand areas and migrated towards the bog's cores, starting on an individual basis only. Here, no drainage canals were dug which initially caused high groundwater levels. Irregular primary peat excavation in combination with the underlying cover sand landscape and successive differential settlement contributed to an undulating morphology in the high moor peatlands.

Fuel scarcity increased values of the remaining, most oligotrophic peat land in the west once the eutrophic fenlands were cultivated for agricultural purposes. There, more and more peat-landowners ordered their lands to be excavated. Special companies (*compagnien*) brought capital of rich Holland people together with local expertise; they closely followed demands from economically active urban populations and associated industries, beginning in Flanders (Brugge, Gent, Antwerpen) in the twelfth to thirteenth centuries AD. That ignited peat bog excavation in Zeeuws Vlaanderen (around Axel and Hulst), and in West Brabant, often initiated by cloisters and rich citizens. Successively, this type of excavation activities moved northward to western Brabant (fourteenth to fifteenth centuries), the delta centre in the sixteenth to seventeenth centuries, to Friesland and Groningen (seventeenth to twentieth centuries), and southward again in the nineteenth and twentieth centuries.

As most peat bogs were depleted earlier, no professional excavation teams have been active in Holland. This was rather done here by individual and independent farmers from the fifteenth to the nineteenth centuries (Leenders 1987) supplying peat to industrialising cities as Haarlem and Leiden, where peat for fuel was needed for beer breweries, laundry services, linen weavers, and brick bakeries as wood was no longer allowed for urban construction, due to the fire risk.

Holland has been self-sufficient in terms of energy for many centuries by using peat as a fuel which significantly contributed to its economic (and cultural) growth in the sixteenth and seventeenth centuries ('The Dutch Golden Age'). Ample availability of peat as a cheap fuel gave the Dutch a comparative economic advantage viz a viz other European nations.

Both excavation and development of excavated peat areas have changed the sixteenth to eighteenth century landscape quite considerably. New landscape types developed in the northern provinces and in De Peel: the *Veenkoloniën*. In contrast to early, irregular peat excavation along the cover sands and ridges, peat excavation in the centres of the peat bogs was done on a quite systematic basis. Canals were dug in perpendicular patterns while farms were built alongside. In the north, 46 such canals with a total length of about 700 km were dug since the sixteenth century (Jongmans et al. 2015). The city of Groningen owned and managed several of these canals, including the *Stadskanaal*, and therefore played a significant role in the regional excavation history (Stol 2010a). Before excavation, peat was burnt there to accumulate mineral content for buckwheat production. In 1880, 8000 ha peat was burnt annually for this reason (in adjacent Germany: 85,000 ha/year), generating abundant smoke pollution. Due to massive peat excavation for fuel purposes very little peat for cereals production remained and burning was almost over by the end of the nineteenth century in this part of the Netherlands whereas peat burning remained allowed in adjacent Germany until 1932.

Since the end of the nineteenth century systematic peat excavation went along with reclaiming the underlying Pleistocene cover sand ridges. These were levelled, top peat layers of the adjacent peat bog were removed and laid on top of this sand and was successively covered with some 30 cm of sand dug out of the canals. Successive ploughing resulted in mixing brought-up sand with peat, resulting in proper soils for agricultural use. Typically, remains of the cover sand morphology were better preserved in peatlands excavated prior to the twentieth century than in later excavations. Landscapes left after complete peat excavation consisted of infertile cover sands. Introduction of artificial fertilizers in the twentieth century made these soils ready for agriculture. Peat remnants left after excavation or by natural causes have also been homogenised with the underlying Pleistocene sands.

Oligotrophic peat from raised bogs/high moorlands are well suited, and have extensively been used, for horticultural purposes, roughly since 1870. That was allowed only if excavators would return at least 40 cm of peat into the soil to be covered with sand to improve soil quality upon excavation. Since 1550, some 100,000 ha (1000 km^2) peat in high moorlands and some 42,000 ha (420 km^2) of peat from fenlands have been excavated or dredged in the north, corresponding with 3.5×10^9 giga-joules of energy. From the onset peat dug in the north supplied 15–40% of the total Dutch demand for energy.

Peat excavation in the southwest of Noord-Brabant and beyond the Dutch border in Flanders was practically over in 1750 when all peat resources were depleted there. In contrast to the north, the sand lands left after peat excavation turned into heathlands since and were not reclaimed before the late nineteenth and early twentieth centuries. The major, higher and flat peatlands situated along the borders of Noord-Brabant and Limburg (*De Peel*) had been exploited since Medieval Times, first on an irregular basis and since the mid nineteenth century also at an industrial scale. As in the north, the topmost peat layers were returned and re-introduced in the remaining sand lands. Initially, such reclaimed lands were almost exclusively used for growing tobacco, later for other crops. Peat was excavated also for horticultural purposes in *De Peel* until 1950. Now, De Peel is a nature reserve where renewed peat growth is stimulated.

With their disappearance in the east of the Netherlands, peat bog's function as water regulators was lost as well. Peat bogs acted as sponges, they absorbed water in wet times and delivered water in dry times. Urbanisation has been another major change in land use since 1900. Excavated, barren sand areas are susceptible to wind erosion and sand dune formation in windy seasons. Excessive use of fertilisers and manure contaminated many of the former peatlands and made these too nutritious (e.g. De Peel) for turning these into proper nature reserves.

Today, former high moorlands in the south belong to the landscape of the higher sand areas as described in Sect. 3.3.2. By the year 2000, only 8400 ha (84 km^2) remained as high moorland/raised peatbog in the Netherlands, mainly used as nature reserve today. Although strongly impacted by humans these landscapes are considered valuable as they still contain parts of the natural biological and geological archive in a country that is otherwise very much affected (and even partly created) by humans. Two areas here have been assigned as National Landscapes.

3.4 The River Belt

The most prominent rivers (Rijn and Maas) and their associated landscapes in the Dutch delta occur in a broad belt from the east and southeast (upstream) to the west (downstream). River Schelde flows from the southwest to discharge further north in the delta while a branch of River Rijn (IJssel) runs northwards from the east to discharge in IJsselmeer in the centre of the Netherlands (Fig. 3.9; Fig. 1.1). River deposits constitute most of the Dutch delta body both in a lateral and in a vertical sense. Dutch river landscapes meet the marine realms in the west and are surrounded by (former) peat landscapes in the centre and in the west. Further east the river belt is bordered by the higher sands areas in the north, the east and the south. In the extreme southeast River Maas cuts through the limestone and loess landscapes.

Dutch river landscapes are quite diverse and depend on their position in the delta. Upstream, river landscape sediments are coarse-grained (sand and gravel), downstream fine-grained (silt and clay). Geographically, the Dutch river belt may be split in an eastern, a central and a western belt.

The Dutch delta developed since 50 million years ago and consists of several kilometres of mainly river sediments, now deeply buried. Except in the east and south most Holocene

Fig. 3.9 IJssel river landscape in the province of Overijssel

river deposits at surface are clays. Pleistocene river sands are generally overlain by wind-blown cover sands generating slightly undulating landforms. The place in the delta where erosion dominates over sedimentation (terrace intersection point) moved eastward due to Holocene sea level rise, from Rotterdam in the Early Holocene to some 15 km east of the Dutch border now. Braided river types characteristic for the Pleistocene ice ages were succeeded and overtopped by meandering river types in the Holocene. In the east and southeast braided river type sands and gravels dominate. They may be covered by younger, fluviatile clays resulting in a topographically varied river landscape with gullies and river banks. Along former river courses, river dunes may have developed in the Late Pleistocene to Early Holocene standing out as topographic heights even if they have largely been covered by younger river deposits.

The central river belt is dominated by deposits and landforms associated with meandering rivers. There, topographically higher and more sandy river banks are alternated by poorly drained, broad flood basins filled with river clays and peat, expanding in western directions. Topographic height differences between sandy river banks and the clay and peat depressions may vary between 50 cm in the west and 1.50 m in the east but generally less pronounced than in the eastern river landscapes. In the centre various generations of river sediments and meandering river channels may cross-cut and be adjacent to each other. River dunes do occur here as well and may be (partly) covered by younger fluviatile clay and peat beds. In the centre clastic river sediments generally dominate and outnumber peat beds.

In the west, the river belt is characterised by indirect influence of tidal movements from the advancing Noordzee. That caused prevailing stagnant and swampy conditions generating abundant peat growth during the Mid and Late Holocene. In vertical sections thinner fluviatile clays alternate with thicker peat layers. Local river dunes may be the only natural topographic heights in the otherwise flat western river belt. Here, all fluviatile sediments are young.

The delta has been a good place for early Dutch to settle as the rivers in the delta provided food and transport facilities which enabled them to trade with people both in the delta's hinterland and overseas. Human settlement in the river belt began on the higher and sandy river dunes from the Late Palaeolithic onwards and on elevated sandy river banks, also in the current west of the Netherlands as the river mouths were far in the present Noordzee until the Late Neolithic. Numerous Neolithic archaeological finds have been reported from such sites. Bronze Age occupation is known from upstream river Maas and their settlements have also been identified in the western river belt, covered by younger river clays which normally hide early settlements there.

More people lived in the river belt in Roman Times. Food had to be produced for the Roman and allied local armies along the *Limes*. Rivers Maas and Rijn then served to transport freight while troops normally moved via roads, e.g. connecting the Roman cities of Maastricht and Nijmegen. Romans have been the first who interfered in Dutch river courses by hydrologic measures (Box 2.1). Early occupation of the downstream parts of the delta is witnessed by rather primitive hydrological measures, including construction of low dikes since about 1000 AD. Numerous small but rather (up to 12 m) deep ponds along current and former river dikes point to former dike bursts. Especially river dikes built on top of buried sandy channels were vulnerable for dike bursts due to piping.

Settlements and towns along the downstream parts of the major rivers were often threatened by river floods from the eleventh to twentieth centuries (Chap. 2). There, flood risk grew both by deforestation in the hinterland and by peat excavation. River flood prevention measures by constructing more solid dikes and dams since the thirteenth century certainly helped to better drain the surrounding fenlands in the central and western river belts. That caused major land subsidence and differential settlements resulting in inversed relief forms: former sand-filled channels turned into topographic ridges and relative high peat lands were drained deeply and turned into flood basins. Another way of flood prevention was by building dwelling mounds, practised in the river belt much later than in the north.

Population grew in early Medieval Times and most of the present river belt towns and cities originate from then. Two types of villages developed there: (1) From scattered farms around a central *brink*, in the east (Merovingian Age or older), and (2) Stretched, along and in between two parallel roads, mainly on inversed sandy creeks in the west (Carolingian Age). The higher river dunes have always been attractive sites for settlements. Most parts of the sandy hinterlands in upstream delta were covered with forests dimming irregular and sudden river water supplies; many of these woods were cut in Roman and certainly in Medieval Times.

Agriculture has been dominant for centuries in the central river belt, both on the river dunes and the levees and on grasslands in flood plains. Cattle manure was applied as fertiliser for agricultural lands. But when population grew further together with more demands for food farmers began to exploit the wetter river forelands for agriculture and cattle breeding too. With the invention of better ploughing techniques even solid floodplain clays were cultivated by digging small canals (*weteringen*) mainly occurring in the province of Utrecht.

By the end of the nineteenth century fruit plantations were introduced and successfully extended on the river banks. Post-WWII parcel re-allotment improved production and hydrological conditions for agriculture in the river landscape (Box 3.1). Coarse ceramic industry producing bricks and roofing tiles, excavated river clays in the forelands and may have had devastating impacts on landscapes along some rivers. Moreover, the typical Dutch river landscape changed by urban expansion which may locally extend into floodplains, and by the construction of highways with their associated noise isolation screens, bridges and by railway lines, mainly for cargo transport, and by large-scale sand and gravel excavation.

Managing river courses in the Dutch delta belonged to the main tasks of Rijkswaterstaat (Chap. 2). Upon its creation in 1798, this public body was faced with three major river challenges: (1) distribution of waters between rivers Rijn, Waal and IJssel; (2) condition of river winter beds; and (3) position of confluent rivers. The first issue has been resolved by construction of *Pannerdens Kanaal*. A new river mouth was made and a new 'river' (*Brielse Maas*) was dug; river Maas was shortened to half by cutting 10 of its meandering river curves, to cope with both last issues. Most of these works were completed during the nineteenth century.

Not least, the recent *Ruimte voor de Rivier* Programme (Chap. 2) has further and locally intensively reshaped the characteristic central and western river landscapes, primarily to increase water accommodation capacities of the rivers. River beds and floodplains were deepened, new side river branches excavated, summer dikes removed and some local dikes strengthened. To safeguard old villages and farmhouses, dike reinforcement works were conducted mainly along their external sides. Some lands were designated as potential flood accommodation basins.

Within the river embankments much has changed as well. High-stem orchards were exchanged for low stem fruit plantations since WWII, parcel re-alottment improved agricultural efficiency, high- and railways considerably fragmentised the river belt and changed its typical character, but also broke its isolation. River belt towns became commuting sites for many people.

As a consequence of current policy views nature areas were significantly extended as part of this Programme. One National Landscape and UNESCO World Heritage site deals with the (western) river landscape: the Holland Water Defence System (*Nieuwe Hollandse Waterlinie*) (Box 2.1) which is basically a man-made system of fortresses, dikes, canals and inundation facilities. Another National Park (*Utrechtse Heuvelrug*) belongs to the central river belt. This region is part of the National Landscape *Rivierengebied* too (Fig. 3.15). Here, rivers and their forelands are the main constituents. Together, these 'blue-green corridors' aim to connect

German and Dutch nature areas. By 2015, some 7000 ha agricultural lands had been transformed into new nature.

River Maas' water discharge is currently being improved in the eastern river belt again by the Project *Maaswerken*, to be completed in 2018. Here, flood risks are being reduced amongst others by deepening river forelands, excavating parallel river courses, deepening and broadening river beds by excavating more gravel from the river bed. Navigation is being improved and large-scale new nature areas are being created. Dikes are constructed also along upstream parts of the river. The eastern river belt hosts two out of the twenty Dutch National Parks: *Maasduinen* (River dunes east of river Maas) and *De Meinweg* (terraces of river Maas, east of Roermond).

3.5 The Marine Landscape

Marine landscapes in the Netherlands have common features but may be quite diverse as well. Common features include their age: sediments in these areas at surface are all relatively young, i.e. less than about 4000 years. Apart from intercalated peat beds they are also distinctly marine and often clayey. Topographically, marine landscapes are mainly at or below present Mean Sea Level (MSL), except for some salt marshes in Groningen which may be slightly above MSL. Coastal and marine sand dunes will be dealt with in Sect. 3.7 while landscapes where peat has gone and have later been reclaimed will be discussed in paragraph 3.6: Polder areas. Marine landscapes occur in a broad zone along and behind the present Noordzee/Waddenzee coast.

In the north, such landscapes occur along the coasts of Friesland and Groningen. While tidal basins in the west of the Dutch delta upon Mid Holocene closure of the coastline declined, tidal basins in the north continued to expand until the twelfth century AD. Although these landscapes are generally flat some higher salt march ridges, roughly parallel to the present coast line, and the main former (inversed) tidal channel courses may give some relief to this landscape.

Western marine landscapes occur in Holland. This landscape had been protected by an almost fully uninterrupted coastline. Here, we discuss the marine landscapes that developed from the few former inlets or estuaries in the coastline which are younger than the peat upon they rest or which they have eroded.

The northernmost of these inlets is the Bergen Inlet, north of Alkmaar, a major inlet active from about 6500 BP until 3500 BP (Vos 2015). Major tidal channels penetrated deep in the north (Kop van Noord Holland and Westfriesland) and deposited large volumes of sand and clay on top of peat that was compressed and occurs as relatively thin peat layers in the marine Holocene succession now (De Mulder and Bosch 1982). On the other hand, these younger deposits made e.g. Westfriesland more resistant to later marine floods and storms (Box 2.2). Upon closure of the Bergen Inlet all tidal channels filled-up with sandy clays followed by peat growth.

A second, rather deep tidal channel: Oer IJ was somewhat south of the Bergen Inlet. Oer IJ supplied the area west of Amsterdam with tidal deposits. Simultaneously, this connection with the Noordzee served as the main drainage system of the vast peat areas and lakes in the centre of the Netherlands. Oer IJ closed about 200 years Before Christ (Vos 2015).

Near the city of Leiden, river Rijn's (Oude Rijn) mouth and estuary was a gap in the coastline. Marine floods deposited mainly clays on top of wide peat areas surrounding the Rijn estuary while river floods were responsible for fluviatile clays in the estuary's hinterland.

The marine landscape south of the Maas estuary extends toward Zeeuws Vlaanderen, at the Dutch – Belgian border. This coastal region covers most of Zeeland and remained a tidal area longer than in Holland. Extensive peat growth began half a millennium later in Zeeland while the closing beach barriers only left open the Schelde estuary in the present Oosterschelde. Peat growth continued here until about 2500 BP when new inlets developed along the south-western coastline causing a relatively early break-up of the peatlands which accelerated after Roman Times primarily due to human influence (Vos 2015).

Mesolithic hunter-gatherers, Neolithic and Bronze Age settlers lived in NE Friesland on top of Pleistocene sand highs. Frisians began to settle in nearby salt marshes during the Iron Age. Initially, they built their settlements straight on the higher and sandy salt marshes but when sea level rose further, low and isolated dwelling mounds were built to protect their farms against high tides. The coastline expanded away from the Pleistocene sands. About three centuries later, in Roman Times, more settlements appeared here and a next series of dwelling mounds was erected further north. Earlier dwelling mounds were not abandoned but rather raised by sods dug from surrounding marsh lands as sea level rose and salt marshes were silted-up further. Subsequently, dwelling mounds expanded both in size and in height protecting more farms and families from aggressive northern floods.

In Late Iron and in Roman Times peat lands were cultivated at significant scale to feed the growing local populations and the Roman armies and their allies. Peat cultivation and exploitation eventually caused subsidence and fragmentation of the peat lands and made the area prone to marine erosion, starting in the southwest. After Roman Times population density decreased everywhere in the marine landscape while storm surges increasingly threatened the north. Marine aggression seems to have slowed down since the seventh century and another generation of settlers occupied the area on higher and wider dwelling mounds. Some of them were connected hosting larger populations. By then, the Frisian flood hill area belonged to

Fig. 3.10 View of salt marshes in the northern marine landscape with dwelling mounds in Groningen province

the most densely populated and the most prosperous areas in northwest Europe (Borger 2010). As the northern coast expanded further wider salt marsh areas became available for cultivation. To consolidate these new lands, time and again dikes were built further seawards in Late Medieval Times. This type of land reclamation was applied mainly in the northern and occasionally in the southwestern marine landscapes (Fig. 3.10).

No human occupation is known from Westfriesland when the Bergen Inlet was active. There, the first new settlers arrived not before 700 AD. Peat was drained and reclaimed. Consequently, the land subsided but remained relatively high and resistant to major flooding. Dwelling mounds were built here as well to keep settlers, their homes and villages dry until completion of the Westfriese Omringdijk. Differential settlement caused landscape inversion and former sand-filled tidal channels were now favoured for settlement and villages developed on these grounds.

Peatland cultivation intensified together with population growth in Medieval Times and inversed sandy tidal creeks and channels were used not only as transport tracks but were densely populated too. Invading Vikings were countered by building wooden fortresses in the tenth century AD. Since the twelfth century people no longer abandoned their homes when storm surges intensified. Instead, they built dams, dwelling mounds and dikes, followed by ring-dikes. This could not preclude that substantial areas were captured by the sea.

In more recent times, major sea walls were built all along the Waddenzee coast as well to protect newly reclaimed salt marshes from periodic flooding. The latest of these northern sea walls was *Lauwersmeerdijk* (1969) which closed the former *Lauwerszee*. But salt marsh development went on beyond these sea walls until Waddenzee including the salt marshes was proclaimed a protected area. Other sea walls and storm surge barriers were built in the southwest (Deltawerken) and in the Holland coastal defence system (Hondsbossche Zeewering) (Chap. 2).

About a century ago, marine landscape geomorphology and soil composition in marine landscapes were still reflected in land use, at least to some extent. Grasslands normally were in the lower and wet areas, and agriculture, infrastructure and urbanisation on the higher, sandy tidal creek ridges. By today, agriculture occupied most of the lower areas as well while urbanisation and infrastructure has grown substantially and anywhere disregarding soil composition and geomorphology. Most of the marine landscapes in the southwest are presently used for agriculture (cereals, potatoes, beets, onions, et cetera) today and 13% only for grassland. Parcel re-allotment almost completely remodelled and reshaped the original marine landscapes. In Westland, south of Den Haag, horticulture emerged since the onset of the seventeenth century. There, Late Medieval settlers improved soil quality by mixing marine clays and peat remains with dune sands. These mixed soils are well-suited for horticulture and many greenhouses were built here since WWII.

In the north, minor height differences occur in the wide and open marine landscape. That has been so for many centuries. Grasslands were in the lower parts of the salt marshes and the higher, sandier parts were and still are used for agriculture (potatoes). The clay region northwest of Groningen turned from agriculture into grassland due to higher demands from the dairy industry. Raised grounds under dwelling mounds were mixed with marine clays to improve their quality. That caused massive flood hill excavation in the nineteenth and early twentieth centuries, demolishing about 60% of them in the region. Marine clays in Friesland were excavated to produce characteristically red bricks and for roofing tiles since the thirteenth century. Land use has not changed much in the northern marine area except for urbanisation and infrastructure.

Soils in Westfriesland are well-suited for crop production, horticulture, grassland and for fruit trees. Hydrological and agricultural conditions of the soils were improved by parcel re-allotment resulting in fully levelling and destruction of most small-scale geomorphological features. Some of the more sandy regions in the west were turned into flower bulb production areas and greenhouses erected everywhere. No major nature development plans were designated for Westfriesland (yet).

Northern marine landscapes contain two National Parks: the island of Schiermonnikoog (which partly belongs to the Coastal dune landscape), and *Lauwersmeer*. Tidal influence disappeared in Lauwersmeer since the sea wall was built that separates this former tidal channel from Waddenzee and water quality in the dammed tidal channel became fresh. However, re-introduction of tidal influence in this National Park is being considered now. Waddenzee is a UNESCO Bioreserve since 2009. This area also has one National Landscape: *Middag-Humsterland*, northwest of the city of Groningen (Fig. 3.15). This National Landscape is open in character with many small, meandering tidal channels and containing some one hundred relatively well-preserved dwelling mounds.

The southwest marine landscape also contains two National Parks: Oosterschelde (a marine estuary) and *De Biesbosch* (a fresh-water tidal area). Once the Oosterschelde storm surge barrier became operational in 1986 (Chap. 2), hydrological dynamics in this region changed dramatically. Tidal channels were gradually filled with clay, water levels rose due to fresh water supply from the hinterland and shoals will probably be flooded by the next century. One of the artificial sand banks for Oosterschelde storm surge barrier construction purposes has been reshaped into a nature area. Five percent in Zeeland's agricultural lands should be transformed into nature before 2019. Experiments in National Park De Biesbosch are supposed to generate more biodiversity by re-introduction of tidal ranges.

Two more National Landscapes were proclaimed here: *Zuidwest Zeeland* and *De Hoekse Waard*. Policy in Zuidwest Zeeland aims to increase nature and forests along the inner dunes, to improve leisure facilities and to preserve and restore cultural historical elements. De Hoeksche Waard National Landscape is a polder with very young tidal sediments deposited after the St. Elisabeth flood (Sect. 2.2), in 1421 AD. National Landscape policy for De Hoeksche Waard aims to reinforce the wet nature component, to maintain characteristic landscape elements in this polder and to make the area more accessible for leisure activities.

3.6 The Polder Area

Polders are land areas with man-controlled water-level systems and include areas of different geneses. Here, we discuss two types of polders, those created by drainage of a former lake or part of a lagoon, and low-lying, excavated peatlands that require strict water management. In contrast to all other Dutch landscapes, polders are fully man-made/controlled. But their topography and geomorphology may still contain natural components. Polder' surfaces often are the floors of former (shallow) lakes which occurred in former wide peatlands that covered the Netherlands for several millennia. Draining the water from such lakes was done by various types of mills and brought into canals constructed around such lakes (Chap. 2). A common feature of polders is that these are all below MSL and will be flooded inevitably unless continuously drained. That has happened frequently in the past, sometimes deliberately (Box 2.1). Impacts of permanent drainage on the land in and beyond polders has been described in Chap. 2.

From a geomorphological perspective, polders are windows showing landforms that were created prior to the start of extensive peat growth. Although many polders have been reclaimed in areas where remnants (or islands) of peat still occurred, most of these peat deposits have since gone by oxidation or by ploughing. However, some of the former peat islands still exist and one of these (*Schokland*) has been proclaimed an UNESCO heritage site (Fig. 3.11).

Except for peat remnants, sediments at the surface of Dutch polders belong to earlier (Early to Mid-Holocene) sedimentation phases than those in the marine clay areas. These mainly consist of sandy tidal channel and clayey tidal basin deposits with some intercalated, thin peat beds. All tidal channels silted-up shortly before major regression of the sea.

Once surfaced such sediments have been disturbed significantly by ploughing, ditch excavation, drainage works and other agricultural, urbanisation and infrastructural activities. However, (satellite) images of polder surfaces often display pre-peat tidal flat landscapes (Fig. 3.12) which may correspond to small topographic height variations caused by drainage driven differential settling. Relief inversion is a quite common feature, also in polder landscapes. For a more technical description of polder construction and (wind)mills one is referred to Chap. 2.

The first Dutch polder was made in 1533 (Grote Achtermeer, south of the city of Alkmaar) and served as a pilot. After several other smaller polders, two more significant lakes were milled dry in 1564: *Egmondermeer* (686 ha) and *Bergermeer* (620 ha). Construction was initiated by singular, often nobility persons. That changed in the seventeenth century when such projects were financed by coalitions of wealthy merchants who were able to jointly collect much more capital needed for larger projects. In the beginning polder construction was primarily done for economic reasons as high prices for agricultural products made development of new land potentially profitable. Another motivation was combatting loss of land. That became more prominent since public authorities became more involved in land governance by the end of the eighteenth century. Prior to that no public money was invested in polder construction. The twentieth century IJsselmeerpolders were entirely financed by public funds (Stol 2010b).

Polder construction was invented and tested in Noord-Holland and most of its lakes were milled dry in the sixteenth and seventeenth century. In Zuid-Holland this mainly happened in the eighteenth century, and in Utrecht and Friesland a century later. Soil formation may take 10,000–15,000 years depending amongst others on climate, vegetation type, soil minerology and soil texture (Jongmans et al. 2015). As a

Fig. 3.11 Schokland: a former peat island in the former Zuyderzee, currently integrated in Noordoost Polder (Note the standing construction in the centre of the picture, a former wooden sea wall)

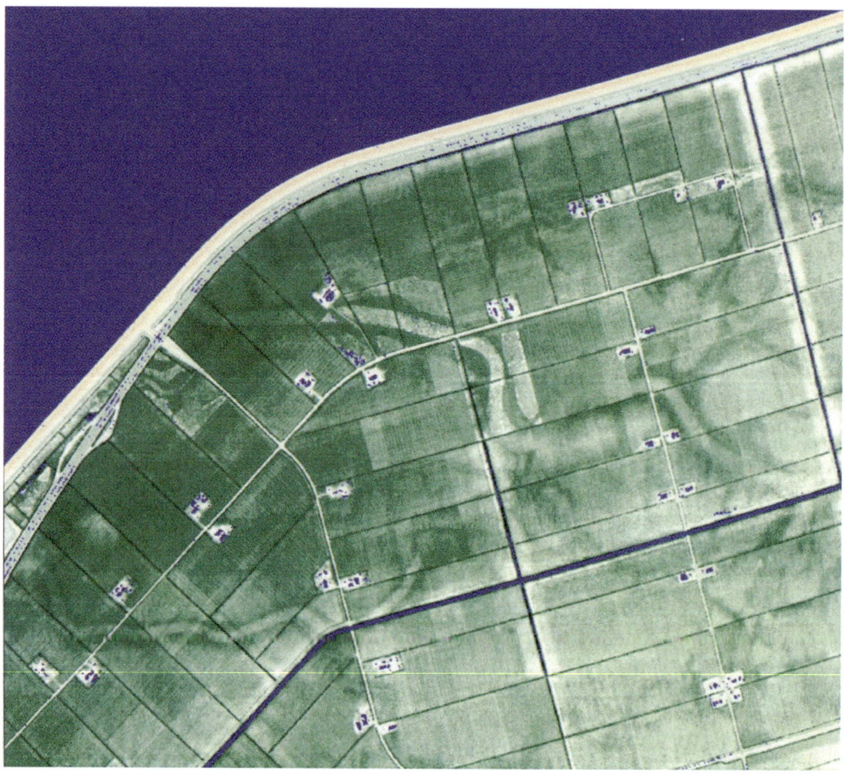

Fig. 3.12 Aerial photograph of Polder Oostelijk Flevoland with height differences at surface by laser altimetry, showing patterns of tidal creeks (lighter coloured). (Source: De Bosatlas van de geschiedenis van Nederland 2011)

consequence, polder floors exposed to the surface for a few centuries at most underwent incomplete soil formation. However, most of these polder floor sediments experienced soil formation processes long before due to Mid and Late Holocene peat growth. Consequently, the topmost beds in polders were often decalcified already and textured. To prepare new land for agriculture most efforts were geared to improve evaporation, drainage potentials of the formerly water-saturated soils, and soil texture, e.g. by sowing reed followed by rapeseed.

Three types of sub-environments may occur in the upper marine sediments of most of the polders: tidal flat deposits, salt marsh lands and reed clays. This last type formed the transition from salt marshes to peatlands. In former tidal flats differential settlement caused relief inversion of up to 150 cm vertically. Such variations may occur between salt marsh ridges and adjacent clay-filled salt marsh basins as well.

In the youngest polders (*IJsselmeerpolders*) reclaimed since the 1930s, processes of physical and chemical maturation had just began. Significant settlement in the clayey deposits occurred directly upon their complete drainage. Depending on texture and presence of peat remains, a further settlement of 30–150 cm may take place (Jongmans et al. 2015). Locally, old river dunes emerged after reclamation of parts of IJsselmeer. Archaeological excavations displayed remnants of early human occupation there.

Although not much time may have elapsed since their reclamation, present land use in the polders may differ quite significantly from their initial land use. Most of the polders were reclaimed for agricultural purposes but not all polder land was suited for profitable crop production. For example, soil quality in Polder *De Heerhugowaard*, reclaimed in 1630, appeared to be rather poor and was used for pasture land and cattle instead. In other polders the area of agricultural land increased over the last century, e.g. in Polders *De Beemster* (reclaimed in 1612) and *De Schermer* (reclaimed in 1635).

Dominantly sandy pasture polders like *De Zijpe* (reclaimed in 1600) in the north of Noord-Holland eventually turned their lands partly into flower bulb production. There, in some other polders where marine sand layers were covered by up to 1 m of clay, the succession may have been turned mechanically to bring the sand on top of the clays. Water tables can be controlled much better in sands which is indispensable for flower bulb production. Also horticulture and greenhouses appear more and more often in today's polder landscapes. The older polders did not escape from large-scale parcel re-allotment neither (Box 3.1). Younger IJsselmeer polders were designed already according to such concepts which varied over time.

Urbanisation and infrastructure expanded quite significantly in polder landscapes since WWII. In the past 30 years, the once small polder town of Zoetermeer expanded to a city of over 124,000 inhabitants now. The first houses in the new city of Almere in Polder Zuidelijk Flevoland were built in 1976; by today, Almere hosts 200,000 people, mainly migrated from nearby Amsterdam (Chap. 8). From their inception, agriculture has been the prime reason for creating IJsselmeerpolders. Nature and leisure were barely incorporated in the original land use plans. But large leisure/holiday parks and camps have emerged there since. The large (56 km^2) nature reserve (*Oostvaardersplassen*) has developed in the deepest and wettest part of Polder Zuidelijk Flevoland, between the cities of Lelystad and Almere. Today, land use in this most recent polder is half for agriculture only, a quarter for urbanisation and 18% for nature.

Box 3.5: De-poldering

In their drive to create more and new nature environmentalists may go quite far. For example, to compensate for deepening the Westerschelde allowing the north Belgian city of Antwerpen to host larger cargo vessels, creation of more and new nature has been a matter of significant debate. As a result of negotiations between Belgian and Dutch authorities the *Hertogin Hedwigepolder*, a small diked part of *Zeeuwsch Vlaanderen* near the Belgium – Dutch border was assigned to be flooded by demolishing the surrounding dikes. The (former) *Hertogin Hedwigepolder* would than become part again of the adjacent tidal marsh lands of *Verdronken land van Saeftinghe*, a current nature reserve. Deliberate flooding of this polder for creating more nature raised considerable opposition by regional people and beyond. For generations they had been fighting to protect their lands against the sea. This conventional attitude is in strong contrast to novel land-management ideas. This issue has been brought to EU Court who decided in favour of the naturalists. In 2012, the Dutch government decided that flooding would be completed by 2019.

3.7 The Coastal Dune Landscape

With some interruptions, this landscape occurs in a rather narrow strip along almost the entire Dutch coast. It consists of wind-blown dune and beach sands and may be split in two types, the Older and the Younger dune and beach sands.

The Older dune and beach sands generally occur in more inland strips, positioned roughly parallel to the present coastline. These normally consist of shell-bearing, yellow-brown to white, calcareous sands which have been active as beach barriers and thus marking fossil coastlines since the main Holocene regression, about 5000 BP. Often these are topped by wind-blown and fine-grained sands. The Older dunes are (much) lower than the Younger dunes.

The Dutch Noordzee coast with its associated Younger beach barriers and dunes appear as one single, stretched SSW-NNE trending zone, in fact this zone consists of two parts. The northern part consists of fine sands poor in calcium and comprises the chain of islands: *Waddeneilanden* and the northern part of the coastline until beach town Bergen, where the orientation of the buried beach barriers and dunes turns to the east (Fig. 3.13). There, the dune coast is broadest, about 5 km. The dunes and beach barriers in the southern part of the Dutch coast are oriented roughly parallel to the present coastline and consist of fine sands which are, in contrast to the northern part, rich in calcium, iron, magnesium and other minerals. From an ecological perspective this difference in mineral content is quite significant as their respective floras and faunas are quite distinct as well. This distinction is associated with different origins of these dune sands. The northern dune and beach sands are derived from eroded, Pleistocene deposits, witnessed by e.g. the Pleistocene core of the island of Texel. These Pleistocene sands are generally decalcified. Their southern counterparts are derived from the Noordzee floor and originate from mineral-rich river sands, transported along the Dutch coast by SW-NE sea currents.

As the Older dunes, the Younger dunes consist of beach sands and dune sands, differing in age and in morphology. These date from the twelfth century onwards and consist of light yellow fine to medium fine sands. The Younger dune and beach sands constitute the present beach and dune landscape and dunes may reach heights and thicknesses of up to 60 m. They were deposited while the Noordzee coast was eroding and while tidal inlets expanded and sand storms have been most severe in Holland between the fourteenth and seventeenth centuries, in Zeeland between the sixteenth and nineteenth centuries. Significant parts of the Dutch SW and NW coast have been eroded by then. Over the centuries successive chains of parabolic coastal dunes migrated landinward, overprinting the former system of alternating beach-plains (with clay and peat-beds) and former beach-ridges (with low dunes). These active coastal dune-fields created hostile landscapes for its inhabitants and interfered with contemporaneous agricultural activities on the Older dune and beach sands; they may cover clay and peat beds in the beach valleys adjacent and parallel to beach barriers. By the end of the nineteenth century the dune areas were almost barren due to lack of proper management (Beckman 2010).

Professional protection and management of the Dutch coastal dunes began in the early twentieth century and was geared at preserving the (Younger) dune landscape, mainly by fixing the sand bodies by dedicated types of vegetation, i.e. by marram grasses and by *Pinus* trees. However, no such nature management occurred in the Older dunes and beach sands. These have been affected by humans from the onset of human occupation of the Dutch coastal zone. Many archaeological remains in the Older dune and beach sands point to almost continuous human occupation and settlement since 4450 BP (van Heeringen 1983). Most settlements of the *Cananefates* and their predecessors were strategically positioned near river and estuary mouths and in elongated areas between forest-covered beach barriers and swampy beach plains. Heavy stone axes for cutting trees and elaborating the sandy lands, were found. Some of these had been produced in England pointing to early (Bronze Age) trade routes across the Noordzee. The dune areas have been rather intensely occupied since early Medieval Times as, amongst others, witnessed by the names of present villages in this area. But many of the original settlements have since been covered by Younger Dune sands or been taken by the Noordzee. Some of the more inland villages, as Haarlem and Alkmaar, have grown into significant cities in the twelfth and thirteenth centuries.

Structure and orientation of the Older beach barriers made these well-suited for Medieval infrastructure. They served as transport routes and several castles, abbeys and mansions have been built at or along these firm sand strips. Bricks were produced from clays dug out of the wet beach plains and the coastal valleys in between beach barriers and used as building stones. Most of these structures built from such bricks were situated on firm beach barrier sands.

Dutch dunes also serve as natural filters to decontaminate fresh surface waters transported to and inserted in the dunes for drinking water supply. In addition, deep fresh water reservoirs occur under the Dutch dunes which are and have been exploited for the same purpose (Chap. 2). Beach barriers were also used as the principal transport networks connecting settlements and later towns and cities. Peat was excavated from coastal valley swamps for fuel. Dutch coastal plains are dominated by peat and clay while sand is a rare construction material. Therefore, in the sixteenth and seventeenth centuries large volumes of Younger dune sands were transported to expanding cities as Haarlem, Leiden and Amsterdam mainly for foundation purposes.

Since the last decades of the nineteenth century Older beach and dune sands were used as producing grounds for flower bulbs, mainly between the cities of Haarlem and Leiden (Fig. 3.14). These sands are particularly well-suited

Fig. 3.13 Coastal dunes poor in calcium near Groote Keeten (Noord-Holland). (Photograph taken by Bas Arens)

Fig. 3.14 Flower bulb production on the older beach and dune sands of former beach barriers, oriented roughly parallel to the present Noordzee coastline

for this purpose as water levels can be controlled quite precisely, critical to the success of this industry (Fig. 3.14). Flower bulb production has now exceeded the beach barriers along the Dutch Noordzee coast and being practised in many natural or man-made sandy regions in and even beyond the west of the Netherlands today.

Except for specific designated areas in the coastal zone in the Netherlands, dunes have been fixed by special grass types to prevent dune sand to be moved too much by the wind. Thus, fixing sand dunes is now part of the Dutch national system of coastal preservation. Another type of coastal protection is to insert large quantities of sand dredged from the Noordzee floor offshore and dumping this into the active coastal zone and to create new beach barriers which might then be redistributed by sea currents. Where such beach

Fig. 3.15 National Landscapes in the Netherlands

barriers emerge above sea level, the sands might be subject to wind and new dunes will be created (Chap. 2).

References

Barends S (1989) Percelen in Nederland. Veranderingen in de percellering tussen 1900 en nu. Pudoc, Wageningen

Beekman F (2010) Duinlandschap. In: Barends S et al (eds) Het Nederlandse landschap. Een historisch-geografische benadering. Matrijs, Utrecht

Borger GJ (2010) Agrarisch veenlandschap. In: Barends S et al (eds) Het Nederlandse landschap. Een historisch-geografische benadering. Matrijs, Utrecht

De Bosatlas van de Geschiedenis van Nederland (2011) Noordhoff Atlasproducties, Groningen

De Bosatlas van Nederland (2007) Wolters-Noordhoff Atlasproducties, Groningen

De Mulder EFJ, Bosch JHA (1982) Holocene Stratigraphy, Radiocarbon dating and Paleogeography of central and northern North Holland (the Netherlands). Mededelingen Rijks Geologische Dienst 36 (3):111–160

De Mulder EFJ, Geluk MC, Ritsema IL, Westerhoff WE, Wong TE (2003) De ondergrond van Nederland. Geologie van Nederland deel 7. Wolters Noordhoff, Groningen

Jongmans AG, Van den Berg MW, Sonneveld MPW, Peek GJWC, Van den Berg van Saparoea RM (2015) Landschappen van Nederland. Geologie, bodem en landgebruik. Wageningen Academic Publishers, Wageningen

Leenders KAHW (1987) De diffusie van een techniek. De vergraving van het veen in de Nederlanden. 1150–1950. In: Tijdschrift van de Belgische Vereniging Aardrijkskundige studies, pp 197–216

Lendering J (2000) De randen van de aarde, de Romeinen tussen Schelde en Eems. Ambo, Amsterdam

Nota Belvedere (1999) Beleidsnota over de relatie cultuurhistorie en ruimtelijke inrichting. Ministerie van Onderwijs, Cultuur en Wetenschap, Ministerie van Volkshuisvesting, Ruimtelijke Ordening en Milieubeheer, Ministerie van Landbouw, Natuurbeheer en Visserij, Ministerie van Verkeer en Waterstaat. VNG Uitgeverij, Den Haag

Poelman JNB (1966) De bodem van Utrecht. STIBOKA, Wageningen

Renes J (2010) Krijt/Lösslandschap. In: Barends S et al (eds) Het Nederlandse landschap. Een historisch-geografische benadering. Matrijs, Utrecht

Stol T (2010a) Turfwinningslandschap. In: Barends S et al (eds) Het Nederlandse landschap. Een historisch-geografische benadering. Matrijs, Utrecht

Stol T (2010b) Droogmakerijenlandschap. In: Barends S et al (eds) Het Nederlandse landschap. Een historisch-geografische benadering. Matrijs, Utrecht

Van Heeringen RM (1983) 's-Gravenhage in archeologisch perspectief. In: de Mulder EFJ (ed) De bodem van 's-Gravenhage. Mededelingen Rijks Geologische Dients 37, 1

Vervloet JAJ (2010) Zandlandschap. In: Barends S et al (eds) Het Nederlandse landschap. Een historisch-geografische benadering. Matrijs, Utrecht

Vos PC (2015) Origin of the Dutch coastal landscape. Long-term landscape evolution of the Netherlands during the Holocene, described and visualised in national, regional and local palaeogeographic maps series. Dissertation, Deltares. Barkhuis, Groningen

Wong TE, Batjes DAJ, De Jager J (2007) Geology of the Netherlands. Royal Academy of Arts and Sciences, Amsterdam

4 Natural Resources & Energy

Eduardo F. J. De Mulder

Abstract

The Dutch subsurface is relatively rich in natural resources. The Netherlands has mainly been self-supporting with respect to its energy needs. Wind energy has been used to drain the polders and thick peat occurrences provided enough fuel for heating for powering industries. Today, the Dutch are significant global players in terms of natural gas and rock salt production and transport. Discoveries of giant natural gas resources in Groningen in the late 1950s generated a major energy transition and construction of a nation-wide network of gas pipes. Moreover, Dutch gas addresses more than 25% of NW Europe's demands. Regardless decades of production, large gas reserves still occur in the subsurface which contains significant oilfields and coal resources as well.

The Dutch are now moving towards a next energy transition, i.e. from fossil to alternative and sustainable energy sources (Sect. 4.3). Such alternative energy sources may include solar, wind, water, geothermal, bio-fuel and nuclear power. Strategies for new technologies and policy directions in energy supply and demand have been developed and political support collected. The Dutch energy market is privatised.

The Netherlands contain sufficient fresh water and construction materials as sand and clay to be self-supportive. Dutch soils are a very important natural resource and were made highly productive by many generations of agricultural engineers (Sect. 4.6). The Dutch surface is used intensively, making underground space – an often forgotten natural resource – of growing interest (Sect. 4.7).

Natural resources exploitation may jeopardise ground stability and cause land subsidence and earthquakes which is a current matter of public concern on gas exploitation in the province of Groningen (Sect. 4.8).

Keywords

Natural gas · Groningen gas field · Earthquake risk · Oil reservoirs · Alternative energy sources · Rock salt · Coal · Energy transition policy · Constructon materals · Soil · Underground space

4.1 Energy Materials

Natural Gas occurs abundantly in the Dutch subsurface and the Netherlands is a gas-exporting nation. These resources mainly occur in giant geological reservoirs deep below the surface of the northeast of the country (Fig. 4.1). The Groningen field is the largest gas field in Europe and the tenth largest in the world (Whaley 2009). This gas field was discovered in 1959 in the Slochteren-1 well at a depth of about 3 km in Permian sandstones, determined as the Slochteren Formation (Geluk 2012). The Groningen gas field is situated mainly onshore in the provinces of Groningen and Drenthe. For a minor part this gas field extends offshore and into Germany.

The multi-coloured sandstones and conglomerates of the Slochteren Formation (Fig. 4.2) belong to the Upper-Rotliegend Group and were formed during the Middle to Late Permian, i.e. about 260 million years ago. These sandstones are porous and permeable, were deposited mainly by rivers and by the wind in coastal dunes (Fig. 4.3). They constitute the main body of the giant Groningen gas field (De Jager and Geluk 2007). The Slochteren Formation which extends to the north into the finer grained Silverpit

E. F. J. De Mulder
Delft Technical University (emeritus), Haarlem, Noord-Holland, The Netherlands
e-mail: e.demulder@planet.nl

E. F. J. De Mulder et al., *The Netherlands and the Dutch*, World Regional Geography Book Series,
https://doi.org/10.1007/978-3-319-75073-6_4

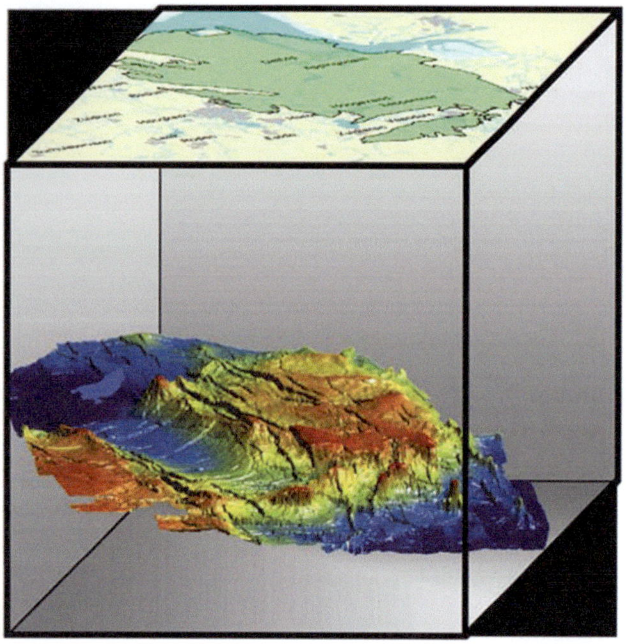

Fig. 4.1 3D subsurface view of the Groningen gas field. Level shown is top of Upper Rotliegend Group (Source: Geluk MC 2012)

Fig. 4.2 Bore hole core of the Slochteren formation. Depth of core: 3580 m. (Source: De Mulder et al. 2003)

Formation, rests with an angle on Westphalian (Late Carboniferous) coal-bearing shales and sandstones. These are the source rocks of the natural gas of the Groningen field. The gas-bearing Slochteren Formation is between 80 and 200 m thick, maximally 235 m, and is covered and sealed from upward gas migration by thick, impermeable salt beds of the Zechstein Group of Late Permian Age.

Gas resources in the Netherlands are not limited to the Groningen gas field only, nor to the Slochteren Formation. Several other major gas fields occur in sandstones of comparable age, i.e. the Annerveen and Ameland Complexes and in the offshore K and L quadrants of the continental shelf (De Jager and Geluk 2007). Other gas fields have been discovered in sandstones of Late Carboniferous Age (Westphalian) in the east of the Netherlands and offshore in the NW part of the continental shelf. In addition, gas fields occur in traps within the Zechstein salt succession in the west of the Netherlands. Gas fields have also been found in the thick Triassic sandstones, but only where no Zechstein salt caprock occurs. These are located in the east, in the West Netherlands Basin and in the north and west offshore. These reservoirs are often sealed on top by Triassic or by Lower Cretaceous shales. In all these cases the gas has migrated from Westphalian source rocks. Gas occurrences in the fields of the Rotliegend are by far dominant in volume (De Mulder et al. 2003).

All such gas fields have been discovered and are being exploited through bore holes. On average these holes are some 3000 m deep. Hydrocarbon exploration both onshore and offshore has been significant during the four last decades of the twentieth century. Seismic methods have been the main exploration tools. Introduction of modern 3D seismic methods in the 1980s resulted in many more natural gas discoveries. Even today, every year bore holes are drilled on- and offshore, both for exploration and for exploitation. For example in 2014, 18 exploration bore holes were drilled, 11 of these encountered natural gas or oil, or both. In the same year, 35 bore holes were drilled for exploitation purposes (Ministerie van Economische Zaken 2015). As the composition and tectonic structures in the Dutch subsurface are better known and understood every year, less and less new gas occurrences have been found, mainly since the late 1990s (Geluk 2012). In 2014, 1.6 billion Nm^3 new natural gas resources have been discovered in the Netherlands.

Although large quantities of Dutch natural gas have been produced since the 1960s, quite considerable resources remained unexploited in the Dutch subsurface. In its Annual report 2014, the Dutch Ministry of Economic Affairs states that commercial natural gas resources in the Dutch subsurface have been determined 883 billion Nm^3, 671 billion of which (some 50% of original volume) are still left in the Groningen field. The remainder of these gas resources occur in some 230 smaller gas fields, both on- and offshore. Natural gas prospects in the Dutch subsurface

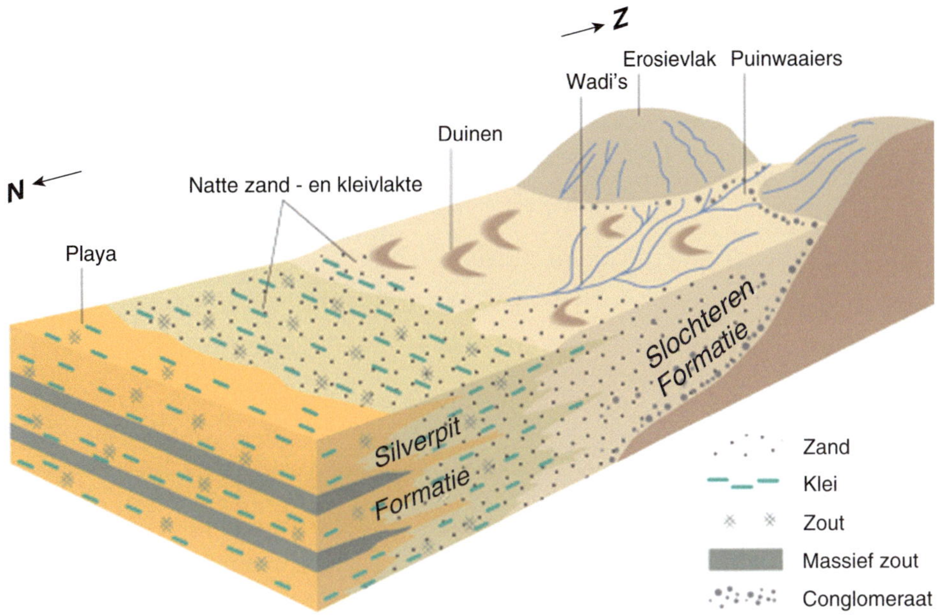

Fig. 4.3. Schematic representation of the deposition conditions of the Slochteren and Silverpit Formations during the Late Permian in the NE of The Netherlands. Grain size of the sandstones decrease to the north while the clay content increases in the same direction. (Source: RGD 1995)

(i.e. geologically, technologically, economically and environmentally feasible) amount to another 113 billion Nm^3 (Ministerie van Economische Zaken 2015). The original gas resources in the entire Dutch subsurface (before exploitation) were estimated at 4349 billion m^3 (De Jager and Geluk 2007).

Production & Consumption When exploitation of this gas began in 1963, the Groningen gas field contained 2.8 trillion m^3 low-caloric gas. A nation-wide network of 15,500 km of pipelines was installed serving 98% of all Dutch households. The natural gas of the Groningen field is composed of 81% methane, 14% nitrogen and 5% other gasses. Atmospheric Nitrogen is added for distribution to and for consumption purposes by Dutch and German households. All existing gas devices had to be transformed to accept the specific Groningen gas quality. Groningen gas supplies about 60% of the entire Dutch electricity production. Gas production rapidly increased until 12 years after its start when production peaked in 1975–1976 and more than 100 billion Sm^3 (is about 100 billion m^3 and an energy potential of more than 3500 PJ) were extracted from the Groningen field together with gas from some smaller fields on land and offshore. Since, total production declined to an average of some 75 billion m^3 annually, while production volumes from (often newly discovered) smaller fields with high-caloric gas increased relatively as large quantities of the Groningen gas field contents remained in stock as a result of the 'smaller fields strategy' (see below and Fig. 4.4). This Figure also shows that production from the Groningen field increased again and almost tripled since 2000 until 2014.

Total gas production from the Groningen field decreased from 53.9 billion m^3 in 2013, via 42.4 billion in 2014 to 28.1 billion m^3 in 2015 (NAM 2016b). This reduction generated a 0.4% loss in economic growth in 2015 (CBS 2016a).

In the Netherlands natural gas and oil are produced by EBN BV, a state-owned enterprise *(Energie Beheer Nederland*; 100% of its shares are owned by the Dutch government). EBN BV is responsible for exploration, production and storage of these hydrocarbons. The Dutch government charged EBN BV to maximally secure its revenues from the production of oil and gas. Through EBN BV the Dutch government participates in joint ventures with private enterprises to explore, exploit and store oil and gas from the Dutch subsurface (EBN 2016). Seventy percent of all revenues from gas and oil exploitation in the Netherlands is attributed to the Dutch State treasures. Since discovery of the giant Groningen gas field the Dutch State received some 211 billion € from oil and gas exploitation.

In 2013, natural gas supplied 43% of the total energy consumption in the Netherlands, the highest amount in NW Europe and a comparable share as oil in Dutch energy consumption. The Dutch and their industries consume about 40 billion m^3 natural gas annually. Dutch gas satisfies more than 25% of the demands in NW Europe (Aardgas in Nederland 2016). Over 50% is exported to Germany, Belgium, France and Italy generating considerable additional income to the Dutch government's treasures. Slightly over half of this gas (51%) is used for fuel and energy in the built-environment (households and small to mid-sized companies), while roughly one third (32%) of it is applied in industry and 15% by gas-fuelled power plants in 2013.

Initially (upon discovery), Dutch policy was focused on exploiting the Groningen gas field as soon as possible as it was commonly believed by then that nuclear energy would make all other fossil fuels redundant. This explains the initial

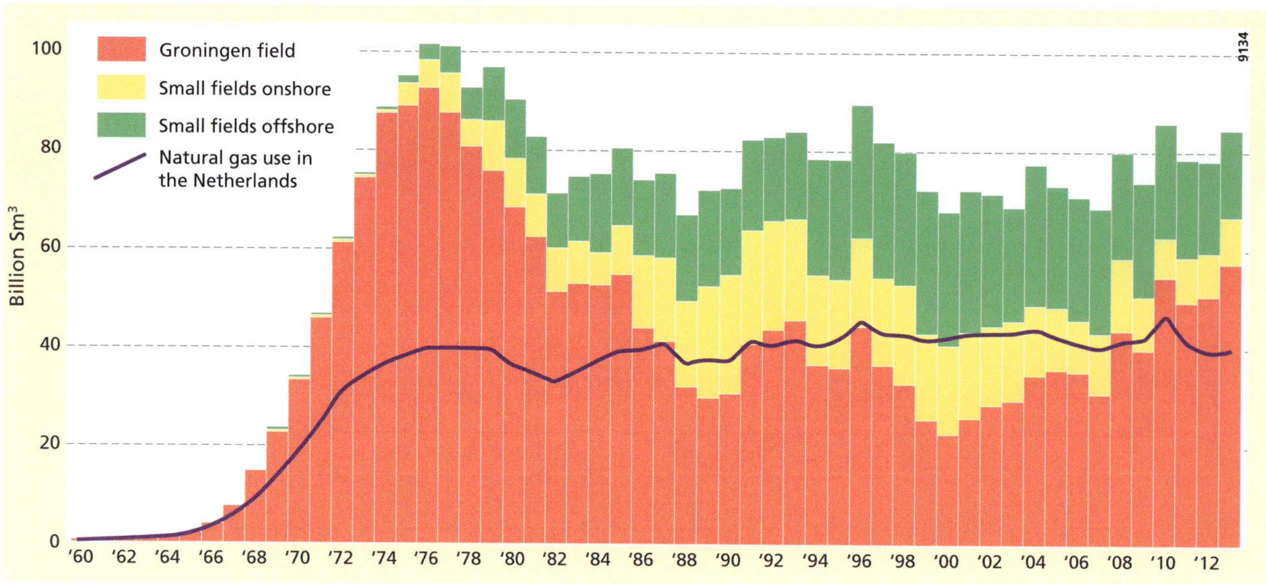

Fig. 4.4 Annual natural gas production in the Netherlands since development of the Groningen field in 1963. Gas production from the Groningen field in red, gas production from the smaller fields offshore (in green) and from the smaller onshore gas fields (in yellow). The blue line corresponds with the total annual gas consumption in the Netherlands which is about half of what has been produced. The other about half is exported, in 2014 this has been 53% of all natural gas produced from the Dutch subsurface

rapid increase in production capacity partly. But there have also been geopolitical reasons for this policy. To avoid some European nations to become (too) dependent on Russian gas resources, the Groningen gas was offered cheaply. This has been an important reason that a considerable share of the Dutch gas resources were exported since. But the 1973 oil crisis forced the Dutch government to review this policy and to be more reluctant in rapid exploitation of its resources. A new policy, launched in 1974, was geared at keeping gas of the Groningen field as much as possible as a strategic resource and instead to give priority to exploiting gas from smaller fields. Simultaneously, exploration of hydrocarbons was intensified resulting in discovery of numerous, mainly smaller gas and oil fields on- and offshore, hundreds of which have been brought in exploitation since. This "smaller fields first" policy together with a policy to spread Dutch energy supply through a multitude of energy sources, has been maintained until recently. But many of these smaller fields began to deplete since about 2000 and could no longer satisfy increasing demands; more and more gas was tapped from the Groningen field until 2014 when parliament forced government to significantly reduce gas production from the Groningen field in response to growing protests by the Groningen population to increased risks of earthquakes. More about current governmental policy on energy is given in Sect. 4.3.

Relatively small gas fields have been found in Lower Tertiary beds in the east of the Netherlands. But most of the younger natural gas accumulations were found offshore in the A and B quadrants (Geluk 2012). Seismic surveys often show chimney-type of structures associated with such occurrences showing that these are leaking gas straight towards the Noordzee floor (De Jager and Geluk 2007). Tiny gas pockets may occur in marine Holocene beds as a result of decomposing organic remains, particularly in places previously covered with thick peat beds. By erosion or excavation of such peat beds, surface pressure may reduce which may explain the rather widespread occurrences of gas in water wells and outcropping sand beds in polders. Such gas occurrences may be trapped in containers and distributed to e.g. a farm houses. Traditionally, they served for small-scale illumination and heating purposes (De Mulder et al. 2003; Fig. 4.5). Today, such shallow gas exploitations have become a matter of environmental concern as such gas-bearing groundwater may contain natural pollutants exceeding norms set for agricultural use.

Oil is produced in the Netherlands but unlike gas, the Dutch are not self-supporting in oil. The first oil was discovered in 1923 when traces of oil were found in Zechstein deposits in a bore hole in the east of the country. A second time oil was discovered has been 15 years later by pure co-incidence when a hole was drilled for demonstration purposes in Den Haag and oil was found at a depth of 460 m only. Following a gravimetric survey in that area in Zuid-Holland, many oil occurrences were identified there. Oil discoveries during the early 1940s near the German border prompted the Dutch to the east of the country where the first commercial oil was found in a field near Schoonebeek, in 1943. These oil resources occur in the Lower Cretaceous Bentheim

Fig. 4.5 Shallow gas container near a farm in polder De Beemster. (Source: De Mulder et al. 2003)

sandstones at some 800 m depth. The oil originates from lacustrine, Late Jurassic shales of the Coevorden Formation (Nedersachsen Group) which widely occur in Lower Saxony (Germany – Fig. 4.6). With a volume of about one billion barrels, the Schoonebeek field proved to be the largest onshore oil field in western Europe. The (less viscous) oil in the Zuid-Holland and offshore fields comes from the Vlieland Sandstones of the Early Cretaceous Rijnland Group. Their source rocks are the organic-rich, Lower Jurassic *Posidonia* shales (Fig. 4.10).

Current (as of 1 January 2015) on- and offshore oil resources (developed and not yet developed) in the Netherlands amount to 34.7 million Sm^3, 27.8 million Sm^3 of which are onshore and 6.9 million Sm^3 offshore. 12.4 million Sm^3 (9.6 million Sm^3 of these resources onshore and 2.8 million Sm^3 offshore) are qualified as conditional, i.e. longer term production plausible (Ministerie van Economische Zaken 2015).

Production and Consumption Oil is one of the most important trade commodities in the Netherlands and oil imports belong to the most significant traded goods in Rotterdam harbour. A relatively small share of the Dutch oil demand is produced from oil fields occurring in the Dutch subsurface (Fig. 4.8). Production of oil from offshore fields started in 1983 and peaked in 1987 with a volume of over five million m^3. The average total production from Dutch fields since 2010 has been about 1,5 million m^3 per year (Olie-import 2016). Production through 'nodding donkeys' began in 1947 and was halted in 1996 (Fig. 4.7). That was due to the very high viscosity and the high paraffin contents of the oil which would keep significant quantities of the unexploited oil retained in the reservoir.

Production from this oil field resumed in 2011 when horizontal drilling techniques, high-capacity pumps and steam injection technology became available. NAM (a Shell & Exxon joint venture) hopes to extract another 100–-120 million barrels of oil before 2035 from this reservoir (NAM 2016a). Oil production from Dutch regions is given in Figs. 4.8 and 4.9.

The Dutch import crude and other oil types from Norway, Russia, Saudi Arabia and many other nations overseas. Such oil imports rose from 66 billion in 2007 to 94 billion kg in 2015, while exports and transit goods increased from 75 to 110 billion kg respectively (CBS 2016b). Crude and other oil types are brought onshore by super-tankers in Rotterdam harbour. From there, pipelines transport about 50% of the imported oil volumes to other destinations in the Netherlands, in Belgium and in Germany. The other 50% is stored temporarily in large storage tanks in Maasvlakte, a reclaimed area in Noordzee where top-sized super tankers can land and discharge their loads of oil. Several other major storage tanks complexes occur in the Rotterdam harbour area but with a storage capacity of four million m^3, the Maasvlakte Oil Terminal (MOT) is the largest. Stored oil is further processed into semi-finished or finished oil products which are successively being transported over land (by rail, pipeline or by trucks) or shipped over rivers and seas.

As oil import not always proved to be reliable, the Dutch government keeps a strategic volume of oil in storage to cope with temporary shortages in supply corresponding with a volume of four million tonnes (Oil policy 2016). In 2012, the Netherlands, together with other nations, installed an oil crisis policy by law geared at joint measures in reducing oil consumption. Moreover, large oil consuming industries are obliged by law to keep oil in stock as well. If necessary, old oil fields might be re-exploited as these have not been entirely depleted although their production is no longer economical. For the longer term the Dutch government is working towards a transition from fossil energy materials to renewable, home-grown energy which would make the nation independent from external energy sources (see Sect. 4.3).

Coal is a combustible black to dark brown rock that occurs in seams or layers. Coal originates from plant and other organic remains. Initially these accumulated as peat that successively was strongly compressed by high pressures and temperatures. Coal contains 75% (lignite) – 91.5%

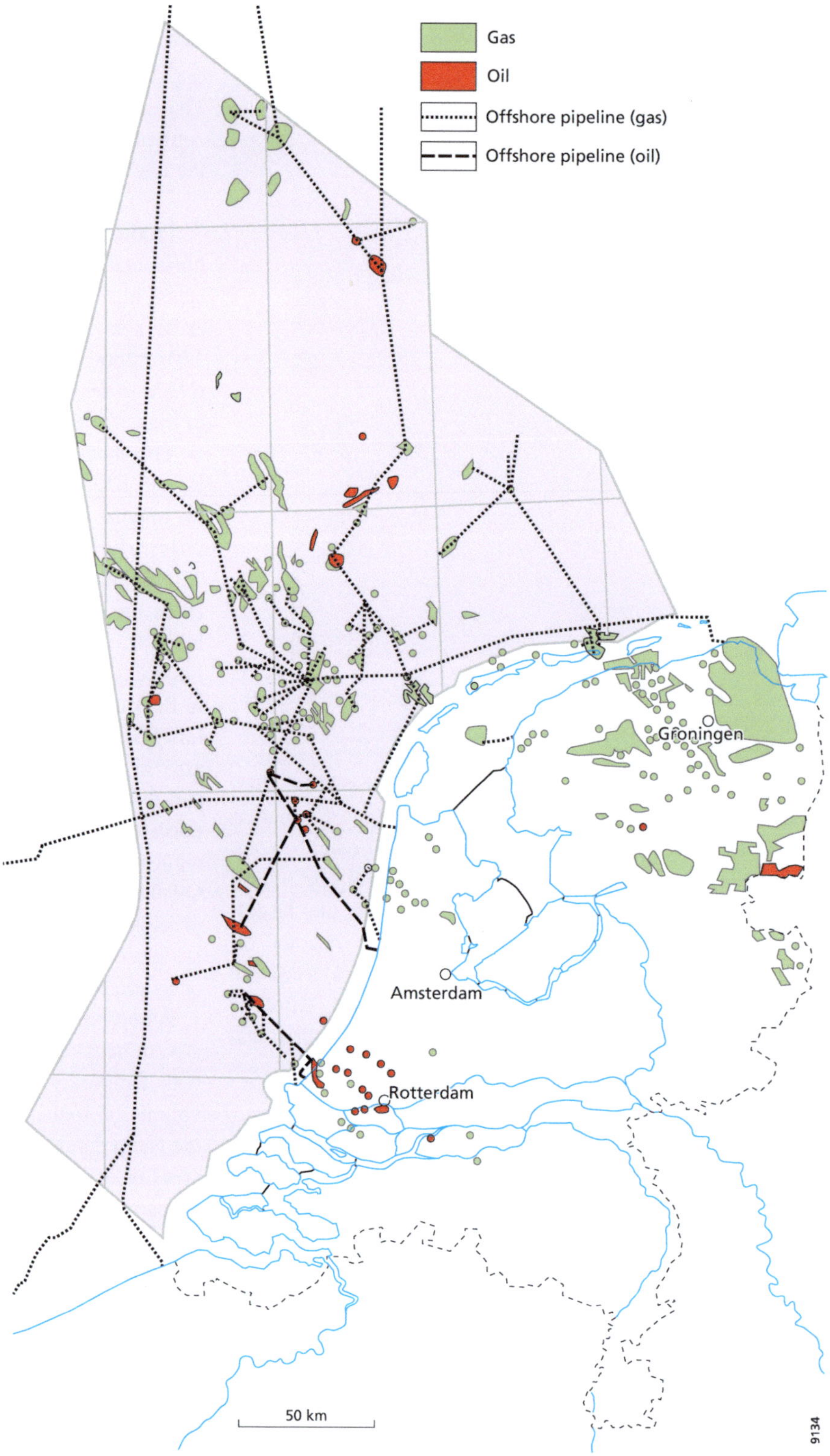

Fig. 4.6 Situation of the oil and gas reservoirs in the subsurface of the Netherlands. The NW-SE orientation of the major tectonic structures is clearly visible. Gas fields dominate over oil fields. (Source: modified after De Mulder et al. 2003)

(anthracite) Carbon (Coal 2016). Coal resources in the Netherlands are of Late Carboniferous (Westphalian, Moscovian) age (318–309 million years ago). Such coal-bearing deposits of that age (Limburg Group; De Mulder et al. 2003) came to rest in a continental environment and occur in the subsurface of most of the Dutch territory. They serve as source rocks for the abundant natural gas resources in the country (De Jager and Geluk 2007). These deposits are exposed at surface north of the city of Aachen (Germany), close the Dutch border, where they have been mined since

Fig. 4.7 Nodding donkeys producing oil from the Schoonebeek field, east Drenthe

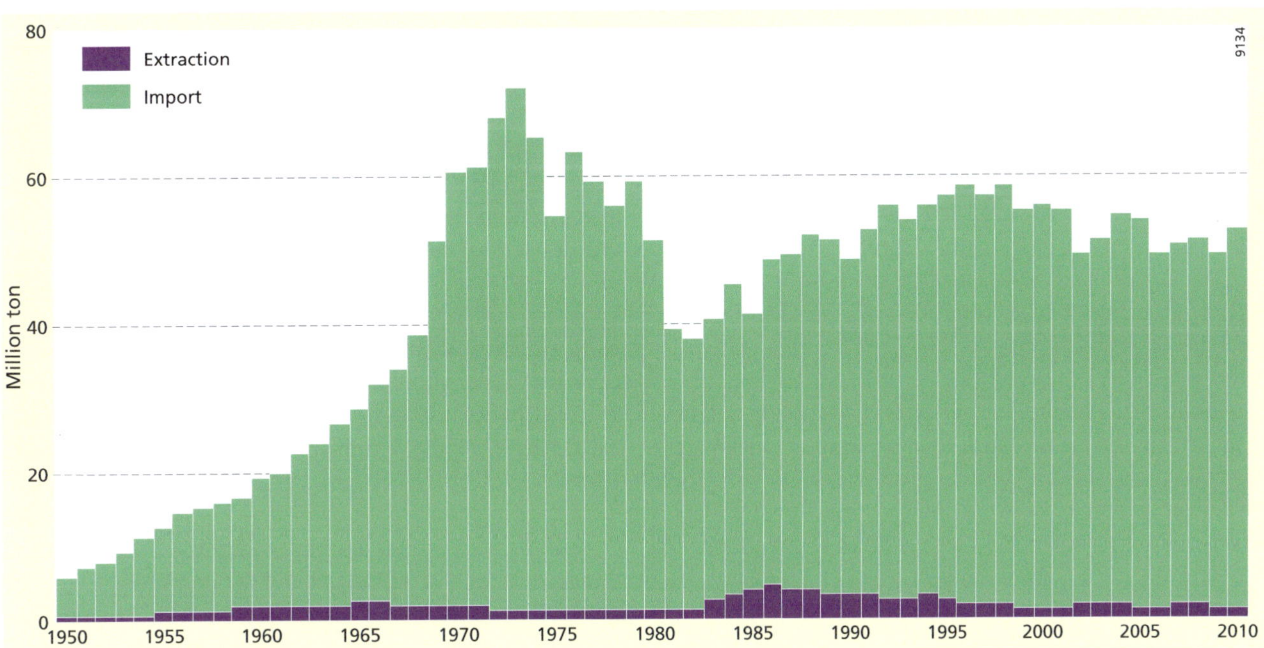

Fig. 4.8 Import of oil in the Netherlands in relation to Dutch oil exploitation from 1950 to 2010. (Source: https://nl.wikipedia.org/wiki/Ruwe_olie_in_Nederland)

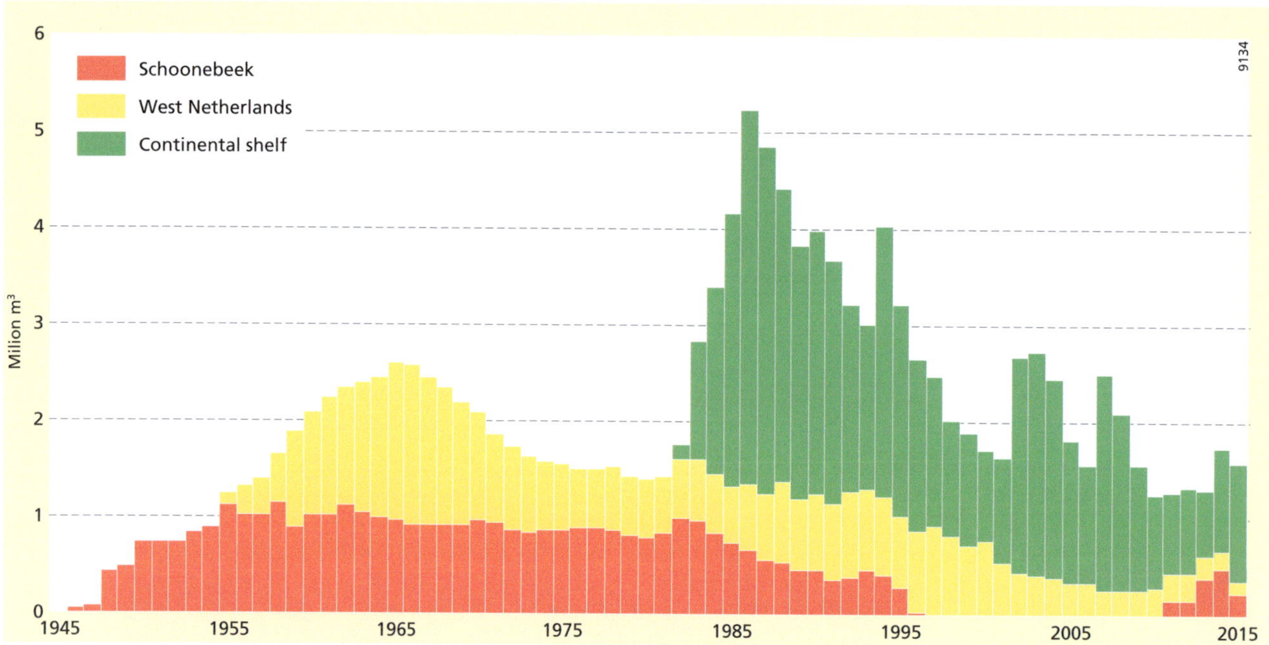

Fig. 4.9 Post-WWII oil production from different regions in the Netherlands (Schoonebeek oil field in red, oil fields in the west of the Netherlands in yellow and offshore (in green). (Source: https://nl.wikipedia.org/wiki/Ruwe_olie_in_Nederland)

Roman Times and dip towards the north and northeast. In the subsurface of Zuid Limburg in the Netherlands, where they have been mined, coal beds lay at depths of roughly between 20 and 1000 m. Individual coal-bearing strata may range in thickness from a few millimetres to some three meters and may alternate in cycles with grey, silty to sandy shales and with grey sandstone beds. The deposits in the coal-bearing Limburg Group are strongly tectonically affected and occur in blocks. Detailed surveys from mining shafts and bore holes revealed 248 individual coal seams with an accumulated thickness of 73 m in the Dutch subsurface (Geluk 2012). According to some authors all Dutch coal beds together correspond with initial peat beds of an overall thickness of 1000 m (Van der Vlerk and Kuenen 1962).

Despite their significant exploitation in the Netherlands, in particular during the twentieth century, remaining geological coal resources in the Dutch subsurface are still some 2900 million tonnes in seams thicker than 50 cm at depths less than 1500 m, 800 million tonnes of which are technically exploitable (Van Montfrans et al. 1988).

Production & Consumption Coal has been the main energy supplier in the late 19th and in the first 65 years of the twentieth century in the Netherlands. As in most surrounding nations, access to coal resources spurred the Industrial Revolution in the Netherlands. Underground coal mining via galleries began in the early eighteenth century near the abbey of Rolduc in Limburg. More significant coal production developed since 1845 when the Dutch government granted mining concessions to private companies. Coal mining began flourishing by the end of the nineteenth century when the region of south Limburg and their coal resources became accessible by rail and water transport. By that time the State also began to exploit coal resources by establishing four State mines which expanded during the early twentieth century (Geluk 2012). A national coal exploration drilling campaign was conducted from 1903 to 1918. They found new prospects in the north of the Province of Limburg and in the east of the country, in *Achterhoek*. Coal mining boomed in the Netherlands from about 1925 until the 1960s and has been responsible for major economic development during the early twentieth century (Chap. 7). Discovery of giant gas resources in the Groningen field (see above), steadily declining coal prices and the ever higher mining operating costs caused the Dutch government and private firms to close all mines ultimately by 1974 (Geluk 2012).

Coal production peaked with more than 14 Mt in 1937, crumbled during WWII to about 5 Mt, revived after the war and with about 12.5 Mt peaked again in 1963, just before all Dutch coal mines were announced to be shut down (De Mulder et al. 2003). In total, roughly 300 Mt coal have been produced from Dutch coal mines over time.

Since 2002, some 12 Mt coal is being imported in the Netherlands due to low market prices and to fuel Dutch coal combustion plants. There are eight such coal combustion

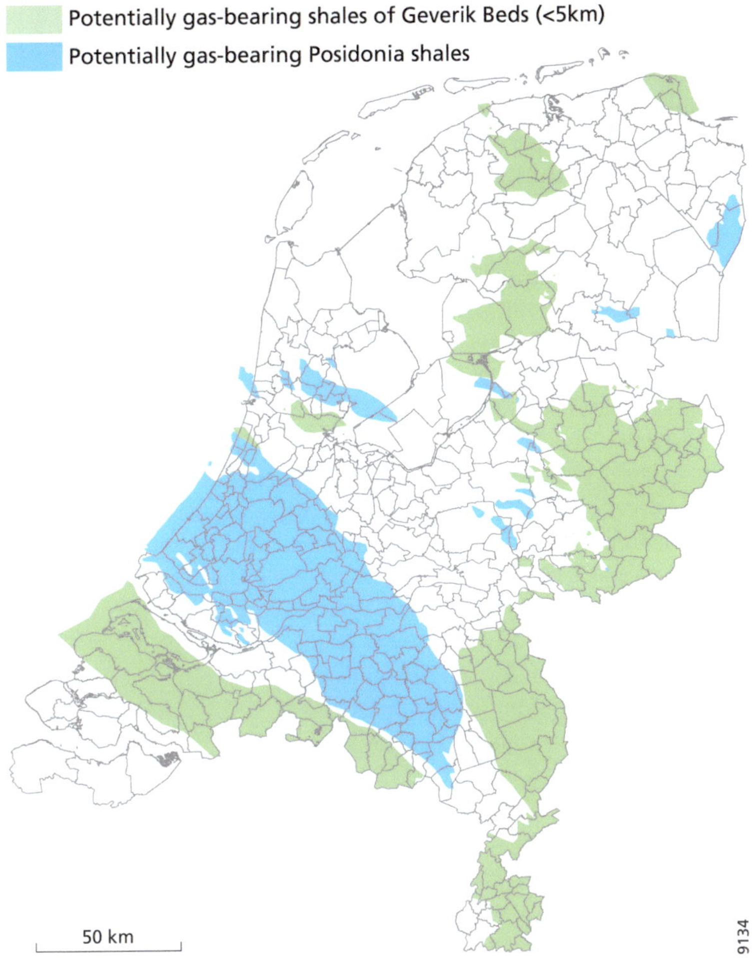

Fig. 4.10 Distribution of Posidonia Shale (in blue) and of the Geverik Beds (in green) to a depth of 5000 m below surface onshore in The Netherlands. (Source: TNO/Mart Zijp)

plants in the Netherlands, often combined with other fuels, as biomass. The two most recently built ones have been equipped with end-of-pipe CO_2 capture installations. Dutch policy concerning coal is to close all coal combustion plants in due time as part of the Climate Agreement on energy transition towards climate neutral, renewable energy supply. Options for Coal bed methane (CBM) exploitation (see Box 4.1) are not feasible economically yet but are being kept open. As all former Dutch coal mines have been dismantled options for renewed coal extraction from former mines do no longer exist.

Box 4.1 Unconventionals: Coal Bed Methane and Shale Gas

In general, renewed interest exists in coal exploration, also in the Netherlands. This interest is primarily focused on unconventional exploitation methods: Coalbed Methane and Shale gas. Coal bed methane is a method to recover methane gas that is retained in the coal beds, by drilling rather than by mining. As many

Box 4.1 (continued)

coal beds and seams still occur in the Dutch subsurface the theoretical potential recoverable volumes for coal bed methane exploitation are high. Currently, coal bed methane production is not considered economically feasible.

Shale gas is gas trapped in consolidated clay beds (shales) in the subsurface. The usual method for extraction is to drill deep bore holes and to disintegrate the shales by pumping a mixture of highly pressurised sands, water and chemicals (fracking). Two types of potential shale gas resource rocks have been identified in the Netherlands, the Jurassic *Posidonia* Shale and the Carboniferous Geverik Beds. These deposits mainly occur in the deep subsurface, in the south and east of the Netherlands (Fig. 4.10). Potential geological shale gas reserves are estimated between 200 and 500 billion m^3 (Van Bergen et al. 2007). Shale gas is not being produced yet in the Netherlands. Exploitation options are in study. However, major environmental concerns for exploitation have been expressed.

Lignite and Peat Lignite is a combustible, dark brownish to black-coloured soft rock. It is a compressed peat that has been exposed to high pressures and temperatures. Lignite has a Carbon percentage between 60 and 75 and a high inherent moisture content (up to 75%). Lignite occurs in the shallow Dutch subsurface and limited mining took place after WWI in Zuid Limburg, near *Brunssum*.

Peat is the first phase in the carbonisation process where plant remains are still well recognisable as pressure and temperature have not yet affected its composition. Peat resources have been widespread in the Netherlands and were easily accessible (Fig. 4.11). Peat has been used for salt collection and for combustion (Chaps. 2 and 3). Large parts of the west, centre, north and the east of the Dutch territory were once covered by thick peat bogs. Major Holocene peat formation in the Netherlands developed since about 4000 years ago and was caused by closure of the Dutch coast line, the Holocene sea level rise and associated stagnation of water discharge via the main rivers in the Dutch delta, a situation that lasted for several millennia (Fig. 2.8; De Mulder et al. 2003). In the west, peat mostly developed as fens while high moor and bog peat is generally found in the higher, often sandy parts of the Netherlands. Peat exploitation declined when coal and later oil became available in the nineteenth and twentieth centuries and not much of this important resource was left by then (Chap. 3).

Production & Consumption Apart from some scattered small-scale mining activities in the mid-eighteenth century, legal lignite mining began in the Netherlands in 1919 but was hampered initially by lack of appropriate transport facilities via the national railway system and by the rapidly declining prices once WWI was over and most lignite mines stopped operations. Overall lack of energy materials in WWII increased interest in lignite and several new mines were opened since, some of these producing until 1962 (Van Rooyen 1989). Geological lignite reserves in the Dutch subsurface are estimated at 1.6 billion tonnes but technically feasible resources are much smaller: five million tons (Van Montfrans et al. 1988).

Lignite only marginally contributed to the Dutch energy demands in times of scarcity. Peat, however, played a dominant role as fuel provider in this respect from about the 13th until the second half of the twentieth century as described in Chaps. 2 and 3. Annual peat production has been three million m^3 on average between 1550 and 1850 (see Fig. 4.12).

Lignite mining halted in 1962 when other and cheaper fossil fuels became available in large quantities. Moreover, most of the directly available lignite resources in the open pit mines were depleted. Large-scale lignite mining to depths of 420 m currently takes place not far from the Dutch border, in Germany, west of the city of Cologne (Rhineland 2016).

Apart from wood, peat has been the sole energy material available in the Netherlands for many centuries. Wide availability of peat did not only heat the houses of Dutch people and powered industries, peat also fuelled flourishing Dutch economies until coal took over as the number one energy material, in the nineteenth century. Peat production has almost directly been related to consumption although trade in peat has been considerable during the sixteenth to nineteenth century. According to De Zeeuw (1978) energy production from peat combustion would have been about 25×10^{12} kJ/year for the nation during the Dutch Golden Age in the seventeenth century, corresponding to on average 12×10^6 kJ/year per capita. These numbers decreased to about 17×10^{12} kJ/year for the nation and 8.5×10^6 kJ/year per caput in the much less flourishing period between 1750 and 1850.

4.2 Alternative Energy Sources

In this Section non-fossil energy resources in the Netherlands are described. Most of these are renewable, i.e. these can be used practically forever and will never become exhausted. Sustainable energy is energy that is consumed in low volumes as compared to its supply and with manageable collateral, especially environmental effects. Another definition of sustainable energy is an energy system that serves the needs of the present without compromising the ability of future generations to meet their needs. Here, we describe the types of sustainable energy that are (or potentially be) used in the Netherlands.

Solar Energy With 1521 sun hours (average) per year the Netherlands does not belong to the world's best targets for solar energy generation. Solar power may be generated via heating systems, Concentrated Solar Power (CSP) systems and through photovoltaic (PV) solar panels. Solar heating systems basically heat fluids that directly or indirectly (via storage tanks) heat domestic hot water, swimming pool water, or for space heating. Such systems may be quite efficient but are rather infrequently applied in the Netherlands. In CPS solar power is concentrated by mirrors and are mostly used in extremely sunny countries as Spain and Israel.

Solar panels are quite popular in the Netherlands and can be seen at many rooftops. Such PV panels, each consisting of energy-generating solar cells, convert sunlight into electricity using semiconducting materials that exhibit a photovoltaic effect. Energy efficiency of solar panels gradually improved from 6% only in 1954 to almost 17% in 2016 for ordinary

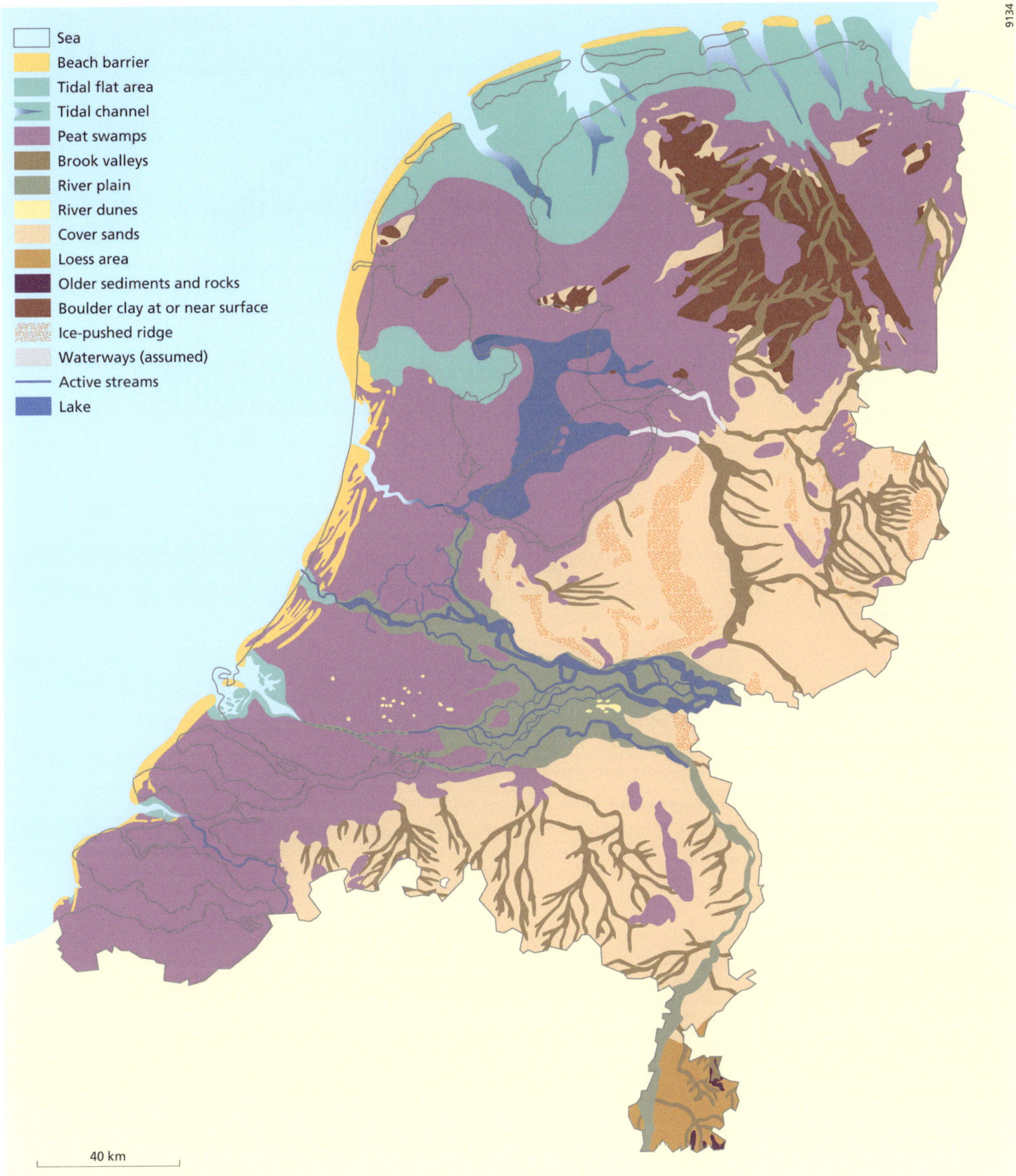

Fig. 4.11 Paleogeographic map of The Netherlands 2600 years ago. Note the extensive peat (purple on the map) accumulation in the west. (Source: De Mulder et al. 2003)

crystalline silicon PV cells. Since two decades more and more PV parks and solar power plants have been installed and many more are in construction. Such parks widely occur now in Germany and in some other (particularly southern) European countries. Plans have been developed for construction of some of such parks in the Netherlands with a total capacity of some 250 MW in 2020. A new (third generation) development in solar panels is application of nanotechnology

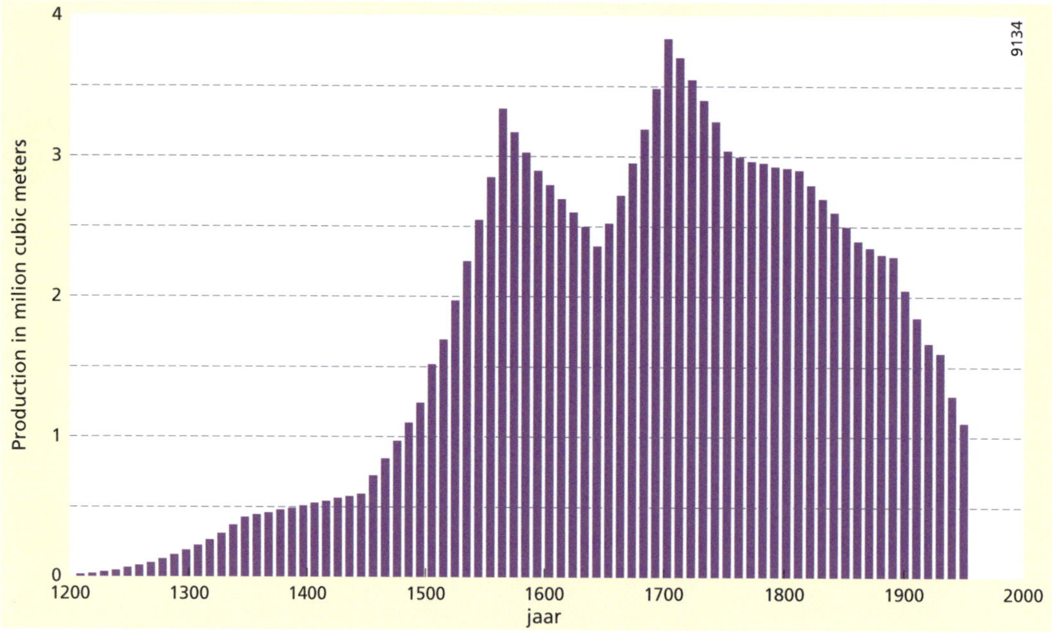

Fig. 4.12 Estimated peat production in the Netherlands between 1200 and 2000. (De Mulder et al. 2003)

creating circuits out of individual silicon molecules. Such cells may cost half as much as traditional photovoltaic cells. Until mid 2013, a total capacity of 347 MW of solar power was installed by more than 93,000 installations while that was 46 MW only in 2003. Most current installations are installed on rooftops of private houses. The Dutch government has facilitated installation of solar panels through subsidies. Private persons may sell surplus energy to the national electricity grid. The Dutch government aims at a solar power capacity of 1500 MW in 2020, serving some 400,000 households (ECN 2016).

Wind Energy The Netherlands is a quite windy country, in particular along the Dutch coasts. Wind power has therefore been used extensively to drain the many Dutch polders since the fifteenth century until steam power took over in the nineteenth century. Thus wind power has significant potential as an alternative, non-fossil fuel for the Dutch. Main concern (as in solar energy) is irregular wind availability per time unit. Even if more than sufficient wind would be available (e.g. during storms) the wind mills will be stopped to prevent damage to the rotation system. Wind mills may be positioned in isolation or together in wind parks. Both types occur in the Netherlands. In rural areas and near urban concentrations isolated wind mills occur and wind parks may be installed in areas with low population densities, in large freshwater lakes and in the Noordzee (Fig. 4.13). Three of such wind parks at sea have presently been installed with a total capacity of 357 MW. A fourth (Gemini, 600 MW, 150 wind mills) is in implementation, and a fifth (Doggersbank) will be a joint project together with Denmark, Germany, UK and has a

Fig. 4.13 Wind park at the North Sea, near Ijmuiden

planned capacity of 9000 MW. The Dutch *Energieakkoord* (Energy Agreement (SER 2013); see Sect. 4.3) aims at 6000 MW of wind power in 2020. The 2015 evaluation shows that without additional measures this ambition will not be met and will barely reach 5100 MW by 2020.

Solar and wind are intermittent energy sources that supply electricity 10–40% of the time. To compensate for losses in production such types of energy are normally paired with already existing powers as, for example, nuclear or natural gas generation. As long as no solid solutions are in place to adequately store energy produced by wind power this type of energy can only be applied as additional to other energy powers. Another disadvantage of wind power is of an aesthetic dimension. Many people oppose building of these tall installations in their immediate neighbourhoods (NIMBY: Not In My Backyard) as they spoil (their views of) the landscapes, even far away from the coast in sea. Others believe that wind parks may be obstacles for migrating birds thus forcing them to deviate from their normal routes. Finally, opposition also occurs against perceived noise of rotating wings.

Water Energy The most significant of this type of non-fossil energy worldwide is hydroelectric power. Hydropower is being used for over 100 years and belongs to the first generation of alternative energies. This type of energy, however, demands for considerable differences in topographic height which do not exist in the Netherlands where the highest 'peak' is a hill of 322 m above sea level, only. Other types of water energy include wave energy and tidal energy. An experimental wave farm is in Portugal and another is planned in Scotland. Two small tidal power plants are operational in Dutch waters, one off the island of Texel and another in the Oosterschelde Storm Surge barrier.

Geothermal Power Geothermal energy is a natural source of renewable energy and gained specific interest rather recently as an alternative for fossil (hydrocarbon) fuels and as a source for sustainable energy production. Geothermal power is independent from weather conditions and has significant potential in terms of capacity, but it is accessible only in limited areas.

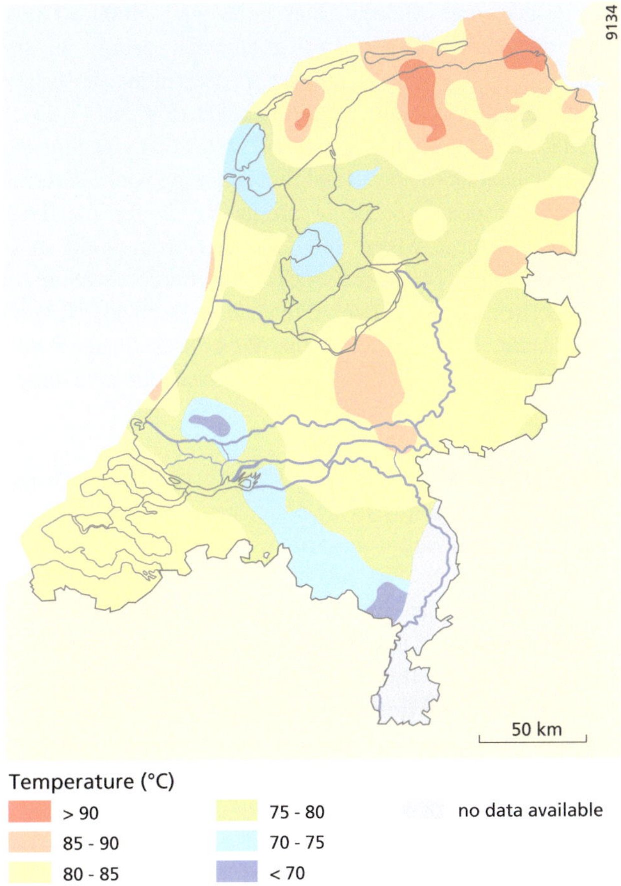

Fig. 4.14 Temperatures at 2000 m depth onshore below the surface. (Source: Lokhorst and Wong 2007)

Two major types of geothermal energy are currently being applied in the Netherlands, shallow and deep geothermal energy. Both types derive energy from relatively warm (or cold) groundwater in aquifers (Chap. 2). Two main boundary conditions determine geothermal energy suitability: temperature of the groundwater source in the aquifer and transmissivity (i.e. product of thickness and permeability) of the aquifer. Both subsurface parameters are determined by geological factors only and are quite variable. A map of temperatures at 2000 m depths is given in Fig. 4.14. Two sorts of geothermal energy systems occur: open and closed systems. Groundwater used in open systems may circulate within the aquifer while liquid in closed systems is never in contact with the groundwater in the aquifer.

Shallow geothermal energy generally is derived from aquifers less than 500 m deep. In an open shallow system (some 2000 in operation in the Netherlands in 2015; Bloemendal and Hartog 2016), relatively warm groundwater is pumped in winter and is used for both domestic and non-domestic heating. After heath extraction the cooled water is re-injected through another bore hole into an aquifer. In summer times, the water flows the opposite way. Thermal power generation in open shallow systems is generally between 200 and 20,000 kW (Lokhorst and Wong 2007). This system is often used in greenhouses for horticulture. In closed systems (close to 3500 put in operation in the Netherlands in 2016 only; Bodemenergie 2017) the fluid is pumped down in winter to a (relatively) warm aquifer and successively heated through heat exchangers. In summer, the system operates in a reversed way for cooling. Closed shallow systems are mainly used for domestic (private) applications. Thermal power generation in closed shallow systems is generally between 50 and 100 kW (Lokhorst and Wong 2007).

Another type of geothermal energy production deals with temporary storage of hot or cold water in the shallow subsurface (KWO 2016). Cold groundwater e.g. is used for cooling a building. Upon use the warm water is temporarily stored in a shallow aquifer through a pipeline system. In winter, this water is used again but now for heating. This method has been applied widely during the last decade in the Netherlands, in particular for climate systems in offices and in greenhouses.

Heat production for electricity generation is the main purpose of deeper geothermal energy applications. In the past decade some 20 deeper bore holes (up to 4000 m) have been drilled to exploit deep geothermal energy in the Netherlands. Hot water locally occurs deep in the subsurface of the northern part of the Netherlands (Fig. 4.14). Deep geothermal energy is mostly applied for heating greenhouses, for block heating of houses and for swimming pools. A recent example of deep geoenergy production is for the Trias Westland Project where 140 °C water from 4000 m is pumped from Trias-aged rocks for future heating of greenhouses near Naaldwijk (Financieele Dagblad 2017). In some cases, for example in the West Netherlands Basin, (small) gas and oil occurrences have been encountered during drilling for geothermal energy.

Energy from Biomass Biomass is organic matter derived from living, or recently living organisms. It can be used as a source of energy and it most often refers to plants or plant-based materials that are not used for food. As an energy source, biomass can either be used directly via combustion to produce heat, or indirectly after converting it to various forms of biofuel. This first type of biomass is solid and normally produced from wood residues, plant remains, garden waste et cetera and pressed into briquettes or pellets. These are often added to the combustion processes in coal-generated power plants.

Biofuels can be derived directly from plants, or indirectly from agricultural, commercial, domestic, and/or industrial wastes. Biofuels (e.g. bio-ethanol) may either be directly used for energy production or be mixed with traditional fuels, for example with gasoline for the transport sector. In 2013, 5% of all gasoline use in the Netherlands were from biofuels. In this country biomass pellets are used in coal-generated power plants. In total, 5×10^9 kWh was produced by biomass in 2015 (BIO 2016).

Sustainability of biomass is a matter of debate as pristine rain forests may be cleared and converted into e.g. sugar cane plantations, which in turn may generate land erosion. New (third-generation) technologies, often based on algae, include advanced biomass gasification and biorefinery technologies, presently are in study or in development.

Nuclear Power does not entirely belong to the group of alternative energies as it is derived from geological deposits ('fossil') and therefore does not meet with the criteria of renewable energy although resources are large on a global scale. However, nuclear energy is non-carbon and this type of energy production does not generate greenhouse gasses. Since the oil crises that hit the Netherlands relatively hard in the 1970s, the Dutch energy policy has always been geared at diversification and spreading of energy materials, definitely including nuclear energy. Successive major failures in international nuclear energy production (Chernobyl in 1986, and Fukushima in 2011), together with fierce debates on safe disposal of nuclear waste materials have made the Dutch reluctant to proceed on nuclear energy production. In 2016, one nuclear power plant (*Borssele*) is in operation supplying 3.9% of the total energy production in the Netherlands. Two more, small nuclear plants occur, an experimental one in Delft, connected to the Technical University and one in Petten (a village in the dunes of Noord-Holland) producing nuclear material for medical applications in hospitals only.

4.3 Dutch Policy Toward a Non-fossil Energy Transition

The Dutch energy market is privatized. Gas and electricity is delivered by the same providers. Consumers are free to choose out of tens of energy providers, mainly based on tariffs and options as green or traditional energy provision. The Netherlands has now set some first steps towards a next energy transition. This is not because Dutch natural gas resources are about to be depleted but rather due to changed political views to give priority to CO_2 emission reduction and thus to transform energy production from fossil fuels to renewables and to focus more on energy saving measures. This switch in political orientation has been inspired by the 2015 Paris Treaty on globally combating climate change (IPCC 2014) to which the Dutch government has committed itself.

Two years before the Paris Treaty, the Dutch government negotiated a national Energy Agreement (SER 2013) with 47 governmental and non-governmental organisations, trade unions, environmental activists and industry in September 2013. That Agreement will be effective from 2013 to 2023 and was based on earlier EU ambitions (EUR-LEX 2016) and has been the result of a typical Dutch 'polder' approach in bringing so many and diverse groups of representatives of the entire Dutch population around the negotiation table for months and in reaching a tangible compromise. In some way the Paris Treaty on Climate (December 2015), almost all nations have jointly set additional international targets: no more global heating than 2 °C as compared to pre-industrial times, and striving for a further reduction to 1.5%. The Paris targets are even more ambitious than those reached in the Dutch Energy Agreement. Ambitions of the Dutch government are 40–50% renewables in 2030 and almost complete CO_2-free power production by 2050 (Ministerie van Economische Zaken 2016). According to Compendium voor de Leefomgeving, renewable energy comprised 11.1% of total Dutch energy production in 2015 (Renewables 2017).

Evidently, implementation of such commitments and ambitions will face major challenges. They require considerable pro-active and creative thinking and operations, including decentralisation of energy supply and linking with societal structures as the built environment and with the construction and transport sectors. That, in turn, will generate new domains, business models and opportunities for market parties and for the public sector. Apart from switching to more nuclear energy, three types of greenhouse gas reductions may be achieved in the Netherlands: (1) large-scale energy savings and efficiency. In addition to what has been accomplished many options still wait for implementation including those in mobility and in technology towards a more sustainable production of goods. (2) large-scale investments in renewable energy production as described in Sect. 4.2. Energy generation from such resources is improving and production costs decline. In some urban areas solar energy production has become competitive with fossil fuels. But none of the renewable energy sources could replace fossil fuels in terms of safe and secure energy production yet (TNO 2015a, b). (3) application of cleaner technologies in fossil fuels. During transition to a non-fossil energy future fossil energy sources will remain necessary to satisfy growing energy demands, in particular in emerging countries. Partly, produced CO_2 might be stored underground (CCUS: Carbon Capture Use and Storage) by capturing Carbon dioxide from the combustion process, using it for e.g. crop production in greenhouses, and to store it successively in e.g. depleted, deep offshore gas fields.

How rapid this energy transition will be strongly depends on production costs of individual non-fossil energy sources and will considerably accelerate once green energy prices will become really competitive, disregarding subsidies, apart from political pressure. But if societal costs as dust-related health issues, mitigation costs of environmental pollution and costs of climate change would be incorporated in this comparison, some renewables might be actually cheaper than fossil fuels are now already. Another important transition criterion is secured and rapid energy delivery which may face challenges for weather-related renewables. Until these criteria have been met, fossil fuels will remain playing an important role in energy production and combinations of fossil and non-fossil energy resources will be applied in the Netherlands ('Hybrid energy systems').

CO_2 reduction is a prime target in Dutch policy today. Carbon dioxide emissions are measured and used to control their production. In 2005, the European Union (EU) introduced an Emission Trading System (ETS), valid for some 11,000 energy plants and energy producing factories in 28 EU and some other European nations. ETS operates as a 'cap and trade' system. Maximum annual emission is determined on basis of the allowed emission that would not exceed global warming of 2 °C. Emission Trading vouchers are distributed among or sold to these plants who are free to buy additional ETS rights from plants that produce less CO_2, or vice versa. As the world economy declined and as cheap fossil and renewable energy supplies flooded the market since, ETS rights sharply declined in value and are very cheap today and not stimulating these plants (in particular not the relatively efficient Dutch plants) to reduce their (CO_2) productions. To counter such developments the Dutch government plans to intervene in this process (Van Geuns and Van Cleef 2016).

The Dutch government has been ambivalent in its ambitions to promote sustainable energy and a clean environment as a significant share of the State income is generated from production and distribution of fossil fuels. Application of green power in the Netherlands has been hampered by capricious subsidy policies by successive Dutch governments. However, early 2018, government decided to terminate natural gas production from the Groningen field, via 7,5 NM^3 in 2022, to zero by 2030. The forthcoming energy transition will also affect position and relevance of major harbours as Rotterdam and oil companies as Shell (stranded assets) which currently strongly rely on fossil fuel production and distribution (TNO 2016; Van Geuns and Van Cleef 2016).

4.4 Construction Materials

Construction materials are natural materials in the subsurface that are generally used for building and construction purposes. These mainly consist of sand, gravel, limestone and clay. Most construction materials occur abundantly in the Dutch subsurface and in the Dutch sector of the continental shelf. Their distribution has been well identified during extensive systematic geological mapping campaigns from the 1960s until the 1990s, and by the successive development of a National database of borehole data by the Geological Survey of the Netherlands.

Gravel is coarse grained (30% of a sample contains components larger than 2 mm) detritus material of rock and is an essential component of concrete. Dutch gravels are of Pleistocene age and may occur associated with sand, clay (as in boulder clay), or, but rarely, with blocks of rock. Gravel quality is determined by size, composition and shape of the coarse components (NEN 1989). Higher silica contents of components generally contribute to higher gravel qualities. Gravel has a quite limited distribution in the shallow (less than 25 m deep) Dutch subsurface and onshore. These only occur in the upstream parts of the River Maas, i.e. in the eastern parts of Noord-Brabant and in Limburg. There, gravels are exposed at surface in the terraces of River Maas. Deeper gravel deposits may be found upstream River Rijn in the east part of Gelderland, in the River IJssel Valley (province of Overijssel), in Noordoostpolder, and in the adjacent area of IJsselmeer. Associated with boulder clays, gravels locally occur in the shallow subsurface in the north of the country and offshore in the Dutch sector of Noordzee

(Van der Meulen et al. 2007). The most common way to extract gravels from the Dutch subsurface is by dredging from artificial lakes.

Sand is also made of rock detritus but has smaller grain sizes (average between 100 and 500 μm) than gravel. It occurs abundantly in the Dutch delta which partly consists of thick layers of Pleistocene sands. At surface or at shallow depths sands prevail in the higher sands areas (Chap. 3) and are predominantly present in the ice-pushed ridges in the central parts of the Netherlands and upstream rivers Maas and Rijn in the east. Coarse-grained sands are being used for concrete and mortar. Their quality is determined by (low) contents of fine materials and iron. Most Dutch coarse sands are used as aggregate for the concrete and asphalt industries while many of the medium-grained sands are used for building purposes, to raise ground or for infrastructural and coastal defence (dikes) purposes. In the west and offshore, sands are excavated by dredging while they are mechanically excavated in dry pits in the central and eastern parts of the Netherlands.

Clay is the finest-grained type of construction material and contains high percentages of platy lutum components (< 2 μm). Large volumes of marine clays at or directly below surface can be found in the marine areas in the west while fluviatile clays extensively occur in the river valleys of rivers Maas, IJssel and Rijn (Chap. 3). Significant quantities of mainly river clays are used for brick and roofing tile production, smaller volumes of in particular older (Tertiary) clays for the production of fine ceramics. Clays are mainly excavated in dry pits.

Limestone is used for the cement industry and only occurs in the south of Limburg and in the east of Gelderland. There, limestone deposits are exposed at surface and are of Late Cretaceous or Jurassic (in Achterhoek) age. Limestone quality depends on its % of $CaCO_3$.

Geological reserves of construction materials exceed national demand. However, certainly not all geological reserves may be extracted. Restrictions apply for economic, technical or environmental reasons. Technically exploitable reserves of construction materials in the Netherlands are estimated in Table 4.1, for the period 2013–2050.

Due to above mentioned restrictions reserves may change over time. Volumes of technically exploitable reserves may increase due to introduction of new and cheaper exploitation techniques while economic conditions and scarcity may determine exploitable volumes in course of time as well. Environmental limitations may have a positive or negative effect on resource volumes due to fluctuations in levels of public awareness, to politics and to legislation.

4.5 Industrial Minerals

Industrial minerals are natural resources required for production of other goods, except for energy and construction materials. (Rock) salt is the most relevant of these industrial minerals in the Netherlands which is Europe's largest exporter of rock salt.

Salt is not only used to flavour food but more importantly for the production of a wide range of chemical products as chlorine, sulphate, sodium, magnesium and calcium. Dutch salt is produced from rock salt. Vast quantities of rock salt occur in the subsurface of the NE part of the Netherlands. Such rock salt deposits are of Late Permian and Triassic age. These developed during large-scale evaporation processes at times that shallow seas underwent extremely dry and hot climatic (desert-like) conditions. Through evaporation thick layers of salt of varying composition eventually came to deposition on the sea floors. When these salt layers, with a relatively low specific gravity, were buried by younger sediments with relatively high specific gravity, (rock) salt eventually began to move upwards into pillar-shaped underground constellations (salt domes, Fig. 4.15). Such structures may reach accumulated rock salt thicknesses of more than 3 km. The top of such pillar structures may reach the surface as they locally do in NW Germany, but the shallowest rock salt occurrence in the Dutch subsurface is found at a depth of 1 m below the Noordzee floor. In total 34 salt pillars and domes have been identified in the Dutch subsurface, 25 of which occur in the north and nine in the east of the country. Due to its impermeability and visco/plasticity rock salt deposits often serve as caprocks to seal underlying, porous Permian host rocks for natural gas in the same regions.

Rock salt deposits were first found in the Dutch subsurface in 1887. However, exploitation did not start before 1919 and was done through solution mining, a technology basically still used today. In the west of Friesland (Barradeel) rock salt is exploited through solution mining at depths up to 3000 m, making that the world's deepest salt mine (Geluk 2012). This mining technique creates large underground caverns which may be a challenge to salt mine stability. Due to the high temperatures at such great depths salt caverns tend to

Table 4.1 Reserves of construction materials in the Netherlands

Construction materials	Geological reserve[a]	Technical exploitable reserves 2013–2050[b]
Gravel	>8 × 10^9 m^3	550 × 10^9 tonnes
Sand	> 590.3 × 10^9 m^3	6.2 × 10^9 tonnes
Clay	1.8 × 10^9 m^3	220 × 10^6 tonnes
Limestone	5 × 10^9 tonnes	154 × 10^6 tonnes

Sources: (De Mulder et al. 2003; CBS et al. 2016)
1 m^3 construction material weighs 1.5–1.8 tonnes
[a]Cf De Mulder et al. 2003, excluding Zilverzand
[b]Cf De Mulder et al. 2003, excluding data CBS et al. 2016

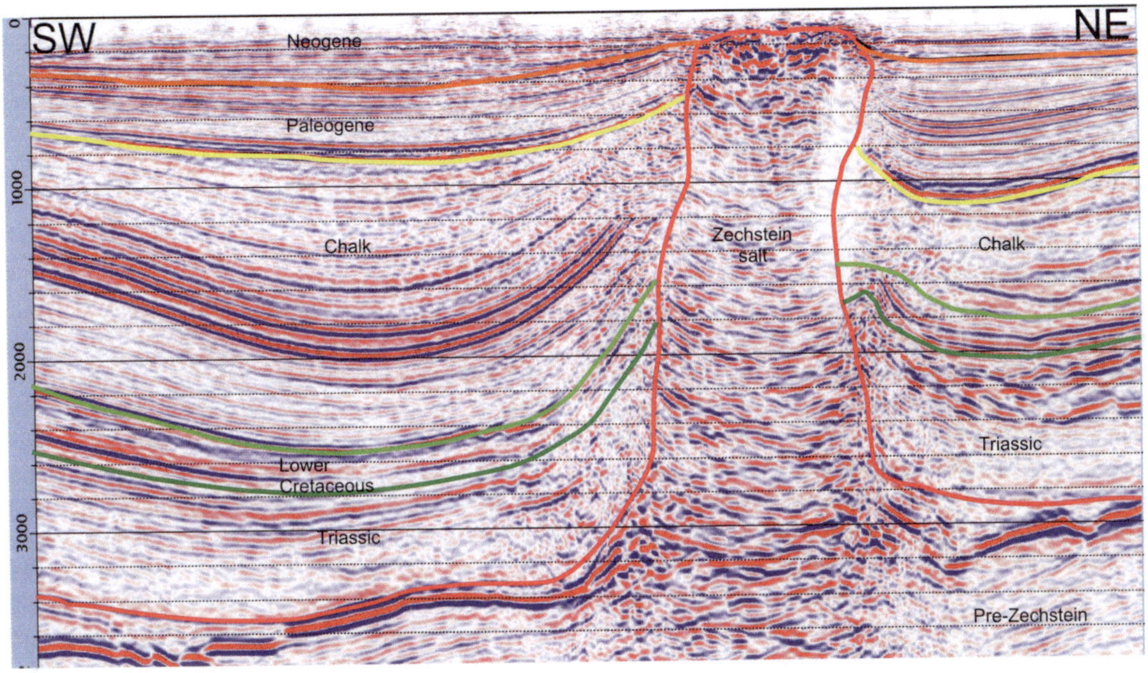

Fig. 4.15 Seismic section of the Zuidwending salt dome. This dome is the shallowest onshore dome in the northern Netherlands reaching to about 100 m below surface. Thickness variations in the adjacent sediments result from periods of main salt movement. (Source: Geluk 2012)

converge much easier, causing deeper and more rapid land subsidence at surface than above shallower salt mines (Geluk 2012). With a quantity of four to five million tonnes/year the Netherlands is one of the largest salt producers in the world. Estimated geological reserves amount to more than 1000 billion tonnes rock salt but technically exploitable is only about 300 million tonnes until 2050 (De Mulder et al. 2003).

Other industrial minerals as metals are very scarce in the Dutch subsurface. Iron oxide occurrences were excavated locally only, e.g. in Achterhoek. In Pre-historic times quite significant volumes of flint (SiO_2) had been excavated from limestone beds in the south of Limburg. Limburg flints were used for high quality tool production in the Stone Age and have been identified in a wide realm in NW Europe (De Mulder et al. 2003).

4.6 Soil

Soil has been a prime natural resource since the Neolithic Revolution which occurred in the Netherlands about 7500 years ago (NEO 2016). Although today this country is densely populated and highly industrialised, its land use still reflects its agricultural roots. About 68% of the Dutch territory is currently in use for agricultural purposes. This number has been slightly fluctuating since the onset of the twentieth century, for example as a result of massive parcel re-allotment (Box 3.1) in the past century. Half of the agricultural land is grassland (CLO 2013).

Three soil types prevail in the Dutch landscapes: sand, clay and peat. Almost all of these soils are of Quaternary age and roughly coincide with topographic differences in the Netherlands. Sands dominate in the higher (i.e. above sea level) grounds in the centre, east and south of the country, clays and peaty soils are mainly situated at or below sea level in the north and west. The sands are mainly of Pleistocene age and that is where agriculture in the Netherlands began. By then, the west and north of the country were still submerged by the Noordzee. There, coastal swamps began to develop which turned into thick peat beds later. The clay beds in the west were deposited by the sea. Clays in the centre of the country often have been deposited by rivers. Normally, peaty soils were used for grassland and for cattle only, while clayey soils are quite well-suited for growing crops. The sandy soils were generally poor in quality but fertilizers turned sands into agricultural highly productive soils as well. Limestones, Tertiary clays and loess deposits have a much smaller distribution and occur in the southeast of the country (Chap. 3).

Soil quality has been improved by more advanced mechanical treatment techniques and in particular by applying the extensive knowledge gained by many generations of Dutch agricultural engineers. That made Dutch soils among the most productive in the world. Soil productivity in the Netherlands is five times higher than the European average (NRC 2014). Composition of Dutch soils, their distribution and regional variation are described in more detail in Chap. 3. Dutch soils are under pressure. Many agricultural soils have been contaminated by abundant use of fertilizers and pesticides. Soil contamination and remediation is discussed in Chaps. 3 and 5.

4.7 Underground Space

Space is an often underestimated natural resource and rather recently gaining some public and political interest. This interest is due to a growing number of residents in this already densely populated and industrialised country. Moreover, people and politics increasingly request for higher quality standards as to the quality on their living environment. Moreover, their high ambitions in terms of nature preservation and development puts surface space more and more under pressure. Remaining spaces become scarcer and thus a more costly natural resource. This is in particular true for city centres, for example prizes of building ground rose 400% between 1995 and 2002 (Grondprijs 2016). In 2001, the governmental organisation for spatial planning (PBL) stated that the demand for space by the Dutch people until 2030 is about 25% higher than its present national territory (Ministerie van Volkshuisvesting, Ruimtelijke Ordening en Milieubeheer 2001). Although the size of the Netherlands has been expanding by land reclamation works in the past millennium (Chap. 2), further options for spatial expansion have become rather limited in this country, but do occur (IJburg, Maasvlakte).

The obvious other directions are building upward or downward. The Netherlands have no significant reputation in constructing high-rise buildings, primarily due to the poor foundation conditions of its rather weak and soft subsurface and its high water tables. Other reasons are the many protected, often Mediaval, townscapes and the often negative experiences of high-rise residences as described in Boxes 8.3 and 8.4. Most high-rise buildings (only 46 of such constructions > 100 m exist in this country) occur in the cities of Den Haag and Rotterdam.

Where options for on- and offshore expansions and skyscrapers fail or are insufficient to satisfy such demands, the space below surface, the subsurface, becomes an increasingly relevant option. Technically, the subsurface may be used for any application. The same geological conditions of the Dutch delta that hamper construction of high-rise buildings, however, may make underground development rather complex in terms of safety and costs. Nevertheless, significant progress has been made in tunnelling techniques in unconsolidated sediments in the past few decades.

Any case of underground development may jeopardise delicate balances between the current complex, large- and small-scale biogeochemical processes in the subsurface. Disturbance of such balances may eventually lead to e.g. changed groundwater quality and land subsidence. Although interactions between such processes are still not well known, several of the larger scale impacts in time of such developments may be predicted through geohydrological and geotechnical modelling techniques (De Mulder et al. 2012). The current political climate in the Netherlands would not allow other interferences in the subsurface than environmentally harmless ones.

Underground development is always more expensive than building at surface. However, in comparing both options the longer term environmental gains generated by underground construction might turn the balance in decision making. Such gains may include strongly reduced air and noise pollution, substantial energy savings and reduced risk for natural hazards (earthquakes). Moreover, underground structures have longer lifetimes and are thus subject to lower depreciation and maintenance costs. Underground development makes multiple land use possible what is particularly relevant in the city centres.

Except for in the topmost 5 m of the subsurface where all water mains, electricity, gas, sewerage and internet infrastructure networks are positioned, the Netherlands does not have a long tradition in underground construction. But that is changing. Underground parking lots in city centres (and beyond) are becoming a rule rather than an exception. Most underground construction works in the Netherlands today are focussed on infrastructure. In total 38 longer (> 100 m) traffic tunnels have been constructed in the past two decades (1996–2016, Fig. 4.16). The longest of these tunnels was a

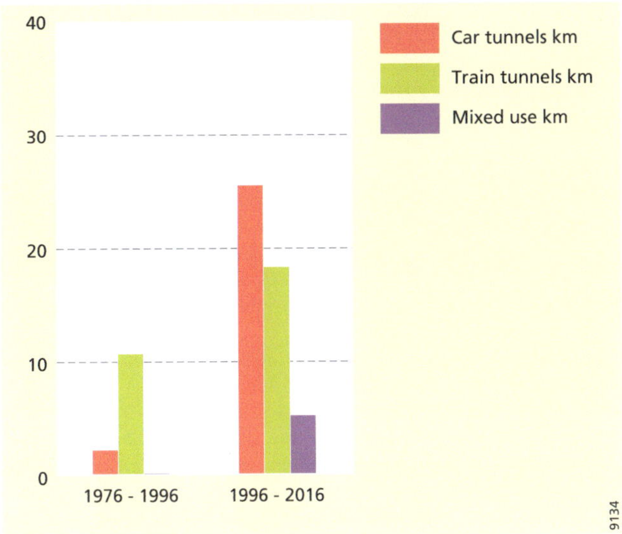

Fig. 4.16 Comparison of tunnel construction in the Netherlands between the period 1976–1996 and 1996–2016 in km

train tunnel, the *Groene Hart* tunnel, with a length of 7.1 km. The longest car tunnel built in that period was the Westerschelde tunnel: 6.6 km. The total length of Dutch tunnels built in the past 20 years has been close to 50 km. Some were built for mixed purposes, including some tunnels solely for bicycles (Tunnels 2016). These data are in sharp contrast with tunnelling data of two decades earlier (1976–1996) when only ten tunnels were built in the Netherlands, just 2.3 km of these for road traffic (Fig. 4.16).

Apart from infrastructure, underground construction works in the Netherlands include cellars under houses, underground office floors, options for storage of wastes, for small-scale household waste in underground containers and for industrial waste as brines and CO_2. Energy materials as natural gas are temporarily stored in the subsurface to meet peak demands. For that purpose several smaller depleted gas reservoirs have been given a new destination. Energy and waste materials may also be stored in former salt mines if such facilities would meet with safety and environmental demands. Permanent storage of high-level radioactive waste materials in salt domes was studied extensively in the Netherlands in the 1980s. Such studies proved that permanent storage may technically be feasible. However, this option has been abandoned for political reasons. Like rock salt, deep and consolidated thick clay deposits of Oligocene age (Boom Clay), widely occurring under the Dutch delta, have excellent thermal isolation properties and may provide good options for underground storage as well.

With growing demands for space and an increasing and more wealthy population, in combination with growing demands for a better quality of the environment, a more extensive use of the subsurface is expected to take place in the Netherlands in the near future. Such development calls for proper, new legislation and a more dedicated management of the underground (De Mulder et al. 2012). The Dutch subsurface may provide a high potential to meet many of the future demands, but not for all. Clearly, there are geotechnical limitations and possibilities that determine the use of the subsurface. Meanwhile, the Dutch subsurface has been explored intensively during the last half century and is now one of the best known in the world. Millions of bore hole descriptions and other subsurface data are stored in databases which are available to the Dutch public, to industries and to the government (BRO 2016) who must legally check existing subsurface data before any development may take place. Cadastres for city centres are moving towards 3D.

Except as a provider of natural resources, a storage place for waste or strategic materials, a place to position infrastructure and other facilities that may disturb the quality of the living environment, the subsurface also serves as an archive containing scientifically valuable traces of the geologic, archaeological and historic past. Moreover, the subsurface is a huge realm where numerous large- and small-scale processes operate that keep the ecosystems sound and running. Awareness of such values urged for protection of the subsurface also through legislation in the Netherlands.

4.8 Exploitation-Related Hazards

Exploitation of natural resources may generate hazards and risks. Where losses of lives were often associated with coal mining, exploitation of natural gas may rather generate earthquakes and losses of properties. Dutch coal mines have been operating relatively safely but still many hundreds of underground workers lost their lives during the 75 years of their operational existence. In addition, many more miners suffered from deathly silicosis. Burning coal by households and industry caused smog and considerably contributed to air pollution during most of the twentieth century. Coal mining also caused sink holes at surface and local subsidence affecting foundations of houses and other constructions.

Natural gas exploitation generates land and seafloor subsidence above gas reservoirs. Moreover, massive gas extractions in the North of the Netherlands (Groningen) has caused seismicity resulting in shallow earthquakes due to sudden settlement in the productive sandstones and/or in the strata above the gas reservoirs. Such earthquakes are anticipated to reach magnitudes of up to 4.5 on the Richter scale (Subsidence 2016). Until June 2018, the highest registered seismic magnitude been 3.6 (Huizinge in 2012). Seismic risk related to gas extraction from 3–4 km deep reservoirs may be amplified and aggravated by the composition of the poorly consolidated sediments in the shallow subsurface. In general, the weaker the subsurface (peat and clay), the stronger intensities of seismic events are being experienced. The composition of the Groningen subsurface has been mapped in detail and been used in the GeoTOP model that served to predict seismic effects at surface (GeoTOP 2007). So far, such earthquakes did not cause physical injuries or casualties yet but did generate significant damage to houses and other constructions, apart from considerable social unrest in the province of Groningen. To reduce earthquake risks in Groningen exploitation rates declined significantly since 2013.

Land subsidence has also been reported in association with solution mining of rock salt, in particular in Friesland. Groundwater extraction may induce settlement of the land as well. In the past centuries this has caused significant problems in the Netherlands, mainly in the peaty areas in the west which surface levels subsided many metres due to oxidation, erosion and excavation (Chap. 3), jeopardising the coastal and river defence systems and triggering major and often catastrophic floods.

References

Aardgas in Nederland (2016) http://aardgas-in-nederland.nl/nederland-aardgasland/aardgas-in-de-nederlandse-energievoorziening. Accessed 23 Nov 2016

BIO (2016) http://www.fluxenergie.nl/economie-2015-productie-hernieuwbare-energie-stijgt-weer. Accessed 30 Nov 2016

Bloemendal M, Hartog N (2016) After the boom: evaluation of Dutch ates-systems for energy efficiency, European Geothermal Conference 2016. EGEC, Strassbourg

Bodemenergie (2017) https://www.bodemenergienl.nl/Actueel/Nieuws/Nieuwslijst/5686.wlid. Accessed 4 Jul 2017

BRO (2016) http://www.briinfo.nl. Accessed 1 Jun 2016

CBS (2016a) https://www.cbs.nl/nl-nl/nieuws/2016/12/lagere-aardgasproductie-drukt-handelsoverschot. Accessed 21 Nov 2016

CBS (2016b) https://www.cbs.nl/nl-nl/nieuws/2016/19/nederlands-record-handel-aardolieproducten. Accessed 24 Nov 2016

CBS, PBL, Wageningen UR (2016) Winning en verbruik van oppervlaktedelfstoffen, 2000–2013 (indicator 0067, versie 15, 21 april 2016). CBS, Den Haag/Planbureau voor de Leefomgeving, Den Haag/Bilthoven/Wageningen UR, Wageningen

CLO (2013) http://www.compendiumvoordeleefomgeving.nl/indicatoren/nl1001-Ontwikkeling-van-het-bodemgebruik-vanaf-1900.html?i=3-18. Accessed 1 June 2016

Coal (2016) https://en.wikipedia.org/wiki/Coal. Accessed 17 Feb 2016

De Jager J, Geluk MC (2007) Petroleum geology. In: Wong TE, Batjes DAW, De Jager J (eds) Geology of the Netherlands. Royal Netherlands Academy of Arts and Sciences, Amsterdam, pp 241–264

De Mulder EFJ, Geluk MC, Ritsema IL, Westerhoff WE, Wong TE (2003) De ondergrond van Nederland. Wolters Noordhoff, Groningen

De Mulder EFJ, Hack HRGK, Van Ree CCDF (2012) Sustainable development and management of the shallow subsurface. Geological Society, London

De Zeeuw JW (1978) Peat and the Golden Age. The historical meaning of energy attainability. AAG, Bijdragen 21:1–31

EBN (2016) https://www.ebn.nl/olie-en-gas-in-nederland. Accessed 21 Nov 2016

ECN (2016) http://www.milieuloket.nl/id/vhurdyxrjtyt/zonne_energie. Accessed 28 Nov 2016

EUR-LEX (2016) http://eur-lex.europa.eu/legal-content/nl/TXT/?uri=CELEX%3A32009L0028. Accessed 30 Nov 2016

Financieele Dagblad (2017) Trias Westland. 11 May 2017

Geluk MC (2012) Exploiting our riches, significance of Dutch industry. In: Boekschoten GJ et al (eds) Dutch earth sciences. Development and impact. Koninklijk Nederlands Geologisch Mijnbouwkundig Genootschap, Den Haag

GeoTOP (2007) Het opslingereffect van de Groninger ondergrond in kaart gebracht. Geo.brief 42(1)

Grondprijs (2016) www.huisprijskompas.nl/article/grondprijs. Accessed 5 June 2016. Accessed 5 June 2016

IPCC (2014) Climate change 2014, synthesis report. Contributions of working groups I, II and III to the IPCC fifth assessment report, summary for policymakers. Intergovernmental Panel on Climate Change, Geneva

KWO (2016) https://nl.wikipedia.org/wiki/Koude-warmteopslag. Accessed 20 Feb 2016.

Lokhorst A, Wong TE (2007) Geothermal energy. In: Wong TE et al. (eds) Geology of the Netherlands. Royal Netherlands Academy of Arts and Sciences, Amsterdam, pp 341–346

Ministerie van Economische Zaken (2015) Delfstoffen en aardwarmte in Nederland Jaarverslag 2014. Ministerie van Economische Zaken, Den Haag

Ministerie van Economische Zaken (2016) Energieagenda, naar een CO_2-arme energievoorziening. Den Haag, December 2016. https://www.rijksoverheid.nl/documenten/rapporten/2016/12/07/ea

Ministerie van Volkshuisvesting, Ruimtelijke Ordening en Milieubeheer (2001) Ruimte maken, ruimte delen. Vijfde nota ruimtelijke ordening 2000/2020. Ministerie van VROM, 's-Gravenhage

NAM (2016a) http://www.nam.nl/content/dam/shell/static/nam/downloads/pdf/biblotheek/redevelopment-schoonebeekoilfield.pdf. Accessed 15 Feb 2016

NAM (2016b) https://nl.wikipedia.org/wiki/Nederlandse_Aardolie_Maatschappij. Accessed 13 Apr 2017

NEN (1989). Geotechniek, Classificatie van onverharde grondmonsters. NEN 5104. Nederlands Normalisatie-instituut, Delft

NEO (2016) http://nl.wikipedia.org.wiki/neolithische_revolutie. Accessed 1 June 2016

NRC (2014) http://www.nrc.nl/handelsblad/2014/11/19/de-landbouwgrond-in-nederland-is-de-beste-ter-wer-1439728. Accessed 6 June 2016

Oil policy (2016) https://www.rijksoverheid.nl/onderwerpen/energiebeleid/inhoud/olie. Accessed 24 Nov 2016

Olie-import (2016) https://nl.wikipedia.org/wiki/Ruwe_olie_in_Nederland. Accessed 24 Nov 2016

Renewables (2017) http://www.clo.nl/indicatoren/nl0517-hernieuwbare-elektriciteit. Accessed 13 Apr 2017

RGD (1995) Toelichting bij kaartblad III: Rottumeroog-Groningen. Geologische Atlas van de diepe ondergrond van Nederland. Rijks Geologische Dienst, Haarlem

Rhineland (2016) http://www.mining-technology.com/projects/rhineland/rhineland2.html. Accessed 17 Feb 2016

SER (2013) Energieakkoord. Sociaal-Economische Raad, Den Haag http://www.energieakkoordser.nl. Accessed 30 Nov 2016

Subsidence (2016) http://www.nlog.nl/en/hazards/subsidence.html. Accessed 1 June 2016

TNO (2015a) https://www.tno.nl/nl/over-tno/nieuws/2015/10/er-is-een-nieuwe-visie-nodig-op-de-toekomst-van-gas. Accessed 23 Nov 2016

TNO (2015b) https://www.tno.nl/nl/over-tno/nieuws/2015/9/flexibiliteit-en-zekerheid-essentieel-in-toekomstbestendig-energiesysteem. Accessed 30 Nov 2016

TNO (2016) https://time.tno.nl/nl/artikelen/het-fossiele-dilemma-van-rotterdam. Accessed 30 Nov 2016

Tunnels (2016) https://nl.wikipedia.org/wiki/Lijst_van_tunnels_in_Nederland. Accessed 29 Mai 2016

Van Bergen F, Pagnier HJM, Van Tongeren PCH (2007) Peat, coal and coal bed methane. In: Wong TE et al (eds) Geology of the Netherlands. Royal Netherlands Academy of Arts and Sciences, Amsterdam, pp 265–282

Van der Meulen MJ, Broers JW, Hakstege AL, Pietersen HS, Van Heijst MWIM, Koopmans TPF (2007) Surface mineral resources. In: Wong TE et al (eds) Geology of the Netherlands. Royal Netherlands Academy of Arts and Sciences, Amsterdam, pp 317–334

Van der Vlerk IM, Kuenen PH (1962) Geheimschrift der aarde. De Haan, Zeist

Van Geuns L, Van Cleef H (2016) Koolstof tot nadenken. TNO November 2016. https://www.tno.nl/media/8558/koolstof-tot-nadenken_20161107.pdf. Accessed 27 Mar 2017

Van Montfrans HM, De Graaff LWS, Van Mourik JM, Zagwijn WH (1988) Delfstoffen en samenleving. Geologie van Nederland Deel 2. Rijks Geologische Dienst. SDU Uitgeverij, Den Haag

Van Rooyen P (1989) Bruinkool, ontstaan en voorkomen. In: Felder WM et al. (eds) Delfstoffen in Limburg. Grondboor & Hamer 43 (5/6): 339–341

Whaley J (2009) The Groningen gas field. GeoExpro Magazine 6(4)

Biodiversity and Nature

5

Jerry Van Dijk

Abstract

Although the Netherlands is one of the most densely populated countries in the world it has several interesting and sometimes internationally unique ecosystems associated with its position on a major European delta. Wet delta conditions and landscape dynamics have shaped the ecosystems of the lowlands while Dutch glaciation history has had significant impact on ecosystem development on the higher grounds. Today, most specific delta dynamics have disappeared due to human use and landscape alteration for flood protection, urbanisation and agriculture. Less than 12% of the Dutch land surface is more or less natural, and is mostly forested. No native forest survived in the Netherlands, current forests were all replanted for timber production or landscape restoration. Natural forests emerged where production forests were abandoned or by natural succession on open terrains. Most Dutch ecosystems and nature reserves are intensively managed because of external disturbances by other land use and to compensate for the lack of natural dynamics. Thanks to such conservation management and to improving environmental conditions in general, Dutch nature now shows the first signs of recovery after a long period of biodiversity loss.

Keywords

Ecosystems · Biogeography · Nature conservation · Environment · Nature policy · Natural history · Natura 2000 · Flora · Fauna

J. Van Dijk (✉)
Copernicus Institute of Sustainable Development, Utrecht University, Utrecht, The Netherlands
e-mail: j.vandijk2@uu.nl

5.1 Biodiversity and Biogeography

The Netherlands hosts almost 46,000 described species, 1100 of which are Eubacteria and Archaea (which is probably just a fraction of the actual number of species), 10,300 are fungi, 3894 are plants and 27,146 are animals, mostly arthropods (Noordijk et al. 2010).

No endemic species occur in this country, but some subspecies are considered endemic. These include a subspecies of Tundra Vole (*Microtus oeconomus* subsp. *arenicola*) and of Large Copper (*Lyaena dispar* subsp. *batava*). Tundra Vole has a boreal distribution and its Dutch populations are considered relicts of Pleistocene glaciations persisting here after retreat of the glaciers. Currently, it occurs in fen peatlands, in the SW delta area and on the island of Texel (NDFF 2015). The endemic Large Copper subspecies is rare in the Netherlands and only occurs in three northern fens. The subspecies of the bee *Epeolus tarsalis* subsp. *rozenburgensis* is also unique for the Netherlands, but its taxonomic status is under debate. In any case, the Netherlands provides most of *Epeolus tarsalis*' habitat together with that of Sea-aster Colletes Bee (*Colletes halophilus*) on which it parasites (Calle and Jacobusse 2008).

Dutch species populations of international relevance have been identified, amongst others, by Schaminée et al. (2010). They consider species with more than 1% of their total habitat area or population size in the Netherlands internationally relevant. Other criteria for international relevance refer to species occurrences along the margins of their distribution range in the Netherlands, or those registered in the IUCN or European Red Lists. Schaminée et al. (2010) conclude that such criteria apply to 44 vascular plant species (3% of all Dutch vascular plant species), 39 mosses (7% of all Dutch moss species), nine mammals (15% of the Dutch mammal species, predominantly rodents and bats), 20 breeding birds and 42 migratory birds using Dutch territory as a staging area (together 21% of the Dutch bird species), 9 reptile species (13% of the Dutch reptiles), two amphibians (13% of the

E. F. J. De Mulder et al., *The Netherlands and the Dutch*, World Regional Geography Book Series,
https://doi.org/10.1007/978-3-319-75073-6_5

Dutch amphibians) and 12 freshwater fish species (27% of the Dutch fish species). Noordijk et al. (2010) applied more strict criteria, considering a Dutch species population internationally relevant only if it would support more than 10% of the European population of a species for mosses, vascular plants or mammals, or 25% for birds, butterflies, freshwater fish, amphibians, reptiles, dragonflies and grasshoppers. Accordingly, these authors consider only one Dutch moss species, 8 vascular plant species, 1 butterfly, 2 mammals and 14 bird species internationally relevant.

Many such species are associated with typical Dutch delta ecosystems, such as fens and coastal dunes. The semi-natural forests in the Limburg loess area and the extensive heathlands provide unique and internationally relevant habitats to Dutch species as well.

The distribution and abundance of Dutch species are mainly determined by physical geographical and climatic conditions, and by the European Quaternary glaciation history. The Netherlands is part of the Atlantic biogeographical region as defined by the EU (European Environment Agency 2016; Fig. 5.1), with an Atlantic climate: mild temperatures and a water surplus due to significant precipitation and low evapotranspiration. Despite favourable regional climatic conditions, biodiversity is relatively low and rather few endemic species occur as compared to other biogeographical regions (European Environment Agency 2002). This is due to the impact of glaciations on European species distributions that drove many species to extinction while forcing others to move to warmer refugia near the Mediterranean, the Caspian Sea and Southern Siberia (Noordijk et al. 2010). Upon retreat of the Pleistocene glaciers such species began to recolonise NW Europe from these refugia while few boreal species persisted under the warmer conditions.

Although the Netherlands is positioned entirely in the Atlantic biogeographical region it shares most of its species with Central European nations. Many species that survived Pleistocene glaciations in their Iberian and French refugia and that migrated north again during the mild Holocene Atlantic climate have their northern distribution limit in the Netherlands. Vice versa, distribution limits of several Scandinavian species of the Alpine and Boreal biogeographic regions do not reach further south than the Netherlands. Usually, they are confined to ecosystems that occur widely in more northern regions but are scarce south from the Netherlands, such as bogs and fens (Noordijk et al. 2010).

Distribution patterns of terrestrial Dutch species are largely determined by soil types and their related hydrology, land forms (geomorphology) and human use. Hence, biogeographical regions in the Netherlands closely follow the landscapes as described in Chap. 3 (Fig. 5.2). Further differentiation is mainly related to local variability in the Dutch climate. It must be noted that this division in biogeographical regions is also largely based on vegetation distribution, while several other species groups, especially animals, may show quite different distribution patterns (Schouten et al. 2010). The main Dutch terrestrial biogeographical regions are: the higher sands area, the loess and limestone area, the river belt, the fen peat area, the marine clay area, and the coastal dunes. Except for the partly human-made, larger inland lakes in the Dutch delta, most freshwater ecosystems are typical for these regions as well. Both major marine areas in the Netherlands are the Noordzee continental shelf and coastal zone, and the Waddenzee and Zeeland tidal areas. Their species and ecosystems are briefly discussed below.

5.2 Terrestrial Biogeographical Regions

The Higher Sands Area This area covers much of the northern, central, eastern and southern parts of the Dutch delta and is characterised by its texture (sands) and by its relatively high topographic position (at least above sea level). This landscape has largely been shaped by glacial and periglacial processes and has sometimes been reworked by Holocene wind erosion (Chap. 3). The most prominent topographical features of this area are a series of ice-pushed ridges in the centre of the Netherlands. Today, such ridges are predominantly covered by dry oak and beech forests, pine plantations or mixed broadleaved and pine stands. Their undergrowth depends on forest types and nutrient status but generally consists of Wavy Hair-grass (*Deschampsia flexuosa*), Climbing Corydalis (*Ceratocapnos claviculata*), ferns or bramble (*Rubus* spp.) when nutrients are abundant, or Lowbush blueberry (*Vaccinium angustifolium*) and mosses under nutrient-poor conditions. Older forest stands in particular contain abundant birdlife and a large variety of mushrooms. Mushrooms include mycorrhizal species living in symbiosis with trees, such as Fly agaric (*Amanita muscaria*) and *Boletus* species, as well as species living on dead wood like the rare Lions Mane (*Hericium erinaceus*), which is mainly confined to the high sands area in the center of the Netherlands (NDFF 2015). Where birdlife in younger forests is generally restricted to a few small songbird species, the older forests are also rich in birds associated with dead wood and older trees, such as Eurasian Nuthatch (*Sitta europaea*), Short-toed Tree-Creeper (*Certhia brachydactyla*) and woodpeckers like Great Spotted Woodpecker (*Dendrocopos major*) and Black Woodpecker (*Dryocopus martius*) (Berendse 2011). The latter make hollows in old trees which are subsequently used as nesting sites by Pine Marten (*Martes martes*), Tawny Owl (*Strix aluco*), and various bat species. Frequent birds of prey in these forests are European honey buzzard (*Pernis apivorus*) and Northern Goshawk (*Accipiter gentilis*).

Fig. 5.1 The main biogeographical regions in Europe. The Netherlands is situated entirely in the Atlantic biogeographical region. (Source: European Environmental Agency 2016)

The ice-pushed ridges have an important function in the hydrology of the surrounding lowlands. Rainwater percolates from the coarse sandy tops of the ridges and exfiltrates later at their lower parts, e.g. in the surrounding fen peatlands of the Utrechtse Heuvelrug ridge complex. These groundwater flows generate favourable conditions for rare plant species that rely on the calcareous and mineral-rich water (Wassen et al. 1990). East of this particular ridge complex, thick clay beds in the subsurface prevent rainwater to infiltrate (Van der Straaten 2008) resulting in numerous springs which feed small brooks in the forests running to surrounding lowlands (Fig. 5.3). Marsh Marigold (*Caltha palustris*) and Alternate-leaved Golden-saxifrage (*Chrysosplenium alternifolium*), indicators for exfiltrating, iron-rich groundwater, are typical constituents of the undergrowth of these forests.

Such brooks feed fluvial grasslands in the lowlands. In the past, such grasslands were often too wet to allow cattle grazing and sustained low productive vegetation only. The high groundwater tables together with regular flooding by calcareous brook waters or exfiltrating calcium-rich groundwater, and low nutrient availability (due to a haymaking regime), generated an exceptionally high plant species diversity. Vegetation is characterised by short sedges, orchids and a wide range of forbs, including some species that are

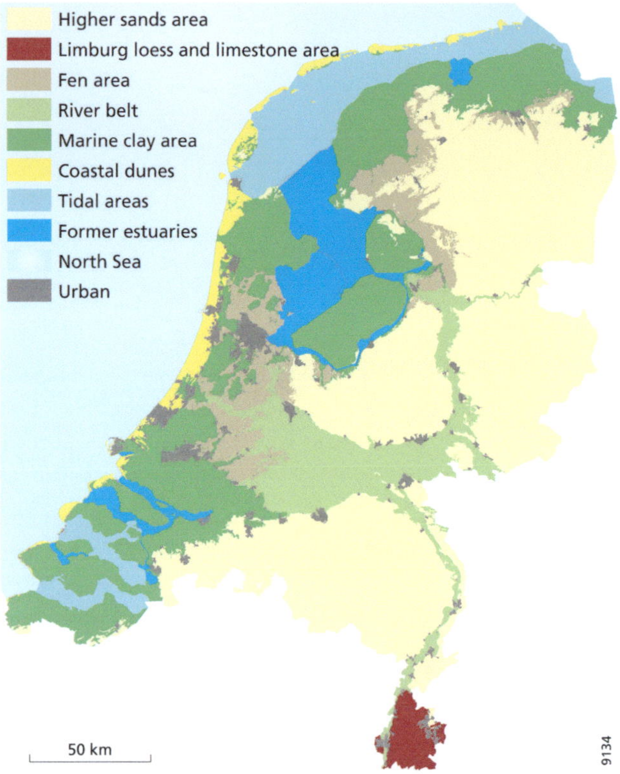

Fig. 5.2 The main biogeographical regions in the Netherlands. (Source: Ministerie van Economische Zaken 2013)

otherwise only found in calcareous wet dune slacks. Due to drainage, stretching of lowland brooks, and eutrophication, these grasslands belong to the rarest and most confined vegetation types in the Netherlands and internationally. Currently, most of these grasslands are nature reserves where haymaking regimes have been re-established as a major conservation management tool together with hydrological restoration measures. Similar springs and brook valleys are found in the east along the German border and in the glacial till plateau in Drenthe.

Peat bogs are rare ecosystems that predominantly occur in the higher sands area. Such peat bogs developed widely in the Netherlands during warmer and wetter Holocene times (Chap. 3). Most bogs were excavated for fuel and burnt for growing Buckwheat (*Fagopyrum esculentum*). Today, only very few more or less intact peat bogs remain, in total about 125 ha. These are all managed as strict reserves together with just over 5000 ha of degraded bogs (Janssen et al. 2014). Conservation measures aim at maintaining sufficient rainwater in the area for *Sphagnum* mosses to grow. Remaining bogs and turf ponds host typical plant species, insects, birds and amphibians. Apart from peat forming *Sphagnum* mosses typical bog plants include Round-leaved Sundew (*Drosera rotundifolia*), Common Cottongrass (*Eriophorum angustifolium*), Small Cranberry (*Vaccinium oxycoccus*) and Bog Asphodel (*Narthecium ossifragum*). These plants host a number of butterflie species unique to peat bogs, including Common Ringlet (*Coenonumpha tullia*), Cranberry Blue (*Plebeius optilete*) and Cranberry Fritillary (*Boloria aquilonaris*) (Berendse 2011). The shallow ponds are home to Moor Frog (*Rana arvalis*) and a range of other, more common frog species. Such ponds also support dragonfly species that are specific for peat bogs, most notably Northern White-faced Darter (*Leucorrhinia rubicunda*). Return of the Common Crane (*Grus grus*) as a breeding bird to the Netherlands in 2001, is regarded as a major success of peat bog restoration. Today, some 14–20 pairs annually breed in the Fochteloërveen and Dwingelderveld bogs. Perspectives for the small peat bog remnants in the Netherlands are uncertain. They belong to the ecosystems most threatened by climate change as they entirely depend on sufficient precipitation (Witte et al. 2012).

Drift sands and heathlands are among the most prominent habitats in the higher sands area. As discussed in Chap. 3, these basically result from anthropogenic forest removal, overgrazing and subsequent erosion. In the past, pine trees were planted to control drifting sands. In contrast present day conservation management aims to removing vegetation to stimulate sands to drift again (Fig. 5.4), because species associated with drift sands are considered rare and rather unique now, even on a European scale. Drift sands provide habitat to thermophilic species, such as reptiles and insects like Sand Lizard (*Lacerta agilis*), Blue-winged Grasshopper (*Oedipoda caerulescens*), Antlion (*Euroleon nostras*) and solitary bee species that build nests in sandy ridges. Specific pioneer plant species that colonise drifting sands include Grey Hair-grass (*Corynephorus canescens*), Bristly Haircap Moss (*Polytrichum piliferum*) and several lichens, all of which are adapted to withstand the harsh conditions of the drifting sand.

Where the higher sands were not fully overexploited or where drift sands stabilised due to succession, dry heathlands developed. The Central Veluwe's heathlands are among the largest in Europe and dominated by Heather (*Calluna vulgaris*; Fig. 5.5). Heather thrives by regular rejuvenation through its use in the extensive agricultural cycle (Chap. 3). Currently, it relies on conservation management by means of sod cutting and controlled burning, or from massive outbreaks of Heather Beetle (*Lochmaea suturalis*). Due to current high levels of atmospheric nitrogen deposition, Heather would be entirely overgrown by grasses, in particular by Purple Moor-grass (*Molinea caerulea*) without active conservation management (Fig. 5.6). Other shrubs that occur together with Heather are Crowberry (*Empetrum nigrum*), Juniper (*Juniperus communis*) and *Genista* species. In wetter areas Heather grows intermingled with Cross-leaved Heath (*Erica tetralix*), Blueberry (*Vaccinium corymbosum*) and Cowberry (*Vaccinium vitis-idaea*) (Weeda et al. 1994).

Fig. 5.3 The Leuvenumse Beek flows over permeable sands of the Veluwe facilitated by the presence of a buried impermeable clay layer. Note the orange color of the brook water caused by high amounts of iron in the exfiltrating groundwater. (Photograph by Jerry Van Dijk)

Fig. 5.4 Hulshorsterzand, a restored drift sand area on the Veluwe. Pine trees, heather bushes, and grasses that proliferated because of high levels of nitrogen deposition were removed in 2009 to stimulate the sands to drift again. (Photograph by Jerry Van Dijk)

Fig. 5.5 Heathland in National Park Veluwezoom with Heather (*Erica tetralix*), Gorse (*Ulex europaeus*) and oak trees. (Photograph by Jerry Van Dijk)

Fig. 5.6 Red deer (*Cervus elaphus*) in the heathlands of National Park Veluwezoom. Note that Heather has been almost entirely replaced by Purple Moor-grass (*Molinea caerulea*) due to nitrogen deposition. (Photograph by Jerry Van Dijk)

Heathlands can support rich insect life. Heather provides nectar to a range of bee species, including domestic Honeybee (*Apis mellifera*) and typical Heather bee species like Heather Mining-bee (*Andrena fuscipes*) and Heather Colletes (*Colletes succinctus*), as well as a range of butterflies and syrphid flies. Heathlands are also important habitat to grasshoppers. Two out of the three Dutch snake species live in heathlands: Common European Adder (*Vipera berus*) and Smooth Snake (*Coronella austriaca*). The Veluwe heathlands and forests belong to the core habitats for Red Deer (*Cervus elaphus*; Fig. 5.6) and Wild Boar (*Sus scrofa*). They also host several rare Dutch bird species including European Nightjar (*Caprimulgus europaeus*), Woodlark (*Lullula arborea*), Northern wheatear (*Oenanthe oenanthe*) and Eurasian hobby (*Falco subbuteo*).

Two Dutch heathland bird species are on the brink of extinction: Great Grey Shrike (*Lanius excubitor*) and Black Grouse (*Lyrurus tetrix*). The latter used to be fiercely hunted and strongly relied on heathlands and their associated extensive agricultural systems. Open heathlands are scarce now because of nitrogen deposition whereas extensive arable fields within the heathlands were replaced by large scale intensive agriculture. However, recreation of extensive arable fields and release of captured Swedish birds did not help Dutch Black Grouse populations to recover. In 2014, only seven breeding pairs were observed in National Park Sallandse Heuvelrug, the only Dutch location where Black Grouse persisted.

The Limburg Loess and Limestone Area The rather typical ecology of the Limburg loess and limestone area is influenced by two principal factors: topography and geology. Limburg is one of the few places in the Netherlands where hard bedrock outcrops, which is often covered by loess (Chap. 3). The area has relatively steep slopes and significant elevation. Geochemical soil conditions are largely determined by the underlying rocks, mostly Cretaceous limestones. These porous, calcareous rocks generally provide well-drained, dry and alkaline conditions. In southern Limburg these may overlay acidic Carboniferous sandstones and shales. In turn, Cretaceous limestones may be covered by Tertiary sandstones and clays, and Pleistocene loess, providing more acidic and wet conditions. These geological conditions result in small-scale patterns of varying soil conditions and concomitant biodiversity gradients along hill slopes. Other relevant topographical features here include Pleistocene river terraces and rather steep, both tectonically and erosion determined hillsides. Such topographical diversity contributes to locally strongly variable micro-climates with higher temperatures on south-facing slopes. Together, these geological and micro-climatic conditions are responsible for distinctively different species assemblages in the Limburg loess and limestone area compared to the rest of the Netherlands, often containing species with a more southern or Central European distribution (Schouten et al. 2009; Noordijk et al. 2010).

Calcareous grasslands are an example of ecosystems with species unique for the Netherlands but with a rather wide European distribution. These occur where limestone beds are at or near the surface. Such grasslands typically contain 30–40 species per m^2 with many orchid species including Longbract Frog Orchid (*Dactylorhiza viridis*), Man Orchid (*Orchis anthropophora*), Military Orchid (*Orchis militaris*; Fig. 5.7), Monkey Orchid (*Orchis simia*) and several Butterfly orchids (*Platanthera* spp.). Other frequent plants in these grasslands are Tor-grass (*Brachypodium pinnatum*), Upright Brome (*Bromopsis erecta*), and herbs such as Dwarf Thistle (*Cirsium acaule*), Slender Bedstraw (*Galium pumilum*) and Chiltern Gentian (*Gentianella germanica*) (NDFF 2015). Such rich flora supports equally rich entomofauna, in particular butterflies. Moreover, Limburg calcareous grasslands provide important habitat for mosses. About 120 moss species occur here, which is almost 20% of all Dutch mosses. The amphibians Common Midwife Toad (*Alytes obstetricans*) and the Yellow-bellied Toad (*Bombina variegata*) benefit from warm microclimates in these grasslands and are mainly confined to the Limburg loess and limestone area.

Where calcareous grasslands were converted into arable fields many rare arable weeds and other plants used to occur. An example is Strap-leaved Earth-moss (*Ephemerum recurvifolium*), that is rare even on a European scale (Schaminée et al. 2010). Most of these species disappeared from the Netherlands because of agricultural intensification. To safeguard such species, conservation organisations own extensively managed arable fields today. Arable fields on loess also host Eurasian Hamster (*Cricetus cricetus*). Limburg is on the northwestern margin of its European distribution. Agricultural intensification almost drove this hamster to extinction in the Netherlands. All animals in the only remaining Dutch wild population were captured for a breeding program in 1999 and their offspring was released in 2002 in special cereal field reserves where wheat was left unharvested. Despite high predation rates by Red Fox (*Vulpes vulpes*), mustelids and birds of prey, the hamster population successfully re-established (La Haye et al. 2010).

Many of the Limburg hills are covered with forests. Forest composition follows the hillside soil gradients discussed above, with Oak-Hornbeam forests under more acidic and wet conditions on the lower parts, to dryer Beech forests on top (Bobbink et al. 2008). In the middle of this gradient, a partially man-made forest type occurs, which is characterised by a mosaic of shaded and open places and the absence of a thick litter layer as a result of traditional coppicing of the undergrowth. This provided optimal conditions for a range of orchid species and butterflies of open forests. Furthermore,

Fig. 5.7 Military Orchids (*Orchis militaris*) in calcareous grasslands in the Gerendal nature reserve in the Limburg loess and limestone area. (Photograph by Jerry Van Dijk)

Hazel Dormouse (*Muscardinus avellanarius*) and Garden Dormouse (*Eliomys quercinus*) are almost completely dependent on this forest type in Limburg. Traditional forest management is therefore now being continued by conservation organisations.

Where geological conditions cause forest soils to be less permeable, springs may occur that feed small forest streams that eventually join in larger valley brooks. Such springs provide optimal conditions for rare Dutch plants like Alternate-leaved Golden-saxifrage (*Chrysosplenium alternifolium*), Oxlip (*Primula elatior*) and Great Horsetail (*Equisetum telmateia*). The streams are home to the larvae of Fire Salamander (*Salamandra salamandra*) while the adults reside in adjacent wet litter forest layers. This amphibian is another example of a more Central-European species with its Dutch distribution limited to Limburg. The Dutch population of Fire Salamander currently suffers heavily from an outbreak of the fungal disease *Batrachochytrium salamandrivorans* that decimated their population size by more than 96% (Spitzen-Van der Sluijs et al. 2013). Rapidly flowing, larger brooks harbour fish species such as Brook Lamprey (*Lampetra planeri*) and Rhine Sculpin (*Cottus rhenanus*), the latter being confined to only a few Limburg brooks.

Quarries, mines and stone walls provide a special habitat in this area (Box 5.1). Subsurface coal mining and open pit limestone quarrying played an important role in southern Limburg. But all coal mines and almost all limestone quarries (except one, in Maastricht) have been abandoned today. Abandoned limestone quarries are important habitat for many bat species now and they host the only known Dutch breeding locations of the European Eagle-Owl (*Bubo bubo*), except for a quarry near Winterswijk, Gelderland. The only natural Dutch population of the Common Wall Lizard (*Podarcis muralis*) lives on an old fortress in Maastricht.

Box 5.1: Limburg's Zinc Flora

On its way downstream from its German source to the Netherlands, a small southern Limburg river (Geul) encounters the Belgian towns of La Calamine and Plombières. These towns used to host lead and zinc mines that strongly polluted brook waters and sediments with heavy metals, which were deposited in the Dutch Geul floodplains as well. Such toxic metals made it hard for plant life to persist, expect for a few species that adapted to heavy metal concentrations. These species are known as Limburg's zinc flora and consist of a subspecies of Mountain Pansy (*Viola lutea* ssp. *calaminaria*), Alpine Penny-cress (*Thlaspi caerulescens* var. *calaminare*) and a Sheep's-fescue subspecies (*Festuca ovina* ssp. *guestphalica*). *V. lutea* ssp. *calaminaria* only occurs on La Calamine's mine tailings and in the Geul floodplains. These species persist only because heavy metal toxicity hampers growth of other, non-tolerant species. Heavy metal tolerance is genetically determined and may evolve to high tolerance levels in a few generations only. Moreover, tolerance for specific metals is not coupled, and is species specific (Van Straalen and Verkleij 1993). Hence, species composition of metal-tolerant vegetation depends on the specific composition of the ores that are being mined and thus may be unique to each individual mining site.

(continued)

Box 5.1 (continued)

While zinc flora used to be quite abundant in the Geul floodplains, it has become rare today. This is mainly caused by a strong decline in heavy metal pollution upon closure of the mines. Moreover, heavy metals were immobilized as metal phosphates in response to phosphate eutrophication, removing the competitive advantage of metal tolerant plant species (Weeda et al. 1994). Limburg's zinc flora species in the Geul floodplains are getting confined to where sediments have been disturbed, thus uncovering older polluted sediments. Recent records of the zinc-tolerant variety of Sheep's-fescue are all restricted to molehills (NDFF 2015). This poses an interesting, yet unresolved conservation dilemma: the zinc flora of the Geul floodplains is unique in the world but can only be conserved by allowing harmful toxins to proliferate in the natural environment.

The Fen Peat Area This area extends in the western and north-western regions of the Netherlands where fen peat occurs at the surface or in the shallow subsurface. This fen peat was mainly formed during the Holocene, when beach barriers began to close the Dutch coast around 5000 BP and extensive mires developed in the regions behind these barriers. Thick peat layers developed in a significant part of the Dutch delta (Chaps. 2 and 3). Most of these peat deposits were reclaimed and successively extracted since Medieval Times. In many places, peat was extracted even below groundwater level. The remaining turf ponds were prone to wave and wind action and expanded successively into larger lakes (Fig. 5.8).

Smaller turf ponds often terrestrialised upon abandonment, starting with the formation of floating vegetation mats. These consist of buoyant individuals of the aquatic plant Water-soldier (*Stratiotes aloides*) and floating organic debris cluttering together, eventually forming substrate for non-aquatic plants. These vegetation mats often have a high plant species diversity, and serve as nesting sites for Black Tern (*Chlidonias niger*), a typical marshland bird. Litter production and ongoing organic debris accumulation kept them growing and eventually filling up these smaller turf ponds entirely. Once their top parts start accumulating rainwater, *Sphagnum* mosses become dominant and a peat bog begins to develop. Next, natural succession develops toward wet heathland and finally to a typical swamp forest or Alder carr, dominated by Black Alder (*Alnus glutinosa*), Downy Birch (*Betula pubescens*) and willow (*Salix* spp.). Since turf ponds were abandoned at different moments in history, all successional stages from open water to swamp forest occurred in the fen peat area, resulting in a high biodiversity. Therefore, current nature management consists of regularly setting back succession by opening-up terrestrialised former turf ponds, thus stimulating all successional stages to occur.

Only few natural lakes remain in the fen peat area, including the Naardermeer, the first nature reserve in the Netherlands (established in 1906). Most natural lakes in this fen peat area were reclaimed for peat harvesting or for agriculture. Both artificial and natural peat lakes are shallow, no deeper than 1.5–2 m. Such lakes are often oligotrophic and well buffered, as they are normally fed by nutrient poor groundwater rich in calcium and bicarbonate. Such conditions are favourable for Stoneworts (*Characeae*). These macro-algae have largely disappeared from the Dutch fen peat areas because of deterioration of water quality. They have been largely replaced by submerged and floating macrophytes, most notably European White Water-lily (*Nymphaea alba*), Yellow Water-lily (*Nuphar lutea*), Water-soldier and a range of Pondweed species (*Potamogetonaceae*). In shallower waters along lakeshores many emergent macrophytes develop, including rushes, sedges and Common reed (*Phragmites australis*).

Measures to improve water quality have been moderately successful. One of the more iconic species that benefitted from increased water quality is the Eurasian otter (*Lutra lutra*). In 1988, this species became extinct in the Netherlands but was reintroduced in 2002 in De Wieden, a large fen peat area in the north. Since then, otter populations have been growing and young otters colonised several other fen peat areas (Koelewijn et al. 2010).

Where water levels drop below the soil surface occasionally, enabling the germination of Common Reed seeds, extensive reed swamps have established. These host a range of typical marshland bird species including Common Reed Bunting (*Emberiza schoeniclus*), Eurasian and Great Reed Warbler (*Acrocephalus scirpaceus* and *A. arundinaceus*), Great bittern (*Botaurus stellaris*) and Bluethroat (*Luscinia svecica*). Spatial succession from shallow water to dryer terrestrial parts basically follows the timewise succession as was described for turf pond terrestrialisation. A species often found in both the terrestrial and aquatic parts of fen peat areas is Grass Snake (*Natrix natrix*), one of the three Dutch snake species. This snake is an excellent swimmer that feeds mainly on frogs, small mammals and fish, all of which are abundant in fen peat areas.

Most terrestrial parts of fen peat landscapes are currently used for dairy farming and have little room for natural vegetation except in ditches and on ditch banks. Intensive use of fertilizers and manure on adjacent fields generally leads to low water quality and correspondingly low natural values. Initially, a group of wader birds now known as meadow birds benefitted from fen peat reclamation for agricultural

Fig. 5.8 Ankeveense Plassen, an extensive shallow peat lake in the west of the Netherlands. Such shallow lakes remained after turf extraction below groundwater. The trees are growing on a remaining shallow strip of land that was used for drying turf. (Photograph by Jerry Van Dijk)

Fig. 5.9 Black-tailed Godwit (*Limosa limosa*) in the Wormer- en Jisperveld nature reserve, one of the remaining strongholds for meadow birds, and therefore appointed as a Natura 2000 area. (Photograph by Jerry Van Dijk)

purposes. Typical species are Common Redshank (*Tringa totanus*), Eurasian Oystercatcher (*Haematopus ostralegus*) and Black-tailed Godwit (*Limosa limosa*; Fig. 5.9). These birds moved from natural coastal and wetland areas to fen peat meadows, because with their high groundwater levels, open landscape and the abundant availability of earthworms and other prey items (because of manure application), these meadows provided optimal habitat (Van Dijk et al. 2015).

The Netherlands is of international importance for meadow birds, in particular for Black-tailed Godwit: an estimated 40% of the global population and 87% of the European flyway population of Black-tailed Godwit breeds in the Netherlands (Birdlife International 2004; Kentie et al. 2016). However, their numbers dropped dramatically since the 1970s due to ongoing agricultural intensification, including lowering of groundwater levels, extremely high levels of manure and fertilizer application, sowing of monocultures of highly productive Perennial Rye-grass (*Lolium perenne*) and earlier and more frequent mowing of the sward (up to 6 cuts/year now; Kleijn et al. 2010). Around 1970, the population size of this species was estimated around 120,000 breeding pairs. By 2015, this number had dropped to about 33,000 breeding pairs. Species that are most sensitive to the changes in agricultural practice, like Ruff (*Philomachus pugnax*), have even almost disappeared from the Netherlands (Boele et al. 2017). So far, conservation schemes have not been able to curb this trend.

On the other hand, geese strongly benefitted from agricultural intensification and the resulting high availability of protein-rich grass to feed on. Some geese species have increased fivefold since the 1970s. In 2015, more than 2 million individuals were overwintering in the Netherlands (Sovon Vogelonderzoek 2017), causing considerable damage to agricultural grasslands and arable crops. In 2016, the Dutch government paid farmers about 18 million € in damage compensations (Beintema 2017). Moreover, flocks of geese pose a threat to aviation, especially around the Netherlands main airport Schiphol. Here, geese have been actively eradicated since 2014. However, following a law-suit by several environmental NGO's, further eradication was prohibited in 2016.

The River Belt The major Dutch rivers have been significantly modified for better use as international shipping routes and for flood protection. Consequently, much of the natural river dynamics, including regular flooding, disappeared and little is left of their original ecosystems and species (but see Box 5.2). Populations of anadromous migratory fish such as Atlantic Salmon (*Salmo salar*) suffered severely from river channelisation, dam and sluice constructions, and pollution. Apart from the aquatic ecosystem, the river belt's floodplains provide valuable habitats in between river dikes. Beyond the dikes, most of the land is used for agriculture as because Dutch river clays are very fertile.

Box 5.2: Tidal Freshwater Wetlands of the Biesbosch: A Unique Riverine Ecosystem

The major Dutch rivers provide unique conditions for freshwater tidal wetlands in their downstream parts. Marine high tides hamper river discharge, thus creating

Box 5.2 (continued)

periodic upstream river water flood pulses. Consequently, river discharges in the Rijn-Maas delta may be reduced by up to 57% at high tides (Barendregt and Swarth 2013).

Although tidal action was strongly diminished by damming many coastal estuaries, a considerable tidal influence still occurs along the shipping route to the Rotterdam harbour, the Nieuwe Waterweg. One of the major freshwater tidal wetlands in both the Netherlands and Europe is National Park De Biesbosch, east of Rotterdam, which emerged when a major polder was flooded in 1421 (Chap. 2). A combination of tidal action and river sedimentation resulted in the development of shallow mudflats. Sedimentation was enhanced by natural settlement of macrophytes and actively stimulated by humans who built small levees and planted rushes (Schaminée et al. 2010). Eventually, such mudflats were cultivated and used for rush and reed production, and willow coppicing. Ongoing sedimentation served as a natural fertilizer. Finally, emerged mudflats were reclaimed and used for agriculture. Willow coppicing and further cultivation of the area was abandoned in 1970 when closure of the Haringvliet estuary strongly diminished tidal ranges. As reclamation of the Biesbosch developed from the northeast to the southwest, today natural freshwater tidal wetlands only remain in the southwestern part of the area.

They are largely covered with alluvial softwood forests dominated by Willow species (*Salix* spp.), and reed swamps. Although biodiversity in De Biesbosch is not particularly high, it contains a unique combination of species that have adapted to the tidal nature of the area. De Biesbosch is of international importance for a subspecies of Marsh Marigold (*Caltha palustris* subsp. *araneosa*) and a number of moss species that are largely confined to tidal freshwater wetlands. It also supports rich birdlife. De Biesbosch was proclaimed a National Park in 1994. The area is now managed as a nature reserve and was expanded by deliberate flooding of some adjacent polders. Success of this conservation management is demonstrated by the occurrence of large numbers of Eurasian Beaver (*Castor fiber*) and the recent establishment of Osprey (*Pandion haliaetus*) as a breeding bird. In 2018, sluices in the Haringvliet dam will be partially opened to enhance tidal action in De Biesbosch, adding to the unique natural values of this tidal freshwater wetland.

Ecosystems typically associated with Dutch river floodplains are wet grasslands and swamps in the regularly flooded zones and former river branches, dry grasslands beyond flood zones and on river dunes, and softwood and hardwood alluvial forests. Many alluvial forests were cleared in the past as they hamper river flow velocities and reduce the water storage capacity of floodplains during high water, all leading to increased flood risk. Forests were also cleared to be converted into pastureland. Restoration projects aim to restore these woodlands, but have only been successful for softwood alluvial forests so far. Wet and dry riverine grasslands are important habitats for marshland birds, meadow birds and geese. Cut-off river branches are important nurseries for fish and serve as habitat for (semi-)aquatic insects such as dragonflies.

The Marine Clay Area This area comprises much of the western and northwest of the Dutch delta where marine Holocene clays occur at the surface because of natural sedimentation, the creation of polders or peat harvesting. Like the fen peat area, the marine clay area is largely dominated by agriculture, making biodiversity largely dependent on agricultural land use as well. Few natural ecosystems remain in the area. Agricultural grasslands in the marine clay area provide important habitat to the meadow bird and geese populations in the Netherlands. A typical example is the Arkemheen-Eempolder, appointed as a Natura 2000 reserve and largely managed by national conservation agencies. It is one of the core breeding areas for Black-tailed Godwit and other meadow birds (Van der Vliet et al. 2015) and one of the few areas where populations are not declining (Sovon and CBS 2005).

Much of the marine clay area consists of polders reclaimed from the Noordzee or the former Zuyderzee (Chap. 2; Boxes 5.3 and 5.4). The once brackish subsurface of these polders sustains salt-tolerant plant species in grasslands and ditch banks, normally found only along the coast and in salt marshes. Moreover, salt water seepage under sea dikes may also provide favourable conditions for such salt-tolerant species. In the Flevopolders, large areas of forest have been planted, which developed quickly on the fertile clay soil. Many of these forests are managed as nature areas and currently harbor a large diversity of especially birds and mushrooms.

Box 5.3: Reclamation History and Biodiversity

The incremental reclamation history in marine clay areas is often reflected by remains of former coastal defenses, specifically old, defunct sea dikes. Such former dikes have a distinct biogeochemical composition due to their calcium-rich construction materials, have specific microclimatic conditions at south-facing slopes and are characterised by the absence of intensive agricultural development because their steep slopes are inaccessible to heavy machinery and cattle, allowing only for sheep grazing or haymaking.

This combination of features turned them into biodiversity hotspots, with a unique plant species composition of species that are otherwise restricted to calcareous dunes and in the Limburg calcareous grasslands. Typical species are Common Agrimony (*Agrimonia eupatoria*), Wild Marjoram (*Origanum vulgare*) and Field Scabious (*Knautia arvensis*) (Provincie Zeeland 2005; NDFF 2015). Moreover, the former dikes provide vital habitat for species that are rare in the Netherlands, such as Corn Parsley (*Petroselinum segetum*), Woolly Thistle (*Cirsium eriophorum*), Spreading Hedge-parsley (*Torilis arvensis*) and Deptford Pink (*Dianthus armeria*) (NDFF 2015).

The abundant flowers attract many insects, especially butterflies, bees and syrphid flies that all are important pollinators of agricultural crops. Bird species breeding or foraging on former dikes are Grey Partridge (*Perdix perdix*), Little Owl (*Athene noctua*) and European Green Woodpecker (*Picus viridis*) (Provincie Zeeland 2005). Because of their connectivity, inland dikes are crucial for dispersal of the European Tree Frog (*Hyla arborea*), the Great Crested Newt (*Triturus cristatus*) and several bat species, and they are one of the few spots where Common Lizard (*Zootoca vivipara*) occurs beyond the higher sand areas (NDFF 2015).

Biodiversity on inland dikes is currently under threat because they are no longer used for extensive grazing or haymaking, leading to dominance of ruderal species and shrub encroachment. Therefore, some dikes are managed by nature organisations and farmers, who even deploy a small herd of sheep to graze them for conservation purposes. (Fig. 5.10)

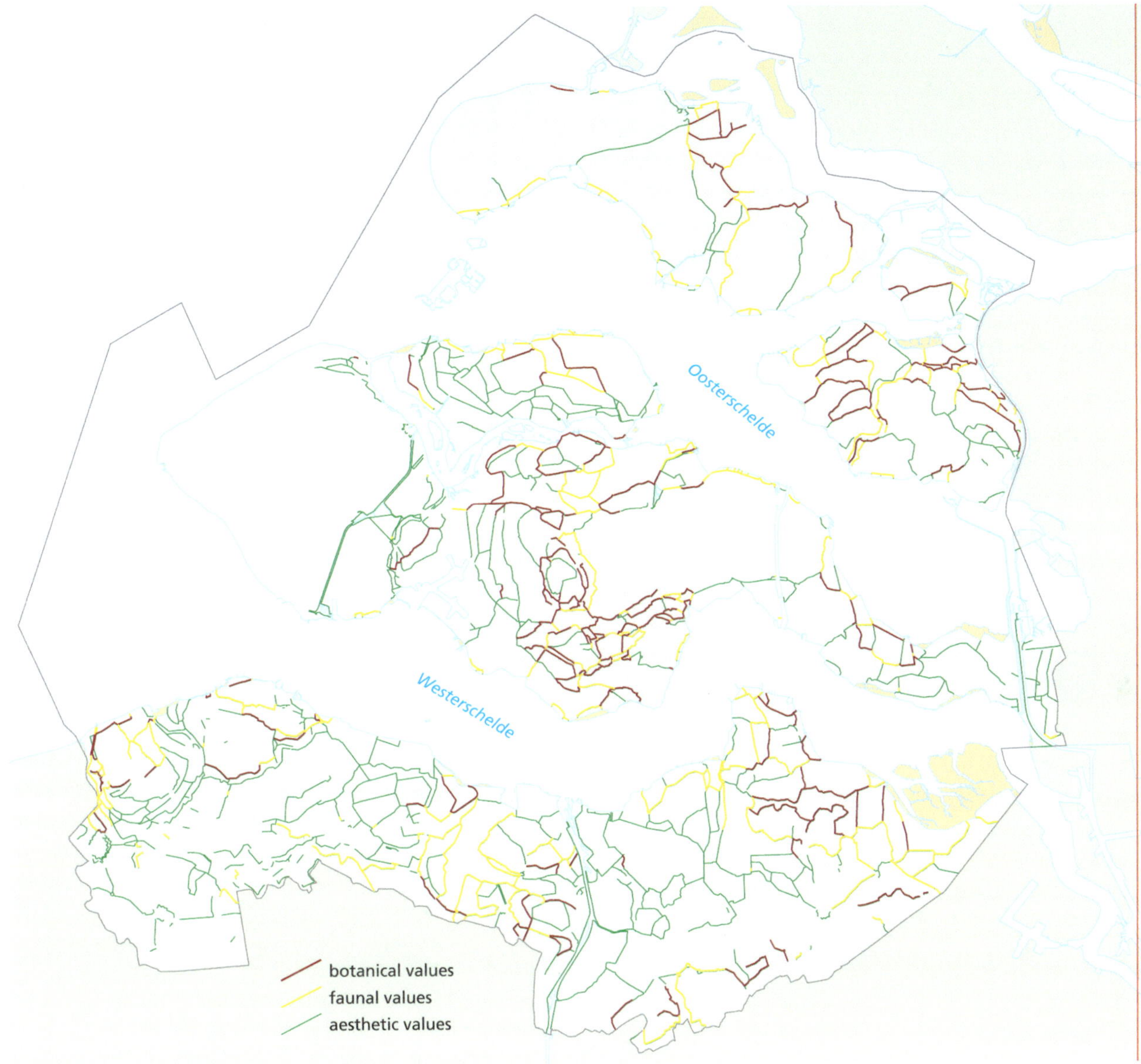

Fig. 5.10 High densities of inland dikes (former sea dikes) reflect the reclamation history of Zeeland. Dikes in red have high botanical values, those in yellow have high faunal values, those in green have high aesthetic values in the landscape. (Sources: Provincie Zeeland 2005, 2017)

Box 5.4: Rewilding the Netherlands

Upon completion of the Flevopolder in 1968, 3600 ha in the northern and deepest part of the polder were left undeveloped because groundwater levels were too high. Natural colonisation turned this area into a marshland that attracted tens of thousands of migratory and breeding birds, including Great Egret (*Ardea alba*) that had been extinct in the Netherlands since 1800 (Berendse 2011). Therefore, this area, now known as Oostvaardersplassen, was proclaimed a nature reserve and was extended by another 2400 ha of grasslands. These grasslands were to become one of the largest and most controversial rewilding projects in the Netherlands.

Rewilding is the deliberate reintroduction of often larger animals to stimulate natural dynamics in ecosystems that disappeared when such animals became extinct. Often such animals operate as 'ecosystem engineers', modifying their environment. Examples are Eurasian beaver (*Castor fiber*) or large grazers and their top predators. Rewilding projects often apply paleoecological and archeological studies in reconstructing earlier natural ecosystems (Marris

(continued)

Box 5.4 (continued)
2009). As many species became extinct, today's rewilding projects apply modern species that are believed to descend from Pleistocene grazers such as Tarpan (*Equus ferus ferus*) and Aurochs (*Bos primigenius*), or purposefully bred animals.

In the 1970s, the rewilding concept was introduced in the Netherlands by biologists and conservationists, united in the 'Working group critical forest management', in response to the highly managed and rather monotonous production forest stands. They proposed to reinstate natural forest dynamics by reintroducing large grazers like the European Bison (*Bison bonasus*) and top-predators such as wolves. Although such concepts were considered outrageous at the time, they sparked discussion on allowing more natural processes in forests and nature reserves, including grazing by large grazers.

In 1983, German Heck cattle (bred in the likeness of the extinct Aurochs) were released in the Oostvaardersplassen in an attempt to arrest natural succession on the grasslands and generate a more open vegetation structure for the many overwintering geese. Later, Konik horses (as a substitute for Tarpan) and Red Deer were introduced (Fig. 5.11). All were treated as wild animals and kept without any human interference. Grazer populations grew rapidly and soon reached the area's carrying capacity at 3500 animals (Cornelissen et al. 2017). This implied that many of the weaker animals died in winter. Carcasses left in the area attracted Raven (*Corvus corax*) and played a crucial role in the return of the White-tailed Eagle (*Haliaeetus albicilla*) as a breeding bird.

The large winter mortality of grazers also triggered strong societal debates whether or not to treat re-introduced animals as truly wild animals. So far, such discussions resulted in a compromise where weak animals are shot in early winter by the managing State Forestry Service to avoid suffering. In addition, some more sheltered areas were built to protect animals from harsh conditions.

More recently, European Bison were introduced in the Kennemerduinen, Veluwe and Maashorst, and a breeding program was started to create a new cattle breed (the Tauros) remeniscent of the famous Aurochs of the Lascaux cave paintings. So far, no top-predators to control herds of grazers have been introduced. However, various documented sightings of the Grey Wolf (*Canis lupus*) in the Netherlands indicate that this species is expanding its territory from Germany.

In the marine clay area arable farming often is too intensive to sustain any wildlife. Eurasian skylark, (*Alauda arvensis*), Grey partridge and Yellowhammer (*Emberiza citrinella*) populations have plummeted with agricultural intensification since the 1970s. Corn bunting (*Emberiza calandra*) is almost extinct now. Birds of arable land may persist only where major agri-environmental measures were taken. A fine example is Montagu's harrier (*Circus pygargus*) that had almost disappeared from the Netherlands in 1990, when only 15 breeding pairs remained. This species is now recovering due to a conservation program in Groningen that

Fig. 5.11 A Konik horse in Blauwe Kamer, a nature restoration area along River Nederrijn. Konik horses are substitute for the extinct Tarpan (*Equus ferus ferus*) in rewilding projects and play a crucial role in vegetation development by dispersing seeds (like the sticky seedpods of Greater Burdock that are completely entangled in the horses' manes and tail in this picture) and opening up vegetation by creating wallows (as visible in the lower right corner). (Picture by Jerry Van Dijk)

includes nest protection on arable fields and establishment of high densities of field margins.

Coastal Dunes The extensive Dutch coastal dunes along the coast are unique in Europe, not just due to their wide extent but also as they are mainly managed as natural areas and actively conserved because of their coastal defense function. Ecologically, a major distinction exists between the dunes in the north and those in the centre and south. This is due to their different origins and compositions: low calcium and mineral contents in the northern dune sands and high calcium and mineral contents in the central and southern dune sands (Chap. 3). Botanically, the dunes belong to the most diverse Dutch ecosystems with more than 700 recorded plant species in the dunes of Voorne alone. Dutch dune vegetation follows a typical zonation related to decreasing marine influence (seawater flooding and salt spray) land inward on the one hand and increasing age of the dunes and beach barriers in the same direction, on the other (Chap. 3).

One of the very few plant species that can establish in the beach zone is Sand Couch (*Elytrigia juncea)*, because it is adapted to salt stress, erosive action of the blowing sand and high levels of potential evapotranspiration. Once these grasses settle, sand accumulates around them, thus forming nuclei of new dunes. Because of their slight elevation, these embryonic dunes accumulate small amounts of freshwater, paving the way for colonisation by less salt-tolerant plant species as Marram grass (*Ammophila arenaria*) (Weeda et al. 1994). Marram grass is vital to dune formation as it adapts to burial by sand through its typical stolon growth patterns (Weeda et al. 1994). Their extensive root systems stabilise dune sands and make dunes less susceptible to erosion. These grasses are insensitive to the calcium content of the soil, allowing them to grow in every Dutch dune region. Therefore, Marram grass is actively planted as part of the Dutch coastal defense strategy and often dominates seaward sides of the outer dunes (Fig. 3.13).

Leeward of the outer dunes where sand dynamics are less pronounced Marram grass is accompanied by a number of other, less adapted plant species, including Rush-leaved Fescue Grass (*Festuca arenaria*) and the Sand-hill Screw-moss (*Syntrichia ruralis* var. *arenicola*), both of which are nevertheless effective in accumulating sand as well (Weeda et al. 1994). In the rather stable areas protected by the outer dunes vegetation development depends on soil acidity and mineral content. Vegetation in calcium-rich sands with low (acidic) organic matter content is dominated by Sea-buckthorn (*Hippophae rhamnoides*) brushwood. Once succession commences, bushes like Common Hawthorn (*Crataegus monogina*), Wild Privet (*Ligustrum vulgare*) and Spindle (*Euonymus europaeus*) and several vine species replace Sea-buckthorn. These late successional bushes finally give way to dune forests characterised by Aspen (Populus tremula), White Poplar (*Populus alba*) and Downy- and Silver Birch (*Betula pubescens* and *B. pendula*), and, in later successional stages, Pedunculate Oak (*Quercus robur*).

Fig. 5.12 Grass-of-Parnassus (*Parnassia palustris*), a rare species in the Netherlands that occurs predominantly in calcareous wet dune slacks. (Photograph by Jerry Van Dijk)

Sea-buckthorn cannot persist in the calcium-poor soils of the northern dune area or in the older southern dunes from where calcium was leached. Here, vegetation is low and more open and mainly consists of Grey Hair-grass (*Corynephorus canescens*) with large swaths of lichen, mosses and ruderal grass species. Succession goes from late successional dry grassland to dune heathland with Crowberry (*Empetrum nigrum*). Here, most soils are not fertile enough to sustain tree growth.

Vegetation is most diverse in dune slacks where fresh groundwater from surrounding higher dunes accumulates. In calcium-rich dune sands biodiversity is quite high and many rare species occur in adjacent slacks. Primary dune slacks form when small primary dunes block parts of the beaches from seawater and start feeding the newly formed small valley with fresh groundwater. Such valleys are dominated by low vegetation often characterised by rare species such as Grass-of-Parnassus (*Parnassia palustris*; Fig. 5.12), Autumn Gentian (*Gentianella amarella*) and various orchid species, including the extremely rare Fen Orchid (*Liparis loeselii*).

Secondary dune slacks may develop where sand is blown out of the older dunes until groundwater level is reached and wet sands prevent further wind erosion. These are characterised by species of more acidic conditions, because older dune sands lost most of their calcium content. Dune slack vegetation is further determined by the soil's nutrient status. Nutrient levels may be high if soils contain much organic matter from established vegetation, leading to species-poor vegetation dominated by Wood Small-reed (*Calamagrostis epigejos*) or Common Reed (*Phragmites australis*) (Van Bodegom et al. 2006).

5.3 Tidal Areas and the Noordzee

The Dutch delta has two main tidal areas: the Waddenzee and the Zeeland Delta. Many other former tidal areas, including the former Zuyderzee were separated from marine and tidal influence by dams for reclamation or flood protection (Chap. 2).

Waddenzee With 4700 km^2 of mudflats emerging twice a day at low tide, the Waddenzee is the largest contiguous areas of emerging tidal flats in the world and extends from the Netherlands to Denmark (Fig. 2.3). Its unique, large size and its crucial role as staging ground for migratory birds prompted UNESCO to proclaim Waddenzee a World Heritage site in 2009. Its mudflats are sheltered from the Noordzee by a chain of barrier islands, Waddeneilanden. The Waddenzee's tidal flats are bordered by salt marshes along the coastline of the islands and the mainland (Fig. 3.10) providing important feeding grounds for the many visiting geese.

As Noordzee waters enter the Waddenzee at high tides with much higher velocities than when retreating at low tides, the Waddenzee functions as a sedimentation basin (Berendse 2011). Active sedimentation sustains an extremely rich benthic fauna that, in turn, supports the higher parts of the food web. The tidal flat benthic fauna is dominated by mollusks, e.g. Common Cockle (*Cerastoderma edule*), Baltic Tellin (*Macoma balthica*) and Sand Gaper (*Mya arenaria*), and a range of polychaete worms. These worms keep the sediment unconsolidated by continuously reworking it and preventing clogging by organic material (Reise et al. 2010). The rich benthic flora and fauna provide food for Shore Crabs (*Carcinus maenas*), shrimp, fish and wader birds at low tide.

With this ample food supply the Waddenzee supports a major part of the East Atlantic and African-Eurasian flyway population of migratory birds as breeding, overwintering and staging area, with an estimated 10–12 million birds visiting the area annually (UNESCO 2009). The Dutch Waddenzee is a habitat of vital international importance to at least 54 bird species, of which 26 also breed here. For some species, such as Brent goose (*Branta bernicla*), as much as 30% of the flyway population is dependent on the Dutch Waddenzee (Van der Jeugd et al. 2014).

Deeper Waddenzee zones are home to biotic mussel reefs (*Mytilus edulis*) and oysters protecting underlying sediments from erosion while actively supporting sedimentation (Baptist et al. 2007). The native European Flat Oyster (*Ostrea edulis*) became extinct because of overfishing. However, its place has been taken by the non-native Pacific Oyster (*Crassostrea gigas*) (Reise et al. 2010). Pacific oyster shells, however, are much thicker than native oyster shells and are thus harder to crack for birds. Therefore, oysters are no longer part of the Waddenzee's wader birds diet (Berendse 2011).

The Waddenzee is of vital importance to fish species, including those in the Noordzee. Eggs and larvae of many species spawning in the Noordzee's coastal waters, e.g. Plaice (*Pleuronectes platessa*), Sole (*Solea solea*), Herring (*Clupea harengus*) and Sprat (*Sprattus sprattus*), drift into the Waddenzee, where their larvae benefit from ample food availability, to return as juvenile fish. Other species use the Waddenzee only as a feeding ground, while some 17 fish species spend their entire life cycle there. A number of anadromous fish species migrate between the Noordzee and the mainland rivers via the Waddenzee.

At the top of the food chain, several marine mammals benefit from the Waddenzee's high productivity, including Harbour Seal (*Phoca vitulina vitulina*), Grey Seal (*Halichoerus grypus*), Harbour Porpoise (*Phocoena phocoena*) and Bottlenose Dolphin (*Tursiops truncates*).

Zeeland Delta This delta is at the confluence and estuaries of the rivers Rijn, Maas en Schelde. Several estuaries were dammed for flood protection and turned into fresh- or brackish water lakes (e.g. Haringvliet and Grevelingenmeer). Two of the estuaries still have open connections with the Noordzee: the Oosterschelde and Westerschelde. The Oosterschelde storm surge barrier is equipped with sluice doors that close during spring tides but remain open under normal conditions, enabling exchange of water and organisms between the estuary and the Noordzee. These doors were installed as a compromise for conservationists who opposed the closure of the estuary because of the impact it would have on its ecology.

Tidal flats and salt marshes in the Zeeland Delta also provide important bird habitats. But in contrast to the Waddenzee, tidal flats and salt marshes only occur along the estuary's fringes due to different geomorphological conditions. Their centres contain very deep (up to 45 m) tidal channels. Tidal flats and channels are separated by large sandy shallows. Reduced flow velocities by the Oosterschelde storm surge barrier cause tidal flats erosion and channel sedimentation, at the expense of bird habitats.

Zeeland's estuaries are characterised by the presence of a more or less natural salinity gradient ranging from fresh river waters to marine saltwater. This is particularly so in the unconfined Westerschelde estuary. Such gradients are especially relevant for anadromous migratory fish like Sea Lamprey (*Petromyzon marinus*), River Lamprey (*Lampetra fluviatilis*) en Twait Shad (*Alosa fallax*) (Schaminée and Janssen 2009).

Due to habitat variety provided by large environmental gradients, the Zeeland Delta has a high biodiversity. There is significant overlap in species that occur in the Zeeland Delta and the Waddenzee. The same bird species use the tidal flats and salt marshes as staging area, and Harbour Seal, Grey Seal and Harbour Porpoise are found in both locations. Seals disappeared from the Zeeland Delta in the 1970s, but a new population has established partly by human introduction of animals from the Waddenzee and partly by natural recolonisation.

Deeper channels and shallow waters in front of the estuary are home to marine fish species, such as Seabass (*Dicentrarchus labrax*) and Cod (*Gadus morhua*). Several other marine fish species use the estuary as a nursery, while Common Cuttlefish (*Sepia officinalis*) and Lumpfish (*Cyclopterus lumpus*) enter the estuary shallows to spawn (Schaminée and Janssen 2009).

The hard embankments of the many coastal defense structures provide habitat to species normally found on rocky coasts, including European Lobster (*Homarus gammarus*) – which is confined to the Zeeland Delta in the Netherlands – and a range of crab species. They also provide substrate for seaweeds such as Lesser Grass-kelp (*Blidingia minima*) and Bladderwrack (*Fucus vesiculosus*), shellfish, sea anemones and even soft corals like Dead Man's Fingers (*Alcyonium digitatum*) (Schaminée and Janssen 2009). Because of the rich underwater life the Zeeland Delta's estuaries have become a popular diving spot.

Salt marshes along estuary fringes are influenced by salinity gradients from the east (with river inflow) to the (marine) west as well. Covering an area of 30 km^2, the '*Verdronken land van Saeftinge*' in the east of the Westerschelde is one of the largest Dutch salt marshes. Positioned at the lower end of the salinity gradient this salt marsh contains brackish rather than saline plant species like Sea Club-rush (*Bolboschoenus maritimus*), Common Reed (*Phragmites australis*) and Sea Aster (*Aster tripolium*). A rare, brackish salt marsh species here is Common Scurvygrass (*Cochlearia officinalis* subsp. *officinalis*). More western salt marshes have many species in common with the Waddenzee, e.g. Sea-purslane (*Atriplex portulacoides*) and Sea Wormwood (*Artemisia maritima*). One of the few Dutch (sub)species considered by some to be endemic that occurs only on the Zeeland Delta's salt marshes is the cuckoo bee *Epeolus tarsalis* subsp. *rozenburgensis*.

Noordzee This is the largest Dutch nature area. Its Dutch sector is located on the shallow continental shelf and covers about 11% of the total Noordzee area. In contrast to the more stratified deeper water, the shallow Dutch Noordzee waters mix rather well (Schaminée et al. 2010). Together with sea floor characteristics this largely determines its underwater ecology.

Today's sea floor used to be land during the ice ages and is as varied in its geological composition and topography as the current terrestrial part of the Netherlands. For instance, the Noordzee's Klaverbank has a variable sediment composition, including large volumes of gravel, and hosts many unique Noordzee species. The coastal zone and its shallow part have a high primary productivity because of the availability of both light and river supplied nutrients. Not surprisingly, this area also hosts most of the 150 Noordzee fish species and the largest populations of marine mammals (Lindeboom et al. 2008). Several other productive areas like the Doggersbank shallows or the Friese Front, where two sea currents meet, are internationally important as feeding grounds for marine birds.

The Noordzee is highly affected by human activities. It contains one of the world's busiest shipping routes, is intensively used for fishing (beam trawling in particular) and is dotted with rigs drilling for oil and gas. More recently, large offshore wind farms were built for renewable energy production. Furthermore, sand dredging for coastal defense purposes further disturbed its sea floors. Although information on the Noordzee's biodiversity is limited, it is clear that fishing and other human activities have altered species composition and the marine food web drastically. As in most of the intensively fished waters larger individuals of most fish species are absent and most large bodied and slow growing species like sharks are severely threatened (Schaminée et al. 2010).

5.4 Environmental Pressure and Nature Conservation

Environmental Pressures Biodiversity and nature are under significant environmental pressure in the Netherlands. The contribution of the main pressures on biodiversity are given in Fig. 5.13. About 70% of the Netherlands land surface is in use for agriculture. The prominent global Dutch agricultural export position is possible only through innovative, intensive and highly mechanised forms of agriculture. Not surprisingly, many Dutch environmental problems, in particular those exerting high pressures on nature, are associated with agriculture. Such pressures include lowering groundwater tables and abundant use of manure and fertilizers resulting in eutrophication and acidification.

Many of the typical Dutch delta species require wet soil conditions. However, in many of the lower parts of the Netherlands groundwater tables are kept artificially low to

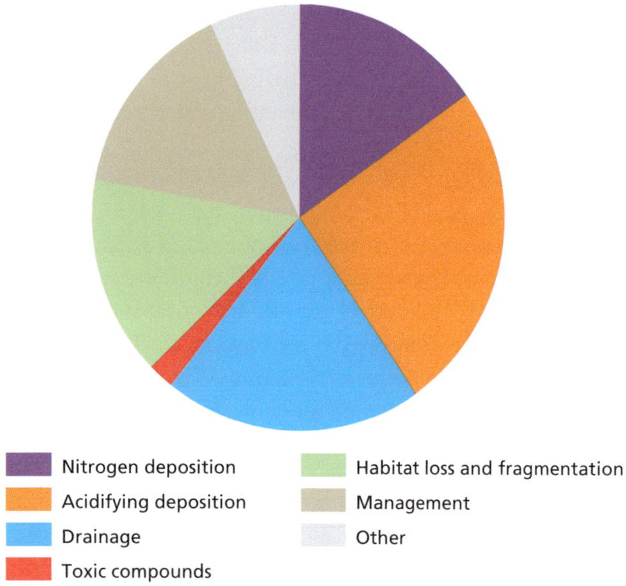

Fig. 5.13 Contribution of environmental pressures to biodiversity loss in the Netherlands since 1950 (in percentages per pressure factor). (Source: Van Veen et al. 2008)

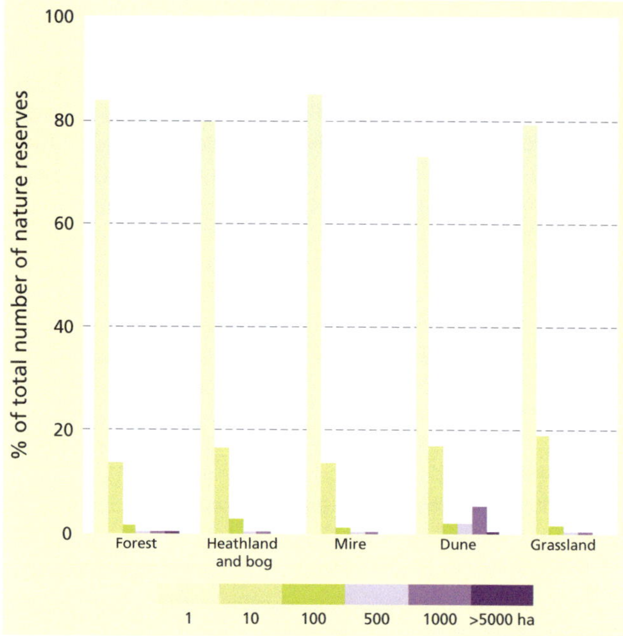

Fig. 5.14 Size distribution of nature reserves for five different ecosystems in the Netherlands. (Source: PBL (2016b))

allow heavy machinery and cattle on the land. Consequently, drought-tolerant plant species significantly have increased in abundance at the expense of species of wetter conditions at least since the 1930s (Tamis et al. 2005). Low groundwater levels also cause problems for meadow birds. Soil organisms, the main staple food for these birds, reside deeper in the soil when are low, beyond reach of birds, and also occur in lower densities when groundwater levels are low. Furthermore, dry soils may become impenetrable for birds. Another, indirect, impact on meadow bird populations is the earlier onset of grass growth at low groundwater levels. Consequently, farmers can cut the grass earlier, during the meadow birds breeding and chick-rearing seasons. Another impact of early grass growth is that much grass has already been cut when the eggs hatch leaving no foraging habitat for the chicks that feed on the higher grass dwelling insects. Finally, where grass has not been cut, e.g. as part of agri-environmental schemes, the sward may become so dense that it becomes inaccessible for the foraging chicks. Climate change impacts since the 1980s have exacerbated this situation (Kleijn et al. 2010). That increasingly exposes meadow bird clutches and chicks to agricultural activities, leading to higher bird mortalities and lower reproduction rates.

Upon derogation from the EU Nitrate Directive, Dutch farmers may apply up to 250 kg/ha of nitrogen on their lands today, almost 50% more than in other EU nations (Van Grinsven et al. 2016). The large numbers of Dutch cattle and livestock, jointly producing nearly 80 billion kg of manure annually (CBS 2017). Despite technological innovations such as biogas production from manure through fermentation (Chap. 4), the main way to process manure is by spreading it over land. Part of the nutrients is taken up by crops but much is left unused and eventually enters groundwater and surface waters, leading to eutrophication. Therefore, much of the Dutch surface waters, particularly in agricultural areas, fail to meet the water quality standards of the EU's Water Framework Directive (PBL 2016a; Chap. 2).

Large ammonia and nitrous oxide emissions from livestock and manure, as well as from traffic and industry, cause high levels of atmospheric nitrogen deposition. Current Dutch nitrogen deposition levels are on average 1600 mol N/ha/yr, but can locally reach up to 4000 mol N/ha/yr. (Velders et al. 2017), the highest in Europe (Working Group on Effects 2004) and far beyond critical deposition levels for most Dutch natural ecosystems (Bobbink et al. 2010). Nitrogen deposition leads to plant species loss through eutrophication and acidification (Stevens et al. 2010). In their analysis of environmental change effects on Dutch flora since 1930, Tamis et al. (2005) showed that plant populations indicative of nutrient-poor soils declined considerably, while populations of more eutrophic species increased, mostly between 1930 and 1980. Since then, more strict environmental policies have been in place that proved quite effective in reducing nutrient emissions to the environment, but still insufficient for providing suitable conditions for many of the Dutch ecosystems (PBL 2016a).

Most of the Dutch natural areas are quite small with only a few larger, contiguous areas present (Fig. 5.14). This is largely due to habitat loss and fragmentation as a consequence of parcel re-allotments and land use change (Rienks et al. 2008; Box 3.1). About half of the semi-natural elements in the countryside, such as hedgerows and trees, were removed from the landscape since 1900 (PBL 2009), thereby significantly reducing connectivity between habitats. However, the Dutch agricultural area is no longer expanding and most current land use change is caused by infrastructure and urbanisation, often at the expense of agricultural land but impacting natural areas as well.

Pesticide application was a huge problem for Dutch nature in the 1970s. Toxins accumulated at the top of the food web, which caused decimation of bird of prey populations. Recognition of this effect resulted in a ban on the most dangerous pesticides and bird of prey populations recovered subsequently. But new types of highly effective pesticides may lead to renewed impacts on the natural environment today. For instance, neonicotinoids have been implicated in honey bee colony losses (Van der Zee et al. 2015) and are so effective in suppressing target and non-target insects that lack of available bird food in the countryside is leading to bird population declines (Hallmann et al. 2014). A more recent environmental problem concerns alien invasive species. Due to globalisation of trade and the Dutch situation on a major European river delta a considerable number of alien species has established here. Not all such species necessarily have adverse effects on Dutch nature but in several cases they do. A most notable example is the near annihilation of the Dutch population of European Crayfish (*Astacus astacus*) due to an invasion of several alien crayfish species carrying a fungal disease to which the native population was not resistant. Only one small population of European crayfish survived in an isolated Dutch pond.

Establishment of exotic species has been accelerated by climate change, thereby actually increasing Dutch biodiversity. Many species from southern regions requiring higher temperatures have now settled in the Netherlands. An example is the Wasp Spider (*Argiope bruennichi*), a Mediterranean spider that was absent in the Netherlands before 1990 but occurs widely in the Netherlands today. Furthermore, thermophilic populations of native plant and animal species are increasing while those that are neutral to temperature or prefer cold conditions (psychrophilic) are decreasing (Fig. 5.15; Tamis et al. 2005).

A potential negative impact of climate change-related temperature rise is a mismatch in phenology, in particular between prey and predator species. Mismatches between caterpillar abundances and hatching of the eggs of Pied Fly-catcher (*Ficedula hypoleuca*) have led to 90% population declines (Both et al. 2006). A further potential impact of climate change is anticipated on water availability for wet Dutch ecosystems due to higher evapotranspiration at higher temperatures and altered precipitation patterns (Witte et al. 2012). However, current climate change adaptation policies

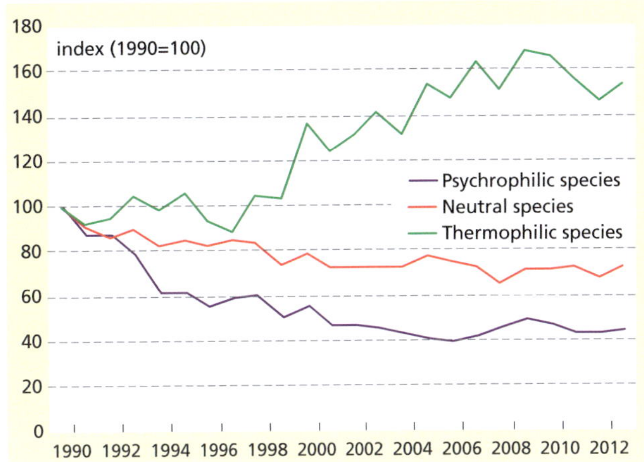

Fig. 5.15 Population development of 20 thermophilic, 20 neutral and 20 psychrophilic animal species in the Netherlands between 1990 and 2013. (Source: PBL (2014))

mainly focus on water safety aspects for humans whereas a coherent adaptation strategy for nature conservation is still largely lacking. So far, such policies seem to be mainly geared at tackling current environmental problems in an attempt to increase resilience of ecosystems to climate change.

Nature Conservation The large environmental pressures on nature and the often small-size of Dutch nature reserves makes that many ecosystems and species do not flourish in the Netherlands and that most nature reserves require active conservation management. A total of 40 Dutch species occur on the IUCN Red List as because they are threatened at a global scale. These comprise mainly fishes (15) and a smaller number of mammals, birds, molluscs, invertebrates and microorganisms (IUCN 2017). With 3389 species, the National Red List of threatened Dutch species is much longer. First, this is because it focuses only on the status of a species within the country borders, ignoring abundance of the species in the rest of the world. Second, the Dutch Red List is dominated by fungi, simply because there are over 10,000 species, of which 60% meet the Red List criteria. In contrast, there are only 20 Dutch *Plecoptera* species, 19 of which on the Dutch Red list (Fig. 5.16).

According to Van der Windt (1995), ideas to protect natural areas began to arise in the late nineteenth century in response to rapid urbanisation, industrialisation and large scale conversion of natural areas into agricultural land which had a devastating effect on environmental quality, biodiversity and aesthetics of the countryside. This led to negative responses by Dutch upper middle class civilians (physicians, schoolteachers and amateur naturalists), writers

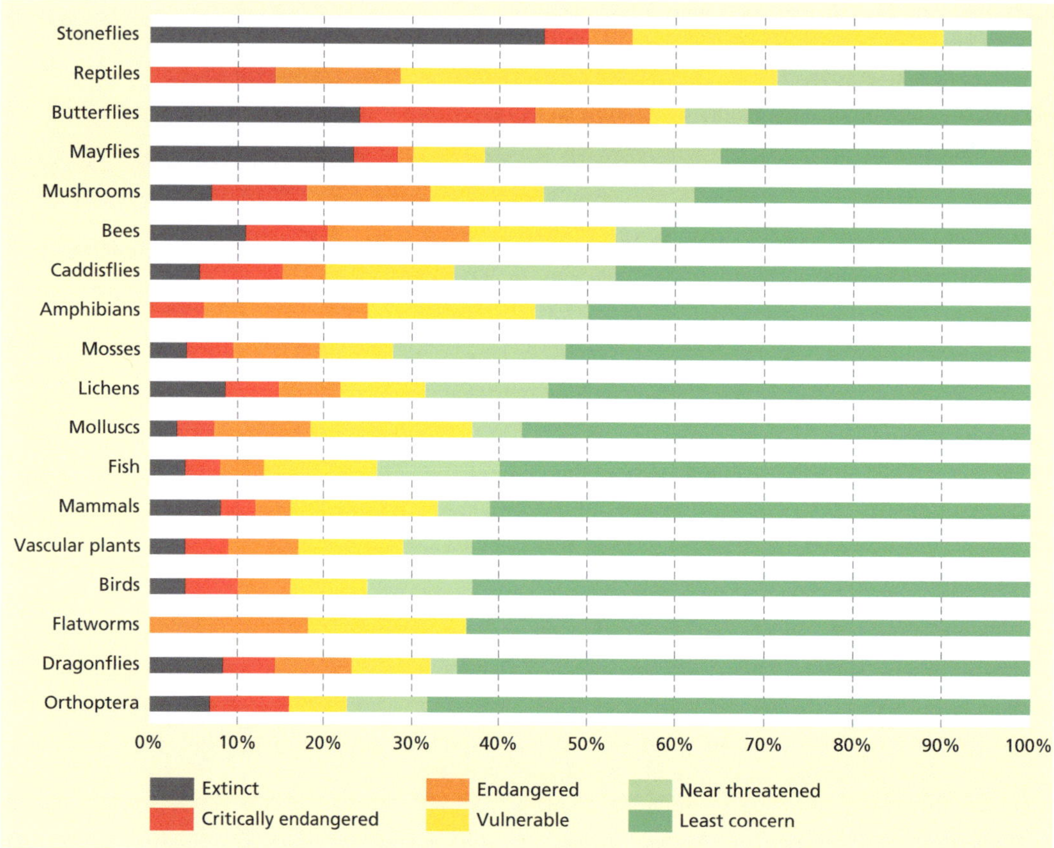

Fig. 5.16 The proportion of each species group in the Netherlands that is threatened according to the Dutch Red List. (Source: PBL (2017b))

and artists who spent much of their free time outside the cities. The first Dutch conservation movement was ignited by a civilian initiative. In 1906, a group of civilians led by schoolteachers and naturalists Eli Heimans and Jac. P. Thijsse bought the peat lake Naardermeer to prevent the municipality of Amsterdam from turning it into a landfill. This acquisition was financed from private funds and initiated the foundation of the '*Vereniging tot Behoud van Natuurmonumenten*' (association for the conservation of natural monuments) in 1905, the largest conservation NGO in the Netherlands today. Its 719,802 members and 107,000 ha of nature reserves (Natuurmonumenten 2016), point to considerable societal support for nature conservation in the Netherlands (Fig. 5.17).

With its acquisition, the Naardermeer became the first nature reserve in the Netherlands. That was the first time that nature was protected for its intrinsic value rather than for its usefulness for humans. Since 1880, the Netherlands already had a law protecting a number of animals because of their usefulness to humans (e.g. game species), but the first conservation law to include intrinsic values for protection was the *Vogelwet*, a law for the protection of all birds, passed in 1912. In 1967, the *Natuurbeschermingswet* (nature conservation law) was passed, that protected a number of rare animals and plants from being captured or harvested in nature. Surprisingly, most of the conservation legislation from this period focused on the protection of species from direct human intervention while private conservation initiatives rather aimed at safeguarding entire areas, including the natural processes and biodiversity therein.

As a result of the ongoing conversion of the last remaining natural areas to farmland in the 1920s and 1930s and continuing attempts by the private conservation initiatives to protect areas with high natural values from such conversions, more and more tensions began to arise between conservationists and farmers, which eventually drew a wedge between both parties. This contributed to a growing belief that agriculture and nature were basically competitors and their territories should thus be physically separated, perhaps with the exception of highly valued semi-natural cultural landscapes (Van der Windt 1995). This finally led to the proposal by conservationists to establish a number of National Parks, like in other nations. Indeed, 20 such National Parks were established in the 1980s, all still existing today (Fig. 5.18).

From the 1970s onward, two developments changed strict division between nature conservation and agriculture. The first was criticism among conservationist on the negative

Fig. 5.17 Red deer drawing a crowd on a weekday evening at an observation screen of Natuurmonumenten in National Park Veluwezoom. (Photograph by Jerry Van Dijk)

social and economic impacts of strict division (especially for farmers) and the need to include alternative views of nature into a more integrative environmental and conservation approach. A second development concerned new scientific insights from studies integrating genetics and ecology, specifically the meta-population theory that was applied to nature conservation by Finnish biologist Hanski (Hanski 1999). This theory stated that individual subpopulations of species should not be considered in isolation but rather in coherence with other subpopulations. Lack of exchange of individuals and genetic material among subpopulations would lead to loss of genetic diversity and insufficient individuals to colonise suitable habitats making the entire (meta)population more vulnerable to extinction. Metapopulation theory gave rise to the idea that Dutch nature reserves should be mutually connected in an ecological network which would require some sort of green infrastructure on agricultural land as well.

This idea was further developed and eventually incorporated in the 1990 National Nature Policy Plan as the Ecological Main Structure, a network of larger conservation areas connected by green or, for aquatic systems, blue corridors. This network was to be realised by new land acquisitions to expand existing nature reserves, by the creation of new reserves and by connecting corridors on former agricultural land through restoration projects. It also included creation of connections between nature reserves through agri-environmental schemes that would lead to more natural infrastructure on agricultural land. The network was aimed to be completed by 2018 and should then comprise about 730,000 ha.

Simultaneously, conservation policy was developed at European level in the form of the Bird- and Habitat Directives, which were later incorporated in the Natura 2000 policy. These Directives use lists of species requiring protection and have a legally binding status. At Dutch national level these policies were legally implemented in the *Flora- en Faunawet* in 2002 and in a new Nature Policy Plan in 2005 (Fig. 5.18). Enforcement of these laws was very strict, often leading to halting major construction projects when protected species were discovered at construction sites. As happened before with agriculture, strong tensions between economic development on the one side and nature conservation on the other emerged. This time, however, these tensions eroded public and political support for nature conservation (De Wit et al. 2011). Further development of the Ecological Main Structure was halted and major budget cuts were introduced by a new government in 2010. The Ecological Main Structure was confined to those areas that had been appointed already as Natura 2000 areas, which meant a reduction of 120,000 ha compared to the original target.

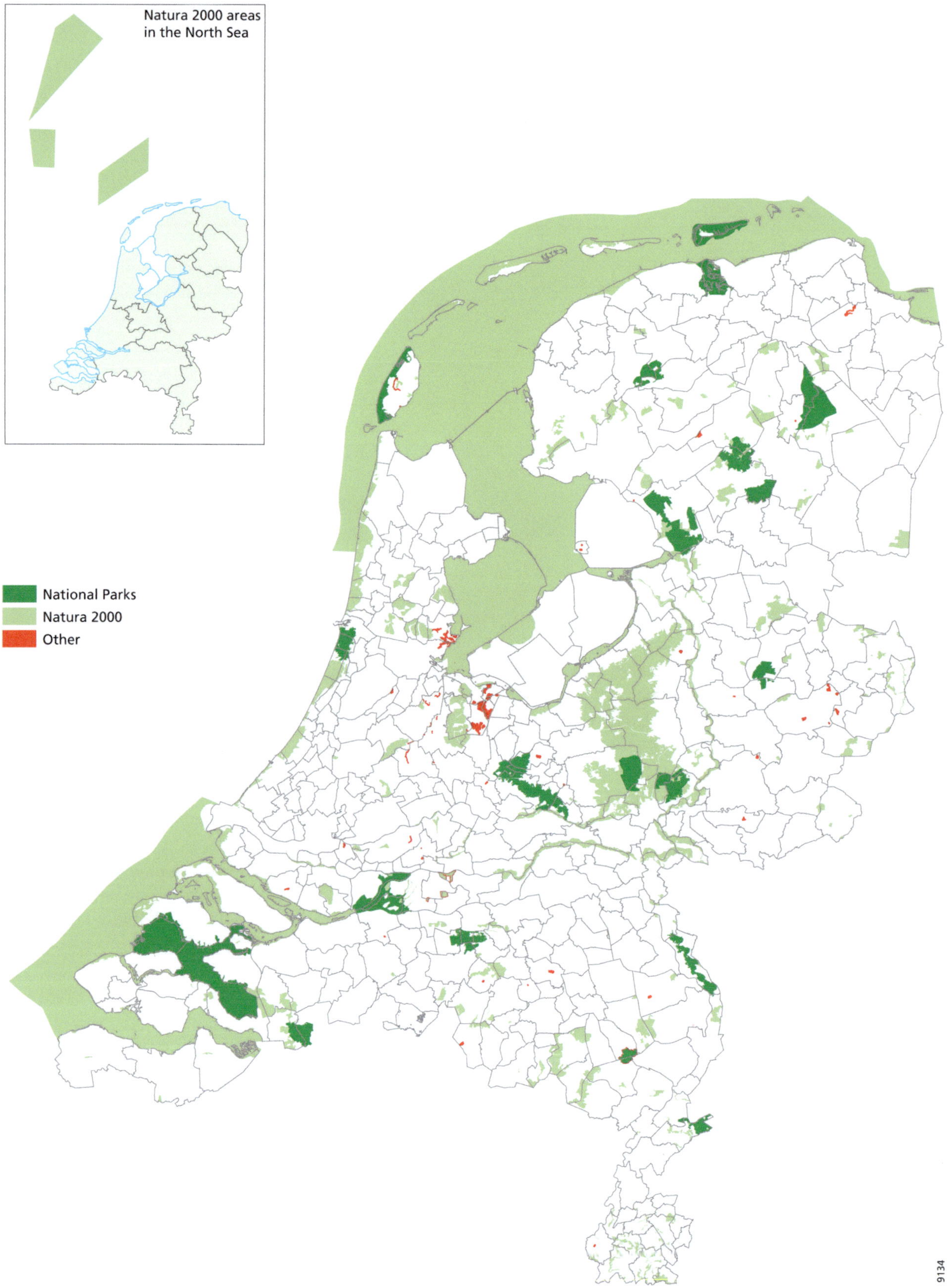

Fig. 5.18 Protected nature reserves in the Netherlands in 2016. The map indicates National Parks, Natura 2000 areas and other protected areas. These partially overlap. (Sources: European Environment Agency 2017a, b)

Fig. 5.19 The currently planned Nature Network Netherlands. (Source: PBL (2016a))

In 2012, yet another government partly reversed this reduction and some of the budget cuts. The revised network is now called the Nature Network Netherlands (Fig. 5.19) which still is some 50,000 ha smaller than the original Ecological Main Structure and has a new deadline for completion in 2027. Simultaneously, the Nature Policy Plan was revised to better reflect the new, less stringent approach to nature conservation. This Nature Policy Plan was implemented in 2017, so the future will learn whether this policy has been effective to curb biodiversity loss and to conserve the delta ecosystems of the Netherlands.

References

Baptist MJ, Dankers N, Smit C (2007) The outstanding universal values of the Waddenzee: an ecological perspective. Wageningen IMARES report C037/07, Wageningen

Barendregt A, Swarth CW (2013) Tidal freshwater wetlands: variation and changes. Estuar Coasts 36(3):445–456

Beintema N (2017) Gedoogbeleid ganzen is mislukt. NRC.nl 10-04-2017. https://www.nrc.nl/nieuws/2017/03/10/gedoogbeleid-ganzen-is-mislukt-7296400-a1549774. Accessed 23 May 2017

Berendse F (2011) Natuur in Nederland. Stichting uitgeverij KNNV, Zeist

BirdLife International (2004) Birds in Europe: population estimates, trends and conservation status. BirdLife International, Cambridge

Bobbink R et al (2008) Preadvies hellingbossen in Zuid-Limburg. Rapport DK nr. 2008/094-O. Ministerie van Landbouw, Natuur en Voedselkwaliteit, Directie Kennis, Ede

Bobbink R et al (2010) Global assessment of nitrogen deposition effects on terrestrial plant diversity: a synthesis. Ecol Appl 20(1):30–59

Boele A et al (2017) Broedvogels in Nederland in 2015. Sovon rapport 2017/04. Sovon Vogelonderzoek Nederland, Nijmegen

Both C et al (2006) Climate change and population declines in a long-distance migratory bird. Nature 441(7089):81–83

Calle L, Jacobusse C (eds) (2008) Bijen en wespen in Zeeland, Fauna Zeelandica, vol 4. Stichting Het Zeeuwse Landschap, Wilhelminadorp

CBS (2017) StatLine: Dierlijke mest; productie en minerale-nuitscheiding; diercategorie, regio. CBS, Den Haag/Heerlen. http://statline.cbs.nl/Statweb/publication/?DM=SLNL&PA=82505NED&D1=0-3&D2=0-1,9,12,17-19&D3=0&D4=a&HDR=G2,G3&STB=G1,T&VW=L. Accessed 28 June 2017

Cornelissen P, Beemster N, Kuypers HE (2017) Vegetatie, vogels, grote herbivoren en recreatie in de Oostvaardersplassen. Staatsbosbeheer, Amersfoort

De Wit AJF, Wieringa K, Hajer MA (2011) Natuurbeleid als strijdpunt. Veranderende politiek-maatschappelijke context en de Natuurverkenning. Landschap 28(4):163–171

European Environment Agency (2002) Biogeographical regions in Europe. The Atlantic region – mild and green, fragmented and close to the rising sea. In: Europe's biodiversity – biogeographical regions and seas. EA report No 1/2002. European Environment Agency, Copenhagen

European Environment Agency (2016) Biogeographical regions. https://www.eea.europa.eu/data-and-maps/data/biogeographical-regions-europe-3. Accessed 28 June 2017

European Environment Agency (2017a) Nationally designated areas (CDDA). https://www.eea.europa.eu/data-and-maps/data/nationally-designated-areas-national-cdda-11

European Environment Agency (2017b) Natura 2000 data – the European network of protected sites. https://www.eea.europa.eu/data-and-maps/data/natura-8. Accessed 28 June 2017

Hallmann CA et al (2014) Declines in insectivorous birds are associated with high neonicotinoid concentrations. Nature 511(7509):341–343

Hanski I (1999) Metapopulation ecology. Oxford University Press, New York

IUCN (2017) The IUCN red list of threatened species. Version 2017-1. http://www.iucnredlist.org. Accessed 12 May 2017

Janssen JAM, RG Jaket al (2014) Habitattypen in Natura 2000-gebieden. Beoordeling van oppervlakte, representativiteit en behoudsstatus in de Standard Data Forms (SDFs). WOt-technical report 8. Wettelijke Onderzoekstaken Natuur & Milieu, Wageningen

Kentie R, Piersma T et al (2016) Estimating the size of the Dutch breeding population of continental black-tailed godwits from 2007–2015 using resighting data from spring staging sites. Ardea 104(3):213–225

Kleijn D et al (2010) Adverse effects of agricultural intensification and climate change on breeding habitat quality of black-tailed godwits Limosa l. limosa in the Netherlands. Ibis 152(3):475–486

Koelewijn HP et al (2010) The reintroduction of the Eurasian otter (lutra lutra) into the Netherlands: hidden life revealed by noninvasive genetic monitoring. Conserv Genet 11(2):601–614

La Haye MJJ et al (2010) Agri-environmental schemes for the common hamster (Cricetus cricetus). Why is the Dutch project successful? aspects of applied biology, vol 100. Association of Applied Biologists, Warwick, pp 117–124

Lindeboom HJ et al (2008) Ecologische atlas Noordzee ten behoeve van gebiedsbescherming. Wageningen Imares Institute for Marine Resources & Ecosystem Studies, Wageningen

Marris E (2009) Reflecting the past. Nature 462(7269):30–32

Ministerie van Economische Zaken (2013) Fysisch Geografische Regio's 2013. http://wwwnationaalgeoregisternl/geonetwork/srv/dut/catalogsearch#/metadata/c8b5668f-c354-42f3-aafc-d15ae54cf170?tab=general. Accessed 28 June 2017

Natuurmonumenten (2016) Natuur laat je leven. Jaarverslag 2016. Natuurmonumenten, 's-Graveland

NDFF (2015) NDFF Verspreidingsatlas. Nationale Databank Flora en Fauna http://verspreidingsatlasnl. Accessed 28 June 2017

Noordijk J et al (eds) (2010) De Nederlandse-Biodiversiteit – Nederlandse Fauna 10. Nederlands Centrum voor Biodiversiteit, Naturalis & European Invertebrate Survey, Leiden

PBL (2009) Natuurbalans 2009. Planbureau voor de Leefomgeving, Den Haag/Bilthoven

PBL (2014) Invloed klimaatverandering op soorten. Planbureau voor de Leefomgeving, Den Haag. http://www.clo.nl/nl142908. Accessed 28 June 2017

PBL (2016a) Balans van de leefomgeving 2016. Richting geven, ruimte maken. Planbureau voor de Leefomgeving, Den Haag

PBL (2016b) Verdeling natuurgebieden naar gebiedsgrootteklasse, 2012, Planbureau voor de Leefomgeving, Den Haag http://wwwclonl/nl158801. Accessed 28 June 2017

PBL (2016c) Percentage bedreigde soorten per soortgroep. Planbureau voor de Leefomgeving, Den Haag http://wwwclonl/nl105213. Accessed 28 June 2017

Provincie Zeeland (2005) Actieplan natuurbeheer binnendijken. Provincie Zeeland, directie Ruimte, Milieu en Water, Middelburg

Provincie Zeeland (2017). Geoloket: Natuur en Landschap. https://zldgwbzeelandnl/geoloket/?Viewer=NatuurLandschap. Accessed 29 May 2017

Reise K et al (2010) The Waddenzee – a universally outstanding tidal wetland. Waddenzee ecosystem no. 29. Common Waddenzee Secretariat, Wilhelmshaven

Rienks WA et al (2008) Grootschalige landbouw in een kleinschalig landschap. Alterra-rapport 1642. Alterra, Wageningen

Schaminée JHJ, Janssen JAM (eds) (2009) Europese natuur in Nederland. KNNV Uitgeverij, Zeist

Schaminée JHJ, Dirkx JGHP, Janssen JAM (2010) Grenzeloze natuur. De internationale betekenis van Nederland voor soorten, ecosystemen en landschappen. KNNV Uitgeverij, Zeist

Schouten MA et al (2009) Determinants of species richness patterns in the Netherlands across multiple taxonomic groups. Biodivers Conserv 18(1):203–217

Schouten MA et al (2010) Defining hotspots of characteristic species for multiple taxonomic groups in The Netherlands. Biodivers Conserv 19(9):2517–2556

Sovon, CBS (2005) Trends van vogels in het Nederlandse Natura 2000 netwerk. SOVON-informatierapport 2005/09. Sovon Vogelonderzoek Nederland, Beek-Ubbergen

Sovon Vogelonderzoek (2017) Numbers of geese in the Netherlands. https://www.sovon.nl/nl/content/aantallen-ganzen-nederland. Accessed 23 May 2017

Spitzen-van der Sluijs A et al (2013) Rapid enigmatic decline drives the fire salamander (Salamandra Salamandra) to the edge of extinction in the Netherlands. Amphibia-Reptilia 34(2):233–239

Stevens CJ et al (2010) Nitrogen deposition threatens species richness of grasslands across Europe. Environ Pollut 158(9):2940–2945

Tamis WLM et al (2005) Changes in vascular plant biodiversity in the Netherlands in the 20th century explained by their climatic and other environmental characteristics. Clim Chan 72(1):37–56

UNESCO (2009) Report of decisions of the 33rd session of the world heritage committee. Report number WHC-09/33.COM/20. UNESCO, Paris

Van Bodegom PM et al (2006) Raising groundwater differentially affects mineralization and plant species abundance in dune slacks. Ecol Appl 16(5):1785–1795

Van der Jeugd HP et al (2014) Geïntegreerde monitoring van vogels van de Nederlandse Waddenzee. Vogeltrekstation rapport 2014–01. Vogeltrekstation, Wageningen; CAPS-rapport 2014–01; Sovon-rapport 2014/18, Sovon Vogelonderzoek Nederland, Nijmegen

Van der Vliet R et al (2015) EU protection is inadequate for a declining flyway population of black-tailed godwit Limosa Limosa - mismatch between future core breeding areas and existing special protection areas. Bird Conserv Int 25(1):111–125

Van der Windt H (1995) En dan: wat is natuur nog in dit land? Natuurbescherming in Nederland 1880–1990. Boom, Amsterdam/Meppel

Van der Zee R et al (2015) An observational study of honey bee colony winter losses and their association with *Varroa destructor*, neonicotinoids and other risk factors. PLoS One 10(7):e0131611

Van Dijk J et al (2015) Modeling direct and indirect climate change impacts on ecological networks: a case study on breeding habitat of Dutch meadow birds. Landsc Ecol 30(5):805–816

Van Grinsven HJM, Tiktak A, Rougoor CW (2016) Evaluation of the Dutch implementation of the nitrates directive, the water framework directive and the national emission ceilings directive. NJAS Wagening J Life Sci 78:69–84

Van Straalen NM, Verkleij JAC (eds) (1993) Leerboek oecotoxicologie. VU Uitgeverij, Amsterdam

Van Veen MP et al (2008) Halting biodiversity loss in 2010. Evaluation of progress. Netherlands Environmental Assessment Agency, Bilthoven

Velders GJM et al (2017) Grootschalige stikstofdepositie in Nederland. Herkomst en ontwikkeling in de tijd. Planbureau voor de Leefomgeving, Den Haag/Bilthoven

Van der Straaten R (2008) De Leuvenumse Beek. Grondboor Hamer 6: 139–143

Wassen MJ et al (1990) Dependency of local mesotrophic fens on a regional groundwater flow system in a poldered river plain in the Netherlands. Landsc Ecol 5(1):21–38

Weeda EJ et al (1994) Nederlandse oecologische flora. Wilde planten en hun relaties. Stichting uitgeverij KNNV, Zeist

Witte JPM et al (2012) An ecohydrological sketch of climate change impacts on water and natural ecosystems for the Netherlands: bridging the gap between science and society. Hydrol Earth Syst Sci 16:3945–3957

Working Group on Effects (2004) Review and assessment of air pollution effects and their recorded trends working group on effects convention on long-range transboundary air pollution. National Environment Research Council, London

6 The Netherlands in Europe and the World

Ben C. De Pater

Abstract

The Netherlands keeps strong and extensive international economic and political relationships with other countries in the world and in particular in Europe. These have certainly contributed to its flourishing post-war economy and its relatively prominent international status. Prior to WWII, however, this international position was significantly more modest. The Netherlands faced rapidly increasing unemployment rates (from 6 to nearly 20% of the workers) since the 1929 economic crisis while the country was politically neutral and international trade was at a much lower level. That position changed fundamentally since WWII when the Netherlands became a NATO member and a founding member of the European Community. This chapter presents a review of the changing patterns of international trade, the international position of Dutch cities and Dutch foreign policy.

Keywords

Dutch international trade · Amsterdam in the golden age · Import-export · Foreign investments · Dutch cities in europe · Dutch foreign policy

Shortly after WWII, in autumn 1945, the Dutch government decided to 'provisionally' maintain Central European (= Berlin) clock time (CET) as enforced by the German occupiers at the onset of WWII, in 1940. However, the Dutch never since returned to the pre-war Amsterdam clock time (40 min behind CET). Following the end of WWII in 1945 all German laws and regulations were abandoned immediately, except for the CET clock time. That decision may be considered symbolic for growing Dutch awareness of being part of a larger (European) world, comparable in dimension with the shift of local times to a central Amsterdam time in 1909 (see introduction to Chap. 9).

This chapter deals with the changing position of the Netherlands as a nation in Europe and in the world. We will focus on international trade, the position of Dutch cities in an international urban context and Dutch foreign policy.

6.1 The Netherlands in the International Economic System

The Netherlands is and has always been a strong trading nation, culminating in the seventeenth century when the Netherlands was a commercial superpower (the Dutch Golden Age). Since the nineteenth century the Dutch share in world trade has been between 3% and 4%. That number is still quite considerable as compared to its current share: Dutch Domestic Product in the world-GDP is about 1% since the 1970s (WRR 2003). In 2015 the Netherlands ranked 17th on the list of the world's largest economies, 9th among the world's largest importers, and the world's 8th largest exporting nation according to international statistical agencies. Here we analyze the geographical-historical development of the Dutch economy in an international perspective.

6.1.1 Dutch Trade Until the End of the Nineteenth Century

In Medieval Times Europe's economic hearth was situated in northern Italy that was economically more powerful than northwestern Europe. But, in course of time this economic power shifted from the northern Mediterranean to the coasts of the Atlantic Ocean and the North Sea. Geographically, the Netherlands was quite well positioned. It was located at about the terminus of an important trade axis connecting northern Italy with the Rhine Basin. The Netherlands was also conveniently positioned between the Flemish trading and industrial

B. C. De Pater (✉)
Department of Human Geography and Spatial Planning, Utrecht University, Utrecht, The Netherlands
e-mail: B.C.dePater@uu.nl

E. F. J. De Mulder et al., *The Netherlands and the Dutch*, World Regional Geography Book Series,
https://doi.org/10.1007/978-3-319-75073-6_6

(textile) cities (Brugge/Brughes and Gent/Gand) on the one hand and the Hanseatic commodities exporting cities (with Lubeck as its core) on the other. In particular, cities in the central part of the Netherlands (Utrecht, Dordrecht, Den Bosch, Nijmegen), in the eastern Netherlands (towns along River IJssel: Arnhem, Zutphen, Deventer, Zwolle, Kampen) and towns along the Zuyderzee (Harderwijk, Stavoren) took advantage of this intermediate position between Flandres and the Hanseac region. Many of them became members of the Hanseatic League. That most ships avoided sailing over the dangerous North Sea has been beneficial to Dutch cities along the inland route via rivers and the Zuyderzee. South-north running rivers (Maas/Meuse, Rijn/Rhine, Weser, Elbe, Oder) connected the interior of NW Europe with the coast. River trade was based on the 'complementarity of successive north to south zones: clay, sand, loess and mountain zones' (Hoekveld 1989). The zones produced and exchanged different goods. Such river trading routes were supplemented by west-east onland trade routes as the Via Regia, connecting the cities of Maastricht and Aachen, via Cologne and Frankfurt to Leipzig and Breslau (now Wroclaw).

In the sixteenth century, the age of discoveries, this economic-geographic structure changed. Spain and Portugal became colonial powers; Northern Italian cities languished. Within the Low Countries the economic hearth gradually shifted from Flanders and Brabant to Holland. A new generation of Dutch towns developed with Amsterdam as their leader. European economy shifted from Mediterranean to Atlantic, from river and coast based to maritime. The 'Dutch seaborne Empire' was born, founding colonies, trade posts and forts in Africa, Asia, and America (Fig. 6.1).

This 'empire' was embedded in an European 'world-economy', an 'economically autonomous part of the planet which could satisfy most of its own needs, a section to which its internal links and exchanges give a certain organic unity', a geographically limited part of the world, not to be confused with 'the world economy' that is global (Braudel 1984). In pre-industrial times a world-economy had, rather vague, external borders, rather sparsely populated zones often with extreme climatic features (drought, cold). A world-economy also had a powerful center, a predecessor of a world city serving as a focal point of a regional world-economy. Such centres shifted geographically and over time: in the fifteenth century such a centre was Venice, in the sixteenth century this position was taken over by Antwerpen, in the early seventeenth century it was North Italy again with Genoa and later in the seventeenth century this position shifted to Amsterdam: the Dutch Golden Age started. By the end of that century Amsterdam's position was taken over by London (Braudel 1984).

Key characteristics of the European world-economy may explain the extraordinary significance of such urban centres. Freight transports were slow, risky and hazardous because of wars, poor weather conditions and criminals. Supply of merchandise often was extremely irregular and may not have been available for months or years even. Other goods were supplied abundantly over some short periods of the year. Demands were quite constant however. Staple markets where merchandise could be stored for some time thus served as a buffer between irregular supply and demand (Van der Kooy 1931). Such staple markets thus served as spots where people were informed about availability, quality and quantity of goods and when these would arrive. On the other hand cargo ship captains could negotiate the best prices and fast payment for their products. European world-economy centres and cities attracted such staple markets and became more and more relevant. Permanent staple markets of the world-economy replaced the classic trade fairs periodically roving from town to town. The world-economy thus replaced the more regionally organised market-economy, the network city the Christallerian central place, the permanent exchange of goods the periodic fair (Braudel 1982).

Apart from the storehouse of the world-economy, the central city also served as the financial center, the place where the most recent and advanced trade-relevant knowledge was available and where raw materials were processed and upgraded in end products or in semi-final commodities, e.g. from timber to ships of from linseed (seed of flax) to vegetable oil. Around Amsterdam such traffic-related industries developed along River Zaan where many hundreds of windmills provided power to these type of industries.

Amsterdam has been such a leader of the world-economy for many decades during the seventeenth century. Its favourable geographic location between northern and southern Europe has been one of the reasons although this was less obvious so on a more regional scale. Sea-going vessels bound for Amsterdam had to sail a long and time consuming route and passed the dangerous tidal channel of the Marsdiep and the shallow Zuyderzee. That changed only in the nineteenth century when new canals directly connecting Amsterdam with the North Sea were dug: first the Noordhollands Kanaal (Amsterdam-Den Helder, open in 1824), later the Noordzeekanaal (Amsterdam-IJmuiden, in 1876). Chance and luck also played a role, e.g. as the elimination of Antwerpen as a rival port city when Spanish troops conquered that city in 1585 and when the Dutch controlled Antwerpen's access to the sea (Zijlmans 2017). On that occasion, many wealthy Protestant and Jewish families left Antwerpen for Amsterdam. Input of their knowledge and capital has been quite beneficial to the city of Amsterdam which grew further. Inventive practices of Amsterdam merchants to create monopolies in certain industries or areas often were successful just because Amsterdam was an economic and financial hotspot by then. All this further strengthened the position the city of Amsterdam as a world-economic centre and simultaneously hampered or

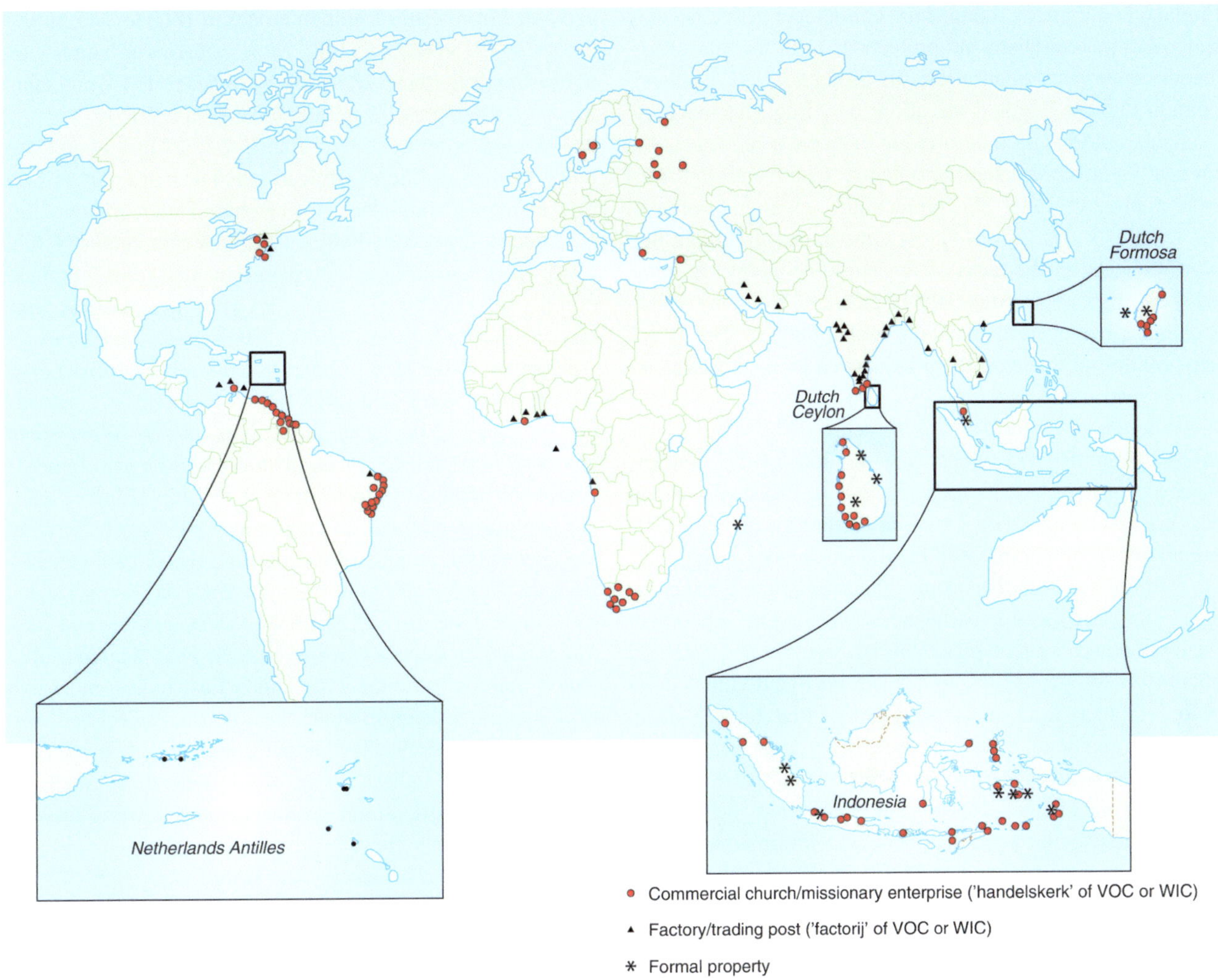

Fig. 6.1 Global Dutch presence during (part of) the seventeenth century. (Source: modified after Hoekveld 1989)

discouraged such urban development elsewhere. In sum, the unprecedented flourishing of Amsterdam as a staple town in the seventeenth century is due its geographic position, to luck and chance and to the success of skilled actors.

The realm of a world-economy may be split into three hierarchically different, regional zones (Braudel 1977, 1984). Around the central city is a core zone. This core zone included Holland, Zeeland and part of Utrecht, Friesland and Groningen in the Dutch Golden Age. Far beyond the core zone is the vast periphery serving as a raw material (e.g. peat and wood) supplier. The zone in between ('semi peripheral zone') supported the core zone. Dominance of the core zone was e.g. clearly demonstrated by the size of its merchant fleet, which was much larger than that of cities and countries in the semi-peripheral zone. On some trade routes, Holland had (quasi-) monopolies, such as the Øresund Strait connecting the North Sea and Baltic Sea. In the Amsterdam-centred world-economy, Scotland, Ireland, parts of Scandinavia, Eastern Europe (east from Hamburg-Venice), southern Italy and the Dutch colonies belonged to the periphery. The countries around the Baltic Sea and North Sea, France, Spain, Italy and Portugal, Rhine- and Elbe regions in Germany may be considered as semi-peripheral according to Braudel (1977).

Such a regional subdivision is dynamic as a central city will not be a leader forever. An area designated as being part of the semi-periphery (as the western part of the Netherlands during the Hanse Times) may turn into a core area, for example due to new transportation routes or to innovations in traffic technology. Old core areas may decline when its leadership is taken over, as happened when London became the successor of Amsterdam by the end of the seventeenth century.

Various reasons and speculations exist about the decline of Amsterdam as a central city, including lack of political organisation and unification of the Republic of the Seven United Netherlands, lack of sufficient human resources (only 2 million inhabitants, much less than rival nations as

England and France), diminishing entrepreneurship, etcetera. Emerging mercantilism and protectionism in the late seventeenth century has been a relevant factor as well. England-bound freight to England was allowed by English ships only. A comparable situation occurred in France where ideas of the French Finance Minister, Jean-Pierre Colbert (colbertism) were applied to trade. This protectionist economic policy at the level of *states* (not of cities) undermined the urban network concept as the basis of the world-economy. In some weaker states as in politically fragmented Germany Dutch merchants managed to maintain their influence. The Dutch kept controlling trade on River Rijn, even today most vessels on this river are Dutch.

Although mercantilism was beaten by liberalism as the ruling economic ideology in the nineteenth century, Amsterdam staple markets never returned. The staple market concept was bypassed by Industrial Revolution technological innovations, comparable with the transition from roving market fairs to staple markets some centuries before (Wagenaar and Van Engelsdorp Gastelaars 1986). The upcoming factories could produce many similar products and potential clients no longer needed to inspect each individual copy before buying. So, goods were transported directly from producer to client, bypassing temporary storage at staple markets. That affected Amsterdam and other cities in Holland as Haarlem, Leiden, Hoorn and Enkhuizen, who were downgraded to the semi-periphery and lost many of their citizens between 1690 and 1870.

6.1.2 Dutch Trade Since the End of the Nineteenth Century

As compared to England, Germany and Belgium, Industrial Revolution began relatively late (in the 1890s) in the Netherland and generated major structural changes. This country was conveniently positioned in between two industrial superpowers, England (active as such since the late eighteenth century) and emerging Germany (unified in an empire since 1871). Both nations and industrialising Belgium were outlets for Dutch agriculture and horticulture. In addition to export, the Netherlands gained a serious transport position by the construction of the Nieuwe Waterweg, a canal directly connecting Rotterdam with the North Sea since the 1860s. Its position in the apex of the Dutch delta gave Rotterdam direct access to the German and further hinterland via Rivers Maas and Rijn. Rotterdam developed as a transit port for the rapidly industrialising German Ruhr area. International river transport was facilitated by abolishing all toll payments along River Rijn in 1868 and by various large-scale physical improvements to this river as dredging since the 1860s (Klemann 2016). Rijn river trade expanded from about 1 million tonnes in 1860 to 37.5 million tonnes in 1913 (Blink 1914b). Most of the river barges went to or from Rotterdam (22.8 million tonnes in 1913) and many less to Amsterdam (1.5 million tons; Blink 1915).

Not only in terms of river transit trade, Amsterdam eventually lost competition with Rotterdam as a sea port as well. That change in dominance between both Dutch harbour cities occurred in the second half of the nineteenth century (Blink 1915; Hoekveld 1972). Rotterdam remained most prominent during the twentieth century as well, e.g. 11,049 ships entered Rotterdam by sea in 1930 and 3590 only in Amsterdam (CBS 1931). Rotterdam bound ships transported wheat, oats, barley and other grains to Germany, and also timber, copper ore and other ores. From downstream Germany iron and steel, natural stones and large volumes of coal were transported to Rotterdam by ship (Blink 1914b). Amsterdam remained an important passenger terminal, together with a major destination for tropical raw materials from the Dutch East Indies (today: Indonesia), as coffee, tea, cane-sugar, tobacco, rubber, copra, spices and rice and some rare mining products as tin and gold. In 1854 a quarter of all Dutch imports came from the Dutch East Indies, more than from any other country. Their share in import declined since to a level well below 10%. The same happened with export to the Dutch East Indies. International trade in colonial raw materials and agricultural products reduced while trade of industrial products grew.

In the nineteenth century, the United Kingdom still was the main trading partner of the Netherlands. That changed around 1900 when Germany took over this position and never left it since. For over one hundred years Germany is the most important trade partner nation of the Netherlands. However, some 40% of this trade is transit/re-export and does not contribute to Dutch economy (WRR 2003). In terms of added value, the share of Germany in Dutch export is 14%, but still remaining the main trade partner. Added value from trade with the USA and China strongly increased while this decreased for Belgium and Luxemburg (ING 2014). The ambition of Rotterdam port trade is to grow in terms of added value rather than in terms of just volumes, making Rotterdam a 'smarter port', not only a 'mass port'.

A spectacular growth in trade (import and export) took place since the 1960s as shown in Table 6.1. In 1880, the Dutch imported for about 70 € per capita, 220 € in 1920, 4900 € in 1980, and for about 22,600 € per capita in 2014. Total import in the Netherlands in 1880 was 0.3 billion € and 382.4 billion € in 2014. Total export grew from 0.3 billion € in 1880 to 433.4 billion € (nor corrected for inflation; CBS 2001; Table 6.1).

Although these growth rates are very high indeed, less spectacular growth rates apply for Dutch trade if calculated in terms of Gross Domestic Product (GDP: total gross added

Table 6.1 International trade of goods in the Netherlands, 1920–2014

	1920	1960	1980	2000	2014
Total imports	1.5	7.8	69.1	206.9	382.4
Imports from Europe	0.9	5.0	44.4	135.6	238.6
As %-age of total imports	62%	64%	64%	63%	62%
Total exports	0.8	6.9	66.7	231.9	433.4
Exports to Europe	0.6	5.2	55.1	196.0	341.4
As %-age of total exports	75%	75%	82%	84%	79%

Since 1980 export exceeds import. Goods in transit (e.g. ore transits in Rotterdam from sea-going vessels via river transport to German clients) are not included. All amounts are in billions €; no correction for inflation
Data: CBS Statline

value created in the Netherlands in 1 year time). The proportion of Dutch trade in goods (imports and exports) as a share of GDP was circa 120% in the 1880s and more than 70% in the 1920s. During the world economic crisis in the 1930s and during WWII that share dropped sharply, but successively rose again to about 90% in the 1990s and to more than 100% in the early twenty-first century. In sum: Dutch international trade strongly increased but GDP as well (WRR 2003).

In addition to import and export of *goods* as shown in Table 6.1, international trade in (non-physical) *services* developed in the Netherlands as well. Export of such services, expressed in terms of added value of services supplied by Dutch residents and companies to non-residents (foreign companies and persons) increases. Such services include international tourism to the Netherlands, telecommunications and computer services, business services as advertising and accounting offices, financial services by banks, et cetera. Values of such imported services reached a level of 147.9 billion € in 2015 and the value of exported services 157.9 billion €. That is almost 40% of the values of import/export of goods. Import and export values in Table 6.1 do not include international energy trades (mainly fossil fuels and electricity) which values were in 2014: 83.2 billion € for importing and 75.5 billion € for energy exports.

Main trade partners with the Netherlands are its neighbouring countries (see Table 6.2). But the same Table shows their declining shares over time, dropping from 53% in 1920 (import) and 54% (export) to 33% (import) and 44% (export) in 2014. Meanwhile, trade shares by other European nations increased due to elimination of national borders in the (expanding) EU. Beyond Europe, the USA remained the leading partner in trade, in particular for imports from the USA. China is upcoming as a prominent trade partner, mainly for imports (35.4 billion € in 2014) which were nearly five times more than in the same year for exports (7.9 billion €). This not only includes trade but investments as well (Box 6.1).

Table 6.2 The five main importing and exporting nations from and to the Netherlands, period 1920–2014

	Import	Export
1920	Germany 27.0%	Germany 24.5%
	United Kingdom 16.6%	United Kingdom 19.1%
	United States 15.7%	Belgium and Luxemb. 10.5%
	Belgium and Luxemb. 9.6%	France 7.9%
	Dutch Indies 7.2%	United States 4.5%
1960	Germany 21.6%	Germany 22.5%
	Belgium and Luxemb. 18.3%	Belgium and Luxemb. 14.3%
	United States 13.2%	United Kingdom 11.0%
	United Kingdom 6.9%	France 5.9%
	France 3.8%	United States 4.9%
1980	Germany 22.3%	Germany 29.9%
	Belgium and Luxemb. 11.6%	Belgium and Luxemb. 15.0%
	United States 8.8%	France 10.6%
	United Kingdom 8.2%	United Kingdom 7.9%
	France 6.7%	Italy 5.8%
2000	Germany 17.8%	Germany 25.7%
	United States 10.2%	Belgium and Luxemb. 11.8%
	United Kingdom 9.6%	United Kingdom 10.8%
	Belgium and Luxemb. 9.3%	France 10.6%
	France 5.7%	Italy 5.9%
2014	Germany 16.4%	Germany 24.1%
	Belgium and Luxemb. 9.7%	Belgium and Luxemb. 11.0%
	China 9.3%	United Kingdom 8.5%
	United States 7.1%	France 8.2%
	United Kingdom 6.6%	Italy 4.3%

Neighbouring countries (Germany, Belgium/Luxembourg, UK) belong to top-five in all years. Transit goods are excluded
Data: CBS Statline, Hoekveld 1989

Box 6.1: Foreign Investment in the Netherlands, Dutch Investments Abroad

International trade and international investment often go hand in hand. Below, some examples are given of foreign investors who acquired several well-known Dutch companies

- DOUWE EGBERTS, coffee and tea factory founded in Joure (Friesland) in 1753, Headquarters (HQ) moved to Utrecht, was acquired by Chicago-based SARA LEE in 1978; today owned by German investment company JAB; current HQ in Amsterdam.
- VERKADE, biscuits and chocolate factory, started in 1886 in Zaandam (near Amsterdam), taken over by British United Biscuits in 1990 and, in turn, by Turkish YILDIZ Holding (Istanbul) in 2014.

(continued)

Box 6.1 (continued)

- SPHINX pottery factory, founded in 1841 in Maastricht, taken over in 1999 by Finnish company SANITEC (HQ in Helsinki), stopped production in Maastricht (Sphinx only has a showroom there). Since 2015 has Sanitec with Sphinx been taken over by Swiss GERBERIT Group, with HQ in Rapperswil-Jona.
- V&D department stores, started with one shop in Amsterdam in 1887 by Vroom and Dreesmann; in 1972, V&D became Vendex that grew into a conglomerate of retail chains; Vendex was sold to Miami-based private equity firm Sun Capital in 2010, that discontinued investing in the loss-making, traditional department store in 2010; bankruptcy forced 62 V&D Department stores to close in 2016.
- HOOGOVENS, steel plant and blast-furnaces in IJmuiden, founded in 1918 along the Dutch coast - convenient for import of iron ore and coal. In 1972, Koninklijke Hoogovens merged with German company HOESCH and regained independency in 1982. In 1999 Hoogovens merged with British Steel, Anglo-Dutch group was re-named CORUS. In 2007, CORUS was taken over by Indian TATA Steel.
- LIPS ship's propellers, founded in 1934 and stablished in Drunen (Province of Noord-Brabant) in 1939 by Max Lips, in 2002 bought by Finnish multinational WÄRTSILÄ (HQ in Helsinki). In 2010 this company closed the Drunen plant, propeller factory moved to China; Lips has been the germ of today's 'Metal Valley' in Drunen.

For many decades, there were no political objections against selling Dutch firms to foreign investors, but in recent years Dutch politicians became more reserved against taking overs from abroad, especially in the case of leading Dutch 'flagship' firms: for instance, PostNL (potential buyer: BPost, the Belgium Post Group) in 2016, Unilever (by Kraft Heinz, Chicago) and AkzoNobel (by PPG, Pittsburgh) in 2017.

Direct foreign investments may comprise real estate, offices, shopping centres and dwellings as well. Prime targets are offices and shops in top-locations, as the Zuidas in Amsterdam and Hoog-Catharijne in Utrecht. Rental incomes have been economically profitable.

Box 6.1 (continued)

Until 2013, foreign shares in office investments were less than 50%, but most recently foreign investors dominate Dutch office markets: their current share is about 75% (Het Financieele Dagblad 2016). In addition, foreign multinationals invest much in new distribution centers, headquarters and production halls and in buying existing firms. In the world list of countries receiving most foreign investments the Netherlands ranked 11th in 2011. Conversely and simultaneously the Netherlands were ranked 9th on the list of countries most investing abroad.

UNCTAD data tell that the Netherlands received 55 billion dollars in direct foreign investment by multinationals in 2000 which was 4.3% of all foreign direct investment worldwide in that year and invested 73 billion dollars abroad (6.4%). These investments rates are quite substantial as compared to the share of the Netherlands in world-GDP (nearly 1%). In 2000, most investments in the Netherlands were from multinationals in EU countries and the United States (Fig. 6.2). Conversely, Dutch multinationals invested in these countries as well. As to sectors, industrial investments have become less prominent while the service sector gained in relevance. In 1986, industry still dominated international investments, since 2000 the service sector was leading (WRR 2003).

Chinese, South-Korean and Arab investors became prominent apart from traditional investors in the twenty-first century. Korean investors bought the iconic skyscraper 'De Rotterdam' along river Nieuwe Maas in Rotterdam from Dutch Rabobank in 2016. In the same year, Chinese multinational COFCO became owner of Rotterdam-based NIDERA, an international agribusiness and trading company with a turnover of 18.5 billion US dollars in 2015 (www.nidera.com).

Likewise, Dutch investments abroad have been numerous. As a consequence, 40% of international trade is *intrafirm*, between companies of the same multinational (WRR 2003). 15% of the Dutch labour force is employed in a Netherlands-based foreign company. Foreign companies contribute 33% to Dutch GDP, according to data of CBS and the Ministry of Economic Affairs.

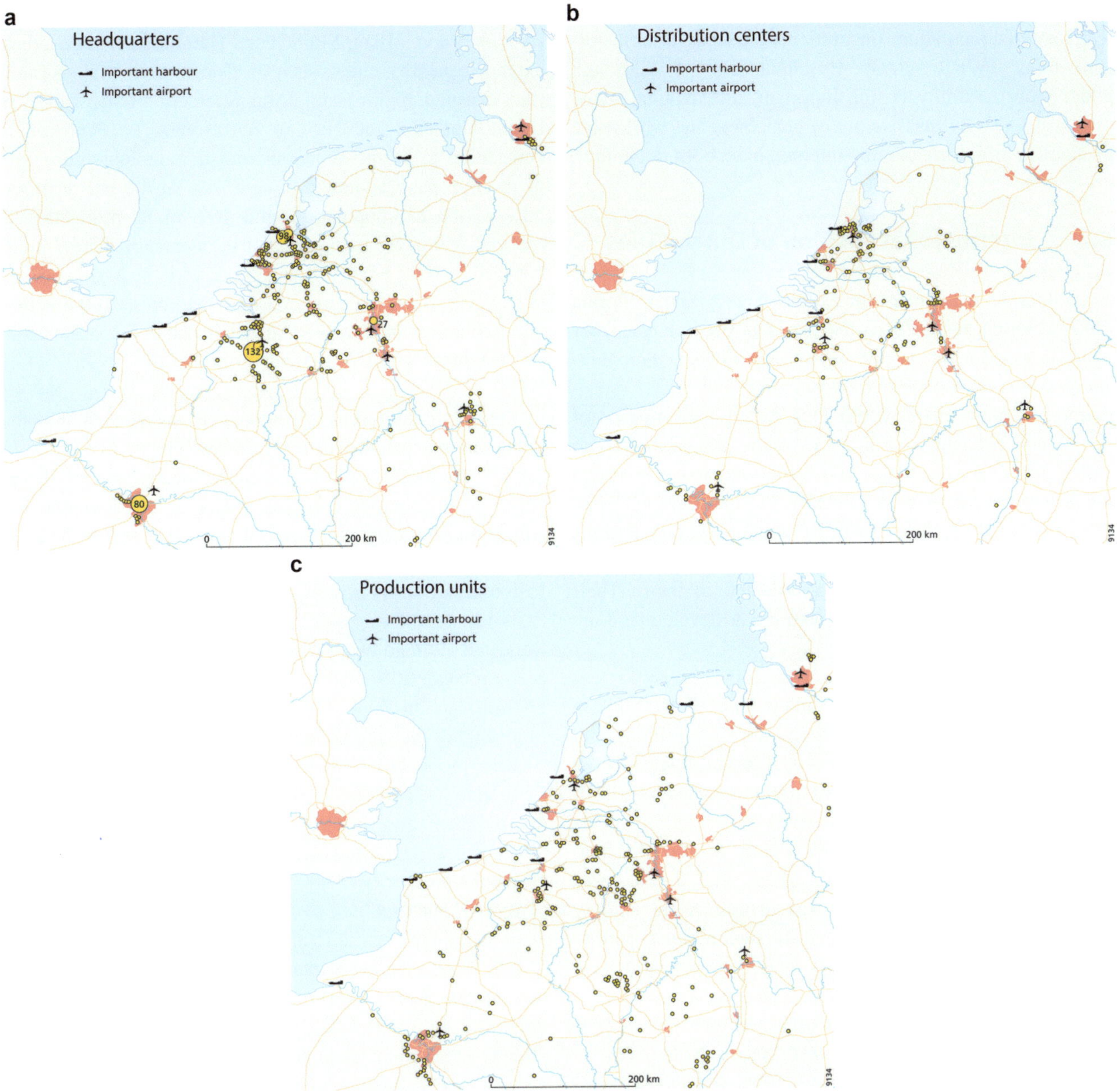

Fig. 6.2 Spatial distribution of American and Japanese firms in NW Europe; Headquarters (**a**), distribution centers (**b**) and production units (**c**) in 1995. (Source: Ministerie van Economische Zaken 1997)

This pattern fits well in the 'gravity model' which may predict trade flows between nations (WRR 2003). This model is based on research by Frankel and Rose (2002; see also Dekker et al. 2007) assuming that trade between two nations is directly proportional to their economic seizes (in GDP) and inversely proportional to their mutual distances. The richer, the more trade; the larger distances, the less trade. By applying an extended version of this gravity model Frankel and Rose were able to explain more than 60% of the trade variances among 186 countries. Dutch trade flows fit well in this model. This holds for nearby economic powerhouse Germany, with nearby Belgium and Luxemburg and with a significant partner in Dutch trade, Great Britain where the North Sea connects rather than separates both nations.

Geographic proximity thus proves to be an important parameter in this model although the true causes for this relevance remain unexplained. Are transport costs a decisive factor or rather acts geographic proximity as a proxy for corresponding values and norms (cultural and social proximity), for equal regulations (institutional proximity), for historical links (WRR 2003; Boschma 2005)? Although geographic proximity is still important, international transport costs, also by massive use of containers and worldwide availability of information by internet, have been declining.

6.2 International Position of Dutch Cities

According to the American economic geographer Florida (2007; Florida et al. 2008) the world is becoming a *spiky world*: a world where people, capital, knowledge, companies and power accumulate in *urban mega-regions*. Such a mega-region is a more or less continuous spatial concentration of people and companies with one or a few metropolises with annual turnovers of at least 100 billion dollars/year around the turn of this millennium. Globally, Florida identified some of such 40 mega-regions, twelve of which in Europe. Realms beyond such mega-regions deteriorate as more and more peripheral features are becoming adhered to them. Their shares in economy, population, research & development are declining.

In the world ranking of the year 2000, the top-3 was Greater Tokyo (with a gross regional product of 2500 billion dollars per year), Boston-Washington (2200 billion) and Chicago-Pittsburgh (1600 billion). Florida assigned the region 'Am-Brus-Twerp' with 60 million people and a GRP of 1500 billion dollars, at 4th position. This region includes Randstad Holland, the Rijn-Ruhr region, the Flemish diamond (Antwerpen-Gent-Brussel-Leuven) and the northern French cities of Lille and Roubaix. This ranking entirely depends on the boundaries chosen for such regions. Further development of this entire urban mega-region would have impact on the future of Randstad Holland and vice versa.

More than 50 years ago the Dutch *Second National Spatial Planning Policy Document* (Tweede Nota over de Ruimtelijke Ordening in Nederland 1966; see also chap. 8) recognized the existence of this vast urbanized area, east of the North Sea. This report not only considered the Am-Brus-Twerp region but the London area and its surroundings (SE England) to belong to this urban zone which encompassed 12% of the European population and covered 2% of the European territory (excluding the USSR) in the early 1960s. The authors of this Second National Spatial Planning Policy Document considered the east and north of the Netherlands beyond this concentration area and would require financial aid and regional support.

Today, spatial planners at EU level think in even larger spatial dimensions than Florida last decade and 1960s Dutch planners. According to them the Am-Brus-Twerp area is part of the so-called *Pentagon*, a pentagon with corner points London, Paris, Milan, Munich and Hamburg. That area may extend to nearby cities, such as Copenhagen-Malmö-Lund (the Øresund region) and from Southern French cities as Lyon, Marseille and Nice to Barcelona. The Pentagon is considered to be the economic and demographic core area of the European Union: it covers 14% of the EU territory, 33% of the EU population, and 46% of its production is realized within the boundaries of the Pentagon. These numbers are growing while, in accordance with the Myrdal model (Chap. 10), shares of rural areas are declining. In the Pentagon, including the Dutch Randstad Holland, a concentration process takes place (European Commission 1999; ESPON 2006).

Spatial distribution of prosperity is unequal. Within the Pentagon less privileged, rural regions occur. According to the EU, people living in such geographically less privileged zones deserve support e.g. in terms of employment and education (the ideal of 'territorial cohesion' within the EU). Table 6.3 lists the top-20 metropolitan areas in the European Union based on their sizes of local economies. London and Paris are numbers 1 and 2 while the Ruhr area, Milan and Randstad Holland occupy positions 3–5. Metropolitan areas accommodate 20% of the EU-population and 31% of EU's total production value. Metropolitan GRP per capita is >1.5 times higher than the EU average.

But production differs strongly among metropoles. Economies in northern European cities are much more productive than those in the south and east. Such inequalities also occur among cities within individual nations. Randstad Holland's prosperity (39,000 € per capita) is in line with that of the EU metropolitan average (38,800 €). But inequality also occurs between individual Randstad Holland cities. Its 'South Wing' (e.g. Rotterdam agglomeration) is less prosperous (34,800 € per capita) than its 'North Wing' (Amsterdam agglomeration) with 42,200 €, or the Utrecht agglomeration: 49,800 € (Manshanden et al. 2014).

6.2.1 Dutch Cities in European Functional Hierarchy

Apart from economic ranking orders (e.g. GRP, company HQs) others, based on a broader set of indicators exist. One of these is the functional urban hierarchy. Figure 6.3 is based on data for 38 urban function indicators (grouped in five dimensions) in 8480 European cities. Cities are thus determined along economy, politics/governance, science, transportation and culture. For these urban parameters Randstad Holland ranks third, following London and Paris, but before Brussels (fourth) and the Rhine-Ruhr cities (fifth). According

Table 6.3 Twenty metropolitan areas in the EU as ranked by their gross regional product (GRP) in 2012

Urban region	Gross regional product (GRP), in billions €	Population in millions	GRP per capita, in thousands €
London	642	14.9	43.2
Paris	620	11.9	51.9
Ruhr area	399	11.5	34.5
Milan	277	7.7	36.0
Randstad Holland	274	7.0	39.0
Madrid	189	6.4	29.4
Flemish Diamond	184	4.3	42.3
Frankfurt/Main	182	4.2	43.3
Berlin	144	5.0	28.6
Barcelona	141	5.3	26.5
Rome	139	4.3	32.6
Hamburg	133	3.3	41.0
Stockholm	124	2.1	58.6
Munich	119	2.1	57.7
Vienna	102	3.0	33.7
Dublin	84	1.8	47.3
Copenhagen	75	1.2	60.9
Lisbon	67	3.2	20.8
Prague	40	1.3	31.8
Budapest	38	1.7	21.8
The total of 20 urban regions	3973	102.3	38.8
EU 28	12,988	508.4	25.5

Also listed are their population and GRP per inhabitant. Source: modified after Manshanden et al. (2014)
Data: Eurostat and CBS Statline

the same criteria, Eindhoven (the former company town of Philips) ranks 43rd and Groningen (the regional capital of the north) 104th. Most such high ranking cities are situated in the northwest, the European Pentagon. Such rankings are all relative and strongly depend on the criteria and geographical boundaries chosen and the attributed weights to each of the parameters. For example, according the Dutch *Atlas Metropoolregio Rotterdam Den Haag* (2014), Randstad Holland's North Wing (with the Amsterdam urban agglomeration) alone would rank 7th and its South Wing (Rotterdam and Den Haag) would rank 13th in this case – much lower than Randstad Holland as a whole (third) (Fig. 6.3).

6.2.2 Amsterdam as World City and a Global City

Research on hierarchies of cities is popular in the UK as well, e.g. by the Globalization and World Cities (GaWC) Research Network. They have an extensive database of 525 cities worldwide. From this database, geographers calculated two indexes (Taylor et al. 2011). First the *city place power* index, dominated by parameters as multinationals HQs, business hotels and business events, comparable with that of Friedmann (1986) on *world cities*. The second index is on *city-network-power* which refers to the *global city* concept (Sassen 1991). That concept includes presence of high-quality service providers in business and banking and the media in a city. Amsterdam ranks 25th on the city-place-power list, i.e. behind (in Europe) London, Paris, Madrid, Stockholm and Moscow, but before Rome, Munich, Zurich, Brussels, Milan and Helsinki. On the city-network-power list Amsterdam is 23th, behind (in Europe) London, Paris, Milan, Madrid, Moscow and Brussels, but before 14 other European cities in the top 50. No other Dutch city occur in the top 50. In conclusion, as a world city and a global city Amsterdam ranks far higher than any other Dutch city. In Europe Amsterdam is seventh on the combined index of city-place and city-network power, worldwide twentieth (Taylor et al. 2011).

6.3 The Netherlands in the International Political Arena

The standing ambition of the Dutch Olympic Committee is a top 10-position. That ambition was realized at the 2000 Summer Olympics in Sydney (8th position) and at the Winter Olympics in Sochi in 2014 (5th) and in Pyeonchang in 2018 (also 5th), but otherwise the Netherlands reach positions

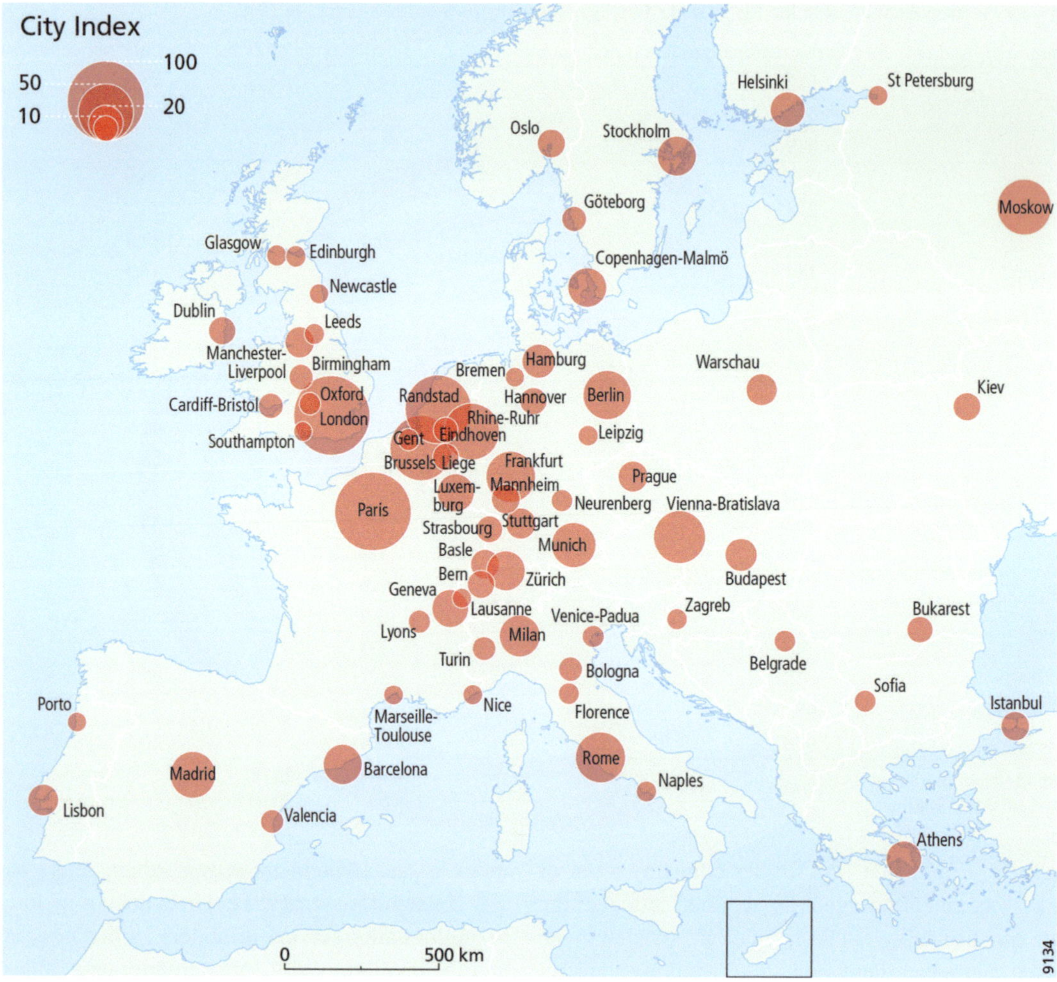

Fig. 6.3 Randstad Holland and 69 other metropoles with the highest functional level in economy, politics/governance, science, transportation and culture. (Source: BBSR 2011)

11–15th. This ambition well reflects the self-image of the Dutch: to be a medium-sized payer on the world podium, and being one of the world's largest small countries. From an analysis in 1979, the later Dutch Defense Minister Voorhoeve concluded that the self-image of the Netherlands is of 'an influential, developed Small Power', a geographically small country but 'yet rather influential, perhaps not in international confrontation but certainly in international cooperation'. In the real world, however, this self-image often appears to be slightly too optimistic.

This positive self-image dates from the Dutch Golden Age (the seventeenth century), when the Netherlands was an economic powerhouse (Fig. 6.1) and from the late nineteenth century when the Netherlands still was a significant empire with large colonies in SE Asia (the present Indonesia) and a smaller one in Latin America (Surinam). An impression of the former size of the Dutch SE Asia colonial empire is given in Fig. 6.4. Awareness among the Dutch that maintaining power in these former colonies has not been without political suppression and violent acts of war, in particular shortly before Indonesia was recognized as an independent nation (1949) came late and is relatively recent.

6.3.1 Three Traditions

Traditionally, the Dutch are champions for world-wide free trade based on their strong trading history. That explains that the Netherlands generally support economic liberalism and, at least for some time, political neutrality. In a political-geographical context, the (maritime provinces of Holland in the) Netherlands adhered to a commercial, maritime, free Atlantic environment rather than to a continental-European environment of protectionism and central authority (Hellema 2009). Voorhoeve (1979) distinguishes three long diplomatic traditions in Dutch foreign policy prior to WWII:

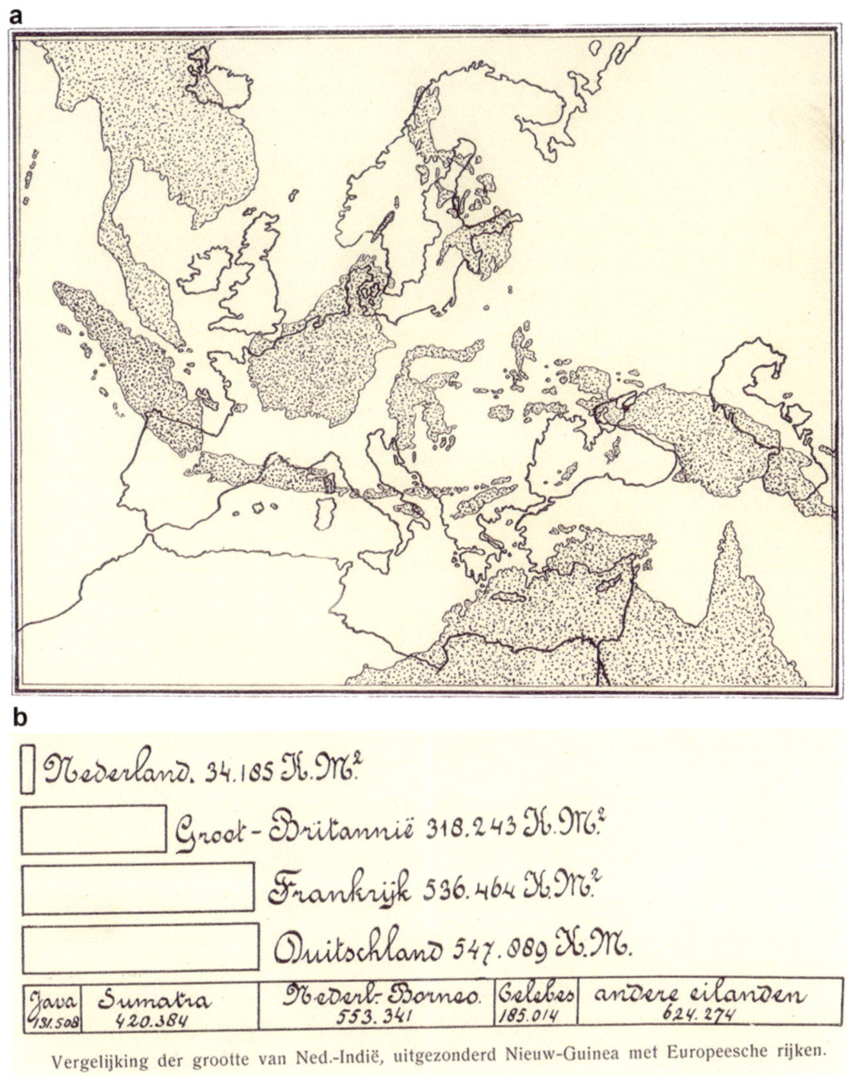

Fig. 6.4 The Netherlands as a global power, illustrated by the size of the colonial possessions of the Dutch in Southeast Asia in maps and charts. (Source: Blink 1914a)

1. The *maritime-commercial tradition*, characterized by commercial freedom, anti-protectionism, free trade, non-intervention, and 'a strong reliance on treaties and orderly legal regulations, an aversion to nationalism, and a craving for international order and peace'. This tradition is strongly linked with its geographical position on the North Sea delta which made 'the Netherlands a natural commercial mediator for NW Europe'. Consequently, all attention and interest was focused on trade cities in the west of the Netherlands, in particular on Amsterdam. 'The eastern and southern provinces' were regarded in this tradition 'as buffers meant only to keep the German and French neighbours out of the blossoming Dutch garden' (Voorhoeve 1979).
2. The *neutralist-non-alignment tradition,* characterized by a self-chosen non-alignment, non-involvement foreign policy and an aversion against power-politics. Based on this tradition and on the initiative of the Russian Tsar, Den Haag was chosen as the site for an international peace conference in 1899. A second peace conference followed in 1907 and in 1913 the Peace Palace - thanks to a gift of the American industrialist and philanthropist Andrew Carnegie – was opened in Den Haag. This Peace Palace hosted the Permanent Court of International Justice in 1922, renamed as the International Court of Justice. Since 1990, Den Haag coins itself as the World's Legal Capital and as the International City of Peace and Justice. This branding will have stimulated the International Criminal Court to be based in this city in 2002 (Van der Wusten 2006; Fig. 6.5).
3. The *international-idealist tradition*, characterized by pragmatist pacifism, 'reinforced by a more principled, religious pacifism based on the Humanist and Christian traditions', legalism, moral qualities, non- or anti-militarism, weak patriotism and strong supra-nationalism, and 'mundialism' (defined by Voorhoeve (1979, Chap. 3) as 'the inclination of the Dutch to identify themselves less with the nation or the

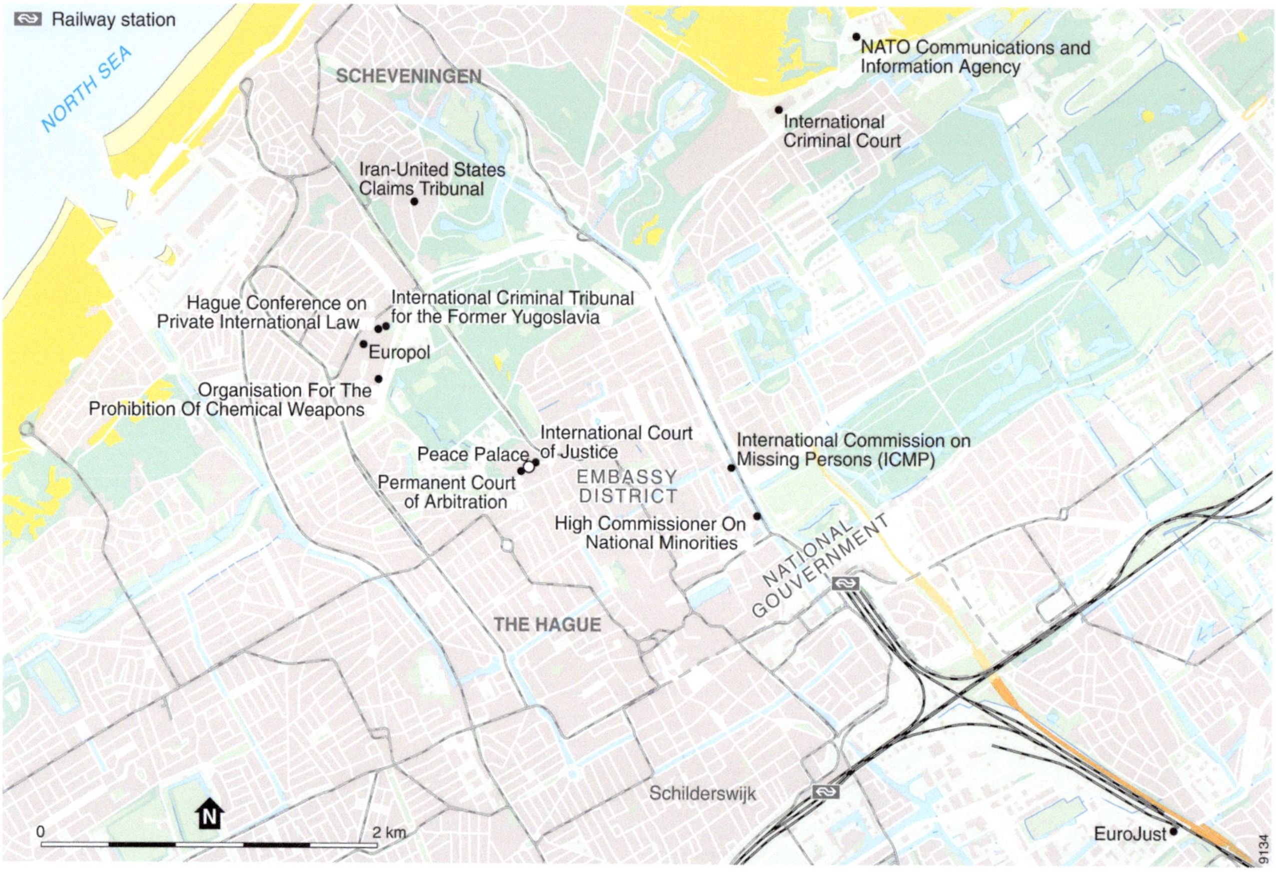

Fig. 6.5 The location of the most important international political institutions in Den Haag, the embassy district and the national government area. (Source: modified after Van der Wusten 2006 and www.thehaguepeacejustice.com)

region and more with mankind as a whole'). Participation of Dutch soldiers in global UN peacekeeping missions, support to poor, Third World nations and Den Haag's ambitions to be the center of international law all fits in this tradition. These traditions, often in combination, played a dominant role in Dutch history.

6.3.2 Neutrality and Non-alignment Until 1940

Non-alignment ruled in Dutch foreign policy since 1839 when Belgium left the Netherlands and became an independent nation (Voorhoeve 1979). Growing rivalry among and nationalism in European nations jeopardized this position of aloofness. Moreover, the geographical location of the Netherlands between two superpowers – United Kingdom and Germany – did not help to keep its neutrality neither. Based on their strong trading traditions the Dutch generally inclined to adhere more to England's maritime tradition however but the mainly continental power Germany became economically more important for the Dutch since the end of the nineteenth century (Hellema 2009).

The Netherlands managed to keep its neutral position during WWI. Non-interference and non-alignment remained the Dutch strategy when WWI was over and the Dutch government tried to mediate between former enemies, e.g. between scientists of both camps by the prestigious Dutch Royal Academy of Sciences (Otterspeer and Schuller tot Peursum-Meijer 1997). However, Dutch non-alignment policy became unsustainable in perspective of the forthcoming WWII. The Netherlands was attacked by German armies in May 1940. After 5 days of battle and German bombing of Rotterdam the Dutch government surrendered and the Netherlands suffered from 5 years of German occupation.

6.3.3 From Neutrality to International Cooperation

Upon liberation in 1945 the Dutch abandoned its long-standing tradition of neutrality and non-alignment and joined the multilateral Northern Atlantic Treaty Organisation (NATO) together with other western European nations. Europe was split in a communistic and a capitalistic block and the 'Cold War' began, manifested by the 'Iron Curtain'.

Western Europe adhered to the democratic, more liberal and more capitalistic world protected by the United States of America. In this situation the longstanding policy of neutrality had become an illusion (Hellema 2009). Voorhoeve (1979) distinguished three major themes in Dutch international policy: international security (NATO), regional economic (and political) integration (EEC, later EU) and worldwide cooperation (UN).

International Security The Netherlands belonged to the 10 European countries and the United States of America and Canada that founded NATO in 1949, in Brussels. Although the Dutch and the USA were allies since WWII, this relationship was jeopardised by tensions about the former Dutch East Indies. In contrast to the Netherlands, the USA favoured rapid independency of the Republic of Indonesia. The Dutch faced considerable pressure before they recognized sovereignty of Indonesia in December 1949.

Regional Economic (and Political) Integration The Netherlands has been one of the six founding members of the European Coal and Steel Community (ECSC), together with Belgium, Luxembourg, West Germany, France and Italy, in 1951. The ECSC aimed to regulate coal and steel production among these nations, in particular between former arch enemies France and Germany. By then, coal and steel were prime commodities for war industry and their supranational regulation should prevent any new war. The ECSC was a precursor of the European Economic Community (EEC), founded in 1957 in Rome. Apart from economic cooperation, primarily in agriculture and abolishment of import duties (Customs Union), these six countries agreed to cooperate on nuclear energy production (European Atomic Energy Community, EURATOM). From then onwards fields and numbers of cooperating European nations expanded. In 1993, the EEC transformed into the European Union (EU).

In its internationally oriented policy the Netherlands used to favour admission of more countries to the EU and to transfer more political power to the EU (HQ in Brussels), if controlled by an elected supranational European Parliament. The Dutch people, however, have been and still are more reluctant to the European ambitions of Dutch diplomats to transfer powers from national to European parliaments. 'Eurobarometer' registered in the 1970s, the heydays of European cooperation, that almost half (49%) of the Dutch population wanted EU member states to have final decision power, not the EEC (Voorhoeve 1979). Euroscepticism and nationalism among the Dutch population has grown significantly since as expressed by negative results (for the EU) of two referendums about European issues and rapid growth of eurosceptic political parties.

Such nationalistic sentiments became more popular in the Netherlands also due to increasing concerns about negative effects of globalisation of the economy, poorly controlled non-western immigration, terrorism, increasing awareness and signals of racism, et cetera. This fueled current impressions of losing control over national issues in policy by Dutch people.

World-Wide Cooperation The Dutch pre WWII internationalist-idealist tradition continued in post-war support for creation of the United Nations, successor of the failed League of Nations, which was founded shortly after WWI to prevent new wars around the world. The Netherlands belonged to the 14 nations on the Executive Committee that chaired the United Nations Conference on International Organization in San Francisco in April–June 1945 (Voorhoeve 1979) and the Netherlands belonged to the 51 nations that signed the UN Charter. This idealism is reflected e.g. in the relatively high financial contribution of the Netherlands to foreign aid and international development cooperation. Together with some Scandinavian countries the Dutch spend at least 0.70% of its GNP for this purpose during decades. Another indication for this internationalist-idealist tradition is the more than average participation of the Dutch in UN peace keeping missions. In 2016, the Dutch were actively involved in 13 such operations ranging between 5 and 400 soldiers each. During the past decades the Dutch international-idealist tradition has been more prominent than its neutralist-non-alignment tradition.

References

Atlas Metropoolregio Rotterdam Den Haag (2014) Versie 2. Gemeenten Den Haag en Rotterdam

BBSR (2011) Metropolitan areas in Europe. BBSR, Bonn

Blink H (1914a) Economische ontwikkeling van Nederlandsch Oost-Indië gedurende de laatste eeuw. Tijdschrift voor Economische Geographie 5(VI):193–320

Blink H (1914b) Ontwikkeling van de economische toestanden in Nederland en zijn betrekkingen met de nabuurlanden gedurende de laatste eeuw. Tijdschrift voor Economische Geographie 5 (IX):345–353 5(X):369–384, 5(XI):416–434

Blink H (1915) De ontwikkeling der economische toestanden in Europa en in Nederland van 1870 tot 1914. Tijdschrift voor Economische Geographie 6(IX):266–276 289–299, 329–344

Boschma R (2005) Proximity and innovation. A critical assessment. Reg Stud 39(1):61–74

Braudel F (1977) Afterthoughts on material civilization and capitalism. Johns Hopkins University Press, Baltimore

Braudel F (1982) Civilization and capitalism 15th–18th century, volume II: The Wheels of Commerce. Fontana Press, London

Braudel, F (1984) Civilization and capitalism 15th–18th century, volume III: The perspective of the World. Fontana Press, London

CBS (1931) Statistiek van de scheepvaart op het buitenland over 1930. Centraal Bureau voor de Statistiek, Den Haag

CBS (2001) Tweehonderd jaar statistiek in tijdreeksen 1800–1999. Centraal Bureau voor de Statistiek, Voorburg/Heerlen

Dekker P et al (2007) Diverse Europe. Public opinion on the European Union & Cultural diversity, economy and policy. European Outlook 4. Centraal Bureau voor de Statistiek/Sociaal en Cultureel Planbureau, Den Haag

ESPON (2006) ESPON Atlas. Mapping the future of the European territory. Bundesamt für Bauwesen und Raumplanung, Bonn

European Commission (1999) ESDP. European Spatial Development Perspective. Office for Official Publications of the European Communities, Luxembourg

Florida R (2007) Who's your city? Basic Books, New York

Florida R, Gulden T, Mellander C (2008) The rise of the mega-region. Camb J Regions Econ Soc 1(3):459–476

Frankel J, Rose A (2002) An estimate of the effect of common currencies on trade and income. Q J Econ 117(2):437–466

Friedmann J (1986) The world city hypothesis. Dev Chang 17(1):69–83

Hellema D (2009) Dutch foreign policy. The role of the Netherlands in World politics. Republic of Letters Publishing, Dordrecht

Het Financieele Dagblad (2016). 18 August 2016

Hoekveld GA (1972) The Netherlands in Western Europe. Tijdschr Econ Soc Geogr 63(3):129–148

Hoekveld GA (1989) Nederland, deel van de wereld. In: De Pater BC, Van Ginkel JA, Hoekveld GA (eds) Nederland in delen; een regionale geografie, vol 1. De Haan/Unieboek, Houten, pp 17–39

ING (2014) Nederland handelsland. Herziening ranglijst belangrijkste exportlanden voor Nederland. http://www.ing.nl

Klemann H (2016) Europa's grootste haven en haar toekomst: Rotterdam 1870-2000. Holland historisch tijdschrift 48 (3–4):157–164

Manshanden WJJ, Bouman-Eijs JM, Koops R, Fischer NM, Chahim M, Rietveld E (2014) De top 20 van Europese grootstedelijke regio's 1995–2012. Randstad Holland in internationaal perspectief. TNO, Delft

Ministerie van Economische Zaken (1997) Ruimte voor economische dynamiek. Een verkennende analyse van ruimtelijk-economische ontwikkelingen. Ministerie voor Economische Zaken, Den Haag

Otterspeer W, Schuller tot Peursum-Meijer J (1997) Wetenschap en wereldvrede; de Koninklijke Academie van Wetenschappen en het herstel van de internationale wetenschap tijdens het Interbellum. Koninklijke Nederlandse Academie van Wetenschappen, Amsterdam

Sassen S (1991) The global city: New York, London, Tokyo. Princeton University Press, Princeton

Taylor PJ, Ni P, Derudder B, Hoyler M, Huang J, Witlox F (eds) (2011) Global urban analysis. A survey of cities in globalization. Earthscan, London

Tweede Nota over de Ruimtelijke Ordening in Nederland (1966) Staatsuitgeverij, Den Haag

Van der Kooy TP (1931) Hollands stapelmarkt en haar verval. Dissertation, Nederlandse Handelshogeschool Rotterdam

Van der Wusten H (2006) 'Legal capital of the world': Political centre-formation in The Hague. Tijdschr Econ Soc Geogr 97(3):253–266

Voorhoeve JJC (1979) Peace, profits and principles: a study of Dutch foreign policy. Martinus Nijhoff, Den Haag

Wagenaar M, Van Engelsdorp Gastelaars R (1986) The rise of the 'Randstad', 1815-1830. In: Schmal H (ed) Patterns of European urbanization since 1500. Croom Helm, London, pp 231–246

WRR (2003) Nederland handelsland. Het perspectief van de transactiekosten. SDU, Den Haag

Zijlmans R (2017) Troebele betrekkingen. Grens, scheepvaart- en waterstaatskwesties in de Nederlanden tot 1800. Verloren, Hilversum

Economic Geography

7

Ben C. De Pater

Abstract

The Netherlands ranks fourth on the list of the world's most competitive economies, after Switzerland, the United States of America and Singapore, and directly followed by Germany (5). This index of the *Global Competitiveness Report 2016–2017* is based on dozens of criteria, arranged in 12 'pillars'. The Netherlands has top 10-rankings for 7 out of the 12 pillars: infrastructure (3rd), health and primary education (4th), higher education and training (3rd), market efficiency for goods (8th), technological readiness (6th), business sophistication (5th) and innovation (7th).

However, this economic competitiveness is distributed quite unevenly over the nation. More than the half of its Gross Domestic Product (GDP) is produced in just 20% of its territory: the Randstad Holland, the Dutch economic core region. An economic upward area is the south, in particular the region of Eindhoven with significant high-tech industry plants. The north is in a more peripheral economic position. Regional policies, initially by the national government and today by the European Union aim at reducing such regional inequalities in prosperity.

Keywords

Regional economic inequality · Randstad holland · Amsterdam as financial centre · Headquarters of companies · Airport schiphol · Eindhoven · Agriculture · Regional policy · Economic clusters

Economists tend to think in terms of national economies. That inclination has been criticized by Jacobs in her book *Cities and the Wealth of Nations* (Jacobs 1984). States are political units, no economic units. If economists would abandon such nationalistic concepts of economies they would (according to Jacobs) discover that 'most nations are constituted of collections or grab bags of very different economies, rich regions and poor ones within the same nation'. But Jacobs failed to recognize that national governments may indeed influence the economic integration and differentiation of regions, e.g. by building infrastructure and by combating inequality. Nevertheless, economic structure and prosperity in almost all countries, including the Netherlands, is quite diverse from region to region. Traditionally, such regional economic varieties were associated with different physical-geographical conditions, with their natural resources contents ('*site* characteristics') and with their geographic location vis-à-vis other areas ('*situation*-characteristics'). The broadest regional distinction applicable to the Netherlands is between maritime and continental regions (Sect. 7.1).

Another, more detailed way to discern regional differences is based on variety in (economic and political) power: core, semi-periphery and periphery. From this approach it appears that part of the maritime regions, the west with Randstad Holland, have become the core region of the nation (Sects. 7.2 and 7.3). Randstad Holland dominates in economic, demographic, in political and in cultural ways. The standard Dutch language that replaced all dialects in the country is rooted in Holland (Box 12.3) and the term Holland which originally refers to the west of the country is often applied (by non-Dutch) to the Netherlands as a whole (Box 3.3). The continental regions in the south (Noord Brabant and Limburg) and in the east of the Netherlands (Gelderland and Overijssel) broadly belong to the semi-periphery while the maritime regions in the southwest (Zeeland) and in the north (Groningen and Friesland, together with continental Drenthe) are mainly a part of the periphery. This is discussed in Sects. 7.4 and 7.5. The attempts initially by the Dutch government and later by the European Union to reduce such regional disparities are described in the last Section of this Chapter (7.6).

B. C. De Pater (✉)
Department of Human Geography and Spatial Planning, Utrecht University, Utrecht, The Netherlands
e-mail: B.C.dePater@uu.nl

E. F. J. De Mulder et al., *The Netherlands and the Dutch*, World Regional Geography Book Series,
https://doi.org/10.1007/978-3-319-75073-6_7

7.1 Maritime and Continental Regions

In his book *A Geography of Europe* (Gottmann 1950) Gottmann distinguished two types of Europe: maritime and continental nations. Maritime nations are situated along Europe's seashores while continental nations are land-locked and in Europe's centre and east. Populations in such nations tend to have different mind sets: people in maritime nations tend to be more open minded and trade oriented than people in continental nations. But much of this is mere speculation. Perhaps continental nations more often face challenges linked to boundaries, custom controls and they always have to rely on other nations for trade while the opposite seems to be true for maritime nations where traffic over water may have been easier and freer while facilitating communication with foreign people. This theme was further elaborated by Fox in his book *History in Geographic Perspective: The Other France* (Fox 1971). Fox distinguished two types of geographical areas, one consisting of regions and cities along sea coasts and navigable rivers, the other without such water transport facilities. Seaport and river port cities were *network cities* (Hohenberg and Lees 1985) or *gateways* (Bird 1983).

Before railway times travelling over land had been relatively easy as compared with travelling overseas, but the opposite was true for transport of goods (Fox 1971). Transport of large volumes of bulky material over water was relatively easy and cheap by then, in contrast to sending messages or transporting soldiers over sea which was easier over land. That made inland territories relatively easy to control but over land transport of goods was expensive. Thus inland cities were good in providing services and facilities, as markets, schools, hospitals, shops and governmental bodies, to their (agricultural) hinterlands. Christaller (1933) referred to such cities as *central places* in his study on spatial distribution of cities in southern Germany, a typical continental area. The same term was used by Hohenberg and Lees (1985). This situation changed after introduction of railway transport and travelling and transportation by train became relatively easy and central places got network functions as well.

From this dual distinction between transport versus travel and land versus sea/river sites two quite different types of societies and economies emerged: (1) areal or territorially organised agrarian cities and states on the one hand, and (2) rather commercially organised port cities with linear sea or river routes on the other. Both types occur at European and at national scales. Similarly, Fox (1971) identified two different France's, each with their own world view and culture: the continental France, an 'agrarian kingdom' with Paris as capital, and the 'maritime republics' at sea or along navigable rivers, as Bordeaux, Nantes, Brest, Rouen, Strasbourg, Lyon and Marseille. A comparable distinction may be made for regions in the Netherlands as well. This distinction was caused by both historical differences and geographical boundaries. However, by today such distinctions have become much less pronounced than in the past. Regional, cultural differences will be discussed in Chap. 12 and economical differences here.

Box 7.1 Large or Small Regional-Economic Differences?

In debates on regional differences in prosperity and production in the Netherlands two view prevail. Experts representing the first view consider such differences modest. Others observe these significant and growing, they consider inequality and uneven distribution strong. Both views are based on political inclination, on the set of indicators applied (as income per capita or per household, unemployment, GRP), and on selected levels of spatial scale. As to spatial scales, five macro-regions may be distinguished (north, east, south, west and southwest), 12 provinces, 40 'COROP-regions' (labour-market areas), the current 388 municipalities (in 2017; 1120 in 1916), and 4051 postal code regions. Criteria meanings are reciprocal to chosen spatial scales. Average scores may hide major internal differences: differences at local levels, e.g. between rich and poor municipalities, are larger than at regional levels (e.g. between provinces).

According to economic geographer Wever (1993), quality differences between regional production environments are quite small in times of advanced means of transport and ICT. According to him, *Achterhoek* (a peripheral area)-based high-tech companies do not suffer from the lack of an university in this specific area. There are three within a radius of just half an hour drive from the Achterhoek. Human geographer Tamsma (1972) points to the modest differences in personal incomes in the peripheral north of the country which were 87% of the national average (and 91% per taxpayer) in 1965 while housing is much cheaper there than in the west. Such percentages remained largely unchanged since (Ministerie van Economische Zaken 2006). According to Tamsma: the north of the Netherlands is (perhaps) a large problem area in a small country, but also a small problem area in a large community. In European perspective regional variation in welfare and prosperity within the Netherlands is modest.

The other group of experts highlight such regional differences. Randstad Holland is the 'engine of the Dutch economy', beyond dominates stagnation. The

(continued)

Box 7.1 (continued)
(large) city is seen as an 'engine of innovation'. More urban dwellers lead to higher population densities and more proximity, more diversity and faster exchange of knowledge and ideas, more creativity, innovations, productivity and higher salaries.

Urban planner Hemel propagates megacities: 'Amsterdam should double its size and become a metropolis of two million inhabitants; that is inevitable in its competition with world cities as London and Paris' (*Trouw*, 19 March 2016). He is strongly opposed to any anti-urban sentiments in the Netherlands and in Europe.

Other experts, as economic geographer Van Oort, however, believe in the power of a multitude of medium-sized and smaller cities in close mutual proximity. They point to 'positive network externalities': cities can 'borrow' each other's size. The benefits of the proximity to neighbouring cities (the 'borrowed size') exceed their disadvantages (of being in their mutual 'agglomeration shadow'). So the whole (Randstad Holland) is more than the sum of its parts (i.e. the collection of cities and towns in the west) (Van Oort et al. 2015; Meijers and Burger 2017).

7.2 Economic Dominance of Randstad Holland

Population distribution and production between the core area and the periphery is uneven (see Box 7.1). In 2015, 45% of the entire Dutch population lived in core area Randstad Holland which covers only some 20% of the country's territory (for a more comprehensive description of Randstad Holland one is referred to Chap. 8, or see De Pater and Van der Vaart 2014). On the other hand, only 10% of the Dutch live in north Netherlands covering 25% of the Dutch territory. Average Gross Regional Product (GRP) of the three northern Dutch provinces is some 28,000 € per capita (excluding natural gas extraction) and 43,000 € in the west.

Randstad Holland's economic profile differs from that beyond (see Tables 7.1 and 7.2; note: Randstad Holland includes only the North and the South Wing, not the three western provinces). Randstad Holland's industrial sector is relatively small as compared to this sector in the Dutch territory beyond Randstad Holland. In contrast, its financial sector and business services are significantly more relevant. Major differences may even be distinguished within Randstad Holland: Amsterdam (with its financial and business services), Rotterdam (with its industry and logistics), Den Haag (with its governmental agencies and international non-profit organisations) and Utrecht (with its more nationally oriented financial and business services).

Randstad Holland hosts many banks, assurance companies and business services and accountancy offices as KPMG, Deloitte, PwC, Ernst & Young. In 1985 e.g. seventeen out of the twenty largest Dutch banks and insurance companies had their headquarters (HQ) in Randstad Holland (eight of them in Amsterdam; Dwarkasing et al. 1988). The nation's capital (Amsterdam) hosts many foreign banks as well (Ter Hart and Piersma 1990; Van Geenhuizen 1993; Box 7.2 and Chap. 6). Due to 'Brexit' the European Medicines Agency will move from London (UK) to Amsterdam by 2019.

Box 7.2 Amsterdam as a Financial Centre
Amsterdam's dominance resulted from a process of spatial concentration. Around 1900, two hundred, often quite small banks occurred. By then, five of them dominated the Dutch financial system: Nederlandsche Handelmaatschappij, Amsterdamsche Bank, Incasso Bank (all with HQ in Amsterdam), Rotterdamsche Bank (Rotterdam) and Twentsche Bank (Enschede). Mergers and acquisitions severely reduced this multitude of Dutch banks. In 1928, a maximum of 381 banks were active in the Netherlands, 35% of which based in Amsterdam. From 1930 onwards shakeout of the banking sector took off. Their numbers declined to 107 banks in 1993, 56% of which in Amsterdam. But the market share of the Amsterdam-based banks is much more than 56% (Boschma and Hartog 2014).

By today (2016), three Dutch banks dominate the sector in the Netherlands: ING Group (HQ in Amsterdam), Rabobank (Utrecht), and ABN Amro (Amsterdam). The largest of these (ING) has been the result of a merger between Postbank and NMB in 1989. Together they possess 67,8% of the total banking assets in the Netherlands in 2016 (thebank.eu). The second largest Dutch bank (Rabobank), traditionally strong in rural areas, was a merger in 1972 between Raiffeisenbank (Utrecht) and Boerenleenbank (Eindhoven). ABN Amro is a merger between ABN and Amro (HQ both in Amsterdam) in 1990. Due to a series of earlier mergers, in fact ABN Amro is composed of all former (1900) large Dutch banks. This bank was about to be acquired by three foreign banks Fortis (Belgium), Royal Bank of Scotland (United Kingdom) and Banco Santander (Spain) in 2007, but was saved and nationalised by the Dutch government in 2008. The Dutch government is currently involved in a process returning this bank to the private sector.

Table 7.1 Added value per sector, in billions of €, in 2012

	Netherlands	Randstad Holland	Neth. outside Randstad Holland	Region Amsterdam	Region Rotterdam	Region Den Haag	Region Utrecht
Total value	538 bill. (100%)	249 bill. (100%)	289 bill. (100%)	90 bill. (100%)	44 bill. (100%)	34 bill. (100%)	30 bill. (100%)
Agriculture	1,7%	1,3%	2,0%	0,5%	0,9%	3,3%	0,4%
Industry	19%	12%	26%	9%	23%	6%	8%
Building sector	5%	5%	5%	3%	4%	4%	5%
Trade, transport & communication	23%	26%	20%	28%	25%	26%	24%
Financial & business services	26%	32%	20%	39%	25%	29%	36%
Government/non-profit	25%	25%	26%	21%	22%	32%	28%

Source: Modified after Manshanden et al. (2014), data: CBS Statline
Randstad Holland North Wing (Haarlem-Amsterdam/Almere-Utrecht) and South Wing (Rotterdam-Den Haag), without the Green Heart
The Netherlands beyond Randstad Holland: The country without North Wing and South Wing
Region Amsterdam: Metropole Region Amsterdam; Region Rotterdam: Stadsregio Rotterdam, Region Den Haag: Stadsgewest Haaglanden, Region Utrecht: Bestuursregio Utrecht (in all cases: the central city and surrounding urban and rural municipalities)

Table 7.2 Employment per sector (number of employees, in millions), in 2012

	Netherlands	Randstad Holland	Neth. outside Randstad Holland	Region Amsterdam	Region Rotterdam	Region Den Haag	Region Utrecht
Total value	8,7 mil. (100%)	3,9 mil. (100%	4,8 mil. (100%)	1,4 mil. (100%)	0,7 mil. (100%)	0,6 mil. (100%)	0,4 mil. (100%)
Agriculture	2%	1%	3%	1%	1%	2%	1%
Industry	10%	7%	13%	6%	9%	5%	5%
Building sector	5%	5%	6%	4%	6%	4%	5%
Trade, transport & communication	28%	29%	27%	32%	29%	24%	29%
Financial & business services	20%	23%	17%	26%	23%	24%	24%
Government/non-profit	34%	34%	34%	31%	31%	40%	37%

Source: Modified after Manshanden et al. (2014), data: CBS Statline
Randstad Holland, etc.: see Table 7.1, footnote

7.2.1 Headquarters of Large Companies

Most larger companies in the industrial, construction trade and transport sectors have headquarters (HQs) in Randstad Holland as well. Out of the 100 largest (in terms of turnover) Dutch companies in these sectors, 72 had HQs in this core area, mainly in Amsterdam, Rotterdam and Utrecht in 1985 (Dwarkasing et al. 1988). For historic reasons other companies were not based in Randstad Holland, e.g. Philips (lighting, medical equipment, home appliances, etc.) was based in Eindhoven until 1997, AKZO (paintings and coatings) in Arnhem and DSM (chemicals and materials) in Heerlen.

Over the past decades Dutch economy internationalised. Dutch companies acquired foreign companies, and vice versa. As early as 1989, 14 Randstad Holland-based HQs controlled 966,000 employees worldwide (Bosman and De Smidt 1993), including Heineken (beverage, 76,000 employees worldwide today) and Ahold Delhaize (retail, 236,000 workers in the USA, the Netherlands, Belgium and Czech Republic today). Within Randstad Holland, Amsterdam dominated with 6 HQs, followed by Utrecht (4), Rotterdam (2) and Den Haag (2) in 1989. In an European perspective Randstad Holland is situated in the shadow of Greater London (with 116 HQs controlling 6.4 million employees in 1989) and Paris (with 64 HQs which controlled 3.7 million employees). The position of Randstad Holland was (and is) comparable with Stockholm (20 HQs, 850,000 controlled employees) and the Ruhr area (23 HQs, 776,000 controlled employees). Frankfurt, Munich, Rome, Stuttgart and Zurich were on positions 6–10 in Europe. By 1989, eastern European cities did not occupy top positions (Bosman and De Smidt 1993).

Since 1989, Amsterdam, 'the core of the core', has strengthened its position. Akzo e.g. moved its HQ from Arnhem to Amsterdam in 1994, after their merger with Swedish Nobel (now: AkzoNobel; 46,000 employees worldwide). Three years later, in 1997, Philips' HQ moved from Eindhoven to Amsterdam, while controlling 23,000 employees in North America, 8000 in Latin America, 40,000 in Asia and 34,000 in Europe/Africa. HQs of global service companies can also be found in Amsterdam, e.g. accountancy and auditing KPMG, with 174.000 employees worldwide, has its HQ in Amstelveen (suburb of Amsterdam), Randstad Holding's HQ, a human resource service provider with 30.000 employees worldwide is in Diemen (another Amsterdam suburb) (Vermeulen 2016).

According to the Fortune Global 500 list of 2011 HQs of eleven top-500 multinationals were based in Randstad Holland, five in Amsterdam (ING, Philips, Heineken, AkzoNobel and in Zaandam, Greater-Amsterdam, Ahold), three in Den Haag (Royal Dutch Shell, Aegon, EADS – Airbus Group, in Leiden, in Greater-Den Haag), one in Rotterdam (LyondellBasell, chemical industry), and two in Utrecht (Rabobank and SHV, a trading company). Beyond Randstad Holland is one more HQ in the Netherlands: GasTerra, in Groningen. The Dutch-British company Unilever has HQs in Rotterdam and London. With 11 HQs, Randstad Holland (Amsterdam included) ranks 7th on the list of world cities based on HQ numbers, after Beijing (48), Tokyo (45), New York (25), Paris (19), London (17) and Seoul (12). Amsterdam ranks 17th, together with Madrid (both with five such companies), and just after Moscow and Zürich (both six companies) (Vermeulen 2016).

Randstad Holland does not only host multinationals but many start-ups as well. Overrepresentation of successful, innovative start-ups in metropolitan areas is illustrated by a survey of 490 Dutch 'gazelles' by *Het Financieele Dagblad* (11 December 2015). Gazelles are small, young, rapidly growing and financially healthy companies. Such companies may not only be found in the high tech sector but may also comprise e.g. restaurants and temporary employment agencies. In north Netherlands (Groningen, Friesland, Drenthe) 30 gazelles are established, in the west (provinces of Noord-Holland, Zuid-Holland and Utrecht) there are 274. Top three gazelle provinces are: Noord-Holland with 123 gazelles, Zuid Holland: 93 and Noord-Brabant: 81. On the other hand, Drenthe has 3 gazelles, Flevoland: 4 and Zeeland: 6. Per capita, one gazelle occurs on 163,000 inhabitants in Drenthe, and 1 gazelle on 22,000 people in Noord-Holland (with Amsterdam). These are considerable differences in a small country.

Beyond Amsterdam and Randstad Holland, many family companies with rather networks and subsidiaries abroad, occur. Bavaria Brewery e.g. is the Netherlands 2nd beer producer, after Heineken. The Swinkels family owns this brewery based in Lieshout, a village in the east of Noord-Brabant already for seven generations. Bavaria's HQ is in Lieshout too. In 2016 Bavaria bought Palm Belgian Craft Brewers, with breweries in Steenbrugge and Roeselaere, Flanders. Conversely, many companies of Dutch origin are now owned by foreign companies or by hedge funds; more information about foreign investments in Dutch companies is given in Chap. 6 (see Box 6.1).

7.3 A Tour of the Dutch Regional Economies: The West

Average population density in the Netherlands is more than 500 persons per square kilometer, but this number varies considerably throughout the nation. Density in the west of the Netherlands has is about 1000, twice the national average. Densities in the south and in the east is about 500, a few more in the south and a few less in the east. The north and the southwestern Province of Zeeland have much lower numbers: about 200. Many urban areas in the west may reach population densities of 5000 and more, many rural areas in the north have population densities of less than 100.

Such regional differences indirectly point to regional economic disparities. The west, or Randstad Holland is known as the core area, the south and most of the east are considered as the overflow area of Randstad Holland, or as the semi-periphery, while the north and southwest is the periphery (Figs. 7.1 and 7.2). Below, the economic geography of the Netherlands will be discussed from this perspective, which is according to Myrdal's theory on relations between core- and (semi-)peripheral regions (Box 7.3).

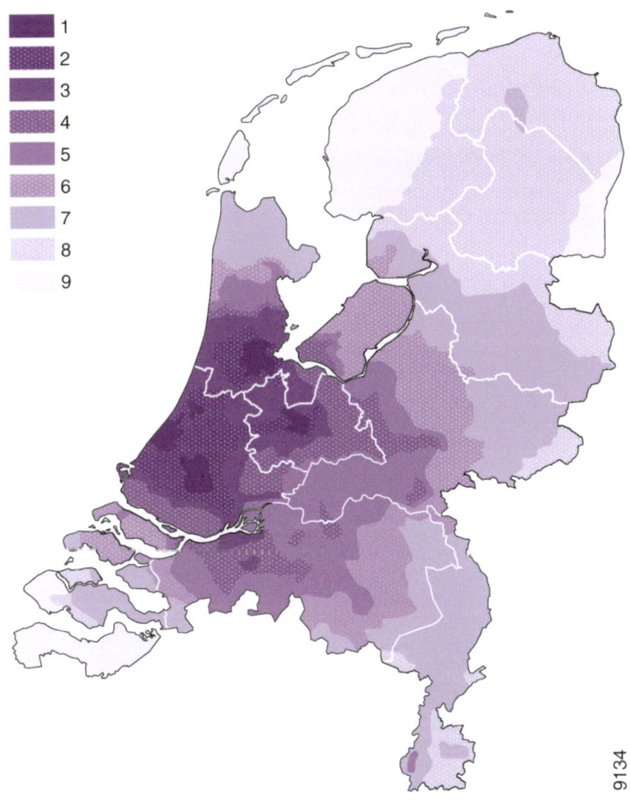

Fig. 7.1 The continuum of core, semi-periphery and periphery in agglomeration economies based on a model using city populations and road distances between cities as proxies. (Source: Dieperink and Nijkamp 1986)

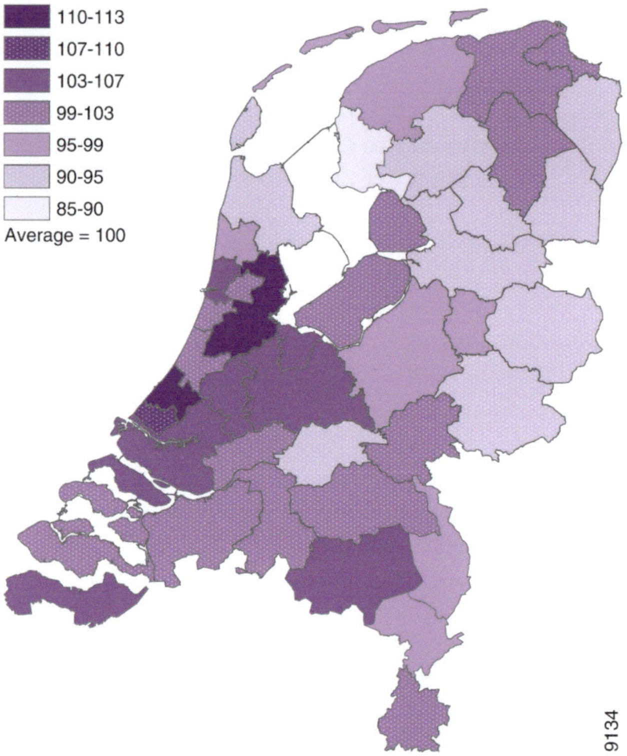

Fig. 7.2 Spatial variety in wages; controlled for individual characteristics of employees. (Source: De Groot 2012)

Box 7.3 The Economic-Geographical Dynamics According to Myrdal

Economic-geographical processes in the Netherlands may be better understood by a theory developed by the Swedish economist Gunnar Myrdal. In his book *Economic Theory and Underdeveloped Regions* (Myrdal 1963), Myrdal investigated development of (in)equalities in economic-geographical, capitalistic systems. Positive feedback effects dominate in such systems while enlarging regional inequalities. 'If things were left to market forces unhampered by any political interferences, industrial production, commerce, banking, insurance, shipping and, indeed all those economic activities which in a developing country tend to give a bigger than average return – and in addition, science, art, literature, education and higher culture generally – would cluster in certain localities and regions, leaving the rest of the country more or less in a back water' (edition 1963).

The core area attracts labour and capital from peripheral areas. The core area more and more

Box 7.3 (continued)

accumulate core characteristics and, consequently, the periphery more and more becomes a periphery. This growing, spatial-economic inequality also has non-economic consequences, e.g. 'periphery' people rather tend sticking to conservative ideas and worldviews and suffer from declining educational and medical facilities than 'core' people. That generates a downward spiral for the periphery: 'poverty becomes its own cause'. On the other hand, core areas increasingly accumulate more wealth. Myrdal points to 'the principle of circular and cumulative causation'. Accumulation of negative effects in the periphery (vis-a-vis core areas: 'backwash effects') is particularly severe in poor nations.

'Spread effects' occur in the semi-periphery, areas geographically situated in the core's proximity. Semi-periphery areas may benefit from the nearby core area (agglomeration advantages), but not consequently suffer from their disadvantages, as traffic jams, lack of space, high rents, pollution, and crime. Such spread effects may be stronger in more developed countries as 'a high average level of development [of the country] is accompanied by improved transportation and communications, higher levels of education, and a more dynamic communion of ideas and values – all of which tends to strengthen the forces for the centrifugal spread of economic expansion or to remove the obstacles for its operation' (Myrdal 1963).

Agglomeration disadvantages in the core may even exceed their agglomeration advantages stimulating spatial decentralisation and expansion of the core area. The geographical range of such drawbacks may be smaller than their range of benefits. A company's site in the proximity of a big city would mean that the company would enjoy (most of) the advantages of the big city nearby but rather avoid (most of) its disadvantages. Eventually, a core area will extend into adjacent areas.

According to Myrdal, 'spontaneous' spread effects reduce regional disparities, but in rich nations only. But welfare state governments tend to start with a nation-wide policy of spatial decentralisation before spread effects begin. Such governments are inclined to reduce backwash and to promote spread effects to the periphery. The paradox is that only richer nations – with their 'spontaneous' spread effects – may afford such costly policies.

7.3.1 Western Netherlands: Core Area

Myrdal's theory may explain post-industrial revolution regional dynamics. In the Netherlands the Industrial revolution began rather late, around 1870, as this nation largely missed the first (steam power-based) industrial revolution. But Dutch entrepreneurs seized their opportunities which emerged from the favourable geographic position of the Netherlands between superpowers Germany and Great Britain, during the second industrial revolution. That revolution was based on electricity, internal combustion engines (cars) and on growing consumer markets.

Dutch industrialisation went along with spatial concentration of population and urbanisation. In the west (the provinces of Noord-Holland, Zuid-Holland and Utrecht) population grew from 39.4% of the total Dutch population in 1860 to 49.1% in 1940. Initially, such growth was most eminent in the four biggest cities in the west (Amsterdam, Rotterdam, Den Haag and Utrecht), where e.g. half of the population in the west was concentrated by 1920. However, the number of people in these cities declined from the 1960s due to a process of suburbanisation with more than half a million inhabitants, a decrease of over 20%. But urban population gradually restored since 1985 due to both a birth- and immigration surplus. In 1980 the share was 44.7%, in 2015 45.1% (Chaps. 8 and 11).

7.3.2 Utrecht and Zeeland Compared

Comparison between the provinces of Utrecht and Zeeland may illustrate how developments in both provinces have deviated. Zeeland, the most southwestern province, had 220,000 inhabitants in 1900 and 380,000 in 2015. In 1900, barely more people lived in the central province of Utrecht (250,000) but there are 1.6 million (five times more) now. Meanwhile total Dutch population grew from 5 million (in 1900) to almost 17 million in 2015. Zeeland, which has been part of the maritime economy in the Dutch Golden Age (seventeenth century), was a 'downward transitional area' by 1900, in terms of Friedmann's (1966) sophisticated centre-periphery model, where cities and trade had been replaced by rural areas and agriculture (Brusse and Mijnhardt 2011).

According to the same model, by 1900 the province of Utrecht was an upward transitional area that became incorporated gradually in the Dutch core area. Spread effects applied, e.g. a major part of machine factory Werkspoor moved from Amsterdam to Utrecht in 1912–1916. Simultaneously, steel plant Demka migrated from Groningen province to Utrecht ('backwash effect'). Utrecht trade fair Jaarbeurs, that developed into a national convention and exhibition centre later, opened in 1917 demonstrating Utrecht's growing national industrial and trade position.

Today, Utrecht province is part of expanding Randstad Holland. Meanwhile, Zeeland underwent significant geographical change. Zeeland had been an archipelago of islands and peninsulas and was situated quite remotely in the extreme southwest and far from Randstad Holland until about 1960. Later, dams and bridges built by implementation of the Deltaplan (Chap. 2), physically connected this province with the Dutch core area and turned the province into a leisure area with holiday parks, beach camping sites and (for poor weather) historic towns. During the Dutch Golden Age towns as Middelburg, Goes and Zierikzee belonged to the maritime network cities. Later, such towns later turned into (Chistallarian) 'central places'. To stimulate regional economy (regional policy, see Sect. 7.6) the Dutch government prompted development of industrial sites along Westerschelde, the waterway to Antwerpen, in particular in Vlissingen and Terneuzen, in Zeeuws-Vlaanderen.

7.3.3 Rotterdam and Amsterdam Compared

Both North Sea facing provinces of Zuid- and Noord-Holland benefited from their geographical position. Since 1872, when a new, wide and deep canal (*Nieuwe Waterweg*), directly connecting Rotterdam with the North Sea, was opened, Rotterdam attracted most of the Dutch transit trade. This port developed as the gateway to industrialising Ruhr area in Germany. Such developments may not only be attributed to Rotterdam's favourable geographic position, inspiring people as e.g. G.J. De Jongh, director of *Gemeentewerken of Rotterdam* by 1900 have played a significant role. He understood Rotterdam's potentials as a transit harbour rather than a commercial port as London. As a result harbour extensions (Rijnhaven, Maashaven) should be different from the London Docks to better address such potentials (De Klerk et al. 2008).

The Nieuwe Waterweg and its related harbour extensions eventually turned the entire Dordrecht-Rotterdam-Schiedam-Vlaardingen region into a maritime industrial area. Hundreds of shipyards, machinery plants, dredging companies and supplying companies emerged along Rivers Beneden-Merwede, Nieuwe Maas, and Noord (Fig. 7.3). These include worldwide operating dredging companies as Boskalis, HBG, Ballast Nedam and Van Oord operate in maritime areas worldwide.

Simultaneously, an industrial landscape developed around the *Noordzeekanaal*, opened just 4 years later in 1876, directly connecting Amsterdam with the North Sea, for the first time in its history. In particular Amsterdam harbour profited from increasing commercial activities in Dutch East India and became the Dutch entry point for tropical agricultural products. However, Amsterdam harbour never reached a comparable prominent position as the port of Rotterdam

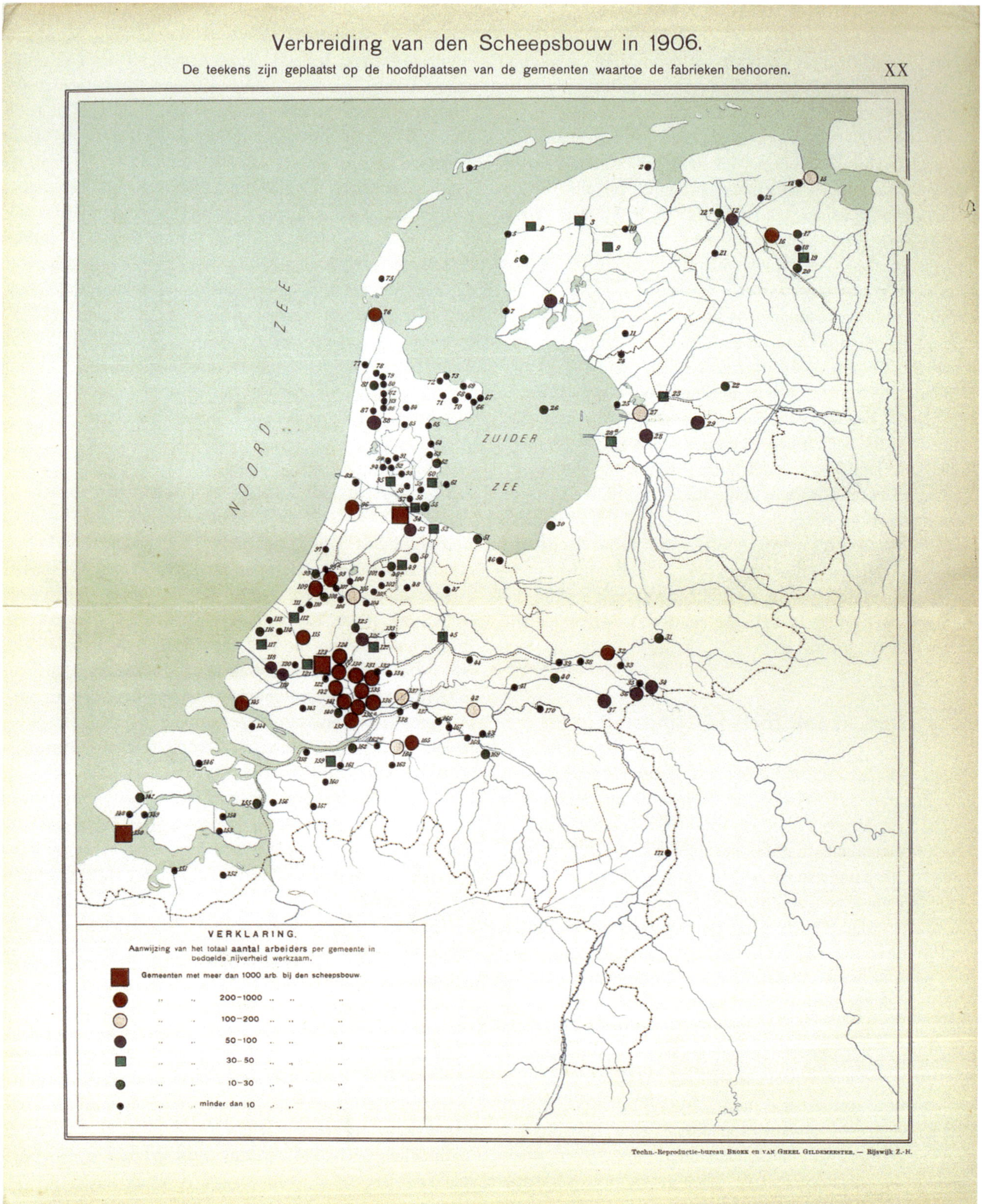

Fig. 7.3 Spatial distribution of ship-building in the Netherlands, in 1906. (Source: Everwijn 1912)

Table 7.3 Employment (number of jobs, in thousands) and added values (in billions) of the Dutch seaport areas, 2014

Seaport areas[a]	Employment (direct jobs)	Employment (indirect jobs, backwards)	Total employment	Added value (direct)	Added value (indirect, backwards)	Total added value
Seaports in the North	9.1	10.9	20.0	1.4	0.7	2.1
North Sea Canal	34.2	34.2	68.3	4.0	2.6	6.7
Rotterdam-Rijnmond	93.8	83.5	–	12.9	7.9	20.8
Other ports in Rijn -Maas region	19.4	21.0	40.4	2.4	1.6	4.0
Westerschelde	15.2	17.0	32.2	3.0	1.9	4.9
Total	171.7	166.6	338.2	23.8	14.7	38.5

Source: Van der Lugt et al. 2015 (condensed version of Tables 2–9)
[a]Seaport areas:
Seaports in the north: Delfzijl, Eemsmond, Harlingen, Den Helder
North Sea Canal: Amsterdam, Velsen/IJmuiden, Beverwijk, Zaanstad
Rotterdam-Rijnmond: Rotterdam, Schiedam, Vlaardingen, Maassluis, other harbours in Rijnmond
Other ports in Rijn-Maas region: Dordrecht, Drechtsteden (as Sliedrecht and Papendrecht), Moerdijk, Scheveningen (Den Haag)
Westerschelde: Vlissingen, Borsele, Terneuzen

(Table 7.3). In the *Zaanstreek*, nearby the Noordzeekanaal and an old industrial area, the numerous windmills were replaced by steam engines around 1900. At the entry point of the NooRdzeekanaal in the North Sea, in IJmuiden, blast furnaces were built in 1918, supplied with raw materials directly by ships from sea. Shortly afterwards, the first foreign investors arrived in Amsterdam, e.g. Ford built a car assembly plant in the Amsterdam port area which have been operational for almost 50 years (1932–1981). Ford has been a frontrunner of an increasing number of foreign investors in the Amsterdam region including its surrounding municipalities and Amsterdam Schiphol Airport.

According to Inbusiness, an Amsterdam acquisition organisation, presently some 2700 international companies occur in the Amsterdam region accounting for 15% of total regional employment. Annually, some one to two hundred international companies arrive (150 in 2015) while 15 such companies leave, on average. Most of them are attracted by Schiphol Airport, by the acknowledged agglomeration benefits of Amsterdam and by Dutch fiscal facilities ('rulings').

In total the Netherlands hosts some 12,000 foreign-owned companies accounting for 16% of total employment in the Netherlands (according to CBS and the Netherlands Foreign Investment Agency, NFIA). This country is popular in particular for American, Asian and European companies. Out of the 11.400 new jobs created by foreign investors in the Netherlands in 2015, some 1200 were in new HQs, 1500 in production facilities, 1100 in service centres and 2100 in distribution centers (NFIA 2016). In this respect, Randstad Holland has been a hotspot for HQs and distribution centres, Amsterdam in particular.

Industrialization has not been restricted to the big cities in Randstad Holland. Historical towns as Haarlem, Leiden, Delft, Gouda, Schiedam and Alkmaar, flourishing in the Dutch Golden Age, but languishing afterwards, revived again by the second industrial revolution around 1900. Often these town were (and are) more or less specialised in a particular product, such as cheeses (Gouda and Alkmaar), textiles (Leiden), food and engineering (Delft), metal (Haarlem) and gin (Schiedam). Here, *industrial districts* with clusters of interconnected small and medium-sized companies flourished (Marshall 1919).

7.3.4 Ship-Building: Rise, Fall and a Modest Renaissance

Growth in industrial employment halted in the Netherlands in the 1960s. Due to rapid and strong increase of Dutch wages since 1963 and to increasing foreign competition many companies no longer were competitive and had to close their doors, e.g. the Demka-Werkspoor cluster in Utrecht ceased production in 1983. That has been quite manifest in the ship-building industry as well. The Dutch have been a global player in the shipping industry. Shipyards in the west were specialised in construction of large ocean-going vessels, for instance for shipping companies maintaining regular services between the Dutch East Indies (until 1948 a Dutch colony) and the Netherlands. These shipyards built passenger and cargo vessels (De Feyter 1982).

Along some canals in Groningen (in the north) coasters were built while construction of river vessels was concentrated along Dutch rivers. The Netherlands hosted Europe's largest fleet of barges, transporting grain, iron ore, coal and other commodities between Rotterdam and the German Ruhr-area, via River Rijn. In 1913, e.g. 65% of the 96,234 ships passing the German-Dutch border in River Rijn was Dutch, both up- and downstream, while 21% was German and 15% Belgian (Blink 1915).

In terms of tons gross, Dutch ship-building was the world's third in the 1930s and fourth in 1956. From then, decline began: fifth in 1962 after Japan, UK, West Germany and Sweden, 15th in 1968, 11th in 1972, and 15th again in 1978 (De Feyter 1982). Subsequently, numbers of employees in the shipyards dropped from 49,000 in 1972 to 26,000 in 1979 (De Feyter 1982). Major shipyards as Wilton-Fijenoord, Verolme (both in Rotterdam-region), NDSM (Amsterdam) and De Schelde (Vlissingen) made structural losses and were forced to cooperate in return for financial support by the State. Such defensive mergers stretched life only temporarily and could not prevent bankruptcies to occur around 1980 as Dutch companies could no longer compete with cheap ship-building nations as Japan, South Korea, China, Turkey and Poland.

While in the 1960s mammoth tankers were built in Dutch shipyards, smaller ship repair companies and smaller yards for complicated and specialised ships are active today, e.g. luxury yachts, dredgers, offshore vessels, patrol vessels for coast guards, tugs, workboats, ferries, et cetera. The largest Dutch company in this sector is Damen Shipyards with 3000 employees and 14 yards in the Netherlands (and 6000 workers and 18 yards abroad). Damen is a family firm founded in 1927, HQ in Gorkum. In total 15,000 people are working in Dutch ship-building industry today and total number of companies is around 1000, most if these building leisure boats. Today, Dutch shipyards have become niche-yards.

7.3.5 Job-Loss by Automation and Relocation

Job losses were also caused by replacement of workers by machines. That occurred e.g. in the food industry in Zaandam, in companies as Verkade (owned by Yildiz), Honig (Heinz), Duyvis (Pepsi Cola) and Albert Heijn (Ahold Delhaize). All manual work has been automated.

De-industrialisation, in terms of employment not (or less) of production value, turned many cities in the west into centres for business and commercial services. Traditional factory cities do no longer exist in the Netherlands. Randstad Holland-based firms are often specialised in products and services at relatively early (explorative) phases in their product life cycles. Companies with products and services in the more mature (so-called Fordist) phase are over-represented in the semi-periphery and in the periphery of the Netherlands (Capasso et al. 2016), except for the science-based industries which are over-represented in the semi-periphery, i.e. in the southeast, in Noord-Brabant (see next Section).

Relocation of industries from (expensive) Randstad Holland to cheaper places in the (semi-) periphery occurred as well. Supported by the government many industrial firms moved from the west to the south, east and north in the 1950s and 1960s (Pellenbarg and Van der Steen 2003). Migration of production units to other nations took place in particular since 1989 (collapse of communism). More recently, relocation of services occurs too, e.g. Dutch ING bank will cancel thousands of jobs soon because of digital innovations and of job relocation to other ING offices in Eastern-Europe and in the Philippines (Haegens 2016).

7.3.6 Amsterdam Airport: Schiphol

In 2016 Schiphol, internationally known as Amsterdam Airport, celebrated its centenary. It started in 1916 as a military airstrip in the reclaimed *Haarlemmermeerpolder*. The first aircraft for passengers touched down on the air strip made of grass in 1920. Slightly before, in 1919, the *Koninklijke Luchtvaart Maatschappij* (Royal Dutch Airlines: KLM) had been established. In 1920 KLM, the world's oldest, under its original name, and still operational national airways, began its operations by a regular service to London, UK. In its first year, KLM transported 440 passengers. Ninety-six years later, in 2016, not less than 479,000 starting and landing aircrafts carrying a total of 63.6 million passengers to or from 322 destinations. Grass airstrips turned into tarmac before WWII. Today, Schiphol Airport has six runways. Schiphol is the fifth largest airport (in terms of passengers) in Europe, after London Heathrow (75.0 million passengers in 2015), Paris Charles de Gaulle, Istanbul Atatürk Airport and Frankfurt, but well ahead of Madrid Barajas (position 6), Munich (7), London Gatwick (8), Rome Fiumicino (9) and Barcelona (10).

Schiphol owes its relatively high-ranking primarily to its position as a global hub in a hub-and-spoke network (Burghouwt 2005). In the top-10 of freight transport airports Schiphol (1.6 million tons in 2015) ranks third, after Frankfurt (2.0) and Paris Charles de Gaulle (1.9), but ahead of London Heathrow (1.5) and Leipzig (1.0 million tons) (Schiphol 2017). Schiphol is a prime example of an airport that developed from an air-side oriented airport to a multi-service airport city (Burghouwt 2002). In earlier times only functions that were strictly related to and indispensable for the air-side of an airfield, as customs offices, desks, and waiting areas were available at the airport. Today, Schiphol Airport contains hotels, conference centers, shops, warehouses, offices, parking facilities as well offering jobs to approximately 65,000 people directly but more indirectly. VNO-NCW (2016), the Dutch employers' organisation, estimates a total of 370,000 jobs for Schiphol Airport (see Box 7.4). All such jobs are not necessarily physically based at the airport but in a wider area and rather are largely in companies that depend on a nearby and accessible airport, an *aerotropolis* (Kasarda 2013; Aerotropolis 2017). For instance, the *Zuidas*, the state-of-the-art banking district in south Amsterdam would belong to Schiphol's aerotropolis (Schaafsma 2014). People employed in the banking, financial

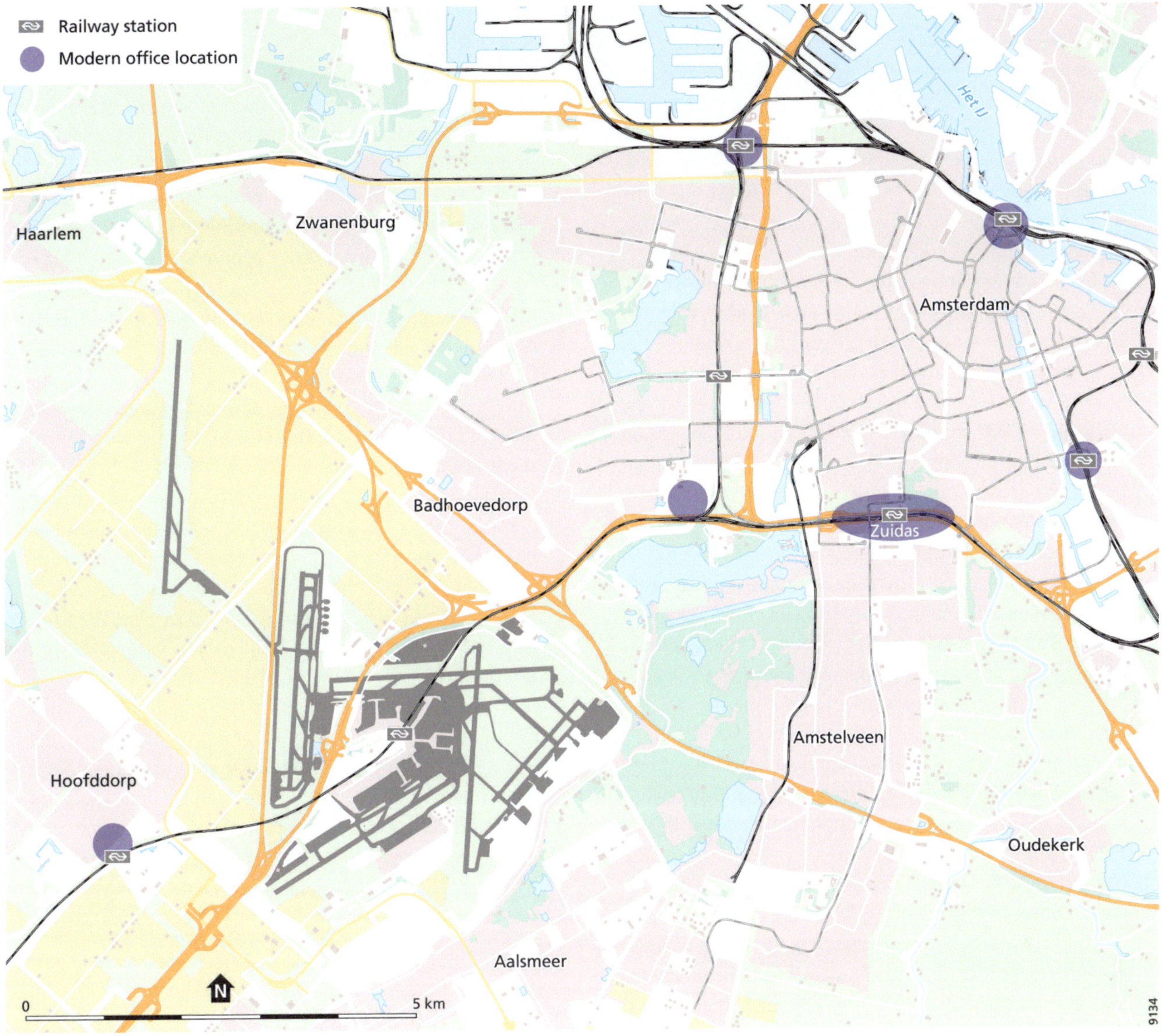

Fig. 7.4 Schiphol, Zuidas and other office locations in the Amsterdam region

and business services sectors working in Zuidas offices may reach Schiphol airport by train in 6 min only. For many internationals an aerotropolis is a more attractive location than the traditional Central Business District downtown (Fig. 7.4).

Despite local and regional protests against noise and air pollution Schiphol expects further growth with another terminal, pier and runway. A local annex, Lelystad Airport, is scheduled to open in 2020 initially for low cost-carriers only. However, predictions in this dynamic sector are hard to make and rather depend a variety of developments as KLM's ability to cope with low-cost carriers competition, emergence of other global hubs, its relation with Schiphol Airport and its position viz-a-viz its SkyTeam partner Air France.

Box 7.4 Economic Relevance of Airport and Seaports

Calculation of the economic relevance of an airport in terms of employment and income is extremely controversial; each new study provides different results. However, a non-controversial calculation of *direct* employment and income, i.e. on the airport e.g. Schiphol, by itself is possible. Assessing *indirect* economic effects, both backward (backward linkages: employment and revenues for suppliers of companies based at Schiphol) and, especially, forward (forward linkages: companies who need airport services) are

(continued)

Box 7.4 (continued)
challenging. The higher the multiplier applied to calculate such indirect economic impacts, the higher total economic relevance of the airport. Based on such multipliers, airport proponents generally arrive at higher total numbers than airport opponents who often stress ecological damage factors by air traffic, pollution and their contribution to global warming.

In a study commissioned by the Dutch Ministry of Infrastructure and the Environment, research company Decisio (2015) assessed for direct effects that 65,000 men and women (55,000 fulltime equivalent: fte) worked at Schiphol Airport in 2015, mainly in the airlines (31,200 employees) and as ground handling staff (13,200 employees). The other 20,000 Schiphol people worked at customs, security, air traffic control, retail, restaurants, et cetera. Their added value is about 5.8 billion €. In addition, there are the indirect effects. The backward effects may be estimated quite well. Schiphol based companies buy products from suppliers who are mostly based near Schiphol, in the Amsterdam metropolitan area. Such products create jobs for nearly 50,000 (indirect) employees (approximately 40,000 fte); their added value is approximately 3.2 billion €. This makes a total of 114,000 employees and an added value of 9.0 € for Schiphol airport, excluding forward indirect effects. In terms of employed persons, this is 1.3% of all working persons (8.9 million) in the Netherlands. A comparable figure applies to Schiphol's contribution to total Dutch GDP.

But forward indirect effects are not yet included. As pointed out before accurate estimates for this parameter are not possible yet. One may argue that several (maybe many?) multinational companies would have not chosen the Netherlands (and particularly Amsterdam) as a business site without availability of Schiphol. The Netherlands Foreign Investment Agency states that 485 European HQs of foreign multinationals (with 25,600 jobs) and 406 European distribution centres (with 24,800 jobs) are present in the Netherlands. Moreover, the nearby city of Amsterdam attracts many foreign business travelers, congress visitors and international tourists. Probably most of these visitors would not have come without Schiphol. HQs, multinational corporations, convention centers, hotels et cetera have their suppliers generating backward multiplier effects. Such double indirect effects (induced effects) should also be included when determining economic impacts of Schiphol. Conversely, without Schiphol Dutch tourists would less frequently travel abroad and would, perhaps, spend their holidays in the Netherlands, providing work and income to national tourists resorts. In short, forward effects are surely present but reliable numbers may not be attached to them.

Another Dutch *mainport* is the port of Rotterdam. Comparable challenges in determining forward indirect effects apply to this port. Economists of Rotterdam-based Erasmus University calculated direct employment and added value, together with backward indirect effects (Van der Lugt et al. 2015). They conclude that the port of Rotterdam-Rijnmond has 94,000 employees (direct) and 83,500 people indirectly employed. In total, 177,000. Out of the 94,000 people working in the Rotterdam-Rijnmond area, 64% works in 'real' port functions (transport, handling and storage) and 36% in harbour related industries as oil refineries, chemical industries and power production. Added value of this area was (directly) 12.9 billion € and indirect added value: 7.9 billion which makes a total of almost 21 billion € in 2014, significantly more than Schiphol (9 billion in 2015).

Other Dutch airports are very much smaller than Schiphol and do not significantly contribute to the economic relevance of Dutch airports in general. But Rotterdam-Rijnmond provides more than half of accumulated direct and indirect employment and added values of all Dutch seaports. Ports along the Noordzeekanaal are second best (Table 7.3).

7.4 A Tour of the Dutch Regional Economies: South and East

Beyond the west of the Netherlands three other regions may be distinguished: the south (Provinces of Noord-Brabant and Limburg), the east (Provinces of Gelderland, Overijssel and Flevoland) and the north (Provinces of Drenthe, Groningen and Friesland). In this Section we focus on the south and the east.

7.4.1 Southern-Netherlands: Overflow Area and Endogeneous Hot Spots

While the western part of the Netherlands industrialised by its accessibility to waterways and the sea and its proximity to markets, industrialisation in the south and east of the Netherlands had been driven by cheap labour. Demographically and economically these regions became more relevant during the twentieth century. Their populations declined from

38.1% of the Dutch total in 1860, to 34.0% in 1900. However, growth began since: 35.7% in 1940, via 41.8% in 1980, to 42.5% in 2015.

This increase has partly bee due to spread effects. Wages has been lower here than in the west, making the south and east particularly attractive for labour-intense industries. An early (nineteenth century) example of spread effects was migration of the textile production from traditional textile industry sites in Haarlem and Leiden to the higher sands in Noord-Brabant. Low wages and local know-how gained in small-scale, domestic cottage farms, were prime reasons for emergence of new factory cities, out of nowhere. Such developments occurred not only in Noord-Brabant (Tilburg and Helmond) but also in the east (Twente region) where textile industry cities as Almelo and Enschede emerged. In turn, these cities attracted a variety of textile supply and support companies, e.g. Stork's machine factory emerged in Hengelo, close to Enschede, together with clothing industries.

But Dutch textile industry rapidly declined since the 1960s. The about 200 textile companies (mainly wool) in Tilburg provided work for 15,000 people. These companies faced increasing competition and did not properly respond to shifting demands. Except for one, innovative company (Innofa), the entire wool industry has disappeared from Tilburg in the twenty-first century (Scheffer 2009). The same was true for Helmond, Noord-Brabant's second textile city, where only the Vlisco company, designing and producing high-quality fashion fabrics for African markets, survived.

Textile industries in Twente underwent a similar process. By 1955, 44,000 people worked in this industry, but only 4200 in 1985 and, simultaneously, employment in Twente's associated clothing industries declined from 6900 to 2100 (Van Zuidam 1989). Only very innovative companies adapting to rapidly changing conditions or opting for new niche products were able to survive, e.g. Royal Ten Cate in Almelo who moved into composites, technical textiles and artificial grass fiber for sports grounds.

7.4.2 Eindhoven: Not Longer a Company Town

The city of Eindhoven, situated in the southeast of Noord-Brabant, has a quite different economic-geography history than above described textile towns. It benefitted from locational opportunities that emerged with the second industrial revolution. New industrial branches were not bound to specific locational requirements and were quite free to settle anywhere (Boschma 1994). By 1891, father and son Philips – living in the river town of Zaltbommel, 45 km as the crow flies from Eindhoven – began a light bulbs factory in the small town of Eindhoven where a factory building was available. It was a quite accidental locational choice.

However, that choice had major implications: today, Eindhoven is the country's fifth city (after Amsterdam, Rotterdam, Den Haag and Utrecht), Dutch high-tech industry centre and delivery room of science-based industries. Greater-Eindhoven is presently housing several Philips spin-offs, e.g. ASML, the world's largest world's manufacturer of computer chips machines, and NXP, a major computer chips producing company. HQs of both spin-offs are in the Eindhoven agglomeration. In 2016, NXP was sold to American peer company Qualcom for 43 billion €. ASML has hundreds of supplying companies, many of these nearby as ASML aims for proximity (De Lange 2016).

High Tech Campus Eindhoven, situated at Philips' former research laboratory, now hosts 142 companies and some 10,000 workers and is a dense network of high tech companies, start-ups, leading research institutes, fast growth enterprises and service companies. This Campus also has numerous links with Eindhoven University of Technology, founded in 1956.

The economic strength of Eindhoven region seems to depend partly on a prevailing regional (professional) culture in southeast Netherlands, geared at co-operation and friendly competition among large companies and suppliers, among engineers and scientists and among firms, local governments and knowledge institutes, the *triple helix* model (Prud'homme van Reine and Dankbaar 2011).

While its spin-offs ASML and NXP developed to global players, the Philips company declined in significance both in Eindhoven and in the Netherlands during the last decades.

While 90.000 people were employed by Philips in the Netherlands (42,000 of which in Eindhoven) in the early 1970s, today some 10,000 jobs have remained in this nation due to *offshoring*, transfer of work to low-wage nations in Asia and Eastern Europe (Atzema and Wever 2004; Jansen et al. 1979; Fig. 7.5). Worldwide Philips controlled some 450,000 workers in the early 1970s, 275,000 in 1996 and 105,000 in 2016. Apart from offshoring this decline is also due to selling parts of their activities which were once (1996) grouped in 14 divisions, three only in 2014. By today (2017), Philips focuses on 'healthtech' only for which about 70,000 employees are working worldwide.

7.4.3 Coal Mining Industry in South Limburg

Large scale production of coal has had rather strong urbanisation effects of south Limburg where both private mines (from 1899) and state mines (from 1906) operated, in particular in the east of that region (*Oostelijke Mijnstreek*). All-time high in coal production was in 1937 (14,3 million ton; see Chap. 4) and the maximum number of miners has

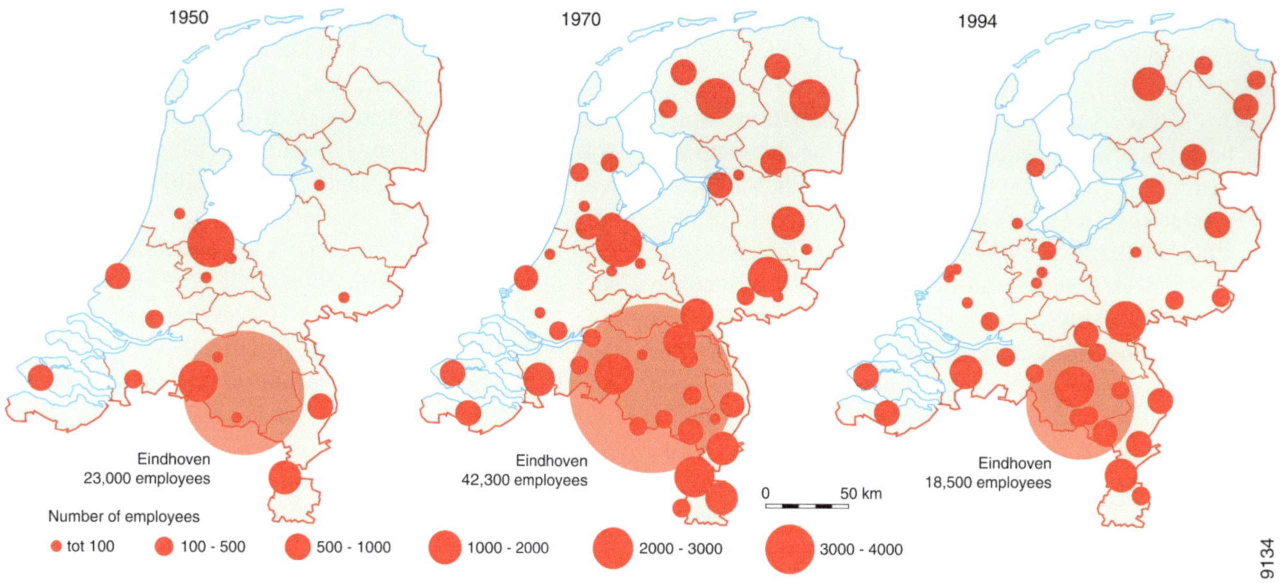

Fig. 7.5 The locations of Philips-plants in the Netherlands, in 1950, 1970 and 1994. (Source: Atzema and Wever 2004)

been 31,500 in 1957 (Mijnsluitingsnummer 1971). Within just a few decades former villages Brunssum, Hoensbroek, Heerlen and Kerkrade developed into significant mining towns (Fig. 7.6).

Unlike much earlier operating mines in Britain, Belgium and Germany there was not much need for positioning power consuming factories nearby Dutch coal mine sites in the Oostelijke Mijnstreek. Coal transport costs had become relatively low in the early twentieth century by the progressing Dutch railway networks (Stohr 1989). Secondary industry remained modest in south Limburg. Moreover, Limburg coal (anthracite) was particularly suited for domestic heating. Other reasons for keeping away new companies in this region was to prevent fights for higher wages and associated social unrest. In addition, the Dutch government blocked initiatives to develop blast furnace plants in Limburg to avoid German annexation of this border area.

In the *Westelijke Mijnstreek* (Geleen) industrial development began with a cokes oven gas (as a byproduct of cokes production from coal) using artificial fertiliser plant, founded in 1930. After WWII, the fertiliser factory developed into a petrochemical complex: DSM, Dutch State Mines. At present DSM (HQ in Heerlen) produces chemical components for food products and plastics. DSM has been the igniter of the Chemelot Campus with 150 companies and some 8000 employees. By 2025, this Campus wants to be 'the most competitive and sustainable chemical and material site of Western-Europe'.

Dutch coal mining was terminated by 1974 and 75,000 direct and indirect jobs were lost (Chap. 4). Although the Dutch government promised alternative employment and despite massive regional subsidies, not more than 18,000 new jobs effectively came to the region (Mijnsluitingsnummer 1971; Toonen 1972; Messing 1988; Luyten 2015). Many of these were in (semi-)government services, as pension fund ABP (in 1969) and the Central Bureau for Statistics (CBS) partly (in 1973) moved from Den Haag to Heerlen. Also, the newly founded Open University was located in Heerlen.

The Dutch government facilitated development of new industries as well, e.g. a passenger car factory in the Westelijke Mijnstreek (Born), where DAF began in 1967. This Eindhoven-based car manufacturer was too small to compete with big brands and was taken over by Volvo a few years later. DAF company still exists but as a successful truck manufacturer now, being part of the American Paccar since 1996. In turn, Mitsubishi took over and owned the factory in the 1990s. They stopped production of Mitsubishi cars in Born, in 2012. However, the factory revived in 2014 when Dutch entrepreneur Van der Leegte (VDL-Group) started assemblage of BMW Mini Cars in Born. This revival has been quite successful; in 2017 the number of employees was 6000, five times more as in 2012. Further growth in employment is announced.

But most former miners could not easily switch to work in car assembly lines and went for early retirement or were provided with long-lasting unemployment benefits. Heerlen, once the center of the mining industry in Oostelijke Mijnstreek, well demonstrates vulnerability of one sector-cities. In 1955, Heerlen was ranked 20th on the list of high income municipalities, also due to high average mining wages. Almost 40 years later, in 1994, Heerlen had fallen to position 555 out of a total of 636 (by then) Dutch municipalities. Similarly, Geleen, the center of Westelijke

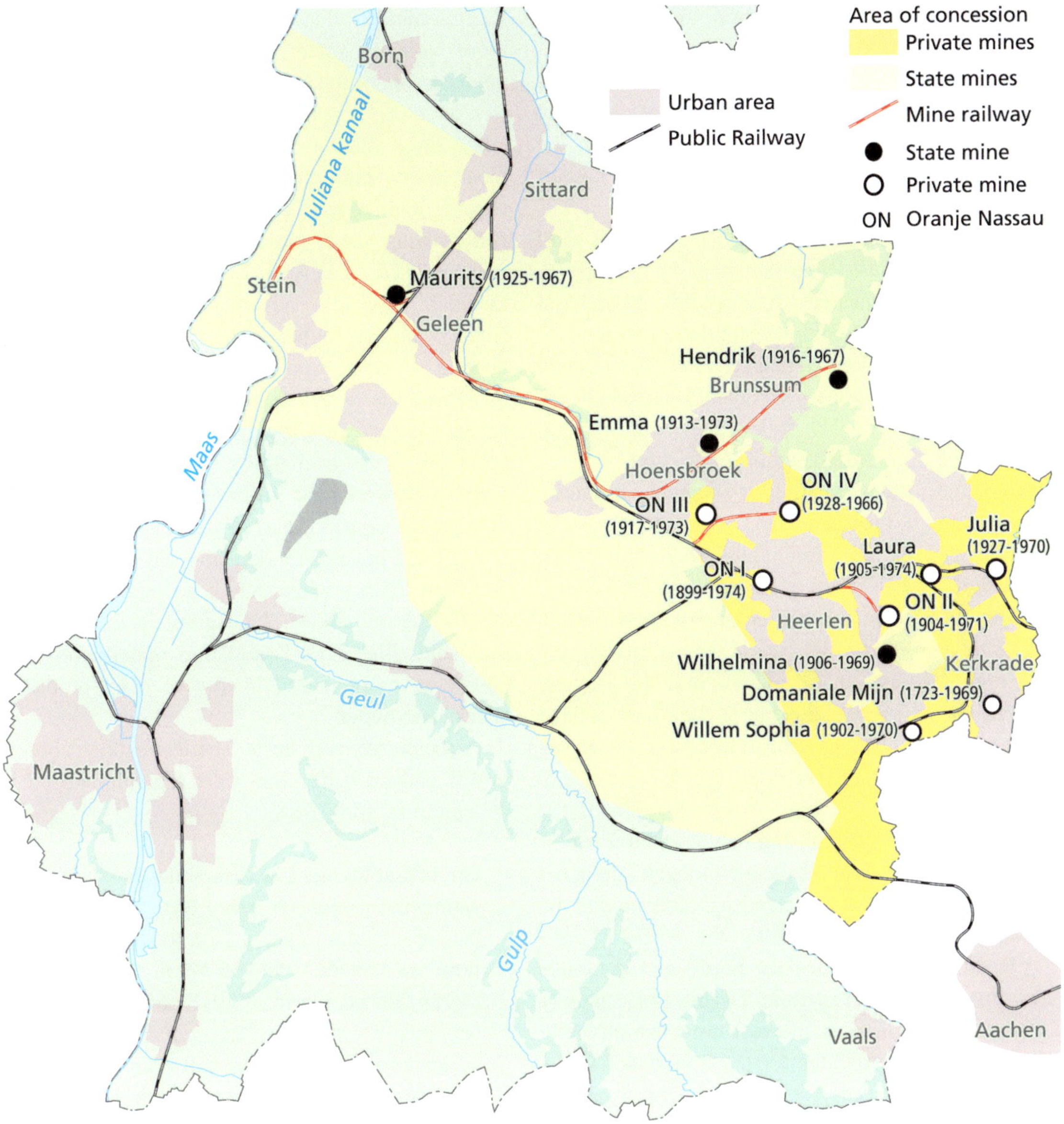

Fig. 7.6 The coal region of Zuid-Limburg. (Based on: Venner 2001)

Mijnstreek, dropped from position 9 in 1955 to 504 in 1994 (see Wikipedia, lemma Parkstad Limburg).

Today, the Oostelijke Mijnstreek, now known as 'Parkstad Limburg', a supra-municipal community of most of the former mining cities with about a quarter of a million people, suffers from an ageing and shrinking population. But there are some successes too. Rather successful promotion campaigns, as 'South Limburg, the bright site of life', helped in attracting many tourists enjoying an atypically Dutch, hilly landscape to this area. In addition, skiing centre SnowWorld, situated on a mining waste site, now serves as a tourist magnet. Even, the region won the Destination Award 2016 of the World Travel and Tourism Council, before fellow finalists Swiss Parks Network (Switzerland) and V&A Waterfront (South-Africa). For 2015, estimates are that tourism would generate some 5800 jobs (in fte) in the region and a turnover of about 368 million €.

7.4.4 East-Netherlands: A Patchwork

As Noord-Brabant in the south, large parts of Gelderland are overflow areas of Randstad Holland, with many commuters. Except for former textile branches in Twente, both eastern provinces Gelderland and Overijssel never hosted large industrial areas. Industry is mostly relatively small-scale and dispersed. Beyond Twente, e.g. in Winterswijk, many small textile companies and iron works occurred. Most other industries rely on natural resources as well, including coarse ceramic industries as brick factories, using clay from river floodplains, agricultural commodities as tobacco for cigar and cigarettes companies in Kampen, Deventer and Veenendaal, jam production plants in the Betuwe in the river belt, and dairy industries and slaughterhouses east of River IJssel.

Today, a transformation process takes place here, toward innovative and smart industries in particular in Achterhoek and Twente. There, many medium-sized high-tech industrial firms supply other industrial companies, often quite innovative in their production processes and products. Achterhoek examples of business-to-business companies include Kaak Group (bakery technology), 247 Tailor Steel (welders) and Pas Reform (incubators) (Atzema et al. 2016).

Such small and medium-sized industrial companies barely affect the rural image of Overijssel and Gelderland, generally known as domains of a quiet, agricultural way of life, with scattered traditional, small farms in nice, rolling landscapes. Agricultural reality is often quite different, e.g. the Gelderse Vallei with abundant bio-industry (see next Section). Nevertheless, the region attracts many tourists and day-trippers from Randstad Holland who may stay in the many cottage parks, (mini) campings and bed & breakfast places sites east of River IJssel: in Salland, Achterhoek, Twente, et cetera. West of River IJssel lies the high, wooded sands of the Veluwe, a traditional area for forestry, nature and tourism.

Apart from the urban area of Arnhem-Nijmegen (along the strategic transport axis connecting Randstad Holland and the German Ruhr region), Zwolle (along the transport axis connecting Randstad Holland and the north) and the cities of Hengelo, Enschede and Almelo in Twente (once mainly textile factory cities), towns in the east are predominantly central places providing facilities for the hinterland. Cities along River IJssel (Kampen, Deventer, Zutphen) have a history as network cities: they once belonged to the Hanseatic network of merchant cities before the main trade focus moved to the western cities in Holland by the end of the sixteenth century. Tourists visit their cultural heritages today.

7.4.5 Agriculture: Intensive and Economically Successful

As cities modernised around 1900, rural areas did as well. Time-space compression was strong. Geographic distances no longer offered farmers protection from the outside world nor did these hamper farmers to explore opportunities beyond their region. More and more farmers faced foreign competition, caused by the import of cheap wheat from the USA and Argentina. They reacted by replacing part of the farm workers by machines. To that end loans were supplied by rural banks, in particular by the Boerenleenbank and Raiffeisenbank, that opened branch offices in more and more villages since 1900. Both banks have been the predecessors of the current Rabobank.

Conversely, modern means of transport as the steamer and cold storage made export possible as well, e.g. demands for meat (cows, pigs, chickens), dairy products as butter and cheese, and vegetables (horticulture) rapidly rose in the fast-growing and more affluent cities in surrounding industrialised nations as England, Germany and Belgium. Dutch farmers took advantage of it, in particular farmers situated in the higher sands (Chap. 3), traditionally the domain of small, self-sufficient farmers. They specialised their production and abandoned food production for themselves as that was bought in stores. That could only be achieved by the availability of artificial fertilisers as originally sands were poor soils which seriously constrained production. In contrast, farmers in the maritime parts of the Dutch delta became market oriented much earlier, since the sixteenth century when they began selling their products on nearby, urban markets. An example is the horticulture in the Westland (Fig. 7.7). Foundation of farmers-run, industrial co-operatives increased quality of dairy products as milk, cheese and butter.

Agricultural and rural modernisation continued after WWII and such developments are summarised below (all data are from the CBS):

- Direct employment in agriculture (including horticulture) significantly reduced, from 34% around 1900 (700,000 people including fishery on a total population of 5.1 million) via 12% in 1960 and 5% in 1980, to about 2% now (=186,000 people in in 2015 out of 8.9 million working people and a total population of 16.9 million). In modern agriculture family labour still dominates (130,000) over outsiders (55,000).
- But indirect employment (and its economic relevance, see below) in this sector agriculture and horticulture is much higher. In 2015, 120,000 companies and 700,000 people were active in Dutch *agribusiness*, the full chain of the food industry, support services, wholesale, food retail, et cetera. Half of these were farms. Relatively few multinationals exist in agribusiness, just 2% of these companies employed more than 100 people.
- Farms strongly decreased in number, mainly after WWII. In 1950, 410,000 farms existed in the Netherlands, in 1970: 185,000, in 1990: 125,000 and in 2015: just 64,000. An example: Around 1970, 37 dairy cattle farms were operational in the village of Twisk (Noord-Holland). By today, only two farms remained (Lengkeek 2015). Villages turned from agricultural settlements into 'dwelling-villages' for commuters and retired persons (Thissen and Loopmans 2013).

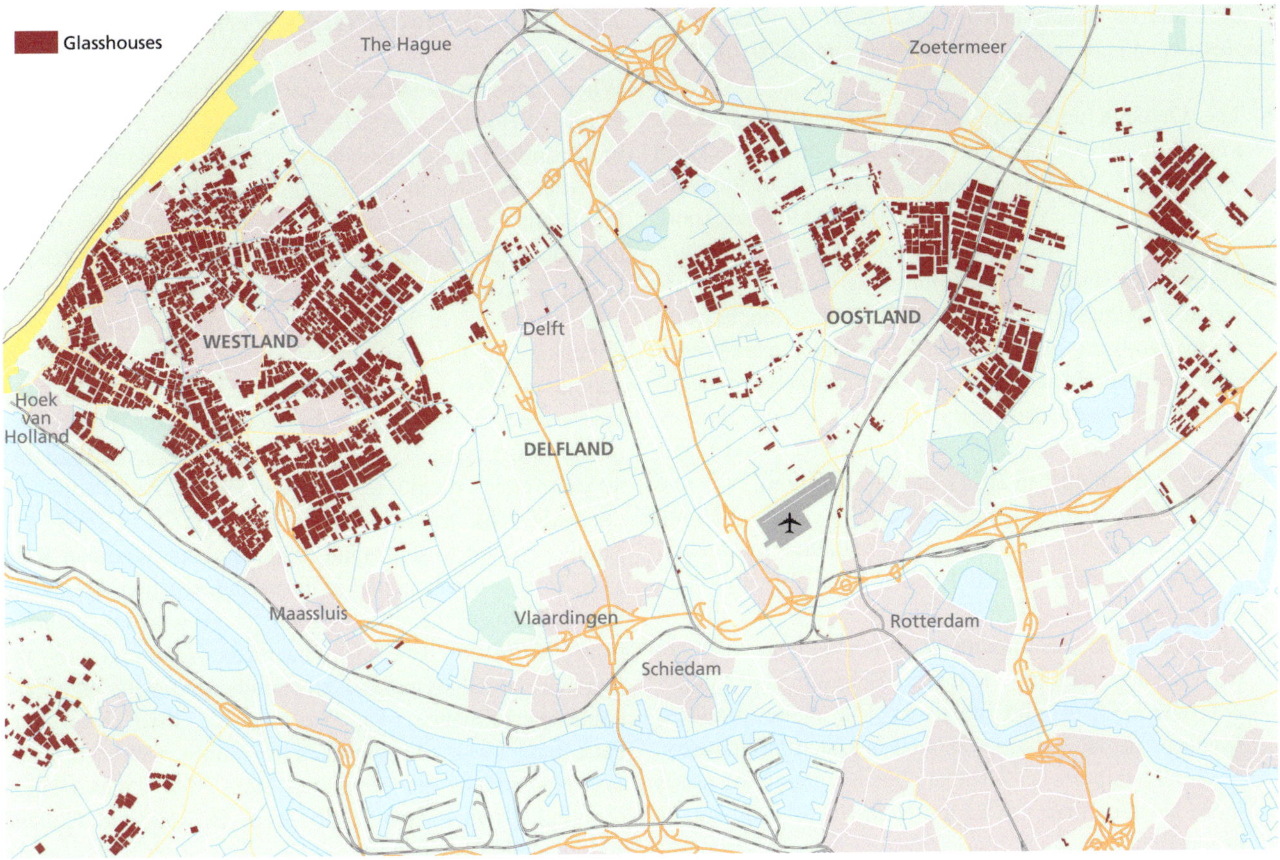

Fig. 7.7 Horticulture plantations depend partly on soil characteristics. Clay and sand are suitable, peat not. In combination with the mild climate (proximity of the North Sea), human agency (expert knowledge was gathered in cloisters and country seats), and local markets (cities in Holland), open-air horticulture developed in Westland since the sixteenth century. Westland roughly covers the sand-clay area within the Den Haag – Hoek van Holland – Maassluis triangle. In the nineteenth century, industrialising and urbanising England became a new and important market. Glasshouse horticulture was introduced. In 1912, the Netherlands had 200 hectares of glasshouses, in 1962: 2000 ha, in 1996: 4200 ha. After the 1970s processes of modernisation, restructuring and scaling-up of greenhouse industry began, including spatial expansion beyond Westland, not expansion to the adjacent Delfland region (a peat area with pastures), but to the polders (reclaimed lakes) of Berkel & Rodenrijs, Bleiswijk and Bergschenhoek (south of Zoetermeer). Together with *Westland* this area *Oostland* forms a so-called *greenport*, a governmental recognised important cluster of agricultural industry (comparable with the 'mainports' Rotterdam and Schiphol and the 'brainport' Eindhoven). (Source: IJsselstijn and Van Mil 2016)

- Total area used by farmers expanded from the end of the nineteenth century until around 1950: from 2.1 million hectares in 1900 to 2.4 million hectares in 1949 (the maximum). Increase was due to the pre-WWII reclamation of natural areas in the south, east and north, and also to reclaimed agricultural polders in the Zuiderzee/IJsselmeer: Wieringermeerpolder and Noordoostpolder. Later, total area used by farmers diminished: from 2.3 million hectares in 1950, via 2.0 million hectares in 1995, to 1.8 million hectares in 2015. This decline was caused by urban expansion, by construction of infrastructure and by nature development. Two-thirds of the agricultural hectares (1.2 million ha) is grassland in use by stock farmers, 0.5 million hectares are fields, 0.1 million hectares is used by horticulture companies. But agriculture is still the largest user of physical space in the Netherlands as a whole (3.4 million hectare).
- Average farm sizes grew from nearly 6 hectares in 1950 to 30 hectares now. In 1950, an average cattle dairy farm had about 10 cows, now about 100. Annual yields per cow (in kg milk) doubled. Production per hectare strongly intensified immensely as well, e.g. wheat production around 1950 was about 3000–4000 kilogram per hectare (weather dependent), by 2000: about 8000–9000 kilogram/ha. Fertilisers, pesticides, better seeds, guaranteed EC/EU prices contributed to the fact that Dutch agriculture is the current world leader in yields per hectare.

In terms of import and export, Dutch agribusiness is impressive and are summarised below (Jaarsma 2016):

- The Dutch export more agricultural goods than they import. In 2016, import of agricultural products valued 57 billion €, which is 15% of the value of all imported goods. In the same year, export value, including re-export, was 85 billion €, which is 20% of total Dutch export value. More than a third of the imported agricultural products was re-exported without value-adding (for example fruit), nearly two-thirds was intended for the Dutch market itself (as a raw material as e.g. cacao, or as finished product as e.g. French cheese or wine).
- Main agricultural export products are flowers and plants (value: 8.3 billion € in 2015), meat (7.6 billion), milk and dairy (7.2 billion) and vegetables (6.3 billion). Agriculture-related products as agricultural machinery, vaccines for animals, and fertilisers are also important export products.
- Most of the Dutch imports of agricultural products are from nearby nations: Germany, Belgium and France; next, the USA (4) and Brazil (5). That nation produces soybeans for cattle-fodder applied in Dutch bio-industry. Soybeans imports shows that Dutch agriculture needs more hectares than just in the Netherlands itself for its production.
- The five most important agricultural export nations are all in the EU: Germany, Belgium, United Kingdom, France and Italy. The Netherlands is the world's second exporter of agricultural products, far after the United States, but before Germany, Brazil, China, Canada and Spain.

Such economic successes inevitably go at the expense of ecological qualities (Chap. 4). Intensive agriculture has strong negative impacts on ecological conditions, both abroad (soybean fields in Brazil, palm oil plantations in Malaysia and Indonesia), and domestically by manure surpluses (phosphate) in intensive livestock farming and in cattle farms. Soil pollution by ammonia emission is particularly significant in sandy soils with low Calcium contents, where many of the pig and poultry farms are located. Diseases are another hazard related to intensive livestock farming, in particular diseases being transferred from animals to humans, as Q fever. Ethical objections against bio-industry are wide spread the Netherlands making demands for organic agriculture growing. However, compared with conventional farming, 'green farming' still is a small sector mainly because of the high prices for organic products which keeps consumer demands relatively low.

7.5 North Netherlands: Peripheral Area

Where parts of south and east Netherlands benefit from Randstad Holland's spread effects, backwash effects prevail in the north. Demographic developments are significant here as well. Population of the three northern provinces of Friesland, Groningen and Drenthe has been 17.4% of the total Dutch population by 1860. From then this ratio steadily began declining: 15.5% in 1900, 12.3% in 1940, 11.0% in 1980 and 10.2% in 2015. In absolute numbers, however, the population in the north tripled, from almost 0.58 million people in 1860 to 1.72 million in 2015. In remote rural areas even absolute numbers declined, e.g. in the rural municipality of Beerta (east Groningen): 4201 persons in 1925 and 50 years later a quarter less: 3174 in 1975 (CBS). People did not only migrate the cities in the west, to Twente and Eindhoven and to the rapidly industrialising German Ruhr area, but also to other cities in the north (see Chap. 11). While Beerta declined, the city of Groningen grew simultaneously, from 96,700 to 164,700 inhabitants. Groningen is the regional capital of north Netherlands, and with high economic performance (Van Steen et al. 2016; Kooij and Pellenbarg 1994). But the relatively thinly populated northern Netherlands as a whole lacks the agglomeration economies which abundantly occur in Randstad Holland and in south Netherlands.

The very modest benefits for this region from gas extraction from the Groningen subsurface may be exemplary for the peripheral position of the north. All State revenues of Groningen gas extraction (about 250 billion € since discovery, see Chap. 4) flow straight into the treasures of the Dutch State. Gas production is by the Nederlandse Aardolie Maatschappij (NAM), a joint venture of Shell and Exxon (50% each) and HQ in Assen, Drenthe. On the other hand, the Groningen population significantly suffers from the disadvantages resulting from gas exploitation, land subsidence and earthquakes (Chap. 4). This resulted in major protests and in successive reduction of gas extraction since 2014. This makes Groningen a clear example of spatial mismatch between advantages and disadvantages. The charges are for the region while most of the benefits are for the entire nation, and in particular for the core area. Calculation show that major shares from these gas revenues have been used to improve rail infrastructure from Randstad Holland to the south (Hogesnelheidslijn from Amsterdam to Brussels) and east (Betuwe Lijn from Rotterdam harbour to Germany), while barely 1% of these revenues returned to the north (Fonds Economische Structuurversterking 2017). For instance, the Zuiderzee line, a high-speed line from Amsterdam to north-Netherlands, has never been materialised, to the dismay of the northerners.

Such developments did generate negative feelings by many northerners in respect to Randstad Holland. Conversely, entrepreneurs valued the north low as a business location for their firms (Fig. 7.8). Large parts of the north have no good reputation as a residential area neither although northerners are more satisfied with their region, also in small and remote villages. Friesland and Drenthe are valued as leisure areas, Groningen not (Meester and Pellenbarg 1989). Negative sentiments concerning 'the north' are not unique in The Netherlands, such sentiments also occur in e.g. England and France.

Regional marketing may be relevant in coping with unfavourable images, e.g. by means of slogans. (Groningen-based) spatial planners promote the north as a residential area full of space and tranquility (Hermans and De Roo 2006). Moreover, opportunities for rural and nature tourism in the north do occur, e.g. water recreation on the Frisian Lakes and the Waddenzee, beach holidays on the Wadden Isles, camping on the sandy soils of Drenthe, and the historic city interiors of Groningen, Leeuwarden (European Capital of Culture in 2018) and of the smaller Frisian cities.

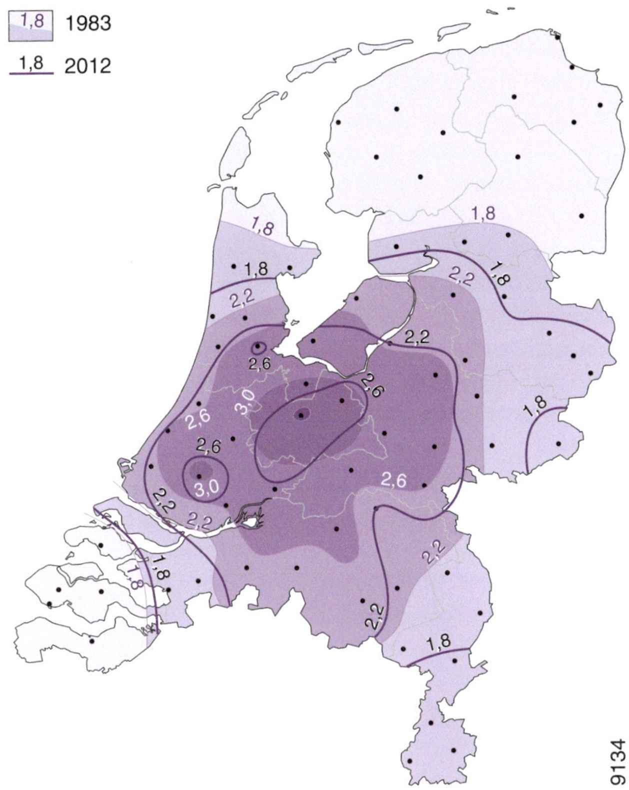

Fig. 7.8 Regional variations in spatial preferences of Dutch entrepreneurs. Changes between 1983 and 2012 are small. The region of Utrecht was and is popular as place of business. North-Netherlands, Zeeland and Limburg were and are considered not popular. (Sources: Pellenbarg and Meester 2006; Pellenbarg 2012)

7.5.1 East-Groningen and the Peat Colonies in Drenthe

Dutch areas with most welfare problems are east Groningen and, in Drenthe, the adjacent peat colonies. Many Dutch consider these areas close to the German border quite peripheral. Any intimacy in these open landscapes with large scale agriculture seems to be missing. Potato – and grain farmers, once prosperous, face rough times now and local industry (strawboard, potato flour, shipbuilding) has nearly disappeared. Social capital is modest, the population low educated and ageing. Young, entrepreneurial people are leaving the region.

When the region became more and more impoverished in the late 1960s, the Queen's Commissioner in Groningen asked the national government for assistance while pointing to risks of communist riots. Some 2 billion € governmental money was supplied to the region and invested in renovation, infrastructure enhancements, land consolidation, environmental measures, water management, home improvement, outdoor recreation, et cetera (Van der Veen 2012). This support has perhaps been the last major exogenous top-down project, in times when the government dreamed they could reshape society and had major funds available for regional restructuring projects, apart from private investors.

Today, the area is in a better state than a few decades ago, e.g. industrial surface water pollution has disappeared and road infrastructure significantly improved. But the government has been unable providing new jobs, still generating high unemployment numbers. And new challenges arrived as mentioned above. Transition to an endogenous proactive policy (see next paragraph) in this area is difficult for various reasons including the perceived unattractive landscape, less favourable geographical location, prevailing negative images and conflicts among administrators. The Blauwe Stad-project is an example of the challenges related to endogeneous regional development. In 1990, Groningen deputy Beukema suggested to turn cornfields north of Winschoten into a lake. By then, grain prices were low enough for the EU to subsidise farmers willing to break their plots. According to the proposal such fields that were thus taken out of production could be turned into a large lake for water sports and leisure, comparable to the popular Frisian Sneekermeer. Along the lake banks 1500 building plots for villa's were planned. Despite massive advertising campaigns not many plots were sold and the project was cancelled prematurely, in 2010, and losses were significant.

But in some cases failed projects might get a second life, e.g. Eemshaven in northeast Groningen. Eemshaven was built for petrochemical industry that could find no place in the Rotterdam harbour area, between 1968 and 1973. However, the Eemshaven harbour and surrounding areas remained empty. A revival occurred in recent years only when Eemshaven developed as an Energy and Data port. Nuon and RWE-Essent built several power plants in Eemshaven and Google with high energy demands for its servers developed a 600 million € datacenter. A transatlantic fiber optic cable will come ashore in Eemshaven together with an electricity cable from Norway. Furthermore, Eemshaven has become a hub for offshore wind farms. Hopefully, existing companies will attract other companies and jobs: Eemshaven as a twenty-first century growth pole, like the city of Emmen in the 1950's (see next Section). So far, Eemshaven is more capital-intensive than labour-intensive, apart from construction activities which have been done mostly by eastern European workers.

7.6 Rise and Decline of Regional Policy

For several decades the Dutch Government has tried to stimulate economic development in northern Netherlands (De Smidt 1981; Tamsma 1980). Regional policy began in 1949 when workers in the peat areas of southeast Drenthe were dismissed as peat extraction was abolished there, leaving the region with almost 4000 unemployed man, about 15% of the male labour force, by far the highest percentage in the Netherlands (average Dutch unemployment rate was about 2.5% in 1949). Regional aid was provided and extended to other areas in the north by next year. There and elsewhere in the Dutch delta agriculture rapidly mechanised generating significant unemployment among agricultural workers. Especially in the higher sand areas a 'small peasant problem' emerged: marginal farms without perspectives.

Not just the peat and the higher sand areas but regions with small-scale industrial enterprises faced difficulties as well (De Nederlandse Ontwikkelingsgebieden 1960). Processes of economies of scale and concentration emerged, e.g. in Friesland's dairy industry (Fig. 7.9).

Regional policy aimed combating poverty and unemployment by creating jobs in industry. This was meant to be implemented through designated 'growth poles', in line with French economist Perroux (Perroux 1961; Fig. 7.10). Accordingly, Emmen, the main town in southeast Drenthe, received support from the Marshall funds to attract companies. That worked out quite well and many labour-intensive industries searching for low wages and subsidies settled there, including Enkalon, a manufacturer of artificial silk for nylon stockings and a subsidiary of AKU with HQ in Arnhem (today: AkzoNobel). By 1956, Enkalon

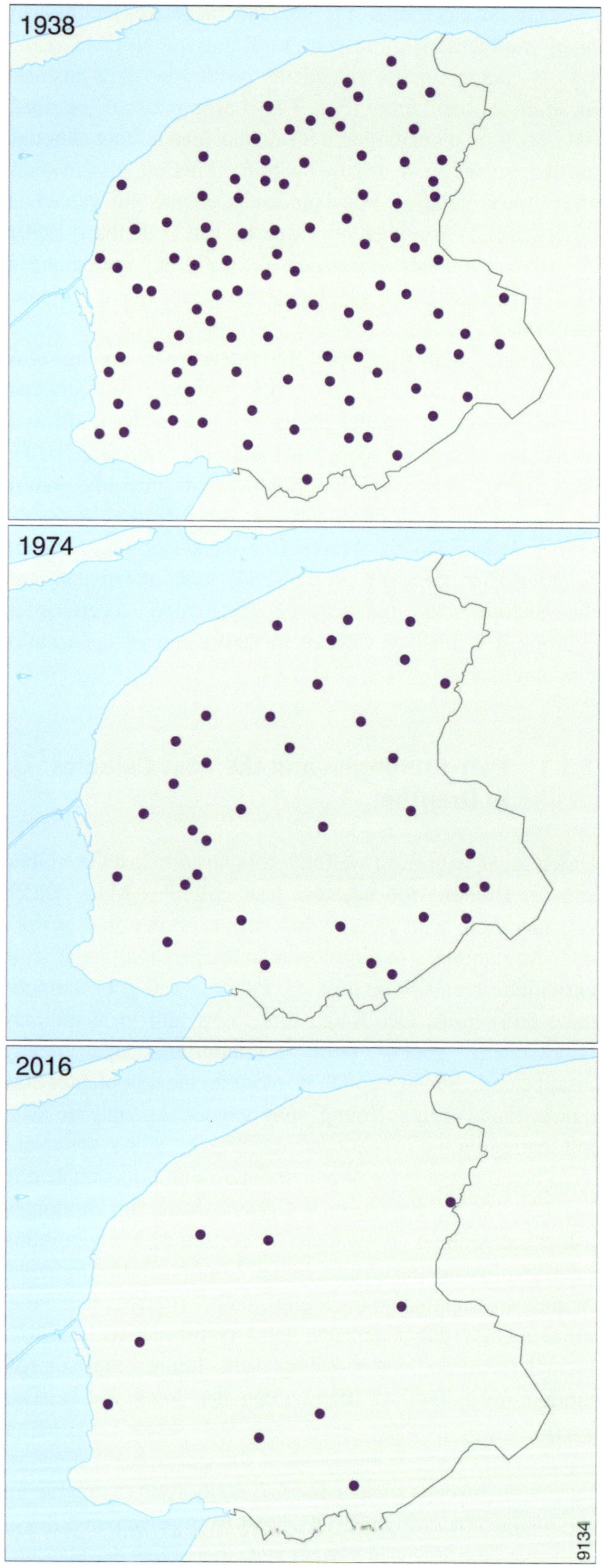

Fig. 7.9 Spatial distribution of the dairy industry (milk and cheese) in Friesland, in 1938, 1974 and 2016. (Sources: Van den Bremen 1977; https://nl.wikipedia.org/wiki/Zuivelfabriek)

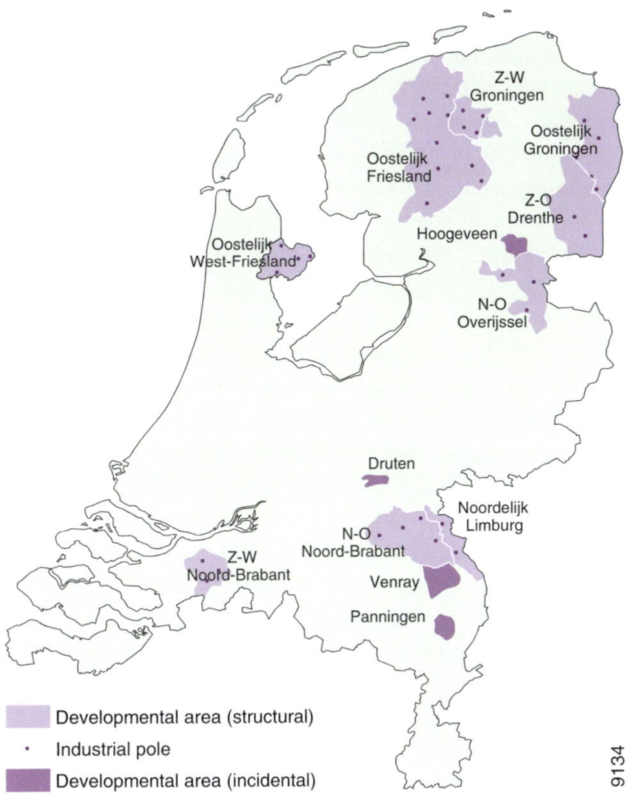

Fig. 7.10 Regional policy regions in the 1950s covering 14% of Dutch territory and containing 6.3% of the total Dutch population. From about 1960, these supported regions extended to entire north Netherlands, and included Zeeland, the Kop van Noord-Holland, east Noord-Brabant, north and central Limburg (except Eindhoven), about a quarter of the Dutch territory. (Source: De Nederlandse Ontwikkelingsgebieden 1960)

employed 1600 people already. But ten years later, Emmen's industrial centre began to decline (www.historisch-emmen.nl).

In the 1950s, Philips company opened assembly plants elsewhere in north Netherlands, in Drachten (razors), Stadskanaal (cathode ray tubes) and in Hoogeveen (domestic appliances), followed by branches in Groningen, Leeuwarden and Winschoten a few years later. Accordingly, employment in the north increased, from 62,500 jobs in 1953, via 94,500 in 1963, to 102,500 jobs in 1972. Such industrial employment growth occurred elsewhere in the Netherlands too, between 1953 and 1963, though less spectacularly. Since 1963, the number of jobs in industry generally began declining but slightly increased still in the north (Jansen et al. 1979).

A second ambition of regional policy was to keep unemployed workers in their home regions as migration to Randstad Holland was considered undesirable. By then, spatial planners disliked expanding big cities and Randstad Holland should never become like London city or a Ruhr area. Moreover, building houses in the higher sand areas was much cheaper than in the swampy west and growing cities in the west would cost much more in terms of infrastructure while job satisfaction would reduce if commuting workers would have to travel for longer times (Winsemius 1949). Aversion against big cities was a widespread phenomenon in the Netherlands in those times (e.g. Stuvel 1956). In sum, regional policy served two purposes in the 1950s: (1) Reducing regional inequality in wealth and well-being by stimulating economic growth in (relatively) deprived areas (2) Reducing growth of the 'over-heated economic engine' of Randstad Holland.

Regional policy and regional devolution culminated in the 1960s. Inspired by French concepts of developing regional metropolises that would counterbalance Paris, Dutch spatial planners adhered believes in a 'bundled deconcentration' of people and economy in and to regions beyond Randstad Holland (De Smidt 1981). According to the model of 'enhanced distribution' of stimulating people and firms moving from the west and south to the east and north, the north was an anticipated dwelling area for 3 million people by the year 2000 (Tweede Nota over de ruimtelijke ordening in Nederland 1966). Actually, only 1.3 million people lived in the north by 1965.

The anticipated Groningen northern metropolis would serve a quarter to a half a million people. HQ of the national post office PTT (now: PostNL) moved from Den Haag to Groningen. Other Den Haag-based public service bodies would be spread to other cities, as Heerlen (the former mining area centre), Zwolle, Emmen and Leeuwarden. As Rotterdam harbour expected insufficient space for further growth, ports in Zeeland (Vlissingen, Terneuzen) and Groningen (Delfzijl, Eemshaven) were extended to accommodate new seaport-bound petrochemical companies (De Smidt 1981).

The 1970s were the heyday of the welfare state and makeable societies. The by then socialist Dutch government focused on spreading power, knowledge and income. Apart from spreading across social classes, spreading comprised a strong spatial component as well. A new law (Selective Investment Scheme) stipulated that investments in Randstad Holland (and later in the Veluwe as well) would only be allowed if special permits were granted. Extending existing companies of establishing new company might thus be refused in (extended) Randstad Holland. For investments in the north and in south Limburg, however, governmental subsidies would be granted.

7.6.1 A New Line of Action

A new, Christian Democrats led Dutch government moved into opposite directions in the 1980s and faith in a mighty government, arranging social, economic and spatial planning, vanished. In reality, households and firms did not move to the north where not 3 million but barely 1.7 million people proved to live by the year 2000. Business parks at the ports in Delfzijl and Eemshaven remained underused. New Dutch governmental ambitions were deregulation, market preference and a smaller government. As in many other western European nations and regions (for instance Flanders; Vantillo and Verhetsel 2012), business was recognised and appreciated again.

Classical regional policy was gradually phased out and this was over by 1990. It was understood that the spreading policy of public services from Den Haag into the periphery resulted in spatial rearrangement of the work only and not to new jobs nationwide. Positive multiplier effects in hosting cities were counterbalanced by (more) negative multiplier effects in their cities of origin, as in Den Haag. Furthermore, incidental successes did not prove to be sustainable. Subsidised, regional companies moved to low-income nations in Asia and later in eastern Europe, e.g. in 1987 Philips company announced to move production and development of irons from Groningen to Singapore leading to angry but powerless reactions by trade unions and local governments. Today, such offshoring is partly compensated by reshoring activities and Philips returned part of their razor production from China back to Drachten where this production began in 1950. However, employees were replaced by robots (Pols 2016).

Such developments learned provinces and municipalities to become more proactive and not to rely fully on the powers of their government. Exogenous top-down policy was replaced by an endogenous bottom-up policy. Available funds were dedicated more to Randstad Holland that was more and more in competition with international urban regions, as Antwerp-Brussels, the Ruhr area, Frankfurt and Munich. 'A regional policy for the periphery' became 'an urban policy for the West' (Van Weesep 1988). In a 'Europe without economic borders', Randstad Holland rather required an 'internationally competitive business environment', to attract international companies and organizations. This new strategy was confirmed in the governmental report *Pieken in de delta* ('Peaks in the Delta'; Ministerie van Economische Zaken 2006). The Dutch government aimed 'to contribute to the ambition to create a competitive and dynamic Dutch economy'. Restoring growth powers and a competitive national economy were the new ambitions which implied focusing on strong economic areas rather than on the weak. Another new inclination was that regions would be leading this process rather than the national government as local and regional authorities best know the problems and opportunities in their areas while directly or indirectly benefitting from their actions.

Peripheral regions re-oriented themselves from Den Haag to the European Union (EU) capital, Brussels. Starting from the 1970s, the European money was inserted into 'less favoured regions' which policy was based on the Treaty of Rome (1957), to 'reduce backwardness of the less favoured regions' and diminishing 'regional imbalances'. In practice, subsidies went to peripheral rural regions and obsolete industrial areas, also in the Netherlands. Such financial support was directly provided by the European Regional Development Fund (ERDF), particularly meant to improve infrastructure. Indirect support was provided via the European Agricultural Guidance and Guarantee Fund (EAGGF). Since the early 1960s, EAGGF supplied income support to farmers providing them a minimum of prosperity in peripheral, rural areas.

7.6.2 Clusters as Lifebuoy

The endogenous bottom-up policy, or the policy of powerful autonomous regions, gained a new focus in the twenty-first century: economic-geographical clusters. One of its designers, American economist Porter (1998) defined such clusters as 'geographic concentrations of interconnected companies and institutions in a particular field'. Such clusters always existed, mostly as 'industrial districts'. Attention for clusters increased since the late 1990s. The EU considers such clusters as powerful means for strengthening the innovative economy. New knowledge may emerge by cooperation of and competition between companies belonging to one and the same cluster, which were seen as 'learning regions'. Local and regional authorities supported creation of new clusters or strengthening existing ones which is considered beneficial to both the economy and employment.

Successful clusters require strong alliances among established and new companies (start-ups), governments (including semi-governments as Chambers of Commerce), knowledge institutions (often technical universities) and investors (banks, venture capitalists), the 'triple helix' model. Silicon Valley, near San Francisco, serves as the prime example of such clusters. Over the past 25 years, thousands of such 'valleys' emerged throughout Europe as a result of local and regional initiatives.

'Cluster fever' also caught the Netherlands where many local authorities dream about creating successful clusters while generating economic benefits and jobs to their region (Anderson 2016).

Het Financieele Dagblad (Zeemeijer 2016) counted 116 partnerships and organisations in the Netherlands aiming to promote innovation, such as the Northern Innovation Board which hopes to bring one billion € of investment to northern Netherlands until 2020 (6500 new jobs). Apart from regional partnerships, local partnerships emerged as well. For instance, there are 13 'valleys' (e.g. a Health Valley in Nijmegen), 32 'campuses' (e.g. Dairy Campus and the Water Campus in Leeuwarden), 9 'delta's' (e.g. Photon Delta, Eindhoven) and 20 'innovation parks' and 'science parks' (e.g. Food Tech Park, Helmond).

Most economic-geographers believe there are too many economic clusters generating power fragmentation over too many geographic locations. According to Ebbekink, all localities/regions have 'their own valley or cluster, and they are all competing with each other, a joint race to the bottom. A few clusters take part at international level but many such initiatives are just empty shells and wishful thinking, only established to compete for central government or EU grants' (in: Anderson 2016). Many such 'Silicon Somewheres' exist all over Europe but they cannot meet the high expectations and get stuck in their start-up phase (Hospers 2006). Necessary conditions for success are: good and lasting cooperation between the parties involved; sufficient critical mass; the use of historical expertise in the region (the idea of path-dependency); and an unique profile, different from competing related clusters elsewhere in the country, Europa and the world. Important is also the presence of capable 'civic entrepreneurs', which can build a cluster (Ebbekink et al. 2015; Ebbekink and Lagendijk 2013).

The economic geography of the Netherlands has been reshaped by forces of 'glocalization', a combined process of globalization and localization. Winning and losing regions, winning and losing cities, winning and losing firms and sectors: together they form the changing economic geography of the country.

References[1]

Aerotropolis (2017) www.aerotropolis.com. Accessed 15 Jan 2017

Anderson R (2016) Elk dorp een Silicon valley. De Volkskrant. 28 May 2016

Atzema O, Wever E (2004) Sluipt Philips weg uit Nederland? Geografie 13(September):6–10

[1] For this chapter we benefited also from internet sources (wikipedia) and articles in Dutch newspapers (de Volkskrant, NRC-Handelsblad, Trouw and Het Financieele Dagblad).

Atzema O, Hospers GJ, Van Oort F, Renooy P, Teisman G, Tordoir P (2016) De kracht van Oost-Nederland. Een economisch-geografische analyse. Werkdocument. Regioplan, IHS, ESD^2, Ruimtelijk Economisch Atelier Tordoir, Amsterdam

Bird J (1983) Gateways: slow recognition but irresistible rise. Tijdschr Econ Soc Geogr 74(3):196–202

Blink H (1915) De ontwikkeling der economische toestanden in Europa en in Nederland van 1870 tot 1914. Tijdschr Econ Geogr 6 (IX):329–344

Boschma RA (1994) Looking through a window of locational opportunity. A long-term spatial analysis of techno-industrial upheavals in Great Britain and Belgium. Dissertation Erasmus University Rotterdam. Tinbergen Institute Research Series, vol 75, Rotterdam

Boschma RA, Hartog M (2014) Merger and acquisition activity as driver of spatial clustering: the spatial evolution of the Dutch banking industry, 1850–1993. Econ Geogr 90(3):247–266

Bosman J, De Smidt M (1993) The geographical formation of international management centres in Europe. Urban Stud 30(6):967–980

Brusse P, Mijnhardt WW (2011) Towards a new template for Dutch history. De-urbanization and the balance between city and countryside. Waanders/Universiteit Utrecht, Zwolle/Utrecht

Burghouwt G (2002) De onweerstaanbare opkomst van de airport city. Geografie 11(September):6–12

Burghouwt G (2005) Airline network development in Europe and its implications for airport planning. Dissertation, University Utrecht

Capasso M, Cefis E, Frenken K (2016) Spatial differentiation in industrial dynamics. The case of the Netherlands (1994–2005). Tijdschr Econ Soc Geogr 107(3):316–330

Christaller W (1933) Die Zentralen Orte in Süddeutschland. English edition: Central Places in Southern Germany, 1966. Prentice-Hall, Englewood Cliffs

De Feyter CA (1982) Industrial Policy and Shipbuilding. Changing economic structures in the Low Countries 1600–1980. Dissertation, Utrecht University. HES Publishers, Utrecht

De Groot HLF (2012) Pieken in de polder?! Oratie. Vrije Universiteit, Amsterdam

De Klerk L, Van de Laar P, Moscoviter H (2008) GJ de Jongh. Havenbouwer en stadsontwikkelaar in Rotterdam. Thoth, Bussum

De Lange R (2016) Inside ASML. Zoektocht naar de succesfactoren achter's werelds grootste chipmachinebouwer. FD, Amsterdam

De Nederlandse ontwikkelingsgebieden (1960) Schets van de sociale problematiek (rapporteur: Van Doorn JAA). Staatsdrukkerij- en Uitgeverijbedrijf, Den Haag

De Pater B, Van der Vaart R (2014) Randstad Holland. In: Besamusca E, Verheul J (eds) Discovering the Dutch. On cultural and society of the Netherlands, 2nd edn. Amsterdam University Press, Amsterdam, pp 69–80

De Smidt M (1981) Ruimtelijke spreiding en regionale ontwikkeling. In: Borchert JG, GJJ E, De Smidt M (eds) Ruimtelijk beleid in Nederland (vierde druk). Unieboek, Bussum, pp 9–41

Decisio (2015) Economisch belang van de mainport Schiphol. Analyse van directe en indirecte economische relaties. Decisio, Amsterdam

Dieperink H, Nijkamp P (1986) De agglomeratie-index: een ruimtelijke indeling op grond van agglomeratieverschijnselen. Planning 27:2–8

Dwarkasing W, Hanemaayer D, De Smidt M, Tordoir PP (1988) Ruimte voor hoogwaardige kantoren. Nederlandse Geografische Studies, Koninklijk Nederlands Aardrijkskundig Genootschap, vol 71. Utrecht University, Amsterdam/Utrecht

Ebbekink M, Lagendijk A (2013) What's next in researching cluster policy: place-based goverance for effective cluster policy. Eur Plan Stud 21(5):735–753

Ebbekink M, Hoogerbrugge M, Lagendijk A, Kerkhof J (2015) Cluster governance. Lessen voor clusters in Nederland. Platform31/ Radboud University, Den Haag/Nijmegen

Everwijn JCA (1912) Historisch-economische atlas. Beschijrving van handel en nijverheid in Nederland. NV Boekhandel, Den Haag

Fonds Economische Structuurversterking 20170 https://nl.wikipedia.org/wiki/Fonds_Economische_Structuurversterking)
Fox EW (1971) History in geographic perspective: the other France. W.W. Norton, New York
Friedmann JRP (1966) Regional development policy: a case study of Venezuela. MIT & Harvard, Cambridge, MA
Gottmann J (1950) A geography of Europe. Revised edition (first edition 1950). Henry Holt and Company, New York
Haegens K (2016) 'Geschrapt' ING-werk naar lagelonenland. De Volkskrant, 22 Oct 2016
Hermans EW, De Roo G (eds) (2006) LILA en de planologie van de contramal. De ruimtelijk-economische ontwikkeling van Noord-Nederland krijgt een eigen kleur. Boekvorm, Assen
Hohenberg PM, Lees LH (1985) The making of Urban Europe 1000-1950. Harvard University Press, Cambridge, MA
Hospers GJ (2006) Silicon somewhere? Assessing the usefulness of best practices in regional policy. Policy Stud 27(1):1–15
IJsselstijn M, Van Mil Y (2016) Atlas van het Westland. 10.000 jaar ruimtelijke ontwikkeling. Thoth, Bussum
Jaarsma M (ed) (2016) Internaliseringsmonitor 2016-II. Agribusiness. Centraal Bureau voor de Statistiek, Den Haag
Jacobs J (1984) Cities and the wealth of nations. Random House, New York
Jansen ACM, De Smidt M, Wever E (1979) Industrie en ruimte. De industriële ontwikkeling van Nederland in een veranderend sociaal-ruimtelijk bestel. Van Gorcum, Assen
Kasarda JD (2013) Airport cities: the evolution. Airport World 18, Issue 2 (April)
Kooij P, Pellenbarg PH (1994) Regional capitals. Past, present, prospects. Van Gorcum, Assen
Lengkeek J (2015) Ik weet nog hoe het was. Dorp tussen nostalgie en nieuwe tijden. Vitruvius 8(January):5–9
Luyten M (2015) Het geluk van Limburg. De Bezige Bij, Amsterdam
Manshanden WJJ, Bouman-Eijs JM, Koops R, Fischer NM, Chahim M, Rietveld E (2014) De top 20 van Europese grootstedelijke regio's 1995-2012. TNO, Delft
Marshall A (1919) Industry and trade. MacMillan, London
Meester WJ, Pellenbarg PH (1989) Het imago van de Nederlandse provincies. Geografisch Tijdschrift 23(2):88–97
Meijers EJ, Burger MJ (2017) Stretching the concept of 'borrowed size'. Urban Stud 54(1):269–291
Messing FAM (1988) Geschiedenis van de mijnsluiting in Limburg. Noodzaak en lotgevallen van een regionale herstructurering, 1955–1975. Martinus Nijhoff, Leiden
Mijnsluitingsnummer (1971) Geologie en Mijnbouw, 2, Spec. Issue (summaries in English), 95-337. Koninklijk Nederlands Geologisch en Mijnbouwkundig Genootschap, Delft
Ministerie van Economische Zaken (2006) Pieken in de delta. Gebiedsgerichte economische perspectieven. Ministerie van Economische Zaken, Den Haag
Myrdal G (1963) Economic theory and under-developed regions. Methuen, London
NFIA (2016) Resultaten 2016. Netherlands Foreign Investment Agency, Den Haag
Pellenbarg PH (2012) De mental map van de Nederlandse ondernemer: 1983-1993-2003-2012. Farewell Lecture Rijks Universiteit Groningen
Pellenbarg PH, Meester WJ (2006) The spatial preference map of Dutch entrepreneurs: subjective rating of locations 1983, 1993, 2003. Tijdschr Econ Soc Geogr 97(4):364–376
Pellenbarg PH, Van der Steen PJM (2003) Spatial perspectives on firm dynamics in the Netherlands. Tijdschr voor Econ Soc Geogr 94 (5):620–630
Perroux F (1961) L'économie du XXème siècle. Presses Universitaires de France, Paris
Pols G (2016) Reshoring: robotarm vervangt Aziatische handen. De Volkskrant 26 May 2016
Porter M (1998) Clusters and the new economics of competition. Har Bus Rev November-December:77–90
Prud'homme van Reine P, Dankbaar B (2011) The dynamic interaction between corporate and regional cultures: the case of Southeast Netherlands. Tijdschr Econ Soc Geogr 102(5):532–547
Schaafsma M (2014) Synergien zwischen Flughafen und Stadt? Das Beispiel Amsterdam-Schiphol. Geogr Rundsch 66(1):28–34
Scheffer M (2009) Fatal clusters. The evolutionary pathway of the Tilburg wool industry. In: Van Boom N, Mommaas H (eds) Transformation strategies for former industrial cities. NAi Publishers, Rotterdam, pp 202–217
Schiphol (2017) Schiphol Group, facts & figures. www.infoschiphol.nl/home/geschiedenis. Accessed 29 Jan 2017
Stohr H (1989) De Mijnstreek: van mijnschacht naar schrijftafel. In: De Pater BC, Hoekveld GA, Van Ginkel JA (eds) Nederland in delen. Een regionale geografie, vol 2. De Haan/Unieboek, Houten, pp 93–120
Stuvel HJ (1956) Het Deltaplan, deel 1. De geboorte. Scheltema & Holkema, Amsterdam
Tamsma R (1972) The Northern Netherlands: large problem area in a small country, small problem area in a large economic community. Tijdschr Econ Soc Geogr 63(3):162–179
Tamsma R (1980) Three decades of regional policy in the Netherlands 1950–1980. Sociaal-geografische reeks 18. Geografisch Instituut, Rijks Universiteit Groningen
Ter Hart HW, Piersma J (1990) Direct representation in international financial markets: the case of foreign banks in Amsterdam. Tijdschr Econ Soc Geogr 81(2):82–92
Thissen F, Loopmans M (2013) Dorpen in verandering. Rooilijn 46 (2):81–89
Toonen WPG (1972) The economic restructuring of South Limburg between 1965 and 1971. Tijdschr Econ Soc Geogr 63(3):180–189
Tweede Nota over de ruimtelijke ordening in Nederland (1966). Rijksplanologische Dienst, Den Haag
Van den Bremen WJ (1977) Verandering en continuïteit. Een studie over de dynamiek in en van de ruimtelijke organisatie van de bedrijvigheid met speciale aandacht voor het stelsel van melkverwerking in het ruraal-urbaan spanningsveld in Nederland. Dissertation, Rijks Universiteit Groningen
Van der Lugt L, Witte JJ, De Jong O (2015) Havenmonitor. De economische betekenis van Nederlandse zeehavens 2002–2014. Erasmus Universiteit – RHV, Rotterdam
Van der Veen C (2012) Een kapitale impuls. De herinrichting van Oost-Groningen en de Gronings-Drentse Veenkoloniën 1969-2011. Van Gorcum, Assen
Van Geenhuizen MS (1993) A longitudinal analysis of the growth of firms. The case of the Netherlands. Dissertation, Erasmus Universiteit Rotterdam
Van Oort F, Meijers E, Thissen, M, Hoogerbrugge M, Burger M (2015) De concurrentiepositie van Nederlandse steden. Van agglomeratiekracht naar netwerkkracht. Platform31, Den Haag
Van Steen PJM, Pellenbarg PH, Groote PD (2016) Economic performance of cities in the Netherlands. Tijdschr Econ Soc Geogr 107 (2):254–256
Van Weesep J (1988) Regional and urban development in the Netherlands: the retreat of government. Geography 73(2):97–104
Van Zuidam JMA (1989) Twente, regio van continuïteit en herstructurering. In: De Pater BC, Hoekveld GA, Van Ginkel JA (eds) Nederland in delen. Een regionale geografie, vol 2. De Haan/Unieboek, Houten, pp 191–203
Vantillo T, Verhetsel A (2012) Paradigm change in regional policy; towards smart specialiation? Lessons from Flanders (Belgium). Belgeo 1-2 (Inaugural Issue)

Venner J (ed) (2001) Geschiedenis van Limburg.Twee delen. Limburgs Geschied- en Oudheidkundig Genootschap, Maastricht

Vermeulen T (2016) De Deltametropool als vestigingsplaats voor grote ondernemingen. Metropool Forum 12–13

VNO-NCW (2016) Nederland voor altijd hét knooppunt van de wereld. NWO-NCW, Den Haag

Wever E (1993) Bedrijvigheid en steden. In: Ottens HFL, Van Weesep J, Wever E (eds) De complexe stad: geografisch onderzoek in actie. Faculteit Ruimtelijke Wetenschappen. Universiteit Utrecht, Utrecht, pp 5–26

Winsemius J (1949) Vestigingstendenzen van de Nederlandse nijverheid, deel 1. Staatsdrukkerij en -uitgeverijbedrijf, Den Haag

Zeemeijer I (2016) Nederland komt om in de valleys, clusters en delta's. Het Financieele Dagblad. 9 May 2016

8 Cities, Suburbs and Villages

Joos C. Droogleever Fortuijn

Abstract

The Netherlands is a densely populated and highly urbanized nation but without megacities. It is a country of towns and medium-sized cities. OECD's Regional Typology identifies no rural and remote areas in the country. Even the smallest and most isolated villages are part of an urbanized or intermediate region. Distances to urban facilities as hospitals are less than a 45 min car drive for almost all Dutch people. Water has always played an important role in urban development of the Netherlands. Most (older) cities are along rivers or the coast and developed as trading centers. Urban expansions may not only occur onshore but on artificial islands in sea or rivers and in reclaimed lands as well.

Keywords

Urban growth · Randstad holland · Suburbanisation · New towns · Urban stagnation · Urban renewal · Rural communities

The Netherlands is a highly urbanized country. It is one of the world's most densely populated nations (Fig. 8.1 and 8.2), with on average 490 inhabitants per km^2. Only average population densities in Bangladesh, Taiwan, South-Korea and city-states as Singapore are higher (Ekamper 2010). At the same time, it is a country without any megacity. Towns and medium-sized cities, all less than one million inhabitants, are distributed over the entire country with relative short distances in between. Even the remotest regions are relatively close to an urban center. The country has no mountains, no deserts, no extensive woods and nature areas (except Waddenzee) and from almost any place in the country a tower of a village church or a skyline of an urban extension is visible.

According to OECD's Regional Typology (2011), based on population density, size and accessibility, i.e. driving time to reach a center of >50,000 inhabitants, no rural areas nor remote areas occur in the Netherlands. However, according to the Dutch national government, one third of the total population lives in an (almost) non-urban municipality and another one-third assigns its own place as rural (Steenbekkers et al. 2008). All Dutch villages are part of an urbanized or intermediate region, and vice versa, urban facilities are within reach of each village. Distance to a hospital e.g. is less than a 45 min car drive for almost all (99%) inhabitants of the Netherlands (Kommer et al. 2015).

8.1 Origin and Growth of Dutch Cities

Most towns and cities developed as market, trade and administrative centers. Towns and cities in the continental east emerged and grew somewhat earlier than those in the maritime west of the country.

In 'Atlas of the Dutch urban landscape', a study of 35 cities with more than 80,000 inhabitants in 2010, Rutte and Abrahamse (2016) distinguish four city types (Fig. 8.3). The *first* type are the oldest towns that originated in the eleventh and twelfth century or even earlier, in Roman and early Medieval Times. These were positioned on topographically high places, the sandy embankment of one of the major rivers in the delta, and usually developed along one of the river sides, e.g. Arnhem, Nijmegen, Deventer and Dordrecht.

A *second* city type, i.e. second-generation towns, originated from the thirteenth and fourteenth centuries and usually are located along inland rivers, several of them as Haarlem, Leiden, Alkmaar and Delft, in the west of the country. Many of the oldest and second-generation towns were prosperous trading centers by the end of the Middle Ages. Several such towns and cities in the east were booming in the fourteenth and fifteenth century as Hanseatic towns,

J. C. Droogleever Fortuijn (✉)
Department of Geography, Planning and International Development Studies, University of Amsterdam, Amsterdam, The Netherlands
e-mail: J.C.DroogleeverFortuijn@uva.nl

E. F. J. De Mulder et al., *The Netherlands and the Dutch*, World Regional Geography Book Series,
https://doi.org/10.1007/978-3-319-75073-6_8

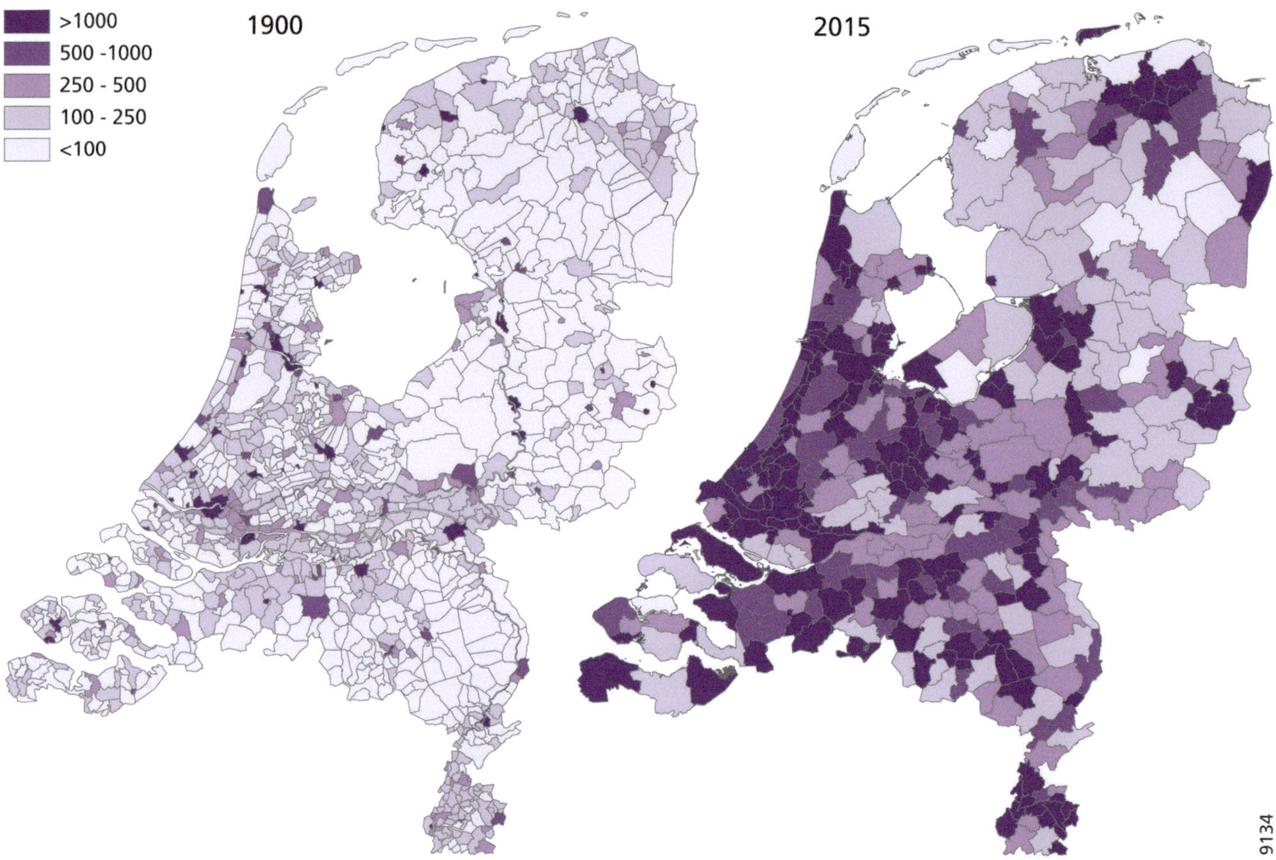

Fig. 8.1 Population density per municipality 1900 and 2015 (number of inhabitants per km^2). (Sources: Ekamper 2010; http://www.metatopos.eu/gemprov.html)

e.g. Groningen, Zwolle, Deventer and Nijmegen, but smaller towns as Stavoren, Kampen, Elburg, Tiel and Venlo as well. Growth in these towns and cities halted until they revived in the second half of the nineteenth century.

The three largest cities in the Netherlands, Amsterdam, Rotterdam and Den Haag, were exceptional cases of this second-generation type. Amsterdam and Rotterdam emerged in the Middle Ages on a river shore. These cities boomed later than most other second generation cities. Amsterdam flourished in the seventeenth century because of trade with Hanseatic, Baltic Sea cities and the West- and East Indies. With 219,000 inhabitants Amsterdam was one of the world's biggest cities by the end of the seventeenth century (Rutte and Abrahamse 2016). Although urban growth stagnated for almost two centuries since about 1670, Amsterdam remained the biggest city in the Netherlands. A next period of rapid growth coincided with the second industrial revolution around 1870 (Wagenaar 1990).

Rotterdam began as a small trading town in Early Medieval Times and rapidly developed in Late Medieval Times. By the end of the seventeenth century Rotterdam was the third biggest city of the country, after Amsterdam and Leiden (Rutte and Abrahamse 2016). Rotterdam took over position of the nation's second largest city upon population decline in Leiden. Stagnation prevailed in Rotterdam during the first half of the nineteenth century but the city revived after 1850 and Rotterdam became the world's largest trading port in the twentieth century (Chap. 7) (De Klerk et al. 2008).

Den Haag emerged in the thirteenth century as a court town of the Count of Holland. After independence of the Netherlands in 1588, Den Haag became the administrative center for the new Dutch Republic and later for the entire Kingdom, founded in 1815. Unlike other cities in the Netherlands, Den Haag faced no stagnation but rather continued to grow from the eighteenth century onwards until 1960 (Schmal 1995). Since 1850 Amsterdam, Rotterdam and Den Haag have been the 'big three'. But with populations of less than one million these 'big three' are not more than medium-sized cities as compared to metropolitan cities in other nations.

A *third* city type includes industrial towns originated by the end of the nineteenth century and developed rapidly from small villages into medium-sized cities. Unlike trading ports of the oldest and second generation cities, these cities were not situated along rivers but rather on sandy soils. Availability of cheap labour had been the main factor for their location.

Fig. 8.2 Red, green and blue spaces in the Netherlands 2012. (Source: CLO 2012)

Infrastructural networks (canals, railways and later motor ways) were developing to accommodate transport of raw materials and distribution of end products. Urban growth in cities as Enschede, Hengelo, Eindhoven, Tilburg and Helmond was caused by some successful entrepreneurs who founded flourishing industries: electric light (Philips in Eindhoven), machines (Stork in Hengelo) or textiles (Van Heek in Enschede). Development of towns and cities in Limburg's mining area, i.e. Heerlen, Kerkrade, Sittard and Geleen, was generated by regional coal mining industry (see Chap. 7).

A *fourth* type, residential towns, developed due to their residential qualities. The oldest residential towns and cities originated by the end of the nineteenth century in attractive sandy landscapes: Hilversum in 't Gooi, Apeldoorn in Veluwe. Spacious suburban villas in woodland areas are characteristic for the oldest residential towns. The newer residential towns, such as Almere, Hoofddorp and Zoetermeer, are rather positioned on clay or peat and developed due to a 'clustered suburbanisation' policy, as initiated by the national government in the 1960s (see Chap. 10).

8.2 Randstad Holland

In his seminal book *The world cities* (1966), British geographer Hall summarizes the main characteristics of the world's metropolitan areas as centers of national and international political power, trade, finances, education, culture and entertainment, and wealth. He identifies 24 such world cities with a population of 3–15 million inhabitants in 1960. Seven out of these world cities were studied more in particular, the six largest (New York, Tokyo-Yokohama, London, Rheine-Ruhr, Moscow and Paris, each with eight million people or

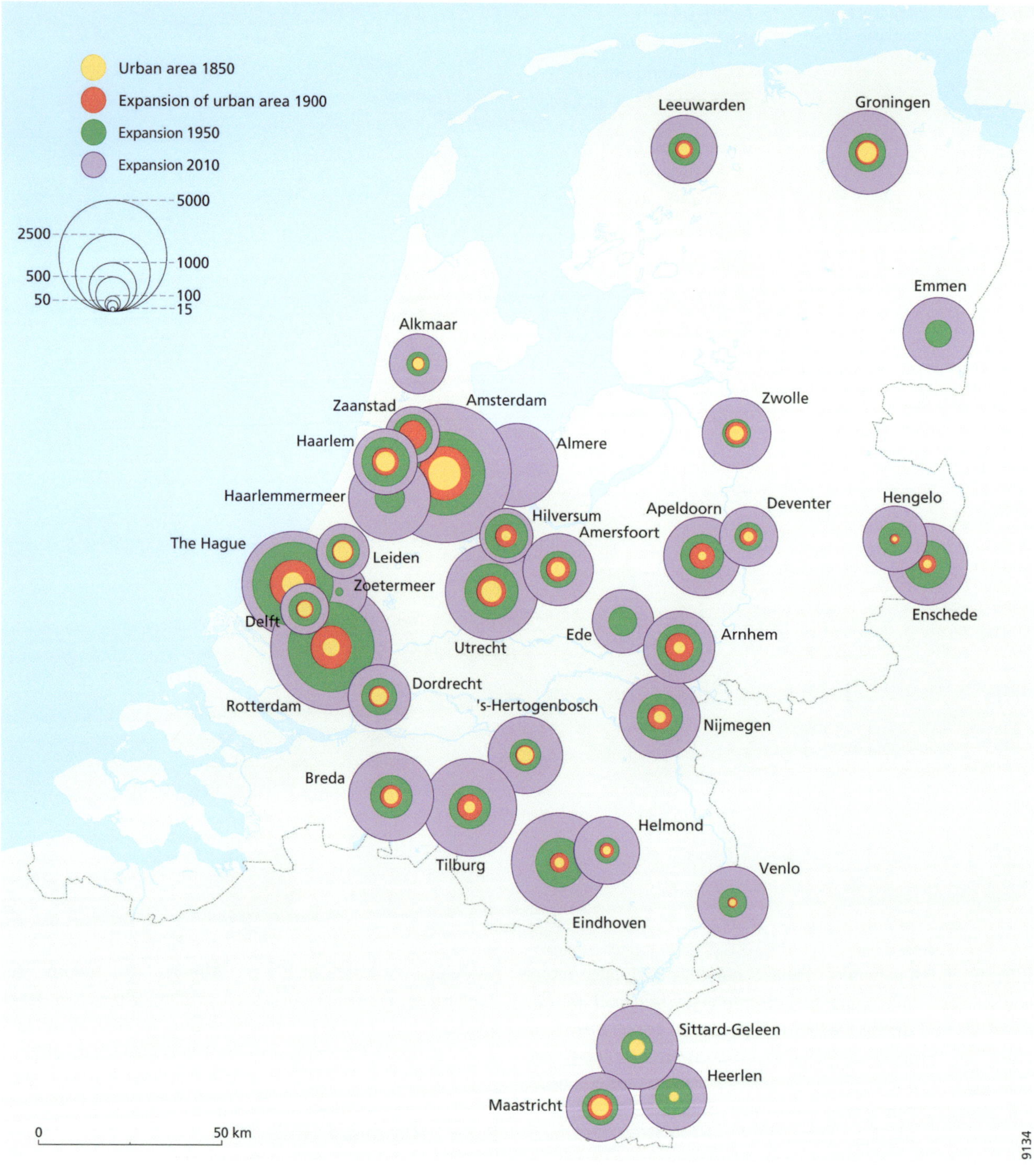

Fig. 8.3 Population growth of major cities 1850–2010. (Source: Rutte and Abahamse 2016)

more) and a special case: Randstad Holland (*Ring City Holland*) with a population of just four million people in 1963. Randstad Holland is a horseshoe-shaped ring of medium-sized cities, running from Dordrecht in the south, through Rotterdam, Schiedam, Vlaardingen and Delft to Den Haag, then via Leiden and a row of densely built villages to the north, to Haarlem and IJmuiden, then eastward to Zaanstad and Amsterdam and then runs SE to Hilversum and southward to Utrecht (Fig. 8.4). Although the Randstad is a more or less unbroken city ring, individual cities may still be recognized.

Unlike metropolitan areas as London and Paris, Randstad Holland is a polycentric metropolis where each city or town has a special function (Chap. 7). Smaller cities, such as Delft,

Fig. 8.4 Built-up area in the Randstad 1900 and 2015. (Source: Bazelmans et al. 2005; PDOK 2015)

Dordrecht, Leiden and Haarlem are specialized in light manufacture, while Hilversum is the media city (radio and television). In 1960, Randstad Holland comprised about one third of the total Dutch population on just 5% of the country's land area.

Randstad Holland towns and cities are positioned around a green area (het 'Groene Hart' – 'Green Heart') with a mixture of dairy farming and nature areas around the lakes that were left from earlier peat extraction (Chaps. 2, 3 and 4). In line with national spatial planning policy, urban extension is intended to take place in outward directions to preserve the 'Groene Hart's green signature. According to Hall (1966), the Randstad Holland model is better equipped to accommodate population growths than monocentral cities as London and Paris.

Whether or not Randstad Holland is one metropolitan area is subject of academic debate. Based on an analysis of relations among companies, Van Oort et al. (2010) conclude that Randstad Holland does not operate as an integrated, polycentric urban network. They did not found any spatial integration within Randstad Holland nor complementarity among their main cities. They rather consider four separate urban regions: greater Amsterdam, Rijnmond, Haaglanden and the Utrecht region, more realistic. These regions constitute the most important economic center where population growth was most predominant.

8.3 Suburbanisation

In the seventeenth and eighteenth centuries, wealthy Amsterdam merchants built country houses outside the city, e.g. on the banks of small rivers as Amstel and Vecht, to temporarily escape from the city's crowds, dirt, smell and noise. By then, commuting on a daily basis was not possible due to long travel times. But rapidly emerging rail infrastructure made daily commuting a realistic option for rich urban residents towards the end of the nineteenth century.

That marked the start of suburbanisation and prosperous families moved from the largest cities in the west to sandy dune areas, to 't Gooi and Utrechtse Heuvelrug. Private developers bought land and developed housing estates with suburban villas and large gardens there. A similar process of suburbanization occurred later in the cities in the north, east and south, where wealthy families built new homes in northern Drenthe, at Veluwe and in the sands around cities in Noord-Brabant. Nowadays, these early suburbs still belong to the nation's richest municipalities (Chap. 13). Bloemendaal (in Noord-Holland's dunes), Wassenaar (in the dunes north of Den Haag), Blaricum ('t Gooi) and Rozendaal (Veluwe) are municipalities with highest average incomes per capita, in 1931 as well as in 2015 (CBS 2015).

In the first half of the twentieth century, suburbanisation remained the exclusive domain of the rich who could afford the costs of a luxury suburban villa and daily commuting from the suburb to the city. Since, travel times and -costs by train decreased rapidly and transport by private car emerged as a commuting option simultaneously (Chap. 9).

Suburbanization processes in the Netherlands changed fundamentally in the 1960s. By then, Dutch economy was booming (Chap. 7) partly due to natural gas exploitation (Chap. 4) and incomes of Dutch employees substantially rose by every new round of salary increases. Wealthier people and improved public and private transport systems enabled middle-class families to live in houses with gardens outside the city. Commuting increased and more and more towns became commuter municipalities: 3.5% in 1947 and 16% in 1971 (Ostendorf 1988). In Noord-Holland and Zuid-Holland even half of all municipalities belonged to this category in 1971.

8.4 Role of the State: New Towns

While suburbanization was driven by private developers in the late nineteenth and early twentieth centuries, mass suburbanization was state-driven since 1960. The Dutch government was alarmed by 'urban sprawl' symptoms in the USA where numbers of suburban housing estates and cars were growing almost infinitely. In reaction on demographic challenges by postwar babyboom (Chap. 11), suburbanization processes and increasing mobility in the Netherlands, the national government introduced a series of planning instruments to control such new urban development (Chap. 10). 'Clustered suburbanisation' became the core concept in urban planning in the 1960s and 1970s. Urban growth was directed into specifically designated, new towns. To preserve the green landscapes, in particular in the 'Groene Hart', existing villages and towns somewhat away from the big cities were designated to absorb their population growths. This steering process was supported by legal and financial instruments.

Growth centers as Groningen, Zwolle, Duiven/Westervoort, Amersfoort, Breda and Helmond, were aimed at stimulating economic development beyond Randstad Holland. New towns in the west were geared to accommodate overspills from the big cities and were positioned away from Randstad Holland. Amsterdam's overspill was directed to Alkmaar, Hoorn, Purmerend, Lelystad and Huizen. Spijkenisse, Hellevoetsluis and Capelle aan de IJssel were assigned to accommodate overspill from Rotterdam, while Nieuwegein and Houten had a similar function in absorbing Utrecht's overspill of people. Due to Den Haag's geographical position near the sea and to preserved nature areas of the coastal dunes, overspill of their citizens could only be accommodated inside the 'Groene Hart': Zoetermeer.

These overspill towns were existing villages and towns with the specific task to grow into cities of 75,000–200,000 inhabitants in a few decades only. Alkmaar, e.g. was a middle-sized historic town with a population of 38,000 inhabitants in 1950. It developed in a new town with more than 100,000 inhabitants in 2017. Nieuwegein was a merger of two villages (Jutphaas and Vreeswijk) where together lived almost 15,000 people in 1950. Today (2017), in the new town of Nieuwegein, situated south of Utrecht, live more than 60,000 people.

Two such new towns were entirely new indeed: Almere built in IJsselmeerpolder Zuidelijk Flevoland and Lelystad in Oostelijk Flevoland. Lelystad is the capital city of the new province Flevoland. In 1967, the first residents arrived there and currently Lelystad has a population of 77,000 inhabitants. In Almere, the first houses were delivered in 1976 and now (2017) this city accommodates 201,000 people.

These new towns were not built immediately adjacent to but rather at a distance of some 10–30 km from their donor cities. The 'clustered suburbanisation' concept was not only meant to absorb urban population growth but also to accommodate any overspill of economic activities. Thus the new towns were not planned as commuting towns but rather as cities with employment opportunities for their new residents. The national government has quite powerful instruments in determining where new residential development will occur but is less effective in assigning sites for new economic activities. This mismatch between residential and economic planning of industries and offices resulted in significant commuting flows between the new towns and their donor cities, e.g. between Lelystad/Almere and Amsterdam.

The 'clustered suburbanisation' policy has been successful in terms of developing new towns. It was less successful in limiting growth of villages and small towns beyond the designated new towns. Towns and villages, in particular in the 'Groene Hart', were growing more than anticipated.

In response to economic stagnation in the 1980s and to population decline of the big cities, Dutch national spatial planners changed their policy toward prioritizing economic growth in the big cities while reducing mobility in the 1990s (Van der Cammen and De Klerk 2012). The Dutch government began promoting 'compact cities' and new urban development was anticipated now in and close to the cities and along main railway lines. IJburg became Amsterdam's main new expansion site in the late 1990s, Leidsche Rijn for Utrecht, Schothorst for Amersfoort, Wateringse Veld, Ypenburg and Leidseveen for Den Haag and Bergschenhoek, Carnisseland and Nesselande for Rotterdam (Box 8.1).

Box 8.1: Urban Expansion on Artificial Islands: IJburg

A major policy shift in the 1990s resulted in new urban extensions near or within cities. One of the challenges of this shift was the lack of building space in cities. Creating artificial islands in cities was an option to cope with this challenge. Amsterdam has a long tradition in urban expansion by constructing artificial islands in former sea arm IJ. By the end of the sixteenth century, the islands Uilenburg, Valkenburg and Rapenburg had been constructed at the north side of the city while creating space for shipyards and timber yards. New islands for shipyards and warehouses were built in the seventeenth century, the 'Dutch Golden Age', and by the end of the nineteenth century.

In contrast to earlier artificial IJ-islands, IJburg is created for residential use, not for harbour or industrial purposes. Amsterdam opted for IJburg in 1996 and after a referendum (with most voters against the IJburg-plans, but the referendum was declared invalid as insufficient voters showed up) construction of the new islands began. Seven islands were planned and three of them realized in 2017. Steigereiland has been the first and part of its residences are 'water'-houses, designed by architect Marlies Rohmer. These houses are accessible via small quays. They were constructed at a shipyard and transported over water to their final destination (Fig. 8.5).

Fig. 8.5 Floating houses in IJburg (Source: http://www.rohmer.nl/project/waterwoningen-ijburg/)

8.5 Stagnation and Revitalisation of Cities

Growth of the three largest Dutch cities peaked around 1960. Amsterdam had a population of 872,000 in 1959 when decline started. Simultaneously, Rotterdam reached its maximum population: 731,000 inhabitants and Den Haag peaked 1 year later with 606,000 inhabitants. Population declines in other Dutch cities began later (in the 1970s or 1980s) and has been less pronounced than in the 'big three' cities.

Dutch cities were faced with a dilapidated housing stock and many urban residents preferred to live in a spacious and modern house with a garden in a suburban environment. Migration from cities to the suburbs was highly selective: it were mainly middle-class families with children who exchanged their urban house for a suburban one. Low-income families could not afford a suburban house and remained in the cities with their relatively high numbers of state-supported social rented housing. But cities also appeared to be attractive for young starters. British geographer Fielding (1992) developed the escalator metaphor to describe new city functions (Fig. 8.6). Young adults from rural areas or from smaller towns moved to cities for entering a college or university or to start a job. Some years later, after graduation, or after getting a job with better career perspectives and starting a family, they leave the city and find a house in one of the suburbs. Increasingly, cities became concentration areas for young, single adults (Latten et al. 2008). The escalator model can be identified in all Dutch cities with higher education institutes (Box 8.2).

Cities became attractive destinations for international migrants. The economic boom of the 1960s stimulated migrants from southern European nations and later from Morocco and Turkey to come to the Netherlands. These labour migrants found residences in the same place where they were employed: the big cities in Randstad Holland and the industrial cities in the east and south (Chap. 11). In the 1965–1985 period of stagnation, urban populations peaked twice: on the one hand by highly educated young singles, and by low-educated, low-income native and immigrant families on the other hand.

Urban municipalities initiated urban renewal programs for their nineteenth and early twentieth century residential areas (De Liagre Böhl 2012). Their sizes, comfort levels and types of construction did no longer comply with modern housing standards. Another problem was subsidence, as many houses lacked proper foundations in Noord- en Zuid-Holland's soft soils. Urban renewal has been conducted in several stages between 1965 and 2015. The period of the 1960s and early 1970s were times of slum clearance and of building new shopping malls and offices, e.g. Hoog Catherijne in Utrecht,

Box 8.2: The City as an Escalator: Groningen
Groningen is the 7th largest city in the Netherlands and the only city in the north having more than 200,000 inhabitants. Groningen is in a region with a stagnating and ageing population; however, the municipality of Groningen has a very young and growing population. Groningen is a typical, although extreme example of an escalator city. Groningen University (founded 1614) is the oldest university of the Netherlands after Leiden, and currently has some 30,000 students. The University of Applied Sciences, Hanzehogeschool, has about 25,000 students. Immigration to and emigration from the city is highly age-specific (Fig. 8.7), even more so than in other university cities. Groningen attracts many young adults, predominantly from the northern provinces, but also from other parts of the nation and from abroad. Some young adults stay in the region after graduation and move to one of the surrounding suburban municipalities. Because of the limited employment opportunities for higher educated people most of the young adults leave Groningen after graduation to make a career or to receive further education in Randstad Holland, Amsterdam in particular (Latten et al. 2008).

Fig. 8.6 City as an escalator. (Source: PBL 2015)

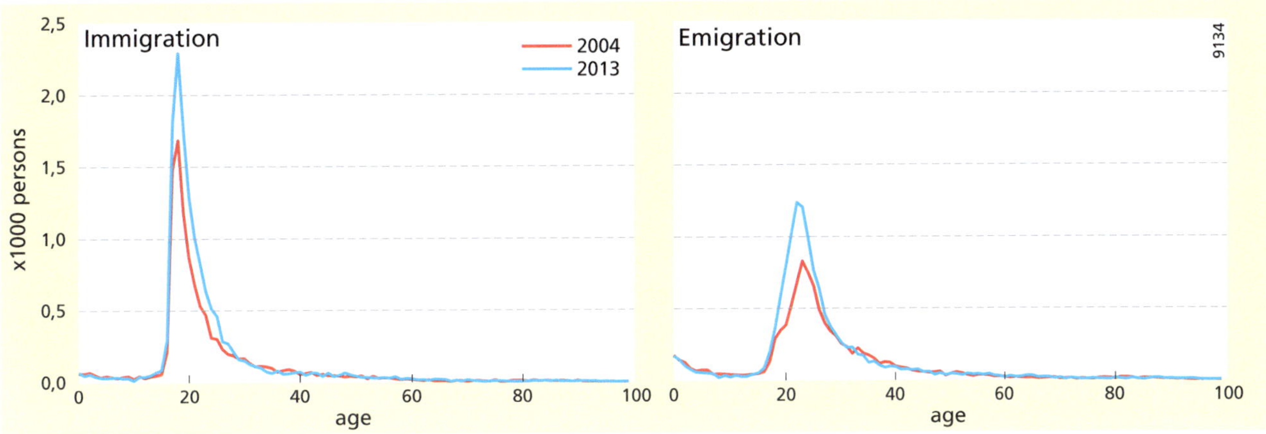

Fig. 8.7 Number of immigrants to and emigrants from Groningen by age, 2004 and 2013. (Source: Rambharos and Beets 2015)

Weesperstraat and Wibautsraat in Amsterdam, and Spui quarter in Den Haag, where nineteenth century buildings were replaced by new governmental offices. Slum clearance and city development generated many urban protests. Action groups of local residents, squatters and student activists protested loudly against this clearance policy and advocated the need of 'building for the neighbourhood', implying renovation of existing houses and building affordable new houses for local residents (Mamadouh 1992; Verlaan 2017).

These urban protests proved successful and generated a policy shift. In the 1970s, national and local governments began with large scale urban renewal focused on renovation and renewal of rental housing where neighbourhoods were preserved according to their original design and where low-income households were allowed to return upon implementation.

This urban renewal policy transformed into 'Grote Stedenbeleid' ('Big Cities Policy'), initially for the four biggest cities and later extended to 25 cities. Together with market parties national and local governments aimed to renew post-WWII neighbourhoods and focused on physical renewal of buildings and public space and on both economic and social aspects. Main aim was to diversify population by means of mixing social rental housing (30%) and owner occupied housing (70%) and a mixture of middle and low-rise apartments and single family dwellings. A total of 105 billion Dutch guilders has been invested in renewal of the late 19th and early twentieth century neighbourhoods in the 1980s. The national government provided one third of the budget and two thirds of the costs were covered by municipalities, housing associations and residents together (De Liagre Böll 2012) (Boxes 8.3 and 8.4).

Box 8.3: Renewal of a Postwar Urban District: De Bijlmermeer in Amsterdam

De Bijlmermeer is an urban extension of Amsterdam and developed since the late 1960s. This new district was built corresponding urban design ideas of French architect Le Corbusier and CIAM (Congrès International d'Architecture Moderne). High-rise, residential buildings in modernist style, positioned in wide green space and strictly separated functions for living, working, leisure and infrastructure constituted the basis of these modernist ideals.

De Bijlmermeer is a polder, reclaimed in the eighteenth century from a peat excavation lake (Chap. 2), in operation as an agricultural area until 1960. The new district was built for family households and contained rather spacious apartments surrounded by large green spaces where children could play safely while motorized traffic was separated from cyclists and pedestrians. The first apartments were ready by 1969 and families moved from older Amsterdam neighbourhoods to De Bijlmermeer. Simultaneously, however, cheaper new family houses were built in Alkmaar, Purmerend, Hoorn, Almere and Lelystad and many preferred such houses with private gardens rather than the modernist apartments in De Bijlmermeer. That resulted in high residential mobility and vacancies in these apartments.

Around 1975, when Dutch colony Suriname became an independent nation, many Surinamese migrated to the Netherlands (Chap. 11). They were

(continued)

Box 8.3 (continued)

entitled to the Dutch social rental sector and to receive rent allowances. De Bijlmermeer, with its many vacancies offered sufficient options to accommodate them. But many of them were unemployed, lived on benefit and were not used to live in high-rise apartments, in a country which was unfamiliar to them.

Instead of the urban planners 'paradise', in a few years only De Bijlmermeer developed into one of the most problematic 'no-go areas' in the Netherlands, a spot where poverty, unemployment, drugs addicts, criminality and vandalism prevailed (Van Kempen 1986). Successively, De Bijlmermeer developed into a 'clearance-house' for any type of immigrant and for native Dutch unable to find a living place elsewhere.

In 1992, municipality and housing associations jointly began reconstructing this urban expansion to better cope with new social conditions and with dilapidation of the otherwise comfortable and spacious apartments. With financial support of the national government (Grote Stedenbeleid) and the European Union (URBAN program), some 7000 out of the initial 13,000 apartments were demolished and replaced by 8000 new, low-rise apartments and single family houses. In addition, 4000 apartments were renovated for the social rental sector and some 1500 renovated apartments were sold to house owners (Wassenberg 2013; Van Kempen and Wassenberg 1996).

In terms of income, De Bijlmermeer population is more diverse and hosts many middle-class immigrants today. Most of the population (83%) has a migration background, 33% is Surinamese-Dutch. Moreover, De Bijlmermeer now hosts many first and second generation immigrants from West-Africa (Ghana) and the Dutch Antilles as well (OIS 2016). Still, local poverty spots occur, but De Bijlmermeer is more diversified than before its urban renovation.

Box 8.4: Renewal of a Nineteenth Century District: Schilderswijk in Den Haag

The city of Den Haag hosts both wealthy and poor districts. Wealthy districts include Vogelwijk and Benoordenhout and have been built on sandy beach barriers (Chap. 3) A poor district is Schilderswijk, built around 1900, on clayey and peaty soils. Schilderswijk's current population is 31,000, which is about half the number before WWII. By today, Schilderswijk belongs to the ten poorest Dutch neighbourhoods (SCP 2016): 39% of the households have an income of less than

Box 8.4 (continued)

105% of the statutory minimum wage (1565 € per month in 2017). This percentage is 14.6 for the entire city of Den Haag. Almost 5000 people in Schilderswijk are unemployed. This district hosts many minorities: 91% of the population is a first or second generation immigrant from Turkey, Morocco, Suriname and other non-western countries (Den Haag in cijfers 2017). Until about 1960, Schilderswijk was a collection of small urban hamlets, each with its own local culture (Klein Kranenburg 2013). The district had a homogeneous Dutch working-class population with a high level of internal cohesion (bonding social capital). Many men worked at nearby industrial estates.

In the decades after WWII, housing conditions in Schilderswijk deteriorated, resulting in dilapidated houses and public spaces. In 1986, Den Haag municipality announced plans to replace existing houses by 28-story high apartment blocks, in line with Le Corbusier's and CIAM's ideas (Box 8.3). But local residents were not consulted and protested fiercely. Successively, plans were cancelled and local residents became involved this time. That resulted in new plans to construct a large number of social rented houses, a mixture of single family houses with gardens and high-density apartments. These plans also included public gardens, open spaces and recreation facilities (sports hall, swimming pool, play grounds, children's farm). Some original houses were saved from demolition and renovated. Implementation was between 1970 and 2010 by means of state funding and funding by housing associations (De Pater 1996).

Schilderswijk was a 'problem accumulation area': an area with both social and economic problems and physical decay. Families who could afford a new suburban house left Schilderswijk and their places were taken by immigrants: labour migrants from Turkey and Morocco, people from the former Dutch colony Suriname, and refugees from several non-western countries. Employment declined, especially in nearby industrial districts. The number of households on benefit rose accordingly. Social cohesion dwindled, vandalism and criminality increased, drugs trade and prostitution flourished.

But housing condition and quality of public spaces in general improved considerably since the 1990s. Today, prostitution is banned but lack of social cohesion, ethnic-cultural segregation and unemployment are still major problems in this district. But overall, Schilderswijk is in a better shape today than 30 years ago.

Population in the four biggest cities, Amsterdam in particular, began growing again after 1985 (De Jong and Daalhuizen 2014). Cities were still attractive destinations for young starters. But in contrast to times of stagnation, increasing numbers of young starters stayed in cities rather than moving to its suburbs. The Second Demographic Transition (Chap. 11) is reflected in household composition of cities. In general, younger generations postpone marriage and family formation while more and more 25–40 years old adults are singles or are part of couples without children, preferring urban residency. More recently, cities are becoming attractive for families with children as well, in particular for symmetrical, dual-career families with both partners working (almost) full-time. They often share domestic work and childcare (Boterman et al. 2010; Karsten 2014). Renewed urban districts and twenty-first century urban expansion areas have become attractive for single, couple and family households with urban lifestyles.

Increased international immigration, including highly educated expats, is a growing share of the urban population, in particular in Amsterdam with expats employed in headquarters of multinational companies and in the financial, educational and cultural sector. In Den Haag, many expats work in diplomacy and in international governmental organisations.

Anticipated population growth for the 2015–2040 period in the biggest Dutch cities is up to 15% (De Jong and Daalhuizen 2015). Nevertheless, future growth will depend on their building capacities, availability of new plots for housing development which more and more compete with preserving green areas.

Revitalization has not been general in Dutch cities. Some industrial cities in the north, east and south of the country rather stagnated or even declined due to de-industrialisation, e.g. Delfzijl in the province of Groningen and Terneuzen in Zeeuws-Vlaanderen (Box 8.5).

Box 8.5: De-Industrialisation and Urban Decline: Terneuzen

In the fourteenth century, Terneuzen was a small market town. In Late Medieval Times, land reclamation began and former islands were united again, including the current region of Zeeuws-Vlaanderen and, accordingly, most of it is 'new land'. During the Dutch Revolt (1568–1648) against the Spanish King and resulting in independence of the northern provinces of the Netherlands, Zeeuws-Vlaanderen was a front area and suffered from inundations and war activities. During the successive Republic of the Seven United Provinces of the Netherlands, Zeeuws-Vlaanderen was an agricultural region with market centres in Terneuzen, Hulst and Oostburg (Thissen 1995).

In the nineteenth century infrastructure improved by construction of the Kanaal van Gent naar Terneuzen (*Canal Gent – Terneuzen*) in 1826 and a railway connection between both cities in 1870. Belgian (Gent) manufacturers invested in new factories around 1900 resulting in a flourishing economy and population growth. A second industrial boom was linked with a Dow Chemical plant in Terneuzen, in 1965. From then, employee numbers rose until 1985, when these began to decline (Dow 2015).

Terneuzen has not been successful in attracting new employment to compensate for employment losses from de-industrialisation. Today, Terneuzen is a shrinking town with an ageing and decreasing population, a stagnating housing market and about 20% empty shops.

8.6 The Urbanised Countryside

In the early twentieth century, almost all Dutch villages were still autonomous, closed communities (Droogleever Fortuijn and Thissen 2002). For their entire lives people stayed in the village where they were born and daily life occurred within the village boundaries. With a few exceptions, men and women were employed in farming or were local craftsmen or shopkeepers. Some of the larger villages hosted local industries, often linked to local agricultural production as canning industries, dairy and sugar factories, or textile industries. Some villages along railway lines in the west were already involved in suburbanisation processes.

In the first half of the twentieth century, rural development was driven by agricultural development, autonomous population growth and migration to cities (Chap. 11). Modernisation and mechanisation resulted in larger scale farms that produced for areas beyond the local market. By the early 1950s, two thirds of the agricultural production was produced for export already (Knippenberg and De Pater 1988). Agricultural modernisation reduced employment in that sector and many young villagers migrated to regional urban centers or to cities in the west to be employed in factories or in the service sector.

Tonckens and Abma (1957) were concerned about declining agricultural employment and, consequently, decreasing population in the rural areas. They expected de-population of the countryside and dying villages in remote rural areas. Similar concerns were expressed by the Noord-Groningen Foundation's report 'Bedreigd bestaan' ('Endangered countryside') published in 1959. In contrast to many other European rural areas, no villages died in the Netherlands. Economies of scale transformed such rural villages from autonomous agricultural into residential villages in the second half of the twentieth century (Droogleever Fortuijn and Thissen 2002). Local shops in small villages shut down and shops in larger villages transformed into branches of store chains. Moreover, local factories closed and many municipal institutions disappeared after several rounds of reorganisation.

In a transition from a productive towards a consumptive order (Van Engelsdorp Gastelaars and Ostendorf 1986), villages became included in regional housing markets, first in the west, followed by the rest of the country. Former farmers and farm workers found jobs in a nearby town or city and commuted to their new work places. Better public transport and more private cars enabled villagers to stay living in their home villages rather than to move to the city. Living in a village became a feasible residential option for people in a wider regional housing market. Villages became attractive residences for more and more external people and village populations increasingly developed as mixtures of locally born and 'import' people.

Simultaneously, a new diversity of villages evolved in Dutch rural areas. Size, distance to urban centers and an attractive residential environment were the main determinants of this new diversity (Thissen 1995; Steenbekkers et al. 2017). National, provincial and municipal governments allocated new housing estates to a limited number of larger (center) villages, equipped with municipal facilities, schools and medical services. These key settlements have growing and diverse populations in terms of age, household composition and income. In general, younger age groups are well represented as these villages provide housing for regional starters. Villages with no such function as regional or municipal centre show rather modest population growth, or no growth at all. Within the context of the regional housing market some of these are quite attractive because of residential amenities.

To discern more from less attractive villages, two factors are relevant. The first factor concerns the physical landscape, i.e. a location near dunes, nature areas, sandy soils, or near lake or river shores. The second factor concerns its cultural-historical quality, i.e. presence of monumental farms and houses and village conservation areas (Fig. 8.8). Population growth is limited in such villages because of regulation by municipal and provincial authorities (Thissen 1995).

Village populations in less attractive landscapes, as e.g. former peat and mining areas and former industrial villages often stagnate or decline and people are relatively low educated and have low incomes. Finally, some very small and remote villages and villages in the 'Bible belt' (Chap. 12) may still have autonomous characteristics with few 'import' people and many working locally.

Population declines have not resulted in dying villages in the Netherlands yet. In many such villages households stabilize in number due to ageing, with the exception of some small, remote villages in unattractive environments with both less people and less households, resulting in empty houses (Steenbekkers et al. 2017) (Box 8.6).

Box 8.6: "The Deeply Rooted": Changes in a Small Rural Community

"The deeply rooted" is the title of an anthropological study about local economy and community life in Anderen, a rural community in Drenthe, written in 1955 by two American researchers, John (biologist) and Dorothy (anthropologist) Keur (Keur and Keur 1955). In the early 1950s, the village of Anderen was a closed community of farmers, some shopkeepers and craftsmen who were deeply rooted in that community, from generation to generation. Daily life was entirely restricted to the local community. People were working in the village, went to the village school, bought groceries in local shops, married with a Anderen-partner or from a nearby village, went to church in a neighbouring village and spent their (rare) leisure times in the local café or in one of the many local clubs: sports, theater, women's association et cetera. Walking and cycling were the dominant means of transport. Only one villager owned a car in the early 1950s.

In 1998, the Department of Geography and Planning of the University of Amsterdam studied local communities of 11 villages in Drenthe, including Anderen. Researchers noticed some remarkable changes since above mentioned study, 46 years earlier. Anderen was still a vital rural community with a vibrant, local community life and even more local clubs (6 in 1952 and 12 in 1998). However, the village school and shops were all closed by 1998. Only a few farmers still lived in the village and most people worked outside the village. Most of the village people was not born in Anderen and 'import' people from Randstad Holland settled in the village. Almost anyone owned a car, daily life occurred at a regional or even national scale and people travelled abroad for their holidays. People were quite satisfied with their village life, felt attached to the village and its local community, but were no longer 'deeply rooted' (Thissen and Droogleever Fortuijn 1998).

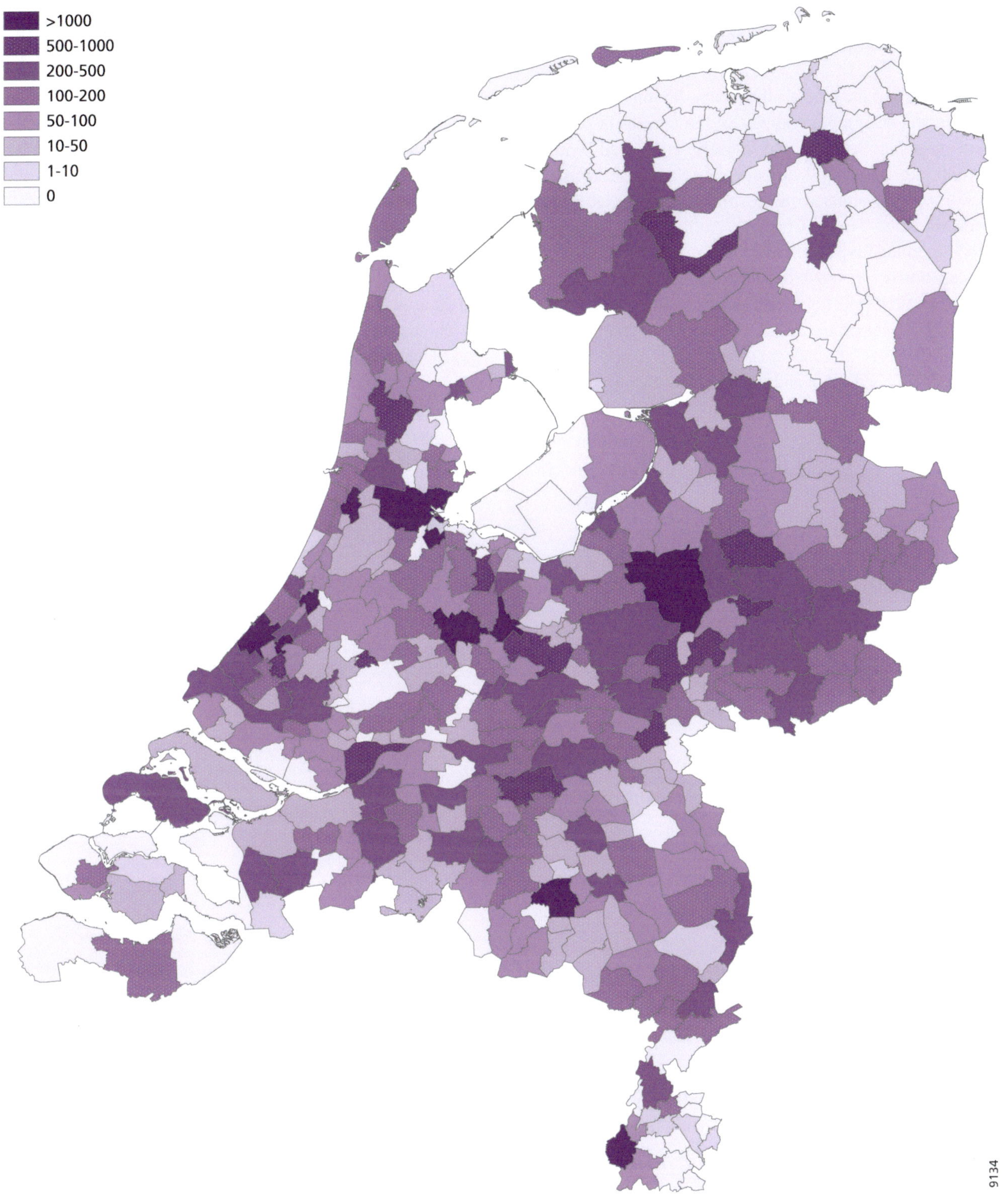

Fig. 8.8 Number of monuments per municipality 2016. (Source: Erfgoedmonitor 2016)

8.7 The Built Environment: From Regional to Temporal Specificity

Until the end of the nineteenth century, lay-outs of towns and villages and design of buildings, of farmhouses in particular, were regional specific and closely connected with the surrounding physical landscape and land use (Fig. 8.9). 'Esdorpen' are situated in the higher sand areas and date from Medieval Times (Chap. 3). Although an 'esdorp' is a dynamic village type with local and regional variation, typical 'esdorpen' are positioned between the topographically higher located crop fields and the lower grasslands along a stream or a river (Spek 2004). The croplands (the 'es') are in the higher parts of sand areas and were raised further as farmers deposited more and more layers of cattle manure on top of the crop fields over the centuries, resulting in the typical, pillow-shaped 'es'. Farmhouses were positioned around a common open field: 'brink'. Most 'esdorpen' occur in the higher sand areas in the east (Drenthe, Overijssel and Gelderland), but they can be found in the sand areas of Groningen, Friesland, Utrechtse Heuvelrug, 't Gooi, Noord-Brabant and Limburg as well.

Villages in the lower parts of the north, west and south-west have different designs. In Groningen and Friesland, most old villages are built on flood hills ('terp'; Chaps. 2 and 3), an artificial round-shaped hill with a church on top and houses around the church and churchyard. Compact, church-centered, nuclear village types often occur in other, lower Dutch regions.

Fig. 8.9 Village types: top left: 'Esdorp Norg in Drente (Source: https://commons.wikimedia.org/wiki/File:Norg_dorpsgezicht.jpg) top right: 'terpdorp' Hogebeintum in Friesland (Source: https://commons.wikimedia.org/wiki/File:Hogebeintum1.JPG) bottom left: 'Dijkdorp' Everdingen in province Utrecht (Source: https://commons.wikimedia.org/wiki/File:Dijkdorp_Everdingen.JPG) bottom right: Village in the Veenkoloniën: Stadskanaal in province Groningen (Source: https://nl.wikipedia.org/wiki/Groninger_Veenkoloni%C3%ABn)

Winding, long ribbon settlements, sometimes more than 10 km, occur on top of creek ridges, river dunes or man-made dikes, protecting people against floods. Straight ribbon settlements developed more recently along roads or canals, e.g. in Veenkoloniën, the vast peat extraction area in the northeast of Groningen and Drenthe, where linear settlements along canals were built in the seventeenth to nineteenth century for peat transport. Such linear villages occur in the peatlands of Noord-Holland, Zuid-Holland and Utrecht as well, e.g. in Westbroek, Tienhoven and Loosdrecht.

8.7.1 Traditional Farmhouses

Designs of old farmhouses are often local- or regional-specific and comply with regional agricultural land use (Fig. 8.10). The northern type, or 'Friese huisgroep', is a square-shaped farm-house ('stolp') with separate spaces for cattle, storage of hay and cereals, threshing space and for residence of the farmer's family, all under one single roof. In some regions (Friesland), the farmhouse is extended with a separate residence in front of a rectangular farmhouse ('kop-hals-romp') (Agrarisch erfgoed 2017).

The 'Hallenhuisgroep', or Low German house, is a typical farmhouse for the higher sand areas in Drenthe, Overijssel, Gelderland, Utrecht and Noord-Brabant, but occur in Zuid-Holland as well. This is a long, rectangular building with a residence at the front side of the building, in some cases transverse, a threshing space and a hayloft in the middle and space for cattle on both sides of the threshing space.

The 'Dwarshuisgroep' is a typical farmhouse type for east Noord-Brabant and for Limburg. The farmhouse is built around an inner court with two separate buildings, or a square of buildings around an inner court.

In Zeeland, the west of Noord-Brabant and on the islands of Zuid-Holland, the 'Zeeuwse schuurgroep' can be found, a

Fig. 8.10 Traditional farmhouses: top left: 'Stolp' farmhouse Beemster in Noord-Holland (Source: https://upload.wikimedia.org/wikipedia/commons/c/cc/Noord-Beemster%2C_Middenweg_44.jpg) top right: 'Hallenhuisgroep'' farmhouse in Gelderland (Source: https://nl.wikipedia.org/wiki/Boerderijtype#/media/File:Rieten_dak_old_-farmhouse.jpg) bottom left: 'Carré' farmhouse Wijnandsrade in south Limburg (Source: https://www.wikiwand.com/nl/Vink_(Wijnandsrade)) bottom right: 'Zeeuwse schuurgroep' farmhouse Poortvliet in Zeeland (Source: http://rijksmonumenten.nl/monument/494110/boerderij-behorend-tot-de-zeeuwse-schuurgroep-en-volgens-jaartalankers-tegen-de-oostgevel/poortvliet/)

long farm house with a residence at the front side and a long shed with typical high, black painted wooden doors on the side. All farmhouse types have local varieties.

8.7.2 Modern Designs

Traditional, regional village designs of villages and farmhouses may still be observed in Dutch landscapes. Several villages are protected and many farmhouses have been declared monuments. But, by the end of the nineteenth century a process of uniformization began. Long rows of similar, cheap houses and apartments were built for working class families in cities. Suburbanization of the more wealthy people around cities resulted in comparable villas with thatched roofs. These types have typical forms depending on when these were built but have no longer regional or local identities.

Between WWI and WWII 'garden villages' have been built by housing associations in their ambition to improve living conditions for working class people, during pillarization of Dutch society (Chap. 12). These small-scale urban districts were designed with winding streets and small terraced houses with gardens. Larger-scale urban districts for the working class were built in different architectural styles in the same time period, e.g. in a modernist style, a functionalistic style with geometrically-shaped buildings in concrete, glass and steel (Bergpolderflat in Rotterdam, 1934) or in an ornamental ('Amsterdam school') style in bricks and red roofing tiles (Spaarndammerbuurt Amsterdam, 1917).

Combined population and economic growth in post-WWII decades caused quite substantial housing shortages. To cope with this tight housing market property developers applied large-scale, standardized, industrial construction methods, rather than leaving this to local contractors. Together with strict Dutch planning and building regulations (Chap. 10) this resulted in rather uniform residential districts without any regional or local identity (Fig. 8.11). Modernism and functionalism, inspired by Le Corbusier and CIAM, have been typical for the urban extensions in the 1950s and 1960s.

Fig. 8.11 Post WWII urban extensions: top left: Modernist neighbourhood in Amsterdam ('Dudokhaken') 1950s (Source: http://www.architect-dudok.nl/dudok-in-amsterdam-west.html) top right: 'Cauliflower' neighbourhood Emmeloord in Flevoland 1970s (Source: https://www.google.nl/search?q=foto+wikimedia+bloemkoolwijk+Emmeloord&rls=com.microsoft:nl-NL:IE-) bottom: Neotraditional neighbourhood Brandevoort in Helmond in Noord-Brabant 2000s (Source: https://commons.wikimedia.org/wiki/File:Zicht_op_de_wijk_-_Brandevoort_-_Helmond_-_20413538_-_RCE.jpg)

They are built in concrete, steel and glass with open, 3–10 story apartment blocks, alternated with green public spaces and ample infrastructure for the rising numbers of private cars (Landschap in Nederland 2017). Rows of houses, all with gardens in front- and in the back, were typical for such extensions built in towns and (even small) villages, while disturbing historical and traditional, local village plans.

In reaction on such monotonous designs of the early post-WWII districts and neighbourhoods, 'cauliflower' type districts were built all over the country, between 1970 and 1985. Last mentioned districts have winding street patterns and houses are positioned around small, green common spaces. Streets end in cul-de-sacs where motorized traffic is forced to slow down (30 km/h zones). Neighbourhoods of the 1970s typically have mixtures of single family houses and small, 3–4 story apartment blocks, built in brick (Landschap in Nederland 2017).

Dominant urban designs of the 1990s and early twenty-first century typically are mixtures of large-scale modernist apartment blocks, small-scale apartment blocks and family houses in ornamental, neo-traditional styles. Some of the most recent districts are based on common themes, e.g. a mediaeval fortress (Brandevoort in Helmond), a Mediterranean district (La Midi in Katendrecht, Rotterdam), an Oriental district (Orient in Transvaal, Den Haag), a water district (IJburg Amsterdam; Box 8.1), or a golf residence (Dronten). However, such themes are invented and disconnected from any local or regional history or from the landscape (Meier and Reijndorp 2010). In contrast to the planned and rigidly designed neighbourhoods of the early post-WWII period, plans in the 1990–2015 period generated considerably more freedom to home owners to design and build their own houses.

Currently the built environment of cities, suburbs, villages, industrial areas and infrastructure makes up 15% of the total Dutch surface. In 1900, this was no more than 2% (CLO 2017). With 62,000 national, 57,000 provincial and municipal urban and rural monuments (Fig. 8.8) and about 400 protected villages, historic towns and city centers, one may still observe specific historic, local and regional designs in the Netherlands (Erfgoedmonitor 2016). However, the 13% of the total space that is covered with buildings and infrastructure since 1900 demonstrate shifts from long-lasting and landscape-rooted, historic, local and regional specificity towards the temporal specificity of the more recently built areas.

References

Agrarisch erfgoed (2017) http://www.agrarischerfgoed.nl/boerderijtypen-2/. Accessed 2 June 2017

Bazelmans J, Bloemers T, Broesi R, Colenbrander B, Visser C, Grootens J (2005) Limes atlas. NAi, Rotterdam

Bedreigd bestaan (1959) Noord-Groningen Foundation, Groningen http://kknnvanmeernaarbeternl/bibliotheek/bedreigd-bestaan-in-noord-groningen-samenvatting-van-onderzoek-uit-1959. Accessed 2 June 2017

Boterman WR, Karsten L, Musterd S (2010) Gentrifiers settling down? Patterns and trends of residential location of middle-class families in Amsterdam. Hous Stud 25(5):693–714

CBS (2015) Bloemendaal en Wassenaar ook al in 1930 rijkste gemeenten. Centraal Bureau voor de Statistiek, Den Haag

CLO (2012) Ruimtegebruik. Compendium voor de Leefomgeving Rijksoverheid, Den Haag http://wwwclonl/onderwerpen/ruimtegebruik. Accessed 2 June 2017

CLO (2017) Veranderingen bodemgebruik, 1979–2012. Compendium voor de Leefomgeving. Rijksoverheid, Den Haag http://www.clo.nl/indicatoren/nl0060-bodemgebruik-in-nederland. Accessed 21 June 2017

De Jong A, Daalhuizen F (2014) De Nederlandse bevolking in beeld – verleden heden toekomst. In: 24 Infographics op basis van de PBL/CBS regionale bevolkings- en huishoudensprognose. Planbureau voor de Leefomgeving, Den Haag

De Jong S, Daalhuizen F (2015) Vier grote steden: toen, nu en in de toekomst. DEMOS Bulletin over Bevolking en Samenleving 31 (10):7

De Klerk L, Van de Laar P, Moscoviter H (2008) G.J. de Jongh: Havenbouwer en stadsontwikkelaar in Rotterdam. Thoth, Bussum

De Liagre Böhl H (2012) Steden in de steigers. Stadsvernieuwing in Nederland 1970-1990. Bert Bakker, Amsterdam

De Pater B (1996) The Hague has awakened, urban renewal and urban revitalization. Aust Geogr 5(5):4–11

Den Haag in cijfers (2017) Gemeente Den Haag

Dow (2015) Weet je het nog? 50 jaar Dow Terneuzen 1965–2015. Dow Chemical, Terneuzen

Droogleever Fortuijn J, Thissen F (2002) Van toepassingsveld tot proeftuin: de ruraal-geografische traditie. In: Knippenberg H, Van Schendelen M (eds) Alles heeft zijn plaats. 125 jaar Geografie en Planologie aan de Universiteit van Amsterdam, 1877–2002. Aksant, Amsterdam, pp 57–76

Ekamper P (2010) De verstedelijking van Nederland. DEMOS Bulletin over Bevolking en Samenleving 26(9):15–17

Erfgoedmonitor (2016) Rijksdienst voor het Cultureel Erfgoed. Ministerie van Onderwijs, Cultuur en Wetenschap, Den Haag

Fielding AJ (1992) Migration and social mobility: South East England as an escalator region. Reg Stud 26(1):1–15

Hall P (1966) The world cities. Weidenfeld and Nicolson, London

Karsten L (2014) From yuppies to Yupps: family gentrifiers consuming spaces and reinventing cities. Tijdschr Econ Soc Geogr 105 (2):175–188

Keur JY, Keur DL (1955) The deeply rooted. A study of a Drents community in the Netherlands. Van Gorcum, Assen

Klein Kranenburg D (2013) 'Samen voor ons eigen'. De geschiedenis van een Nederlandse volksbuurt: de Haage Schilderswijk 1920–1985. Verloren, Hilversum

Knippenberg H, De Pater B (1988) De eenwording van Nederland. SUN, Nijmegen

Kommer GJ, Gijsen R, Lemmens LC, Kooistra M, Deuning C (eds) (2015) Beschikbaarheid, specialisatie en bereikbaarheid van spoedeisende hulp; analyse gevoelige ziekenhuizen. Rijksinstituut voor Volksgezondheid en Milieu, Bilthoven

Landschap in Nederland (2017) Rijksdienst voor het Cultureel Erfgoed. Ministerie van Onderwijs, Cultuur en Wetenschap, Den Haag

Latten J, Das M, Chkalova K (2008) De stad Groningen als roltrap van Noord-Nederland. Bevolkingstrends 56(2):52–59

Mamadouh V (1992) Three urban social movements in Amsterdam: young households in the political arena between 1965 and 1985. In: Van der Wusten H (ed) The urban political arena, geographies of public administration, Nederlandse Geografische Studies 140. Koninklijk Nederlands Aardrijkskundig Genootschap/Instituut voor

Sociale Geografie Universiteit van Amsterdam, Amsterdam, pp 105–127
Meier S, Reijndorp A (2010) Themawijken; wonen op een verzonnen plek. Thoth, Bussum
OECD (2011) Regional typology. Directorate in Public Governance and Territorial Development. OECD, Paris
OIS (2016) Gebiedsanalyse Bijlmer Centrum Stadsdeel Zuidoost. Onderzoek, Informatie en Statistiek. Gemeente Amsterdam
Ostendorf W (1988) Het sociaal profiel van de gemeente. Woonmilieudifferentiatie en de vorming van het stadsgewest Amsterdam: het ruimtelijk beleid van een achttal gemeenten na de Tweede Wereldoorlog. Dissertation, University of Amsterdam
PBL (2015) De stad als roltrap. Planbureau voor de Leefomgeving, Den Haag http://wwwpblnl/infographic/de-stad-als-roltrap-0. Accessed 2 June 2017
PDOK (2015) Publieke Dienstverlening Op de Kaart. Topografische Dienst, Den Haag wwwpdoknl. Accessed 26 June 2017
Rambharos N, Beets G (2015) Groningen. In: Beets G, Van Dam F, De Jong A, Manting D (eds) De stad: magneet, roltrap en spons. Bevolkingsontwikkelingen in stad en stadsgewest. Planbureau voor de Leefomgeving, Den Haag
Rutte R, Abrahamse JE (2016) Atlas of the Dutch urban landscape. A millennium of spatial development. Thoth, Bussum
Schmal H (1995) Den Haag of 's-Gravenhage? De 19e eeuwse gordel; een zone gemodelleerd door zand en veen. Dissertation, University of Amsterdam. Matrijs, Utrecht
SCP (2016) Armoede in kaart 2016. Sociaal en Cultureel Planbureau, Den Haag
Spek T (2004) Het Drentse esdorpenlandschap. Een historisch-geografische studie. Matrijs, Utrecht
Steenbekkers A, Simon C, Vermeij L, Spreeuwers W (2008) Het platteland van alle Nederlanders. Hoe Nederlanders het platteland zien en gebruiken. Sociaal en Cultureel Planbureau, Den Haag
Steenbekkers A, Vermeij L, Van Houwelingen P (2017) Dorpsleven tussen stad en land. Slotpublicatie sociale staat van het platteland. Sociaal en Cultureel Planbureau, Den Haag
Thissen F (1995) Bewoners en nederzettingen in Zeeland: op weg naar een nieuwe verscheidenheid. Dissertation, University of Amsterdam. Nederlandse Geografische Studies 191. Koninklijk Nederlands Aardrijkskundig Genootschap/Instituut voor Sociale Geografie Universiteit van Amsterdam, Utrecht/Amsterdam
Thissen F, Droogleever Fortuijn J (1998) Sociale cohesie en dorpsverenigingen op het Drentse platteland. Verslag van een onderzoek in de gemeente Aa en Hunze. Amsterdam Study Centre for the Metropolitan Environment, University of Amsterdam
Tonckens NA, Abma E (1957) Verdwijnende dorpen op het Groninger Hoogeland. Afdeling Sociologie en Sociografie van de Landbouwhogeschool, Wageningen
Van der Cammen H, De Klerk L (2012) The selfmade land. Culture and evolution of urban and regional planning in the Netherlands. Spectrum, Houten/Antwerpen
Van Engelsdorp Gastelaars R, Ostendorf W (1986) The Dutch system in transition: from a productive to a consumptive order? In: Borchert JG, Bourne LS, Sinclair R (eds) Urban systems in transition, Netherlands geographical studies 16. Koninklijk Nederlands Aardrijkskundig Genootschap/Faculteit Ruimtelijke Wetenschappen Universiteit Utrecht, Amsterdam/Utrecht, pp 231–242
Van Kempen E (1986) High-rise housing and the concentration of poverty: the case Bijlmermeer. The Netherlands Journal of Housing and Environ Res 1(1):3–26
Van Kempen R, Wassenberg F (1996) Amsterdam's Bijlmermeer: trouble in highrise paradise. Aust Geogr 5(5):20–23
Van Oort F, Burger MJ, Raspe O (2010) On the economic foundation of the urban network paradigm: spatial integration, functional integration and economic complementarities within the Dutch Randstad. Urban Stud 47(4):725–748
Verlaan T (2017) De ruimtemakers: projectontwikkelaars en de Nederlandse binnenstad 1950–1980. Vantilt, Nijmegen
Wagenaar MF (1990) Amsterdam 1876-1914. Economisch herstel, ruimtelijke expansie en veranderende ordening van het stedelijk grondgebruik. Dissertation, Historisch Seminarium, University of Amsterdam
Wassenberg FAC (2013) Large housing estates: ideas, rise, fall and recovery. The Bijlmermeer and beyond. Dissertation, TU Delft

9 Infrastructures

Ben C. De Pater

Abstract

The World Economic Forum classifies the Dutch infrastructure as excellent. Out of 138 countries, the Global Competitiveness Report 2016–2017 ranks Dutch infrastructure third. That position is a combination of scores on more specific criteria: first on port quality, fifth on overall infrastructure, fourth on roads and air transport, fifth on energy supply, seventh on railroads. Nevertheless, the Netherlands is not devoid of infrastructural challenges, more specifically the daily traffic congestions in Randstad Holland. This chapter presents a review of the development of (inter)national infrastructure, transport and communication networks since the 1900s.

Keywords

Road traffic · Railroad networks · Beach resorts · Mainport rotterdam · Mass media · Electricity networks · Commuters

Before May 1st, 1909 no common time existed in the Netherlands as a nation, only local times. Each community chose its own time generally based on the top position of the sun (12 am), consequently making clock times in the east later than in the west. That situation well qualifies the weak Dutch interregional connections before the twentieth century when the nation was not more than a gathering of local territories and communities, relatively isolated from the outside world.

All surrounding nations assessed national times earlier. Great Britain in 1880 (Greenwich Mean Time), France in 1891 (Paris time), Belgium in 1892 (Greenwich mean time) and Germany and Austria in 1893 (the time of the 15-degree Meridian East: Central European time). The shift toward national clock times was driven by the telegraph and even more so by the railway companies. Both types of infrastructure mutually connected local societies. Dutch railway companies used the Amsterdam time as the Standard Railway Time for all railway stations in the Netherlands. Beyond these stations, church towers showed the valid local (clock) times.

Although common clock times became inevitable in a modernising nation, accepting the Amsterdam time as the national time caused long debates in Dutch Parliament in the beginning of the twentieth century (Knippenberg and De Pater 1988; Knippenberg and Nauta 1989). As a consequence of the German occupation of the Netherlands in 1940, the Amsterdam time was changed for the Berlin time – the Central European Time. However, after liberation in 1945, Central European Time was maintained for the Netherlands. That has been one contribution to connect the Netherlands with the rest of the world (see introduction to Chap. 6).

Construction of modern infrastructures has been the main driver behind this kind of integration. Transport was time consuming and expensive until deep in the nineteenth century. In terms of travel times and transport costs the Netherlands was quite a large country by then (Fig. 9.1). Geographical distances proved to be barriers which occasionally protected the population as well. A Cholera epidemy in 1832 in the fishing town of Scheveningen spread quite slowly to surrounding towns (Fig. 9.2). In today's mobility conditions the disease would have reached all remote corners of the country within 1 week time, given its incubation period of several days. In particular in the higher parts of the Netherlands people rarely or never left their home regions, unless they either belonged to the upper or the lower classes. Upper class tourists often made a *Grand Tour* through their home country or even through Europe. Poor, unemployed people were forced to leave their roots to find work in cities or in commercial agricultural areas (permanent or seasonal migration).

Not only transport but communication – transmitting news and information between locations – was slow as well. In the nineteenth century mass media did not occur yet; messages

B. C. De Pater (✉)
Department of Human Geography and Spatial Planning, Utrecht University, Utrecht, The Netherlands
e-mail: B.C.dePater@uu.nl

E. F. J. De Mulder et al., *The Netherlands and the Dutch*, World Regional Geography Book Series,
https://doi.org/10.1007/978-3-319-75073-6_9

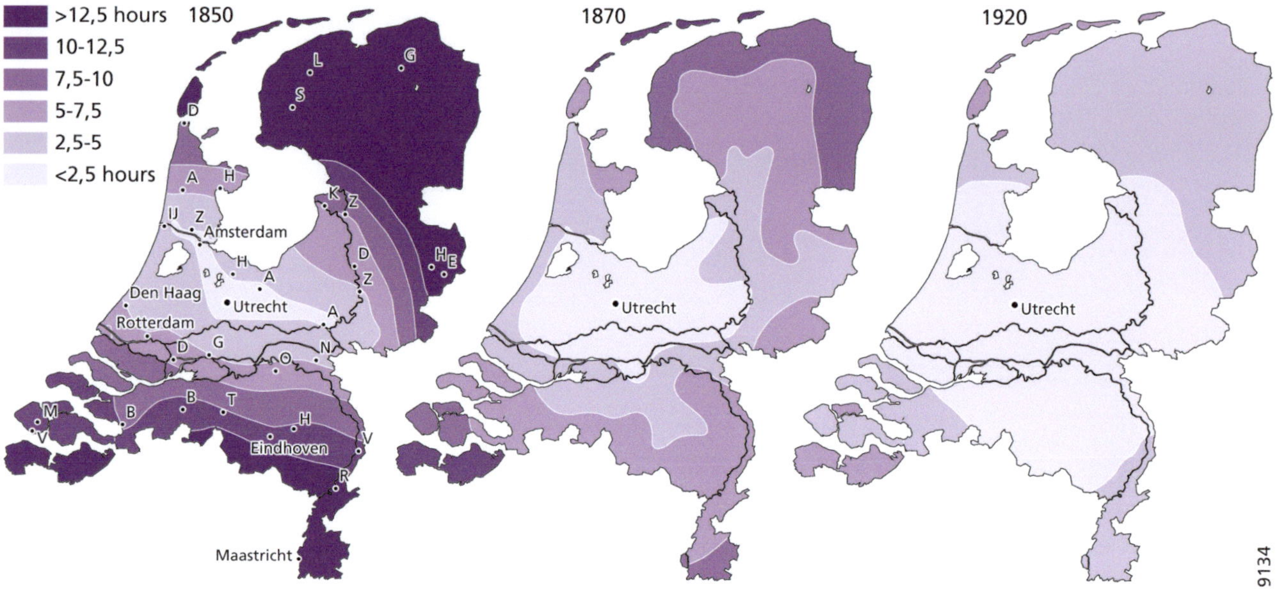

Fig. 9.1 Area of reach from the city of Utrecht for a given travel time, in 1850, 1870 and 1920 respectively. (Source: Thurkov et al. 1984)

were often disseminated via church services. Pastors read proclamations of the King. Death of Royalty members was announced by ringing church bells, three times daily. Regardless societal classes chimes calling for national mourning generated a sense of community feeling and awareness thus contributing to nation-building.

This chapter describes the development of infrastructure in the Netherlands. Section 9.1 deals with transport freight and persons: railways, roads and canals. Section 9.2 discusses communication networks: newspapers, magazines, radio, television and the internet.

9.1 Transport: Rail, Road and Water Links

As discussed in Sect. 7.1, the distinction between the maritime southwest, west and north on the one hand, and the more continental south and east has always been relatively large and these regions were mutually disintegrated. Socio-spatial integration of relatively isolated regional societies may be achieved only through coherent networks of transportation and communication lines.

From the seventeenth century until the onset of railways, regional transport was mainly through horse-towed barges in the maritime parts of the Netherlands where dedicated canals (so-called *trekvaarten*) were dug to facilitate such means of transport. Transport by such barges was slow, about 7 km/h, but relatively comfortable and cheap. Frequency was high and reliable, wind conditions seldom disturbed travel schemes, only heavy frost could disrupt these services.

Connections over water (rivers and the Zuyderzee ('IJsselmeer' upon separation from the North Sea, in 1932)), were initially addressed by sailing ships, later by steam- and motor ships. *Beurtschepen* sailed according to approximately fixed routes and timetables between towns, loaded with freight, cattle and passengers, whereas tramping ships sailed irregular routes. *Beurtschepen* were normally owned by local shipping companies while tramping ships were often family-owned (Hanrath 1938). Transport velocity of these ships strongly depended on wind conditions, delays at locks and bridges, etcetera. Amsterdam for example was connected over water with many towns along the Zuyderzee borders. In 1913, beurtschepen (steamers and motor ships only) made 46,319 trips from Amsterdam while 50,874 such trips were registered from Rotterdam in the same year, connecting the city with villages and towns in the hinterland, as Gorkum, Dordrecht and Gouda (Blink 1914).

In the higher parts of the Netherlands (in the east and south) stagecoaches and diligences moved forward slightly faster than the horse-towed barges (*trekschuiten)* but coaches were more expensive, less frequent and less comfortable as most of the roads were unpaved. No more than 500 km of paved roads existed in the early nineteenth century in the Netherlands. In general, people in the east and south were less mobile and rather preferred not to leave their home towns. This geographic distinction in mobility is supported by calculations on public transport services of ten million passenger-kilometers for 0.8 million inhabitants in the continental part of the country viz-a-viz 105 passenger-kilometers

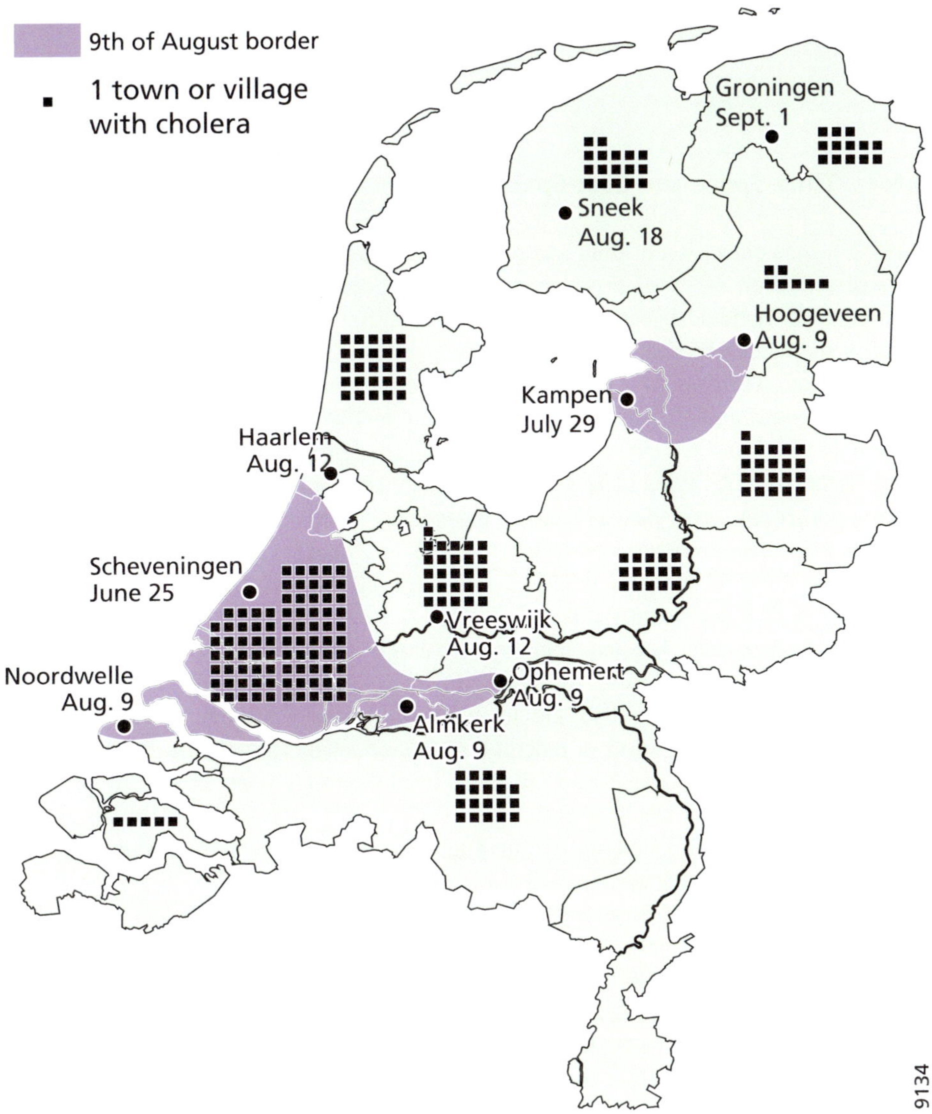

Fig. 9.2 The slow spread of cholera brought onshore on 25 June 1832 by Scheveningen fisher men; they drunk water contaminated with Cholera germs. Primary data: *Staatscourant*, 4 March 1833. (Souce: Knippenberg and De Pater 1988)

for the 1.2 million people in the maritime regions per year before the arrival of the railway (De Vries 1978).

9.1.1 Railway and Tramway Development

Mobility patterns significantly changed by the introduction of rail connections, gradually first but quite radically later. The first train connection was between the cities of Amsterdam and Haarlem, in 1839. This 17 km long railway route was exploited by the *Hollandsche IJzeren Spoorweg Maatschappij* company. Initially, railway construction and exploitation was private, the Dutch state did not enter this business before 1860. *Staatsspoorwegen* opened its first line in 1863. From then it took 20 years only to build an integrated network of private and state railways to connect all major cities, including the villages in between. Total length of the rail network reached approximately 2000 km by the early 1880s.

From 1880s onwards, a fine-meshed grid of local and regional tramway connections began to develop. In 1900, 31 tram networks were in operation, in 1914 ten more. However, these were not interconnected and their track widths varied greatly. In 1930, this regional and local tramway network was 3100 km long, 2400 of which regional. That point in time has been the climax of tramways in the Netherlands which successively began to decline due to the upcoming road trucks for on land freight transport, of busses and bicycles for transport of people. In the 1930s, very few private cars occurred (10 per 1000 inhabitants), but bicycles were quite common (400 per 1000 inhabitants). In the 1960s all regional tramways were gone but the rail network for trains survived.

9.1.2 Time-Space and Cost-Space Compression

Apart from the expansion of train and tramway networks, a massive reduction of travel time and travel cost certainly contributed to regional integration and spatial organisation in the Netherlands. Physical distances have been swallowed by travel time. Average travel time between Amsterdam and one of its 59 surrounding Dutch towns was 142 min in 1880, 89 min in 1930 and 61 min in 1980, including changing trains. Relative travel costs went down as well: in 1880, a factory worker paid more than 14% of his net-weekly salary on average for the cheapest one way ticket from Amsterdam to one of the surrounding towns, in 1920 5%, and in 1980 less than 2% (De Pater and Schmal 1982).

Both developments boosted mobility. Average train distances travelled per person in the Netherlands were 385 km per year in 1913, 680 km in 1948 and about 1000 km in 2015. This growth in mobility is also reflected by the number of train passengers: from about 40 million in the 1930s to 140 million per year today.

Even more impressive in terms of time time-space compression has been developments in road traffic. In 1939, only 11 passenger cars existed for every 1000 inhabitants in the Netherlands. This number is in sharp contrast to six car trucks/1000 people and with Germany: 17 passenger cars, Denmark: 28, England and France: 40, USA: 196 passenger cars per 1000 people in 1939 (Ramaker 1946). Fifty-three passenger cars existed per 1000 people in 1960, 260 in 1975, and 458 passenger cars in 2010 in the Netherlands.

Current (2016) mobility per person is 11,000 km per year within the Netherlands, about half of this distance as a car driver, 22% as a car passenger, 9% as a cyclist and 9% as a train passenger (other ways of transportation, including bus, tram and metro: 10%) (KiM 2016).

9.1.3 Rail Connections in the Twenty-First Century

Evolution of most transport networks occurred in four phases:

- loose ends (companies began constructing potentially profitable connections),
- integration (connecting loose ends, often together with the emergence of a national organisation),
- densification (more connections per unit of space),
- selection (removing redundant or uneconomic connections)

The tramway network never intended to go beyond regions as interregional connections were served by the railways (trains) and thus bypassed the integration phase. But densification did occur until 1930 before fierce competition by busses, bicycles, trucks and passenger cars (selection phase) made regional tramways in the Netherlands redundant, eventually.

However, new, regional tramways came into operation in some agglomerations from the 1980s onwards. These included the suburbs of Nieuwegein in Utrecht and Amstelveen in Amsterdam. Some other plans for new tramway connections were never implemented. Elsewhere, light-rail systems replaced regular train connections. The former Hofplein line (an old railway line between Rotterdam and Den Haag) turned into *Randstad rail* in 2006, connecting Rotterdam, Zoetermeer, Pijnacker and Den Haag.

Today, local tramway networks occur in Amsterdam, Den Haag and Rotterdam only. Further expansion of local tramways in (more) Dutch cities stagnated probably due to massive presence of bicycles (on average 1.3 bicycle per person) in the relatively small and flat Dutch cities. As in many other European cities and New York, Dutch cities encouraged biking and walking in the city centres while banning cars since the 1990s (Box 9.1).

Box 9.1: Cycling

Perhaps apart from China, in no country on Earth bicycles are more popular as in the Netherlands where 17 million inhabitants own 23 million bicycles. Total length of special bicycle lanes is approximately 35,000 km. Bicycles are particularly and increasingly popular in cities while cycling in rural areas is decreasing (Harms et al. n.d.).

Especially in Amsterdam the share of bike trips out of all trips increased significantly: from 33% in 1989 to 53% in 2012. Simultaneously, car trips in Amsterdam declined from 39 to 24% and trips by public transport dropped from 27 to 21%. On average a bicycle trip in Amsterdam covers a distance of 3 km, a car trip 6 km and a trip by public transport 7 km (Harms et al. n.d.).

Bicycles are typically used by commuters to cover the distance between their homes and the nearest railway station. During recent years, many modern bicycle parkings were built in subsurface of station areas, with a capacity of thousands of bikes. The largest bicycle shed can be found in Utrecht Centraal railway station, its capacity is 12,500 bikes (see also Jordan 2013; Oldenziel et al. 2016).

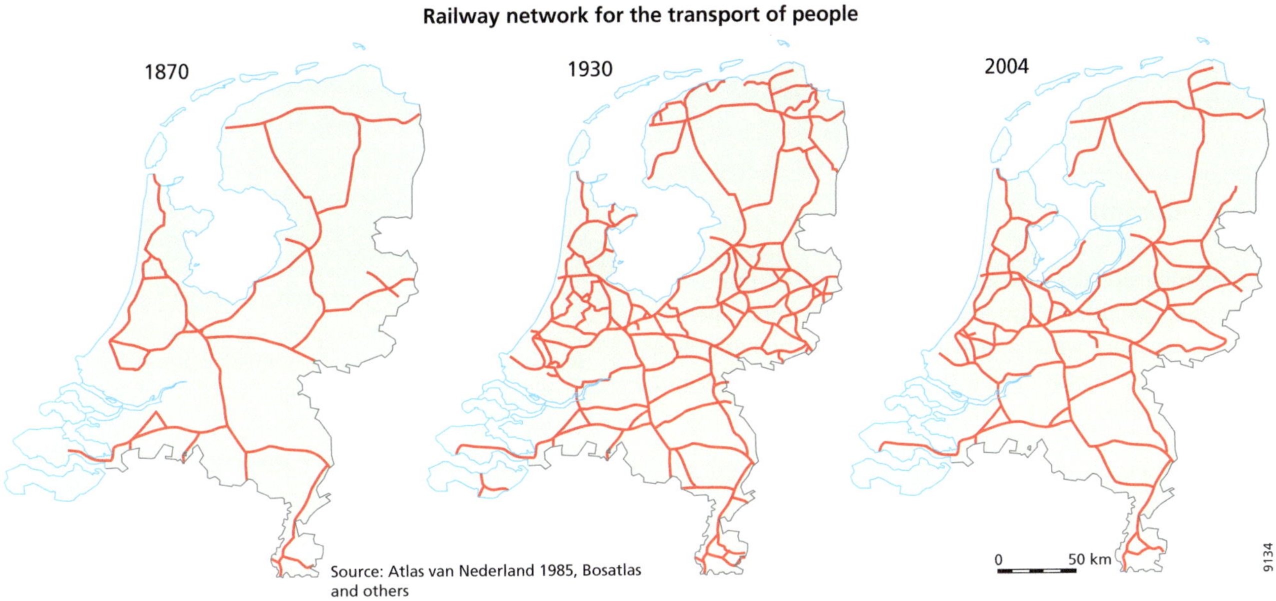

Fig. 9.3 Development of railway network for passengers. (Source: Kranenburg and Meijer 2001)

Over time, the Dutch railway network passed the first three phases as given above but largely escaped from the selection stage (Fig. 9.3). The total Dutch railway network of 3000 km has not changed significantly since the 1930s but several local, uneconomic connections were abolished and new railway lines have been opened. Some of the closed lines turned into tourist attractions with steam trains, others became e.g. bicycle lanes. New, post WWII rail connections were built in Randstad Holland including one between Amsterdam and Leiden via Schiphol Airport, and another one connecting Amsterdam with Lelystad and Almere in the new IJsselmeerpolders (extended to Kampen and Zwolle in 2012: the so-called Hanze line). A planned high speed line – the *Zuiderzeelijn*, desired by the northerners – between Randstad Holland (Amsterdam and Schiphol) and the north (Drachten and Groningen) never materialised, however, to frustration of people in the north.

Despite closure of smaller railway stations the net number of Dutch railway stations grew. According to a list on the Dutch wikipedia, 133 new stations were opened between 1970 and 2016. Most of these railway stations are positioned along existing lines connecting residential suburbs with older cities, mainly used by commuters, students and pupils. Dutch planners always have stimulated commuting by train rather than by car. However, their attempts have been marginally successful only as proved by the many and increasing traffic congestions on Dutch highways which may occasionally range up to 900 km in rush hours and bad weather conditions. Trains may also be packed to capacity as students may travel for free.

9.1.4 International Railway Connections: Betuweline and HSL-Zuid

The Netherlands have railway connections with cities in Belgium, France and Germany since a long time. During the Cold War (1946–1989) an international train connected Hoek van Holland (where ships from UK landed) with Warsaw and Moscow. In 1993 this line was abolished.

Two more international railway lines have been constructed in recent times. The *Betuwelijn* connects Rotterdam harbour with the German hinterland, more in particular with the industrial Ruhr area. This connection aims transporting 300 container freight trains daily. Pushed by Rotterdam lobbyists the Dutch government agreed in 1993 to build this connection which would reduce polluting road traffic through the Netherlands. Construction began in 1998 regardless fierce local protests against disturbing the river landscape, demolition of houses and noise. In response, the line was built parallel and close to a highway. In 2007, the *Betuwelijn* was opened but the anticipated freight transport ambitions could not be met, as the connecting railway lines in Germany were not yet built and as transport rates proved to be rather costly (Filarski 2016).

Another recent international railway line was the *HSL-Zuid* (High Speed Line South), connecting Amsterdam, via Schiphol Airport and Rotterdam, with Antwerp and Brussels, Paris and London (but not Den Haag). This line was believed to be essential in a pan-European railway network connecting all major city centres. In 1994 the Dutch government decided to construct this high-speed connection. To preserve the protected Green Heart (*Groene hart*), a 8.7 km long tunnel was built in the HSL tracé and the line was opened in 2009.

This line suffered serious setbacks as well, including delivery of technically unreliable train coaches. Travel times of 3 h and 18 min between Amsterdam and Paris are normally realised by another user of this line, Thalys.

9.1.5 Commuters

Before train connections greatly reduced travel times and costs and contributed to societal modernisation, people lived where they worked and vice versa. Commuting as a regular way of transportation is associated with the rise of train connections which enabled people to work in another place than they lived. With the expanding rail connections the number of Dutch towns reachable by public transport in 75 min time (considered as a maximum for commuting) from Amsterdam was two only in 1880; eight towns in 1905; nineteen in 1930; thirty in 1955 and 38 in 1980. Around the year 1900, wealthy people travelled between e.g. Amsterdam and their homes in attractive nearby areas as the dunes of Bloemendaal or the forests and hills of ’t Gooi (residential commuters). As travel costs went down further commuting became affordable to middle class people as well. Commuting to and from Amsterdam has grown quite significantly during the twentieth century: from less than 1000 in 1900 to some 60,000 in 1971. In 1900 there were 13 municipalities with over 50 commuters working in Amsterdam, in 1971 115 municipalities (De Pater and Schmal 1982).

Apart from residential commuter municipalities as Baarn and Bloemendaal, municipalities with many work commuters developed*:* towns where many people worked who lived elsewhere. Some municipalities served both groups, e.g. Hardinxveld, a small town along the Merwede River with a large shipyard attracting many workers from surrounding towns while simultaneously supplying workers to surrounding cities as Sliedrecht (Den Breejen 1984).

Initially, most residential commuters were *immigrant* commuters who used to live and work in the city and who continued working in the city after their move to a new residence town. Later, also more *indigenous* commuters entered the stage: people who initially lived and worked outside the city and who found new work in the city. They did not move to the city but rather stayed in their old residence. Indigenous commuting replaced migration, resulting in a slower growth of the city and in a faster growth of the urban surroundings, the expanding *daily urban system*.

In 1925, some 6% of the working population worked in another town than the municipality of residence (Ter Heide 1965). In 1960 this number was 27% and in 1986: 52%, which means that more than half of the working population was a commuter. In addition, commuting distances increased from 3 km in 1960 to 15 km in 1986. Most commuters travel within their own urban regions; especially female commuters are more often working within the city region where they live too (Burger et al. 2014). In the early 1950s commuters generally travelled by train, (company) busses, moped, scooters or by bicycle, later more commuters traveled by car (Box 9.2).

Box 9.2: Beach Resorts: Scheveningen, Bergen aan Zee, Zandvoort

Many of the Dutch beach resorts owe their existence to infrastructural developments as the construction of rail/tramway lines and successive travel cost reductions by the end of the nineteenth century. Hinterland connections were vital for their emergence and their survival. Scheveningen, a former fishing village near Den Haag, started flourishing from 1870 when an international rail connection between Den Haag and Germany was completed. By then, wealthy, sea-loving Rhineland Germans had options to travel either to Ostend (Belgium) or to the beach resort of Scheveningen near Den Haag. That resort remained a popular destination until its decline after 1965. Already since about 1910, the resort was flooded by middle and lower-middle class people who were able to visit this former elitist resort too due to better regional public transport facilities and higher incomes. For ordinary people in and beyond the region trains became affordable for occasional use in summer times.

From the 1920s onwards workers earned some additional payment and free time for their holidays. They normally stayed in the homes of villagers e.g. in Scheveningen. The elite tourists felt less comfortable and moved to other, high-end beach destinations, e.g. Noordwijk, with its chic hotel Huis ter Duin. Further north, in Bergen aan Zee, building plots along the coast were sold since 1908. The landowner understood that sales would go up once the beach would be connected with the hinterland. He built a 3 km paved road and a steam tramway line through the dunes from the village of Bergen to the coast. Bergen was already connected with the city of Alkmaar by a tramway line since 1905, and from Alkmaar by railway with Amsterdam. This made Bergen aan Zee accessible for an (inter)national public.

Since the 1960s, Dutch beach resorts faced significant competition from sunny Mediterranean beach resorts, in particular due to more private cars, package deals and recently low-cost airlines. Since, Dutch beach resorts became more weekend or day stay destinations for common people. Many hotels closed.

That scenario did not hold only for Scheveningen but for the former, poor fishing village of Zandvoort as well. That site was since 1884 the terminus of the Basel-express, a railway line that ran weekly from the

(continued)

Box 9.2 (continued)
Swiss Basel, via some major German cities, to Zandvoort. Initially, Zandvoort attracted mainly wealthy German tourists, but when rail connections with Dutch cities, including Amsterdam, improved and tickets became cheaper, Zandvoort developed in a popular destination for everyone. The so-called Blue Tram brought more than 1.5 million passengers to this beach in 1905. This tramway line closed in 1957, long since the elite tourists left Zandvoort.

Periods of growth, consolidation and stagnation are normally followed by decline, decay or rejuvenation (Butler 1980). Scheveningen and Zandvoort declined since the 1960s. The downturn in Scheveningen was fueled by Reinder Zwolsman, a business man who owned various hotels in Scheveningen since 1962, including the Scheveningen Pier. He replaced some of the grand hotels by apartment buildings making this resort much less attractive. Demolishment threatened the grand hotel Kurhaus (built in the 1880s) too, but the building became a national monument just in time. Significant opposition arose eventually forcing Zwolsman to sell his possessions. After restauration, the Kurhaus was reopened in 1979. The Scheveningen Pier (1961) decayed but rejuvenated and re-opened in 2016. But the golden days of the *belle époque* were gone forever.

Today, Zandvoort is promoted as 'Amsterdam-Beach', in the hope that the (too) many tourists who go to Amsterdam not only visit the IJsselmeer-fishers villages Volendam and Marken (traditional tourist traps since the early 1900s) or Giethoorn (a historical water-village with the nickname Dutch Venice, especially popular with Chinese groups) in the province of Overrijssel, but also spend one or more days in Zandvoort. The city of Amsterdam tries to spread international tourists from the overcrowded tourist-historic inner city to relatively unknown interesting places in the region.

9.1.6 Road Traffic

In the 19th and early twentieth century, rail and tramways consumed most of the public infrastructural investments. That postponed road improvements. Total length of paved roads increased from 12,034 km in 1873 to 15,624 km in 1905, but decreased in terms of per capita of the growing population (Blink 1914). Investments in roads went up after WWI, primarily due to emergence of the tramway-competing busses and trucks which replaced horse-towed wagons, tricycles and pushcarts. In 1912, for example, the Willemsbrug in Rotterdam was crossed by 18,800 horse-towed vehicles and 28,600 hand carts in 1 week time. Twenty three years later, in 1935, those numbers were reduced to 4600 and 3600, respectively. But the number of trucks crossing this bridge increased from a few hundred in 1912 to over 25,000 per week in 1935. However, bicycles were far more popular, both for commuting and for recreation. 77,921 cyclists, 28,835 tram passengers, 17,203 pedestrians, 13,615 car passengers and 10,336 bus passengers were counted crossing this bridge on June 15, 1939 (Ramaker 1946).

The Dutch government encouraged road traffic in the first decades of the twentieth century. All tolls on national and provincial roads were abolished in 1899. Toll stations were still operational on local roads, e.g. 298 local tolls existed in 1916, 164 in 1928, and 12 toll stations left in 1947 (Bos 1951). Between WWI and WWII, plans for bridge and highway constructions were developed and materialised in the first *Rijkswegenplan* (1927). Several river-crossing railway bridges existed by then already but cars had to use ferry-boats. For example, since 1871 trains moved over the *Moerdijkbrug,* the longest bridge in Europe by then, across Hollands Diep. But it was not before 1936 that cars, carts and bicycles could cross Hollands Diep without a ferry-boat. A next *Rijkswegenplan* was published in 1932 and planned for a network of highways, mutually crossing on roundabouts outside cities.

Before WWII tiny fragments of the planned highway network were open for traffic. In 1952, 200 km of highways existed, in 2016 almost 2500 km. In the last decades, growth in length has been quite modest. The main twenty-first century focus was on capacity increase of existing highways. The connection between Amsterdam and Utrecht e.g. has been a narrow road connecting all villages and towns in between both cities until 1954 when a highway of 2 × 2 lanes was opened. Since, this highway (A2) has been widened towards 2 × 5 lanes now.

Today, any highway construction or widening will face, sometimes fierce, opposition from environmentalists, local residents, and sometimes local authorities. Notorious is the construction of a second highway between the cities of Den Haag and Rotterdam in a rural area (Midden-Delfland). The first designs of that highway date back to the 1950s and construction has been contested since. In 2015, this 7 km highway track has been opened for traffic finally while many (expensive) adjustments had been made addressing most of the objections, including preservation of the rural landscape and reducing traffic noise as much as possible.

Despite all investments to improve Dutch road infrastructure pressure for more mobility did not reduce. On the contrary, traffic congestions became an increasingly regular issue. The first such congestion in the Netherlands occurred on a roundabout near Utrecht, in 1955. Since, traffic jams increased both in length and in duration. Figure 9.4 shows the top-25 of traffic congestions in the Netherlands while Fig. 9.5 displays congestion gravities as registered during the last decade. The 2008 dip in traffic congestions is attributed to the economic recession and (according to the Dutch government) to the results of road works to increase capacity as well. Since 2014, traffic congestions are increasing again.

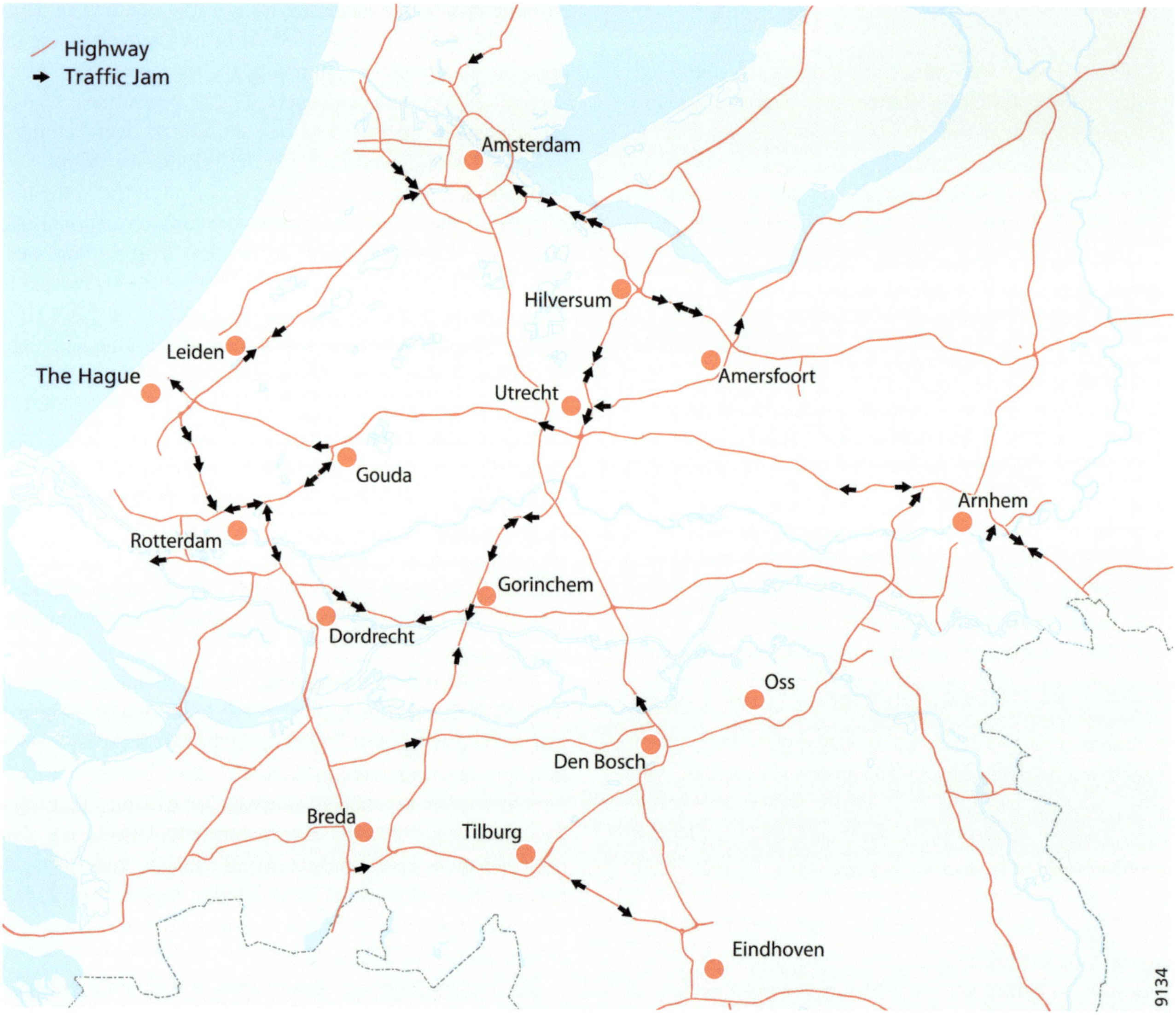

Fig. 9.4 The top-25 of traffic congestions in the Netherlands, measured by *filezwaarte* (length × minutes of delay). Shown are the headers of the congestions; travel directions are indicated by arrows. (Source: Traffic Information Service (Verkeers Informatie Dienst VID))

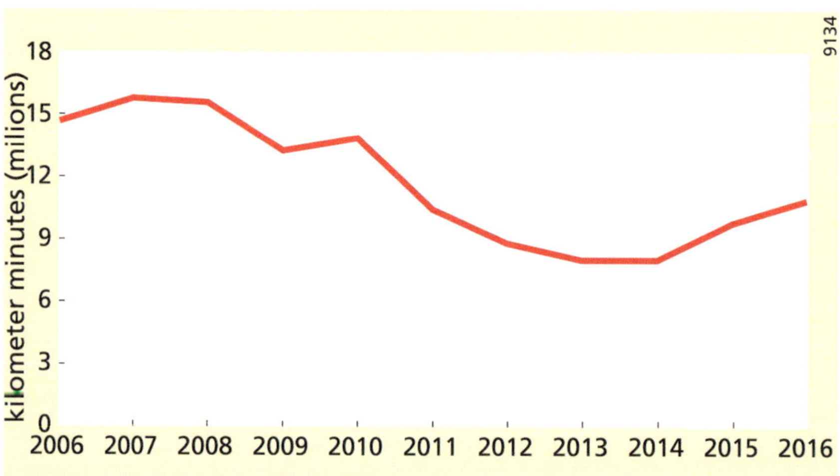

Fig. 9.5 Total gravity of traffic congestions in the Netherlands, expressed in *filezwaarte* (length × minutes of delay), in the period between 2006 and 2015. (Data on earlier years are available but measured according to different standards). (Source: VID)

9.1.7 Canals and Ports

The four phases of network development as distinguished above (loose ends, integration, densification and selection) are valid for canal networks as well. Until late in the nineteenth century, regional water networks occurred as loose ends in addition to the major rivers in the Dutch delta. Lack of integration has not been caused by lack in ambition and enthusiasm for digging canals. King William I (1813–1839), nicknamed 'Canal King', not being aware of the forthcoming railway infrastructural revolution, ordered to dig numerous canals eventually meant for introducing more prosperity in the country. One of these canals in the south of the country is named after him: the Zuid-Willemsvaart (1826).

Canals were also built for water managerial or for peat extraction and transport purposes. The Bergsche Maas has been dug between 1888 and 1904 to drain River Maas. Peat extraction and transport, together with regional development, were the main drives for digging the Dedemsvaart for which permission was granted in 1809. This canal allowed peat transport to the town of Hasselt and further, via Zuyderzee, to other Dutch towns and cities. The city of Zwolle was against the construction as ships might then bypass the city's own river Overijsselse Vecht. Eventually, new factories developed alongside of the Dedemsvaart bringing some prosperity to the region (Varwijk and Seinen 1983).

Water management measures, including development of new canals was a main item during the reigns of King William's successors (Willem II and Willem III). In 1892, e.g. the Merwede Canal connecting Amsterdam with River Rijn and its German hinterland was opened. This canal was meant to strengthen economic power of the port of Amsterdam vis-a-vis Rotterdam harbour. However, the Merwede Canal soon proved to be too narrow and a new, broader canal was dug, the *Amsterdam-Rijnkanaal* which was opened in 1952 and partly coincided with the Merwede Canal.

Interregional integration of the numerous Dutch waterways developed around 1900. Such integration was a major incentive for inland navigation. In 1930, e.g. 83,000 loaded cargo barges sailed from Germany and Belgium to the Netherlands; inland harbours, as those in e.g. the city of Den Haag, flourished. Thousands of barges transported vegetables, cattle, coal etcetera to Den Haag's warehouses and slaughterhouse in Laakhaven, located at the fringe of the city. These goods were brought to shops in Den Haag by trucks and carts.

In 1939, the Dutch inland navigation fleet consisted of 18,500 vessels, their average load capacity was 200 tons. After WWII, their numbers remained about constant for three decades but their average load capacity increased considerably. Since 1975, the number of vessels decreased from 20,000 to some 7500 ships today. The average load capacity per inland vessel increased six times, to 1200 tons. Big ships cannot use all waterways. In 1953, the total length of navigable rivers and canals in the Netherlands was 6920 km but for 1000 tons+ vessels only 1615 km of the waterways was navigable; 1900 km of the network was open to ships of less than 80 tons.

After WWII, smaller canals lost their navigation purposes due to rising numbers of trucks on the road and to increasing load capacity of ships. Some of such canals were reclaimed and turned into roads. That marks the (last) selection phase in infrastructural development. Re-appraisal of smaller canals occurred since the 1980s due to emergence of their ecological, cultural, historical and leisure values, e.g. the *Turfroute*, operational in summer, in the southeast of Friesland, a historic peat-excavation area (Buiter and Raat 2016).

9.1.8 Mainport Rotterdam – Gateway to Europe

Post-WWII extensions of harbour cities at river mouths (as Rotterdam) or along sea-connecting canals (as Amsterdam) typically occur in a seaward direction. This creates a physical separation between city and harbour (see the development of the harbour areas from year to year – from 1815 to 2015 – and in much detail on www.topotijdreis.nl). The British geographer James Bird (1971) designed a model for this development, the *Anyport* model. Old city harbours became inaccessible for larger post-WWII vessels. Successively, these often polluted and barren areas turned into brownfields. Larger and deeper harbours were built, closer to the sea, as in Rotterdam, where the new harbours were surrounded by petrochemical industry plants, mainly oil-refineries. Here, seaward harbour expansion occurred in successive stages: Botlek (1950s), Europoort (1960s) and Maasvlakte 1 (1970s), each new phase closer to the sea and the last one even *in* the North Sea. Most of these harbours are used by bulk carriers: dry bulk (ores, grains, etcetera) and liquid bulk (mainly oil).

Rotterdam's most recent harbour extension is Maasvlakte 2, built since 2008, and situated even further into the North Sea (Koppenol 2016). Maasvlakte 2 is meant for containers and is used by major container handling companies as APM, RWG and ECT. Close interaction between larger international companies in the Rotterdam municipality has significantly contributed to the success of Rotterdam harbour, between 1962 and 2004 the largest in the world, measured in throughput (461.2 million tonnes in 2016). Since that year the throughput in Chinese harbours is (much) larger, but in Europe Rotterdam is the number one (Table 9.1).

With a throughput of 12.4 million containers (in number of TEU's: 20-feet Equivalent Units) in 2016, Rotterdam is the main container port in Europe and the 11th globally,

Table 9.1 Top 5 of European harbours in millions tonnes throughput and top 5 of European container harbours

Harbour	Millions tonnes (incoming + outgoing) in 2015	Harbour	Millions containers (TEU) in 2015
1. Rotterdam	466.4	1. Rotterdam	12,235
2. Antwerp	208.4	2. Antwerp	9654
3. Hamburg	137.8	3. Hamburg	8821
4. Novorossiysk	128.4	4. Bremerhaven	5547
5. Amsterdam	96.5	5. Valencia	4615

Source: Port of Rotterdam (2016)

following mainly Asian (Chinese) harbours. Competitors of Rotterdam harbour sharing the same European hinterland, are between French Le Havre and German Hamburg. Significant Chinese investments makes Piraeus harbour (Greece) a serious, new competitor for Rotterdam as navigation distances and time reduced for Asian ships due to recent widening of the Suez Canal and the geographical position of Piraeus.

Inland Terminals and New Destinations for Old Harbours From the deep sea terminals in Rotterdam containers and other cargo are transported to smaller seaports elsewhere in Europe ('short sea'). Trucks, trains (e.g. the *Betuwelijn*), and river barges bring this cargo to destinations In Europe, or to inland terminals, which have been expanded as well, e.g. Venlo in the province of Limburg, near the River Maas, highways and rail connections. In the last 20 years, TCT Venlo developed into a prominent 'inland logistic centre' (Raimbault et al. 2016).

As in many other traditional harbour cities obsolete pre-WWII harbour areas in Rotterdam and Amsterdam were recently rediscovered and rejuvenated. Old warehouses were renovated, but also spectacular skyscrapers were built. So on the Kop van Zuid 'de Rotterdam' (a high-rise complex of star architect Rem Koolhaas) can be seen, but also the traditional office of the Holland America Line, now hotel New York. Where until the 1960s emigrants to the United States boarded the passengers ships of the HAL, now cruise ship passengers disembark. The old harbour areas of the IJ shores in Amsterdam received new functions as residential areas, office districts, and facilities for leisure and tourism too (Box 9.3).

Box 9.3: Electricity Networks

The Second Industrial Revolution, around 1900 in the Netherlands, has partly been due to electrification of the country. Comparable to the First Industrial Revolution (transition from horse and wind power to steam power) in the mid nineteenth century, electrically powered machines began to replace steam-driven

Box 9.3 (continued)

machines in factories in the early twentieth century. Initially, electrical machines were powered by dynamo's or by energy from local power stations. In 1900, 20 such power stations existed in the Netherlands, in 1913 there were 82. Most of these power stations were private but some were owned by the municipality and served as local networks, as long-distance electrical power transmission was not yet feasible. Electrical power stations did not deliver to factories, offices and households only, they also provided electrical power to railways, tramways, and to cities for street lightning. As many cities have built gas plants in the late nineteenth century, not many of them were eager to invest again in (competing) electrical power stations.

The tangle of local networks was gradually replaced by larger, provincial power plants and networks since the 1910s when provincial governments decided to provide such services. Meanwhile, long-distance electrical power transport became operational. A decade later, in 1920, almost two-third of all Dutch municipalities were already incorporated in the provincial electrical networks. Despite about 10,000 km of high-voltage grid in the Netherlands, interprovincial integration was not yet completed by 1927. The 200 municipalities in 1927 with no electrical power were sparsely populated. The number of municipalities with no electricity went down to eight in 1935.

Electrification of the country is reflected in the number of users and their power consumption. In 1919, only 400,000 houses and buildings were connected with the electricity network. That number rose to 1.9 million in 1939. Over the same time period power consumption per capita increased from 58 to 354 kWh (Groote 1996; Hesselmans 1990, 1993).

Soon after WWII, production and consumption of electricity exceeded the pre-war level. In 1950: about seven billion kWh, 60 years later, in 2010: 120 billion kWh. In 2010, production was lower than consumption as the Netherlands imported often cheaper and sometimes cleaner electricity from neighbouring countries since the 1980s. Between 1995 and 2013, 87% of the supplied electricity was domestic and 13% imported from Germany, Belgium, Norway, and the UK. In the Netherlands, most electricity is consumed by industry. In 2013, industry and business used 91 billion kWh and the 7.6 million households used 23 billion kWh. Electricity consumption per caput was over 6000 kWh in 2012, 1500 kWh of which was for domestic use, about 18 times the consumption in 1939 (Van Wezel 2015).

9.2 Communications Networks

In contrast to transport networks physically moving people, freight and energy, communication networks transfer information from one place to another, without physical movement. Communication systems developed in the course of the twentieth century, comparably as spectacular as transport networks did.

Two types of communication may be distinguished, individual and collective modes of communication. Individual modes of communication include newspapers and radios whereas cinemas and theatres are meant for groups of people and are considered collective communication modes (Dibbets 2014). Panorama paintings, which were quite common in the nineteenth century, may be considered an intermediate mode. Inside buildings with rounded walls realistic panoramas were displayed against an admission fee. Such panorama paintings aimed to communicate views on the world or historic events. One such Dutch, nineteenth century panorama painting displayed in a dedicated building survived: the Panorama Mesdag in Den Haag showing the fishing village and beach of Scheveningen (Rombout 2006). Around 1900, most of these were replaced by (travelling) cinemas.

9.2.1 Telegraph

Apart from communication by semaphore or by carrier pigeons, traditional surface mail faced large scale competition by other, more rapid communication and information types around 1850 in the Netherlands. The first Dutch telegraph line (*Rijkstelegraaflijn*) connecting five cities was opened in 1852. That number of telegraph connected cities expanded to 90 in 1867. Telegraph offices were gradually combined with post offices. Such combined post and telegraph offices increased and telegrams could be sent from 477 offices in 1893. This rapid increase may be explained by a drastic cost reduction, growing prosperity and larger demands caused by migration from rural areas to cities. In 1880, two million domestic telegrams were sent, in peak year 1919, six million. Since, domestic telegrams dropped but telegrams sent to foreign destinations went on rising. In 1924, more telegrams were sent abroad than domestically in the Netherlands. That was caused by the emergence of telephones (Ten Brink and Schell 1954; Groote 1996).

9.2.2 Telephone

The *Nederlandsche Bell Telephoon Maatschappij* was the first company that got permission building local telephone networks. In 1881, the first local network was operational with 49 subscribers in Amsterdam. Only 4 years later, in 1895, 31 local telephone networks, operated by seven companies occurred in the Netherlands and some cities began building their own networks as well (Groote 1996). Once local networks were interconnected the Dutch state began to interfere and started telephone services at a national and international level since 1897. From 1913, all private local networks were nationalised while densifying existing networks and rapidly expanding these networks even into rural areas. By then, 1134 PTT (Post- Telegram and Telephone) offices occurred in the Netherlands (Fig. 9.6).

More and more individuals subscribed for a home connection: in 1939 323,000. Similar to other infrastructural networks the growth was spurred by major cost reductions. Still, telephones were not common in Dutch households. Around 1935, no more than 15–20% of the urban households possessed a telephone (5–10% of houses in rural areas).

The more connections to the telephone grid, the more calls were made. In 1902, the one million calls barrier was passed, in 1909 more telephone calls were made than telegrams sent, and in 1931, 28 million calls were made (Santing 1933; CBS 1937; De Wit 1998). After WWII the number of telephone connections grew further, 0.8 million in 1950, via 2.2 million in 1970, seven million in 1990, to 9.9 million in 2000 (blafhert.wordpress.com). By then, mobile phones began to emerge connecting people rather than places. For several years now more mobile phones occur in the Netherlands than inhabitants.

9.2.3 Mass Media: Newspapers and Magazines

Until late in the second half of the nineteenth century newspapers were scarce and expensive in the Netherlands. In 1850, seven copies of daily newspapers occurred per 1000 inhabitants, and including weekly and other newspapers: 18 copies. However, regional differences were large. Figure 9.7 shows major differences between the lower and the higher parts of the Netherlands. Many newspaper copies circulated in cities in the west but rather few in Brabant and Limburg where illiteracy was relatively high (Knippenberg 1986).

During the last 50 years of the nineteenth century many new newspapers were issued: 92 in 1850, 760 in 1894. Abolition of newspaper taxes in 1869 facilitated this process. Numbers of informative magazines rose as well. Roman catholic families subscribed to *Katholieke Illustratie* (founded in 1867), people who were interested in popular science to *Vragen van de Dag*, founded in 1886 by the geographer Hendrik Blink.

Not just the number of titles, their circulation increased as well. In the 1930s about 200 copies of newspapers were distributed per 1000 people in the Netherlands, which

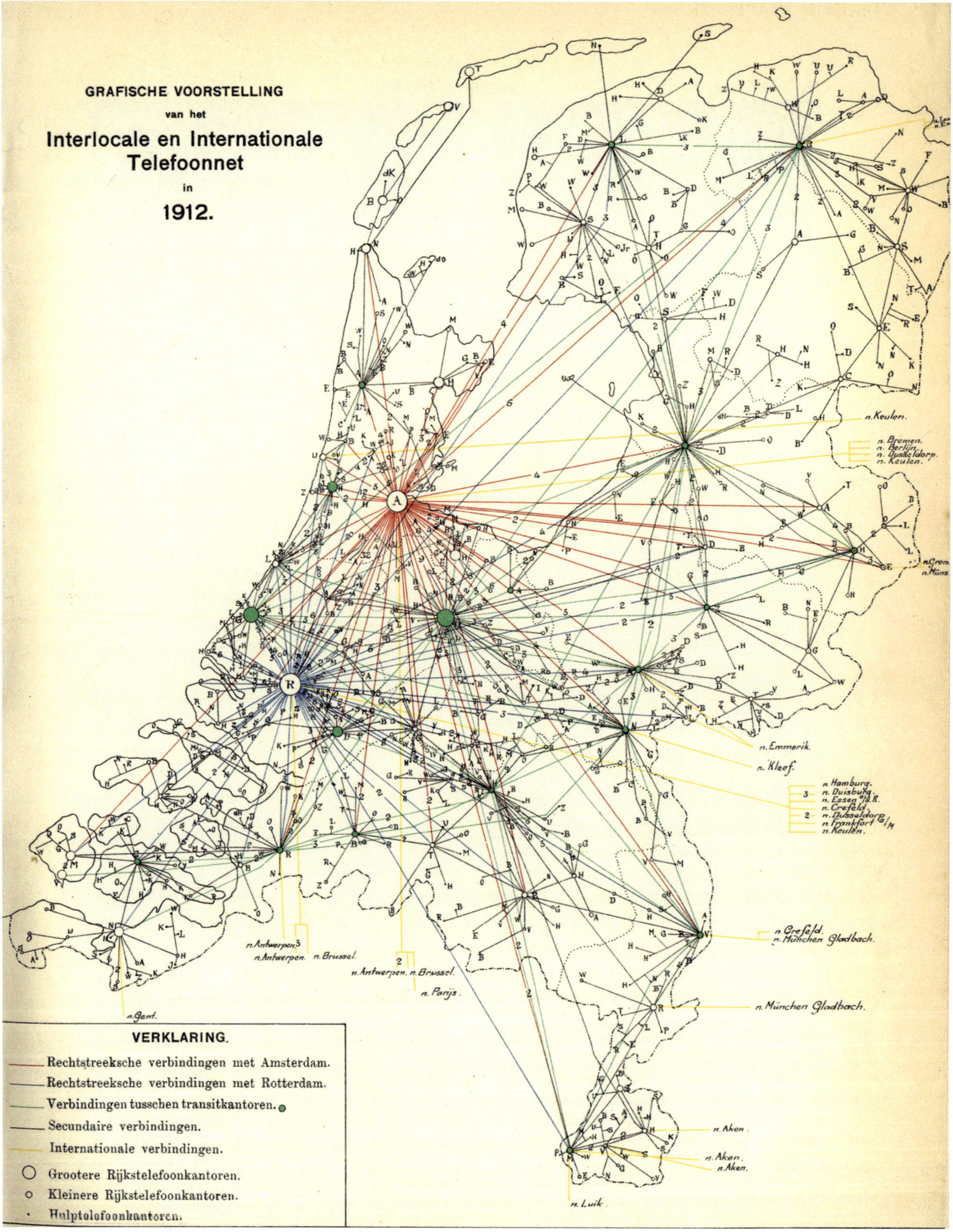

Fig. 9.6 The long distance telephone network in the Netherlands in 1912. (Source: Tijdschrift voor Economische Geografie, August 1912 (annex))

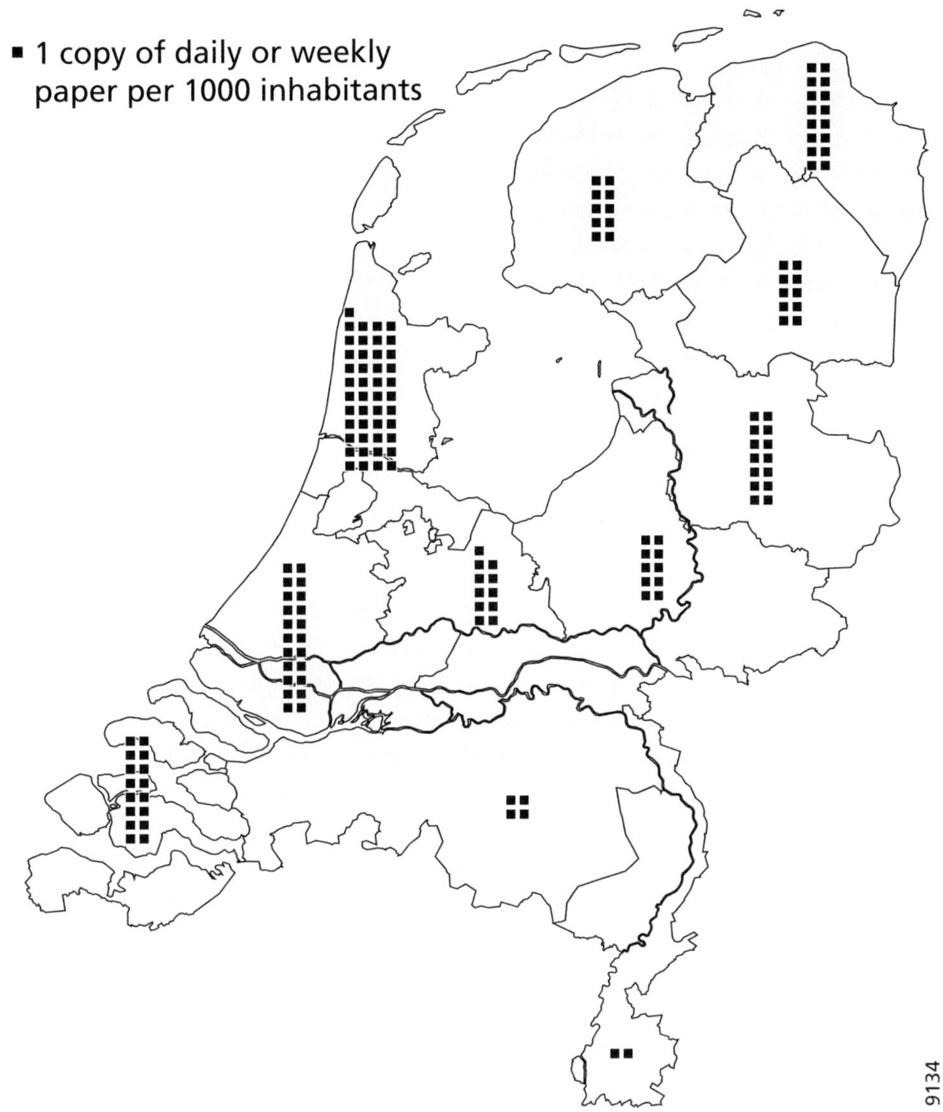

Fig. 9.7 Number of copies of newspapers per 1000 inhabitants per province in 1850. Primary data: Hemels 1969, Census 1849 (From: Knippenberg and De Pater 1988)

number grew to 300 copies around 1990. Since, these numbers are shrinking due to the rise of internet (Hemels 1969; Luykx 1978; Schneider 1968).

9.2.4 Mass Media: Radio and Television

Since its origin, radio rapidly spread in the Netherlands. The first broadcasting associations were founded around 1925, neatly according to the Dutch pillar culture (Chap. 12): a Roman-catholic, a protestant, a socialist, and a more liberal broadcasting association. By coincidence, Hilversum developed as the Dutch broadcasting city. By 1935, almost one million radios were spread and radio density greatly exceeded private telephone density. More than half of the urban households was able to listen to the radio, in rural areas: 10–40% (Fig. 9.8). After WWII, the numbers of radio's rose quickly: two million in 1950, three million in 1960.

However, the postwar period was particularly characterised by the emergence of television. In 1955, 16,000 television sets occurred, 1.8 million in 1965, and more than 3.3 million in 1975. Except for orthodox Calvinists, TV sets were found in every Dutch household. As with radios, TV sets were bought by urban households first, followed by rural households. Regional differences became more diffuse during the final phases of TV distribution (Van der Knaap 1978).

9.2.5 Amsterdam Internet Exchange

The first window of the historical *Canon of Amsterdam* (used for instance in class rooms) is dedicated to Dam Square, the final and fiftieth window to the Amsterdam Internet Exchange (AMS-IX), developed from the mid-1990s, and with 799 affiliated companies, among the leading ones in

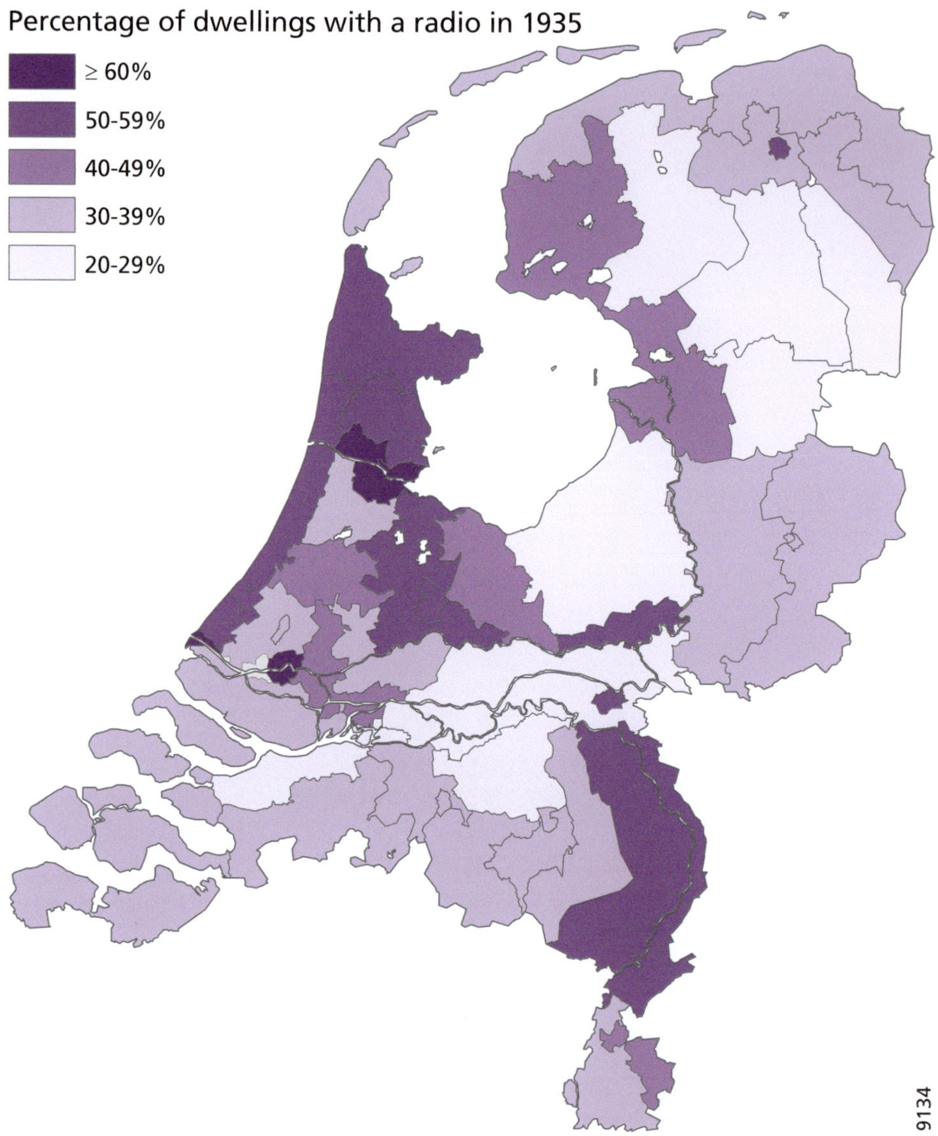

Fig. 9.8 Regional distribution of radios in 1935. Primary data: CBS 1937. (From: Knippenberg and De Pater 1988)

the world. AMS-IX' origin is in Amsterdam Science Park, in the eastern part of the city, where the first fiber optics cable between the United States and Europe ended. In 2012, six out of the 11 currently existing transatlantic cables landed in the Netherlands. Today AMS-IX consists of a web of locations in the Amsterdam agglomeration. The web serves as a 'roundabout' that connects IP networks. Through this exchange points data are forwarded from one network to another (Hemel 2012).

Just as Schiphol Airport, the presence of Amsterdam Internet Exchange stimulates foreign investments. The Advisory Council on the Environment and Infrastructure (Rli 2016) advises the national government to pay extra attention to further developing this new main port rather than to continue investing in former main ports as Schiphol Airport and Rotterdam harbour, under the banner of 'Amsterdam: the most connected city of the world'.

9.2.6 A Highly Developed Information Society

The *Eurostat Regional Yearbook, 2016 Edition* states that the 12 Dutch provinces, together with five British regions, four German regions and Luxembourg, contain the highest share of households with broadband connectivity in the EU: more than 90% in 2015. On the NUTS level 2 the Dutch province of Overijssel enjoyed the highest share of all NUTS level 2-regions in the EU: 97% of all households in 2015 (the EU average was 80%).

The Dutch are heavy internet users. But, again, spatial discrepancy exists between urban and rural areas. High-speed internet infrastructure is more developed in urban areas. Private companies hesitate investing in costly fiber optic cable networks in rather thinly populated regions. That caused local pressure to combat 'digital exclusion' and to improve access to broadband networks. It is estimated that in 2015 circa 200,000 households and 60,000 companies (farms, for instance) had no access to high-speed networks (Salemink and Strijker 2016; Salemink 2016).

References

Bird J (1971) Seaports and seaports terminals. Hutchinson, London

Blink H (1914) Ontwikkeling van de economische toestanden in Nederland en zijn betrekkingen met de nabuurlanden gedurende de laatste eeuw. Tijdschrift voor Economische Geographie 5(X): 369–384 and 5(XI): 416–434

Bos CJ (1951) De tollen op landwegen in Nederland. Dissertation, Utrecht University, Utrecht

Buiter H, Raat R (2016) Nederland kanalenland. Een reis langs twaalf kanalen. Stokerkade cultuurhistorische uitgeverij, Amsterdam

Burger MJ, Meijers EJ, Van Oort FG (2014) Multiple perspectives on functional coherence: heterogeneity and multiplexity in the Randstad. Tijdschrift voor Economische en Sociale Geografie 105 (4):444–464

Butler RW (1980) The concept of a tourist area cycle of evolution. Can Geogr 24(1):5–12

CBS (1937) Marktanalytisch handboek voor Nederland. Centraal Bureau voor de Statistiek, Den Haag

De Pater B, Schmal H (1982) Tijdgeografie en de suburbanisatie rond Amsterdam tussen 1855 en 1980. Stedebouw & Volkshuisvesting 63 (1):37–46

De Vries J (1978) Barges and capitalism. Passenger transportation in the Dutch economy, 1632–1839. HES, Utrecht

De Wit O (1998) Telefonie in Nederland 1877–1940. Opkomst en ontwikkeling van een grootschalig technisch systeem. Otto Cramwinckel, Amsterdam

Den Breejen P (1984) Hardinxveld en Giessendam, van vissers- en hoepmakersdorpen naar industriegemeente. Historische Vereniging, Hardinxveld-Giessendam

Dibbets K (2014) Paul Kruger als toneelheld en filmster. De verbeelding van de Boerenoorlog en de opleving van het nationalisme, 1899–1902. De Negentiende Eeuw 38(4):225–268

Filarski R (2016) On the right track? Decision-making on the Betuwe Route. In: Arts J, Filarski R, Jeekel H, Tousaint B (eds) Builders and planners. A history of land-use and infrastructure planning in the Netherlands. Rijkswaterstaat, Eburon, Den Haag, pp 279–343

Groote P (1996) Infrastructure and Dutch economic develoment. A new long run data set for the Netherlands 1800–1913, Netherlands Geographical Studies 211. Koninklijk Nederlands Aardrijkskundig Genootschap/Faculteit Ruimtelijke Wetenschappen, Utrecht/ Groningen

Hanrath JJ (1938) De Nederlandsche beurtvaart. Tijdschrift voor Economische Geographie 20(11):41–53

Harms L, Te Brömmelstoet M, Bertolini L (n.d.) Ontwikkelingen in fietsgebruik en fietsbeleid (internet). Centre for Urban Studies, University of Amsterdam http://www.verdus.nl/upload/documents/Lucas%20Harms%20Ontwikkelingen%20in%20fietsgebruik%20en%20fietsbeleid.pdf. Accessed 20 June 2017

Hemel Z (2012) Het Amsterdamse mirakel. Plan Amsterdam 2012 (2):4–9

Hemels JMHJ (1969) De Nederlandse pers voor en na de afschaffing van de dagbladzegel in 1869. Van Gorcum, Assen

Hesselmans T (1990) De elektrificering van Nederland. Intermediair 26 (20B):39–45

Hesselmans AN (1993) Elektriciteit. In: Lintsen HW (ed) Geschiedenis van de techniek in Nederland, Volume III. Walburg Pers, Zutphen, pp 135–161, 301–304

Jordan P (2013) In the city of bikes. The story of Amsterdam cyclist. Harper Collins, New York

KiM (2016) Mobiliteitsbeeld 2015. Ministerie van Infrastructuur en Milieu, Den Haag

Knippenberg H (1986) Deelname aan het lager onderwijs in Nederland gedurende de negentiende eeuw, Nederlandse Geografische Studies 9. Koninklijk Nederlands Aardrijkskundig Genootschap/University of Amsterdam, Amsterdam

Knippenberg H, De Pater B (1988) De eenwording van Nederland. Schaalvergroting en integratie sinds 1800. SUN, Nijmegen

Knippenberg H, Nauta B (1989) Naar eenheid van tijd in Nederland 1835–1909. Tijdschrift voor Sociale Geschiedenis 15(4):324–346

Koppenol D (2016) Lobby for land. A historical perspective (1945–2008) on the decision-making process for the Port of Rotterdam land reclamation project Maasvlakte 2. Boom Publisher, Amsterdam

Kranenburg RH, Meijer H (2001) A compact geography of the Netherlands. KNAG Royal Dutch Geographical Society, Utrecht

Luykx T (1978) Evolutie van de communicatiemedia. Elsevier, Brussel

Oldenziel R, Emanuel M, de la Bruhèze A, Veraart F (2016) Cycling cities: The European experience. Foundation for the History of Technology & Rachel Carson Centre for Environment and Society, Eindhoven

Port of Rotterdam (2016) https://www.portofrotterdam.com/nl Accessed 15 Nov 2016

Raimbault N, Jacobs W, Van Dongen F (2016) Port regionalization from a relational perspective: the rise of Venlo als Dutch international logistics hub. Tijdschrift voor Economische en Sociale Geografie 107(1):16–32

Ramaker JG (1946) Verkeer en vervoer in de Nederlandsche stad. Een economisch-stedebouwkundige studie. Dissertation, Erasmus University Rotterdam. Holdert, Amsterdam

Rli (2016) Mainports voorbij. Raad voor de Leefomgeving en Infrastructuur, Den Haag

Rombout T (ed) (2006) The Panorama phenomenon. Panorama Mesdag, Den Haag

Salemink K (2016) Digital Margins. How spatially and socially marginalized communities deal with digital exclusion. Dissertation, University of Groningen, Groningen

Salemink K, Strijker D (2016) Rural broadband initiatives in the Netherlands as a training ground for ne-endogeneous development. Local Econ 31(7):778–794

Santing R (1933) De telefoon. Tijdschrift Koninklijk Nederlands Aardrijkskundig Genootschap 50(May–June):645–660

Schneider M (1968) De Nederlandse krant. Van 'nieuwstijdinghe' tot dagbladconcentratie. Van Kampen, Amsterdam

Ten Brink EABJ, Schell CWL (1954) Geschiedenis van de Rijkstelegraaf 1852–1952. Staatsbedrijf der PTT, Den Haag

Ter Heide H (1965) Binnenlandse migratie in Nederland. Dissertation, University of Amsterdam

Thurkov AJ, Harten JDH, Knippenberg H (1984) Bewoningsgeschiedenis. Atlas van Nederland, volume 2. Staatsuitgeverij, Den Haag

Van der Knaap GA (1978) A spatial analysis of the evolution of an urban system: the case of the Netherlands. Dissertation, Erasmus University Rotterdam

Van Wezel B (2015) Elektriciteit in Nederland. Centraal Bureau voor de Statistiek, Den Haag

Varwijk GH, Seinen JR (1983) Stap voor stap langs de Dedemsvaart. Boekhandel Rooseboom, Dedemsvaart

The Governance of Order

10

Len A. De Klerk

Abstract

God created the world, but the Dutch made their own land. They had to, since a quarter of the country lies below sea level and another 30% along rivers, all endangered by spring tides and temporarily high water levels. As from 1000 AD the Netherlands have been shaped and reshaped, controlled and vehemently protected against natural forces, its main natural enemy being the *water wolf*. Water management is part of the DNA of Dutch culture, and to mastering nature has been the rule rather than the exception throughout Dutch history. Building dikes, reclaiming land from lakes and the sea, shaping polders and to construct orderly cities required collective efforts and investments. These activities, driven by collective gains and compelled the cooperation of private organisations and public authorities, are reflected in the Dutch landscape breathing optimisation, order and symmetry, straight lines and geometrical patterns, as if designed by Dutch painter Piet Mondriaan.

Governance is a too big container of policy, private and public parties, juridical frames, common values, formal and informal rules of government, age-old practices of social and political deliberation, and changing attitudes to all these characteristics. This chapter traces some headlines of the culture, methods and instruments of governance of what may be called, *geography made by governance*, with a focus on spatial planning.

Keywords

Water board · Urban planning · Railway system · Housing act · Land consolidation act · Postwar spatial planning

L. A. De Klerk (✉)
Department of Geography, Planning and International Development Studies, University of Amsterdam, Zaandam, The Netherlands
e-mail: L.A.deKlerk@uva.nl

10.1 Self-Government in a Fragmented Country

Amongst many travellers, the Portuguese Alphonse Esquiros (1858) was impressed by the way the Dutch had created their swampy delta into a human ecotype: 'This country is their work, their creation, as well as the God of the Bible they judge it rightly, that what they have done is well done.'. His statement is extended by the cultural-historians Bank & Van Buuren that 'order is not a characteristic of a specific era of Dutch history [but] the main characteristic, the existential condition of a civilisation which had nested and had to survive in the delta of three great rivers…' (Bank and Van Buuren 2004).

Although widespread over Europe, feudal landownership did not fit to the geography and the mentality of the Low Countries (the Netherlands). After extraction of large volumes of peat by private developers in the eleventh to thirteenth century a landscape of lakes, ditches and small parcels remained. These lakes were successively reclaimed and converted into polders for agricultural land use (Chaps. 2 and 3). A common interest was born: maintenance of dikes, locks, canals, ditches and controlling groundwater levels. Once started as private organisations in hamlets and villages, became the country's oldest public institution: the Water Board (Fig. 10.1). In 1255, the first official Water [Management] Board (Dutch: Waterschap) was founded by William II, Count of Holland.

Water Boards adopted principles as *proportional cost allocation* and *strict compliance to regulations*. Financial means required for maintenance were levied in accordance to the share of the total area under control of each landholder (landlords and lessees), therefore proportionately sharing the burden of ensuring collective safety. Similar to their policies to grant city rights, market rights and toll rights, the Counts of Holland had to enter into partnerships with polder authorities in charge. Actually, a Water Board was a meeting of joint landholders, which fostered a 'collectivist mentality, which

E. F. J. De Mulder et al., *The Netherlands and the Dutch*, World Regional Geography Book Series,
https://doi.org/10.1007/978-3-319-75073-6_10

Fig. 10.1 Gemeenlandshuis of Delft expresses the importance of the Water Board of Delfland. (Source: Hoogheemraadschap van Delfland 2017)

aimed at a rational balance of ends and means, the conscientious governance of the achieved goods and the right of consultation and liability for maintenance of all those involved. This mentality was internalised by large groups in the Netherlands; [and] perhaps constituted an additional precondition for the successful expansion of the Dutch trade.' (Blockmans 2010).

Apart from Water Boards, independent cities are the second historical Dutch source of governance. As from the twelfth century cities were founded for trade, military reasons or as religious centres. The oldest cities, Nijmegen and Maastricht, were founded as Roman colonies. Most cities in Holland (present provinces of Noord- and Zuid-Holland) were founded between 1200 and 1400. In the sixteenth century Holland was the most urbanised region of Europe. By mid-seventeenth century, about 60% of Holland people lived in cities (Chap. 8).

The history of public intervention in land use is as old as the cities. Expropriation was already applied in the thirteenth century. According to Peteri (1913) the concept of private ownership dissociated from collective ownership as a sort of right of use. By the end of Medieval Times local authorities used a combination of private law and public law regulations as expropriation to implement public works aiming at improving traffic infrastructure. The ultimate goals of such measures generally were in public interest: defence against floods (*water wolf*), construction of bridges or military defence systems and generally to facilitate economic growth by improving internal and external traffic accessibility. In the fifteenth and sixteenth centuries public intervention in urban development had been institutionalised in Dutch cities already (Fig. 10.2).

Although City Councils and Water Boards made their own local laws, their autonomy concerning urban development and reclaiming lakes had never been absolute. Both activities, regardless private or public or in mutual cooperation, needed approval (Octrooi, a formal public patent or permission) from the States General, which subsequently requested advise from the High Council of State, the oldest state institution, founded in 1531 by the Habsburgian emperor Charles V and still existing, which supervised state-owned land, rivers and lakes. In 1769, e.g. the city of Rotterdam needed permission (consent) from the States of Holland to construct a second reservoir (overflow) to relieve the existing overflow, which was too small to collect water to prevent the city from flooding. City councils were autonomous, but always needed consent, because most (re)development plans had implications for either the provincial water system or for the military defence system. Such administrative encumbrance is not dissimilar to what is presently experienced (Van der Cammen et al. 2012).

City councils decided autonomously on *building regulations* which were issued as local *Ordinances*. Generally, governance of public works and new urban development was ruled by a combination of private law (buying, exchanging, leasing and selling land) and public law (compulsory purchase; expropriation). In the twenty-first century, buying and renting land, or selling and leasing out land by public authorities are still private law actions which, according to public law, need approval by the municipal council and by the Provincial Executive.

In the early fourteenth century the city of Utrecht enacted an *Ordinance* prescribing maintenance of the public character of streets. Decision making and implementation concerning urban expansion was settled as an act of public-private cooperation. Private corporations, as guilds, influenced public decision making which, for their part, depended on these guilds for implementation of their decisions.

In terms of political structure, the *Republic of the Seven United Netherlands* (seven united provinces, 1588–1795) was quite unique as compared to other European nations. The Republic had no central power, no head of state and no capital city. On all levels, urban, provincial and national, governance was a *consensus-seeking* method of policy and decision making by deliberation. Eventually, decisions were

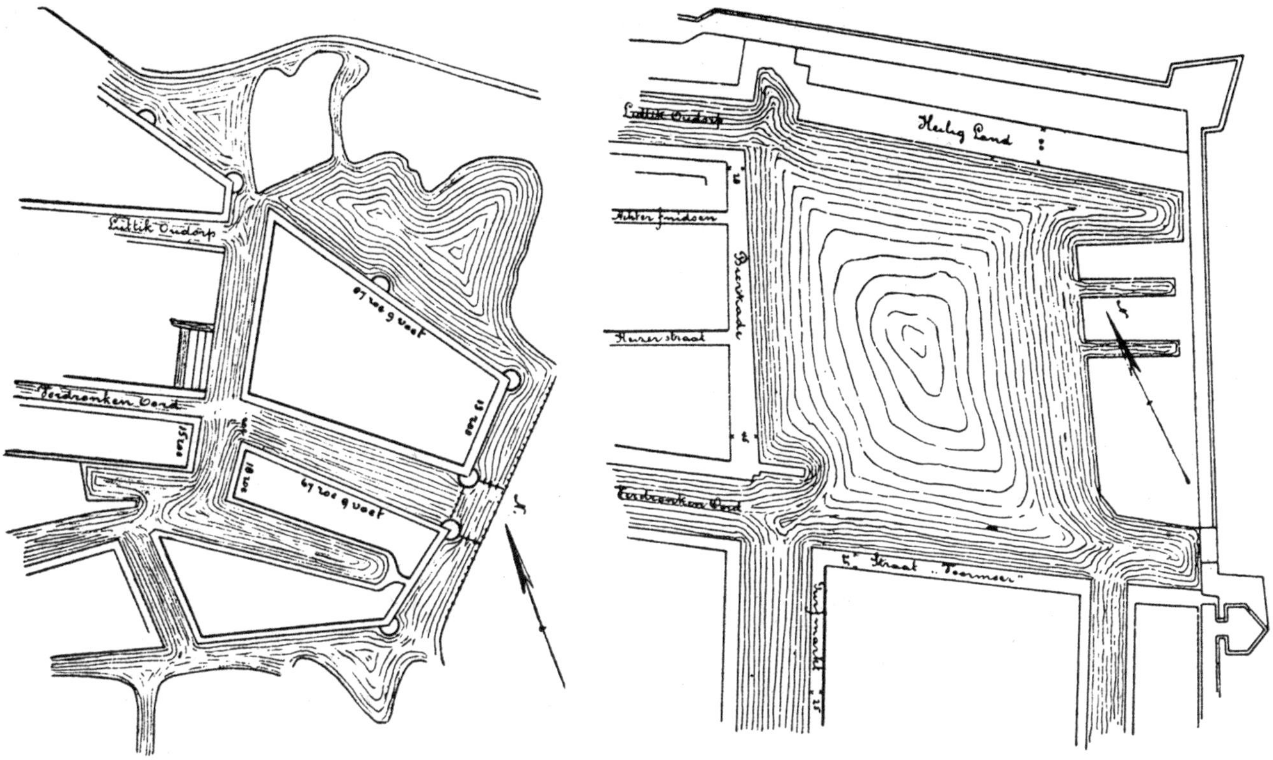

Fig. 10.2 Sixteenth century development planning by the city of Alkmaar: the expensive development plan, not realised (left); the cheaper and realised plan (right). (Source: Peteri 1913)

drive by practical objectives maintained political balances (Frijhoff and Spies 1999). In most rural areas land lords were in charge and dominated the provincial states. Urban governance was self-government that relied on private interest groups of which the guilds and prominent merchant families were most relevant. In the seventeenth century the Dutch Reformed Church had the status of state church, and became the most important public institution, the *Res Publica* which controlled public government. The church became involved into numerous social and cultural issues, as almshouses and even establishment of theatres. On a national level the Republic was ruled by the States General, a meeting of 20 independent city and of seven province representatives, inevitably leading to cumbersome, slow central decision making.

10.2 Renaissance Legacy of Urban Planning

The seventeenth century (the Dutch 'Golden Age') was an era of unprecedented prosperity, urbanisation and migration. Cities expanded, and major lakes were reclaimed. In 1680, about two million people lived in the Republic. In Holland, even 54% of the population lived in cities of 10,000 or more inhabitants (Table 10.1). Between 1570 and 1665, Amsterdam had grown from 30,000 to 160,000 inhabitants, Leiden from 12,000 to 70,000 people.

Two major geographical developments occurred in the Netherlands during the seventeenth century: substantial urban expansion and some large reclamation projects, the latter emerged from technical innovations by engineer and mill constructor Leeghwater (1575–1650; Box 2.4). Between 1610 and 1631, land reclamation increased arable lands in Holland by 40%. Such essentially private land reclamation projects still required approval by city councils and the States of Holland. From a governance perspective such approval was provided rather easily as the States of Holland were dominated by these urban merchant families, often in their capacities as burgomasters and aldermen.

A particular heritage of the Golden Age were the urban development plans for cities like Amsterdam, Haarlem, Leiden and Groningen. Generally, initiatives for such plans originated from the city council which commissioned a plan by its city architect or by a private architect. The city of Leiden, for example, needed urban expansion to accommodate the many Flemish immigrants (Fig. 10.3). After the 1581 census, the Leiden city council agreed on three principles to stimulate urban (re)development and housing conditions (Van der Cammen et al. 2012):

- To stimulate construction of private houses by imposing a *building obligation* on every landowner who possessed a

Table 10.1 Population growth in the province of Holland, 1350–1800

Year	Total population	Urban population	As percentage of total population	Urban demographic figures 1500–1800
1350	290,000		23%	Death surplus: ± 800,000
1514	275,000	127,000	46%	Emigration: ± 225,000
1622	672,000	363,000	54%	Immigration: ± 1400,000*
1680	883,000	540,000	61%	
1750	783,000	475,000	60%	
1795	783,000	462,000	59%	

*By origin urban immigration comprises three streams of people: from rural Holland (about 30%), from the rest of The Netherlands and foreign immigration (more than 50%). Source: Lucassen (2002)

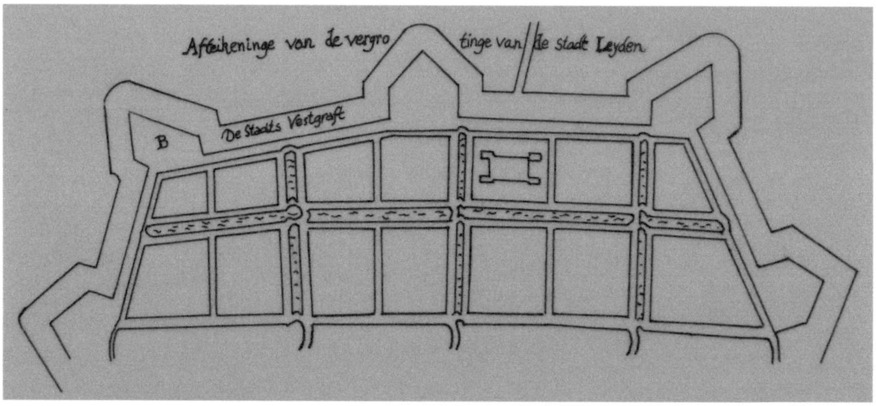

Fig. 10.3 Urban development plan for the fourth (northern) extension of Leiden by Jan Dou 1611. (Source: Van der Cammen and De Klerk 1986)

non-built lot. In the event of non-compliance, the land was surveyed and had to be sold to an interested party;

- To introduce a new *ordinance* obliging landlords turning their buildings that were converted to barns, warehouses or stables over the past 30 years, into houses for human habitation;
- To stimulate *spatial concentration* of the textiles industry in Leiden's industrial district Maredorp, with the dual objective of reducing water contamination and preventing decay in the city centre.

Leiden's urban expansion was mainly inspired by the city's fear that it would lose its competitive advantages viz-a-viz other cities and would decline. That fear was Leiden's most powerful argument in to request States of Holland's consent to expand the city.

In terms of governance urban planning had to resolve three major and fully intermingled problems: cost-benefit balance of a development plan (especially investments in military defence works), compensation for expropriated land owners, and balancing internal and external water systems. Costs of such large investments were covered otherwise. For its first (1612) major development plan for the western part of the Canal District, the city of Amsterdam applied the *Right of Preference* for expropriated land owners who were allowed to buy or lease back the parts of their land not needed by the city for public use (canals, streets, bridges or military defence).

Application of the Right of Preference caused problems like land speculation, in particular amongst councillors who deliberately frustrated perspectives for a beautiful city. To prevent such problems, and despite heavy protests, the city council adopted the so-called *Leiden model* for its 1664 development plan for the eastern part of the Canal District. The Leiden model prescribed a *public auction* to simply sell expropriated land not needed for public use and higher land values would thus be to the benefit of the city's treasury. Another advantage of this model was that Public Works would better control completion of the plan with stricter building regulations on maximum densities and bans on industries in the new residential neighbourhoods. Such regulations created the foundations of the nineteenth and twentieth century functional city.

10.3 Towards Central Decision Making in the Nation-State

In 1769 Adam Smith characterized the eighteenth century Republic as a *stationary state*, a state of economic standstill and decline which hit the cities by then. In 1825, the population of Leiden had shrunk from 70,000 to 25,000 and Amsterdam spent a quarter of its public funds to social relief.

After the French Occupation (1795–1813), Dutch economy slowly recovered. The nineteenth century saw the birth of a modern, central governance system in the Netherlands. Centralisation began with the Batavian Revolution of 1795, supported by the French who occupied the country until the end of 1813. Centralisation was intensified upon foundation of the Kingdom of the Netherlands under King Louis Napoleon (1806) and the successive incorporation of the Netherlands in the French empire (1810). Centralisation was to be continued and enhanced by the Kingdom of the United Netherlands (1815), including Belgium until 1830.

In 1848, afraid for a revolution, the Netherlands rather suddenly adopted a modern constitution, as developed by Thorbecke (1798–1872). The nation became a constitutional state where all citizens are equal by law and governed by a parliamentary democracy at all levels: nation-wide, provinces and municipalities. The 1848 constitution finally disconnected the Dutch from their long history of local and provincial autonomies and their local juridical frameworks. Not only all citizens, also all local and provincial authorities became equal by law. The Netherlands became a *decentralised unitary-state*.

King William I (reign 1815–1840) believed in technological progress for national prosperity. He belonged to the few who noticed the relevance of the Industrial revolution which already occurred in neighbouring nations. Helped by self-founded institutions as Nederlandsche Bank (a banker's bank), Nederlandsche Handel-Maatschappij to foster trade on the colonies and (secret) financial state funds, William I, nicknamed 'the Canal King', initiated the construction of canals and railways to improve Dutch traffic infrastructure. Moreover, William I supported new industries, often with personal financial contributions. Eventually, his policy ruined the exchequer, one of the reasons he was forced to abdicate in 1840. More than once William overruled Parliament. When Parliament rejected his proposal to construct a railway between Amsterdam and Arnhem in 1839, he outmanoeuvred Parliament by an executive *Royal Decree* (Koninklijk Besluit) to construct that railway, financed by a state loan and a state guarantee on rent, which he completed with a personal investment. Out of many, this decision reveals the difference between decision making in the pre-modern constitutional era and after 1848.

In 1821, King William I introduced a model for *public-public cooperation* to expand the number of land roads, between the provincial capitals in the first place. He discerned national, provincial and local roads. National roads were paid by the national state, provincial roads by provincial authorities and local roads by the municipalities. However, maintenance costs would all be borne by the national state.

> **Box 10.1 Central Decision Making on Dutch Railway System**
>
> The Dutch Industrial Revolution began by constructing new canals and improving navigability of rivers in the Dutch delta by eliminating meanders and shoals. Between 1800 and 1850, about 450 km of canals and rivers were constructed and improved. As to railways the Dutch system was far behind internationally. Dutch investors did not expect much profit from railways in a small and water transport dominated country. In 1855, the Netherlands had 314 km railway track only, Belgium 1349 km, Germany 8352 km and Britain 13,411 km. In 1860, Dutch Parliament passed the Railway Act to construct a state railway network. That Act was considered a violation of the free market policy. To arrange for a majority of votes in Parliament, the Dutch cabinet proposed starting railway construction works by connecting all provincial capitals. In 1883, about 1350 km had been constructed (Fig. 10.4).

Most new roads were exploited as toll roads. In 1798, a similar hierarchical distinction had been introduced for waterways and (great) lakes: a national water system (coast, large rivers), regional and local water systems. A long history of local independence, lack of money, lack of maintenance and lack of coordination, pride and stubbornness had made decentralised water control systems obsolete. The 1810 national Dike Act had been a response to the 1804 and 1809 large floods that caused many victims and much damage. The Act met significant resistance from the Water Boards, was suspended in 1813 and withdrawn in 1835 as a victim of Restauration policies (Knippenberg 2013).

A third model of nineteenth century governance was the *concession model*, a detailed contract between a public authority and a private company providing the company an exclusive right – a monopoly – to construct a railway, canal, road, to develop a piece of land, or to exploit a service, e.g. a tramway (De Klerk 2008). Concessions always concerned a specific geographic entity, a city, a region, the entire country. The concession model emerged by introduction of the market economy, around 1865. But the model only worked for a short time for the construction and exploitation of railways, and not at all reclamation of large peat lakes as Zuidplas (between Rotterdam and Gouda) and Haarlemmermeer. Eventually, return on capital for such large investment projects was too uncertain. The concession model failed as well for constructing modern infrastructure, a railway system and a new seaway. In 1860, Parliament finally passed the

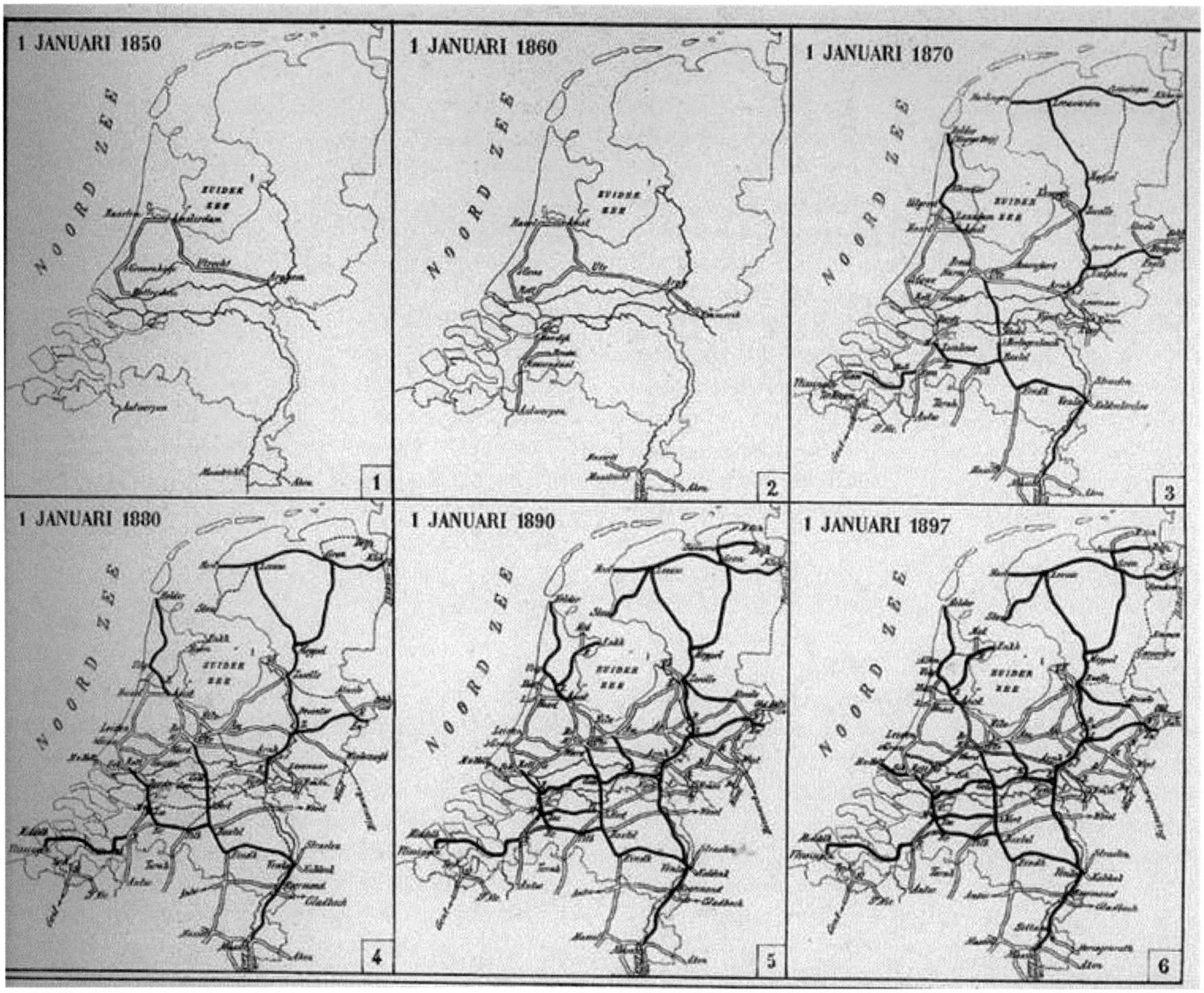

Fig. 10.4 The railway station of Den Bosch, opened in 1896, in substitution for the first station of 1868. It was heavily damaged during WWII and replaced by a new building in the 1950s. The present station is built in 1998. Breda's (1863) was a wooden building and could thus be demolished rapidly in war times. The first station in Breda was opened in 1855. It was replaced by this wooden building in 1863 and by a new building in 1932. (Source: Maatschappij tot Exploitatie van Staatsspoorwegen, 1863–1913)

Railway Act (1860) and Acts to grant a concession to a private (English) company to construct the *Noordzeekanaal* between Amsterdam and the North Sea, and to construct the *Nieuwe Waterweg,* between Rotterdam and Hoek van Holland. The Nieuwe Waterweg was considered a new river estuary, which, by law, was and is a state competence not being eligible to private enterprise. In 1880, the Noordzeekanaal Company got serious financial problems and had to be taken over by the state.

10.3.1 Planning by Dealing and Wheeling

In contrast to large public works the concession model proved successful during the first stage (1870–1900) of urban growth. To allow urban growth, two problems had to be overcome: (1) dismantling the many urban fortifications, which seized the cities in an iron grip; (2) annexation of land from neighbouring municipalities, villages. Fortifications were not owned by municipalities but rather by the national. The Fortifications Act (*Vestingwet* 1876) enabled dismantling. The Act successfully induced a policy to sell land for urban development to cities. The Department of Finance (Domains) hired architect Van Gendt, who designed about 20 city plans to calculate potential land prices for these fortification grounds and began negotiations with city councils. That has been the first large scale intervention by the central state in urban planning.

Regardless the successful seventeenth century Dutch tradition of planned urban development, the Dutch lacked a national law on (urban) land development until 1901. Parliament and city councils, dominated by supporters of the market economy, considered urban development as a private enterprise. Municipalities were not obliged to prepare development plans, but most cities did. In most cases such plans were adapted to existing land property configurations. To force land owners to cooperate city councils operated along two ways:

- Private law: buying, leasing and selling land under specific development conditions.
- Public law: applying building regulations, before 1900 prescribing just elementary urban design rules as minimum street width and obligations to connect new housing to drinking water and sewer networks.

Most cities applied combinations of both private and public law instruments. Prior to 1900, urban development planning had been a matter of knitting together series of (small) street plans by private concessionaries, land owners, building constructors, most of them speculators. Municipal authorities marginally examined private development proposals against Building Regulations and at the same time used the instrument of buying and selling as a means to enforce land owners to municipal planning. The city of Rotterdam bought 'strategic parcels' on development sites to coerce developers to adapt their plans to the municipal street and sewage planning. It was in the city's interest that the costly sewage system would be paid back by as many connections as possible. If a developer refused to cooperate with the municipality his future property would not be connected to the sewage system, or his street was declared unfit to be adopted by the municipality. The developers' risk was to be saddled with housing he could not rent or sell (De Klerk 2008). Thus, urban planning has been highly a matter of dealing and wheeling in times of capitalism between 1870 and 1914.

10.3.2 The Housing Act 1901

Around 1900, the free capitalist order step by step drifted into the phase of *orderly capitalism*, characterised by introduction of numerous rules on socio-economic regulation: rules for (international) economic trade, rules to guarantee free market access by entrepreneurs, legal criteria for minimum quality of products or accountability of enterprises and public authorities, and sanctions in case of violation. Not only the public sector increased and professionalised, numerous private organisations as accountants did as well. Accountants introduced exams to certify professional quality as a certification of trust. The twentieth century began with the first (modest) waves of social laws to protect labourers from risks of the market economy. Town and country planning followed this trend, which found a firm basis in the innovative Housing Act 1901 (Box 10.2).

The Housing Act 1901 and its successive amendments, revisions and supplements had a great impact on urban and regional planning. Its 1931 supplement e.g. introduced regional planning by provincial authorities, as urban growth could no longer be contained by urban planning, but had to be covered by planning on regional scales.

The Housing Act made urban planning a national concern and a local task, *obliging* cities and fast growing municipalities by law to settle urban development plans. A second major innovation in the Act was a breakthrough of the time-consuming parcel-by-parcel procedure of expropriation by introducing a new entitlement for public expropriation: land for housing according to an approved (by city council),

Box 10.2 Dutch Housing Act, 1901

The Dutch Housing Act, endorsed in 1901, comprised two sectors (Van der Cammen et al. 2012):

Urban development:

- Municipalities are statutory obliged to implement building regulations, regulating, among other things, minimum distances between buildings, building heights and building lines.
- It is forbidden to construct buildings without a building permit.
- Local authorities are statutorily allowed to order proprietors to repair or refurbish their buildings (houses) at their own costs, and to declare houses unfit for habitation.
- Municipalities with 10,000 or more inhabitants and a population growth of 20% in the last 5 years are obliged to prepare a development plan with grounds which 'in the near future are assigned to construct streets, canals and squares' and to approve development plans showing future streets, squares and canals.
- Introduction of an approved plan as a legal entitlement for expropriation in the interest of housing and planning.
- For citizens, the possibility of appeal in three stages: municipality, provincial parliaments and the Crown (in fact the Cabinet).

Social housing:

- Introduction of subsidies for 'legally admitted' (operating according to the law) private housing associations.
- Introduction of cheap loans for admitted housing associations: interest must be paid on advance loans (annuity loans) for new housing and slum clearance by housing associations, which were at the expense of the municipality and the central state (50/50).

confirmed (by provincial executive) and irrevocable (after procedures of public hearings on objectives and appeals) municipal development plan. Amsterdam have set the first step on this new path by expropriating almost all the land of the annexed municipality Nieuwer Amstel by deciding that it was to be used for housing. Architect Berlage (1856–1934) designed the development plans for Amsterdam-South in 1915.

10.3.3 Shaping Efficient Agricultural Landscapes

By 1800, about 30% of the Netherlands was waste land, mainly in the east and south. Lands in the west were almost completely used for agriculture. After some heavy storms and damaging floods (1809, 1820, 1825) Amsterdam, Haarlem and Leiden felt threatened by the ever growing peat lakes Haarlemmermeer and Zuidplas (north of Rotterdam) was feared to merge with some other peat lakes jeopardising Rotterdam's safety. In both cases King William I took initiative to reclaim the lakes, but de state was unable to sell concessions as the private sector did not expect a profitable exploitation. Therefore, Dutch Parliament passed special laws to make such reclamation possible. This case revealed the main problem in Dutch economy until 1870: a lack of money to invest.

In the nineteenth century rural areas faced three major problems:

- Fragmented land ownership preventing efficient agricultural production methods.
- A historical burden of common land ownership.
- Heavy competition of cheap American agricultural products, calling for new products, new methods of financing and organisational structures.

Fragmentation of ownership and parcelling of agricultural resulted from specific inheritance principles. According to Franconian right of succession, in the southern provinces of the Netherlands inheritance was divided amongst all sons. In the east, Saxon law prescribed so-called *primogeniture* by which the eldest son inherited the farm. By the end of the nineteenth century, in the former Franconian area a 10% loss of productive land for enclosures, ditches and access roads was calculated. In addition, the common lands, mainly waste lands (woods, sandy grounds, peat lands), were commonly owned by private farmer associations or by municipalities. An early Napoleonic reform act (1810) had been suspended after liberation from the French in 1813.

At the turn of the nineteenth century, state committees urged for defragmentation by land consolidation, regardless of their status as privately owned or as long-term leased land. It took another quarter of a century before Dutch Parliament passed the first *Land Consolidation Act* (1924), more or less copied from Germany. The objective of this typically procedural act was 'to promote agriculture by consolidation according to the criteria and procedures of this Act' (Van den Bergh 2004). This procedural character has been characteristic for the pre-WWII governance structure by not prescribing content criteria, but by supporting the primate of private initiative by inviting to a consolidation initiative, and leading such initiatives along lines of democratic decision making by farmers in their own interest. A consolidation procedure could be started if 25% or more of the land owners submitted a land consolidation plan to the Provincial Executive. Upon agreement, a detailed consolidation plan might be prepared. For implementation of such a plan, a *double majority* was required: the majority of land owners voting in favour would need to represent at least 50% of the concerned land as well. After this referendum a Local Commission with regional representatives was supported by a staff of agricultural and nature preservation experts. Until 1985, these procedures have framed land consolidation plans in the Netherlands.

After 1945, the pre-WII Land Consolidation Act was revised more than once while upscaling land consolidation and adding new objectives, as improving farming technologies, fostering agricultural production, landscaping, tourism, traffic infrastructure and room for industry. In 1954, a third *Consolidation Act* aimed for more inclusive forms of consolidation to stimulate regional economic growth and asked for a comprehensive planning concept. This type of consolidation has been applied in larger parts of the delta between rivers Rijn and Maas.

Another post-WWII trend is regionalisation. This trend addressed the fact that problems may differ from region to region and was applied first in the *Land Consolidation Act Walcheren* (1947) That Act also implied reconstruction of farming. In the 1960s and 1970s, large-scale land consolidation encountered protests against destroying historic landscapes with meandering small rivers, twisting dikes and cosy corners. The *Herinrichtingswet Oost-Groningen en de Gronings-Drentse Veenkoloniën* (1979) shows how improvement of agriculture reduced in priority and became just part of a wider range of goals and objectives 'to foster good living and labour conditions and economic and social development of the region concerned.' (Van den Bergh 2004).

Damen (2002) calculates that about 5 billion € has been spent on land consolidation improving about 3.3 million ha (= 5.7 million acres) during the second half of the twentieth century. A new *Land Development Act* was introduced in 2007, for two reasons: (1) a shift of competences and responsibilities (implementation and finances) from the central to provincial governments, and (2) decreasing popularity of land consolidation (Leenen 2014). To reduce time consuming procedures maximum surface of consolidation projects was restricted to 2000 hectares.

The post-WWII arena shows two successive trends: centralisation and shifting priorities. The first: continuing central decision making on major public works, as the Zuiderzee Works/IJsselmeerpolders (Chaps. 2 and 3) and the Delta Works. Both Works had been responses on disastrous floods, the Delta Works on the disastrous 1953 storm surge, affecting the provinces of Zuid-Holland, Zeeland and Noord-Brabant. In both cases early warnings on poor dike maintenance had been neglected. Implementation of the Deltawerken (Delta Works) lasted 25 years but meanwhile this plan underwent remarkable adaptation due to shifting political priorities in the 1970s in favour of protecting historical landscapes and the natural environment, occurred. Political flexibility and the adaptation of plans to shifting priorities toward environmental protection is the second trend of the post-WWII era, as demonstrated by changing the Oosterschelde plan, a major part of Delta Works. Initially, the Oosterschelde sea arm was planned to be closed entirely and to become a fresh water basin. In response to fierce public protests this plan was abandoned and replaced by constructing a storm surge barrier that would be open normally but be locked in case of heavy storms.

10.4 The Rebuilding Era, the Heyday of Planning

The Second World War and 5 years of German occupation have influenced post-war governance in the Netherlands. In summary: substantial up-scaling of public intervention, centralisation of decision making and public planning as an accepted means of policy making.

During WWII about 4% of all houses were destroyed, 16% of traffic infrastructure (bridges, port installations, railway tracks) was damaged, blown-up or bombed, and about 16% of the land (375,000 hectares) required recovery. Deliberate inundation by the Germans had made more than 200,000 hectares of agricultural temporarily unproductive. Until 1955, more than 6000 farms had to be repaired or rebuild.

The scale of repair, replacement and development was unprecedented and between 1950 and 1970 fuelled, additionally by the baby boom and a booming economy, an almost 4% annual economic growth, two decades long. Unsatisfied with political fragmentation during the 1930s Great Depression and considering the enormous post-WWII reconstruction tasks, centralised decision making was seen as the only way to go ahead for proper implementation. Therefore, the Dutch governance structure of the Rebuilding Era differed quite considerably as compared to the pre WWII period:

- Central state took over numerous tasks from the private sector. In general, the state decreed that all private agencies and institutions, e.g. state-subsidised housing associations were obliged to give access to all Dutch people regardless their religious denomination.

- Central state maintained a strict central allocation planning of investment and labour forces according to economic policy priorities.
- For urban and regional planning, comprehensive planning concepts were introduced.
- A concept of equal prosperity at national level, meaning that peripheral geographic regions would be involved in regional industrialisation schemes through comprehensive land use planning, subsidies and tax reliefs for private firms.

In all political parties and on all governmental levels planning was accepted as an obvious, non-disputed means of policy making and implementation. Planning meant *coordination* of investments in different economic sectors and regions. Spatial planning even received a role of coordinating public and private investments. In the optimistic post-WWII period coordination was strongly supported by *cooperation* as an attitude of public and private institutions, supported by trade unions as well as by employers organisations. Together, they shared responsibility for constructing a welfare state, comparable to the USA and Canada as examples of prosperity which certainly would affect car ownership and land use, or with the UK for decentralising urban growth by developing new town schemes.

The nation's reconstruction policy was ruled by a clear set of national socio-economic priorities as proposed by the *Central Planning Office* (1948) and its influential director Jan Tinbergen, the first Nobel Prize Laureate of Economics in 1969. The priority ladder was supported by all relevant parties: (1) redevelopment and repair of the transport infrastructure, which had always been so vital for the merchant economy of the Netherlands; (2) rebuilding of the industry and agricultural production and (3) housing and consumption. Increasing economic growth made possible to introduce a more or less complete system of public social security, starting with state pensions at the age of 65 years. Thanks to discovery of the enormous natural gas field in Groningen in 1959, and the political decision to spend no less than 85% of the state gas dividends (more than 210 billion € between 1959 and 2009; Chap. 4) on social security, rent on debts and education, the Netherlands could develop a most favorable social security system. Whilst Norway stored state gas and oil revenues exclusively in its Sovereign-wealth Fund (a national investment fund), the Dutch government deliberately allocated 26.5 billion euro in the *Economic Structure Enhancement Fund,* only between 1995 and 2009. Through that fund the Dutch government gave priority to invest in new infrastructure as the Betuwe Line, a freight railway from Rotterdam to Germany, and the High Speed Railway from Amsterdam via Rotterdam to Antwerp-Brussels-Paris, with a link to London (Aardgas in Nederland 2017).

Box 10.3 Principles of Decision-Making: The Art of Agreeing to Compromise

Decision-making in Dutch pacification democracy is a complicated process of deliberation, which in the twentieth century emerged along the following rules (Lijphart 1968):

- A business-like approach to policy-making: pragmatically gaining results above debate about principles;
- Pragmatic tolerance: respect for other fundamental (religious) convictions leading to a pragmatic culture for handling sometimes contrary meanings and convictions;
- Élites rule: important issues have always been discussed and decided in advance by the leaders of different pillars. This process has been supported by the pillarised Dutch press for a long time;
- Proportionality: proportional allocation of positions like mayors, provincial governors, presidents of committees et cetera;
- De-politicisation: the art of turning fundamental political goals into technically (financially, legally) feasible proposals;
- The executive power rules relatively independent and, always based on a coalition, acts somewhat like an independent referee concerning the different goals and objective of political parties.

In general terms, the Netherlands have adopted the *Rhineland model*, a guided market economy comprising a broad range of social laws protecting individual workers and their families from risks of the (full) market economy. As a consequence, the public sector takes a relatively large share of the Gross National Product (GNP), to no less than 72% in 1982. All these measures reinforced a tendency since the early 1900s that citizens shifted their socio-economic commitments step by step from local to the national level where key decisions on income and social welfare policy were taken. Amongst others, that resulted in a gradual reduction of the powers of political-religious pillars and their grip on their ranks and files.

10.4.1 Urbanisation by Planning

During the Rebuilding Era the urban and regional planning system has been accomplished by the 1941 *Basic Decision* (Basisbesluit) to start spatial planning at a national level, during German occupation. This Decision was followed by the establishment of the National Agency for Spatial Planning (Rijksdienst voor het Nationale Plan, 1941). In the next decades national planning built content and form along three lines:

1. A successive series of national planning documents and (regional) planning schemes;
2. A specific legal framework, the Spatial Planning Law (1966);
3. A system of subsidies to resolve specific planning problems, as urban renewal, location of industrial sites, and planning of traffic and transportation.

Together these instruments elevated planning onto an institutional status, almost unrivalled by other countries. Dutt and Costa opened their study on Dutch planning as follows: 'The Netherlands is decidedly the most planned country among the European nations. Only a few democracies of the world can match the planning apparatus of the Dutch government. Such a state of affairs is a product of circumstances created by harsh environmental constraints, a challenging history, socio-cultural forces, hard economic necessities, and the size of the country' (Dutt and Costa 1985).

According to the old Dutch tradition of local independency the local *land use plan* or *zoning plan* (*bestemmingsplan*) was, and is, on top of the planning hierarchy. After legal procedures of public openness (in town hall or in another public building), objections by citizens, decree by the municipal council, approval by provincial executive, and eventually after a High Council of State procedure, a land use plan gains the status of an legally irrevocable and therefore mutually binding citizens and public authorities. The law also framed contents of different types of local and regional plans, generally consisting of (1) a policy section including descriptions and analysis of the character of the plan area; (2) a map projecting future land use: spatial allocation of functions and activities ranging from separate parcels to large sites; and (3) binding building codes, land use regulations (prescriptions and restrictions) and the character definitions for buildings and other accommodations, and in some cases zoning codes (Van der Cammen et al. 2012). Dependent on aim and content, plans are quite detailed (e.g. for urban conservation areas) or more generally describ ing an anticipated future planning policy.

Similar to land consolidation law, planning law has always been a procedural framework, only prescribing abstract on who is the planning authority at stake, which kind of rights citizens have in various stages in the planning procedure, and what kind of information a land use plan should contain. National, regional and local planning have shaped the Netherlands as it looks in 2017. The first Report on Spatial Planning (1960) proposed landscape zoning between urban agglomeration, still to be experienced between Leiden and Den Haag, and between Den Haag and Rotterdam. New towns and contained suburbanisation are the result of planning policies which began in the late 1950s and were continued until 2015. Between 1996 and 2015 about 1.1 million new houses and apartments have been built, 36% of which in new urban districts and 15% in inner city districts. The new districts are called 'Vinex-districts' (Vinex is an abbreviation the Fourth Report on Spatial planning Extra, 1991). 'Vinex-district' expresses the planning-mindedness of the Dutch and perhaps a lack of phantasy as well. In no other nation in the world a series of new urban districts is called after a government report.

10.4.2 Democracy vs Efficiency

Dutch municipalities and provinces have a legal status as determined by the Constitution. At the introduction of the Municipality Act in 1851, 1209 municipalities existed in the Netherlands. A century later, in 1950, there were still 1015, but their numbers rapidly decreased since: from 672 in 1990 to 388 municipalities in 2017. This decrease is due to two different developments: annexation by cities and an emerging public management concept, only able to cope with larger municipalities, i.e. more than 25,000 inhabitants, to maintain proper and basic social, educational and medical facilities. At this point, a major distinction between Dutch and municipalities in other nations must be stressed. Income-wise and since 1929, Dutch municipalities collect about 85–90% of their incomes from both the Municipal Fund and from direct transfers by different ministries, and not more than about 10–15% from local taxes. The Dutch state operates as an income redistributor and municipalities have limited room for decision making on local priorities.

Land annexations, liquidation of independent municipalities and mergers of municipalities are subject of national law, and therefore subject of political debate. After the 1960s, political debates were dominated by two issues. First, migration of the middle classes from cities to suburbs, and second, increasing requests for more democracy and participation by the general public. The trending idea is that both are better off with smaller than larger government units. Not so much in villages and suburbs, but in cities new experiments emerged, mostly connected to urban renewal. It was the first time in history that renovation and reconstruction plans had to be set up for existing residential areas, built between 1880 and 1920. In the late 1960s, tenants began protesting against replacement housing for the twilight zone 1880–1920. But for most tenants new housing would be too expensive and nobody was waiting to see their old neighbourhood being turned into some sort of high rise outskirt.

History welded the upcoming democratic participation movement of 1968 together with the urban renewal challenge. Large-scale and top-down blue print planning, successfully applied to expand cities during the Rebuilding Era, turned into small-scale and bottom-up process planning, adapted to the quest for basic democracy and participation. Generally, this model, with some local variations, was adopted by all cities and supported by new national and local subsidies, which culminated in the Urban and Village Renewal Act of 1985. That Act decentralised policy making to municipalities and streamlined an endless series of administrative orders, but was suspended 10 years later and replaced by a national policy to stimulate middle class (owners occupiers) housing in inner city districts.

10.4.3 Centralisation vs Decentralisation

In the early 1980s, Dutch policies turned into alternative directions. Culmination of an economic crisis (second oil crisis, 1979–1984), structural mass unemployment, an expensive social security system, diminishing tax income and a growing understanding of the second demographic transition (decreasing birth rates since the late 1960s; Chap. 11), put a heavy financial burden on public finances. Economists coined the term 'Dutch disease' to explain Dutch economic dependency, its public finances and its welfare system on natural gas revenues. Slowly, politicians became aware of the long-term impacts of the second demographic transition on the Dutch social security system, which had been designed for a population structure of the Rebuilding Era, focused on traditional nuclear families, the 4-persons household.

Comparable to the USA and the UK, Dutch cabinets adopted a neo-liberal socio-economic policy, in the early 1980s, containing three main issues: an almost continuing policy of austerity and reorganisation of the welfare state; shaping more room for private enterprises by deregulation and privatisation and stimulating the economy by lifting international trade barriers. Expectation of more governance transparency by deregulation was blurred by two parallel trends: centralisation and decentralisation. Central government transferred competences to the EU whilst common citizens tried to increase their influence on decision making concerning their immediate social and physical environment: their own neighbourhood or village.

The ideological foundation of political neo-liberalism told modern citizens amongst others that they generally are well-educated, generally earning a reasonable income and thus are able to bear responsibility for their own housing accommodation and should no longer rely on the drip of state subsidies. As a consequence, all housing subsidies, including cheap state loans for housing associations, were abolished in 1995. Two types of subsidy remained: an indirect contribution to land exploitation (part of the lump sum subsidy system to foster the Vinex-urbanisation policy), and individual rent subsidy for people whose income is too low for renting a decent house in the social housing sector. A consequence of these decisions was the loss of the coordinating role of spatial planning on a national level. Spatial planning lost both the housing sector and the agricultural sector which were often decisive for implementation of plans. As a result, planning procedures involved a confusingly larger burden of non-spatial goals, shareholders and stakeholders within the public sector (separate ministries enhancing their own sector planning competences) as well in the private sector (private landowners and developers who were invited to participate and invest in urban development).

Consequently, the once powerful Ministry of Housing and Planning was reduced to only the Directorate-General Space and Water, since 2010 a Department of the Ministry of Infrastructure and Environment, a merger product of several trends: the loss of allocation power concerning housing, the loss of political priority for housing, and the perception that spatial planning should be an integrated part of environmental planning.

Central state redistributed competences to the EU, but to provinces and municipalities as well. The EU, collected new responsibilities on, amongst others, environmental issues, laid down in *EU Directives*. Concealing unpopular decisions behind EU Directives became part of the political game, although EU-law making is being prepared and negotiated by national governments. Dutch government and parliament did hardly ever use their veto rights to stop EU-law making. On the contrary, the Netherlands are amongst the few member states, which have properly integrated EU-regulations. 'Dutch planners have always seen their country as part of a northwest European delta, well connected and dependent on its hinterland, with a competitive position that needs to be watched and sustained by means of infrastructure investments but also tax policies' (Faludi 2005). A difference with other EU-nations is that the Netherlands have often integrated EU-Directives as a part of spatial planning procedures, whilst other nations have handled the Directives as separate policy issues (Chap. 6).

This international orientation, which is familiar to a trade-oriented country by tradition, made the Netherlands responsive to EU Directives like the *Natural Habitat Regulations* (1997) and *Natura 2000* to protect Europe's most valuable and threatened species and habitats (see Box; Chap. 5). In the Netherlands about 160 areas have been designated as *Natura 2000* protection areas, the Waddenzee being one of the most important ones. These protection plans have set up according to the juridical framework and procedures of the Spatial Planning Act. In many planning procedures entrepreneurs are involved, because most protected areas are important for recreation.

Box 10.4 Natura 2000

Covering over 18% of the EU's land area and almost 6% of its marine territory, Natura 2000 encompasses the largest coordinated network of protected areas in the world. It offers a haven to Europe's most valuable and threatened species and habitats. Natura 2000 is a network of core breeding and resting sites for rare and threatened species, and some rare natural habitat types which are protected in their own right. It stretches across all 28 EU countries, both on land and at sea. The aim of the network is to ensure long-term survival of Europe's most valuable and threatened species and habitats, listed under both the *Birds Directive* and the *Habitats Directive*.

Natura 2000 is not a system of strict nature reserves from which all human activities would be excluded. While it includes strictly protected nature reserves, most of the land remains privately owned. The approach to conserve and sustainable use of the Natura 2000 areas is much wider, largely centred on people working with nature rather than against it. However, Member States must ensure that the sites are managed in a sustainable manner, both ecologically and economically. Source: (http://ec.europa.eu/environment/nature)

10.5 Planning Simplified

At the turn of the millennium complaints about planning became louder concerning endless procedures, too detailed rules and too much bureaucracy. Besides, planning became discredited of curbing economic growth during the neo-liberal growth decade 1995–2008. Why should it take 12 years or more to construct a new high way? And have options for submitting objections for anyone all over the country against all planning stages everywhere not become unrealistic? The new Dutch Spatial Planning Act of 2006 aimed to address calls for speeding-up procedures and simplicity: 'Fewer rules, less central control where possible, and an implementation-oriented approach' are the new guiding principles, as announced by the Ministry of Environment and Spatial Planning (Ministerie van Volkshuisvesting, Ruimtelijke Ordening en Milieubeheer 2007) boastful, including claiming an action-oriented approach, more efficient decision making and simplified legal protection.

Another objective of the new Act was to enhance transparency of the blurred competences within the public sector between planning authorities at different governmental levels. Coordination responsibilities were transferred from the national state to provincial authorities 'in such a way that each tier of government can represent the interests entrusted to it to the best of its ability.' To conclude, the Act makes a clear distinction between *policy making* and *policy implementation*, which in the 1966 Act were closely interconnected, and which consumed so much procedural time.

The 2006 Planning Act preluded to a full new concept of steering spatial development by a broad framework which is laid down in the *Environmental Act* of 2016, a bill which will become a law in 2019. Its aim is to substantially simplify rules and procedures by integration of 26 existing Acts on planning, development and environment, such as contamination of water, air, noise and protection of nature areas. Rules for spatial development are integrated too, from no less than 5000 statutory provisions to 350, from 120 Ministerial Rulings to 10, and from 120 National Administrative Orders to only 4 (Factsheet Omgevingswet 2014). The general aims of this bureaucratic cleansing are to gain transparency for citizens and procedural time for both the private and the public sectors. Spatial planning is becoming an integrated part of environmental planning, diminishing the number of bumps and problems to construct houses and apartments on brown sites, or to construct wind mill parks. Local public authorities will receive more competences and responsibilities to adapt planning to their own goals and objectives. Private parties will receive more room as a result of the exchange of detailed planning licenses by general rules. Initiatives by citizens or entrepreneurs, fitting in the local or regional *Environmental Vision*, may apply on line for a license to construct and go into rather smooth procedures, which will be decided by one single planning authority.

References

Aardgas in Nederland (2017) Aardgas en de economie. http://aardgas-in-nederland.nl/nederland-aardgasland/aardgas-en-de-economie

Bank J, Van Buuren M (2004) 1900: the age of bourgeois culture. Tome 3 of Dutch culture in European perspective. Royal Van Gorcum/ Palgrave Macmillan, Assen/Basingstoke

Blockmans WP (2010) Metropolen aan de Noordzee. De geschiedenis van Nederland, 1100–1560. Bert Bakker, Amsterdam

Damen J (2002) Development of land consolidation in the Netherlands from project objective to project instrument, Prepared for the International Symposium by FAO, GTZ, FIG, ARGE Landentwicklung and TUM Land fragmentation and land consolidation in CEEC: A gate towards sustainable rural development in the new millennium, 25–28 February 2002,Munich

De Klerk LA (2008) De modernisering van de stad 1850–1914. De opkomst van de planmatige stadsontwikkeling in Nederland. NAi, Rotterdam

Dutt AK, Costa FJ (eds) (1985) Public planning in the Netherlands: perspectives and change since the Second World War. Oxford University Press, Oxford

Esquiros A (1858) Nederland en het leven in Nederland. Amsterdam 1858

Factsheet omgevingswet (2014) Ministerie van Infrastructuur en Milieu, Den Haag

Faludi A (2005) The Netherlands, a culture with a soft spot for planning. In: Sanyal B (ed) Comparative planning cultures. Routledge, New York/London, pp 285–307

Frijhoff WTM, Spies M (1999) 1650, Bevochten eendracht. Sdu, Den Haag

Hoogheemraadschap van Delfland (2017) https://nl.wikipedia.org/wiki/Hoogheemraadschap_van_Delfland. Accessed 15 Mar 2017

Knippenberg H (2013) De fysieke kant van het land: grenzen, grond, water en wegen. In: De Haan I, Den Hoed P, Te Velde H (eds) Een nieuwe staat. Het begin van het Koninkrijk der Nederlanden. Prometheus – Bert Bakker, Amsterdam, pp 76–103

Leenen H (2014) Land Development in the Netherlands, *zfv* – Zeitschrift für Geodäsie. Geoinformation und Landmanagement 139 (3):166–172

Lijphart A (1968) The politics of accommodation. Pluralism and democracy in the Netherlands. University of California Press, Berkeley/Los Angeles

Lucassen J (2002) Immigranten in Holland, 1600-1800. Een kwantitatieve benadering, Working Paper 3. Centrum voor de Geschiedenis van Migranten, Amsterdam

Ministerie van Volkshuisvesting, Ruimtelijke Ordening en Milieubeheer (2007) The new spatial planning act gives space. Ministerie van Volkshuisvesting, Ruimtelijke Ordening en Milieubeheer, Den Haag

Peteri WB (1913) Overheidsbemoeiingen met stedebouw tot aan den vrede van Munster. Dissertation, TU Delft

Van den Bergh S (2004) Verdeeld land. De geschiedenis van de ruilverkaveling In Nederland vanuit een lokaal perspectief, 1890–1985. Dissertation, Wageningen University

Van der Cammen HA, De Klerk LA (1986) Ruimtelijke ordening. Van plannen komen plannen. Het Spectrum, Utrecht/Antwerpen

Van der Cammen HA, De Klerk LA, with Dekker G, Witsen PP (2012) The Selfmade land. Culture and evolution of urban and regional planning in the Netherlands. Unieboek, Houten/Antwerpen

Waddenzee (2017) http://www.waddenzee.nl/fileadmin/content/waddenZeewerelderfgoed/island.jpg. Accessed 21 June 2017

11 Demography

Joos C. Droogleever Fortuijn

Abstract

By the end of Medieval Times the population of the Netherlands was one million people and grew to five million in 1900. Today's (2018) population of this country is 17.2 million. According to the Second Demographic Transition model the Netherlands currently is a nation with an ageing population due to low mortality, low birth rates and an ageing post-WWII baby boom generation. Moreover, Dutch society is characterized by a growing diversity in household types and a growing ethnic diversity. Today, no more than 26% of all households are two-parent families with children. One-person Dutch households are growing rapidly because of more and more young singles, divorced people and older widows and widowers.

International migration from neighbouring countries, labour migrants, refugees and migrants from former Dutch colonies contributed to an ethnic diverse population in the Netherlands where 22% of the population is a first or second generation immigrant. Some groups of immigrants and refugees integrate better in Dutch society than others.

Keywords

Demographic transition · Population growth · Birth rates · Death rates · Household diversity · International migration

In March 2016 the 17th million inhabitant of the Netherlands was welcomed. In contrast to the 16th, 15th and 14th million, the 17th 'millionaire' was not a newborn baby but an immigrant. This is exemplary for the so-called First Demographic Transition that occurred in European and other developed countries in North-America, Oceania and in Japan (Lesthaege and Van de Kaa 1986). The demographic transition from high fertility and high mortality to low fertility and low mortality is completed now and further population growth is more due to international migration rather than to the balance between births and deaths. Unlike some European countries and Japan, 20% of the Dutch population growth has been still natural until 2016 (CBS 2017a). In the first part of 2017 natural growth was even negative (CBS 2017b). But migration has become a more important factor in population growth than natural increase. In 2014, the number of immigrants in the Netherlands exceeded the number of births for the first time in more than a century, and in 2016 80% of the total population growth was the result of immigration (CBS 2017a).

11.1 A Completed Demographic Transition

According to the First Demographic Transition model populations of industrialised countries develop in similar stages. The first stage has high birth rates, high death rates and no natural growth. In the second stage, high birth rates go together with diminishing death rates and a population increase. The third stage combines low death rates with decreasing birth rates. The final stage in this model is characterised by low birth and death rates and no population growth (Van den Brekel 1986).

By the end of Medieval Times (around 1500 AD), the Dutch population was one million inhabitants. Population grew about continuously to almost two million in 1660, partly due to immigration in the prosperous seventeenth century (the Dutch Golden Age). A period of economic and population stagnation lasted to the end of the eighteenth century and population began to grow again in the nineteenth century. By the end of the nineteenth century, Dutch demography had arrived in the second stage of the demographic model with high birth rates, declining mortality and a population growth from two million people in 1795, via three million in 1850, and about five million in 1900 (Fig. 11.1).

J. C. Droogleever Fortuijn (✉)
Department of Geography, Planning and International Development Studies, University of Amsterdam, Amsterdam, The Netherlands
e-mail: J.C.DroogleeverFortuijn@uva.nl

E. F. J. De Mulder et al., *The Netherlands and the Dutch*, World Regional Geography Book Series,
https://doi.org/10.1007/978-3-319-75073-6_11

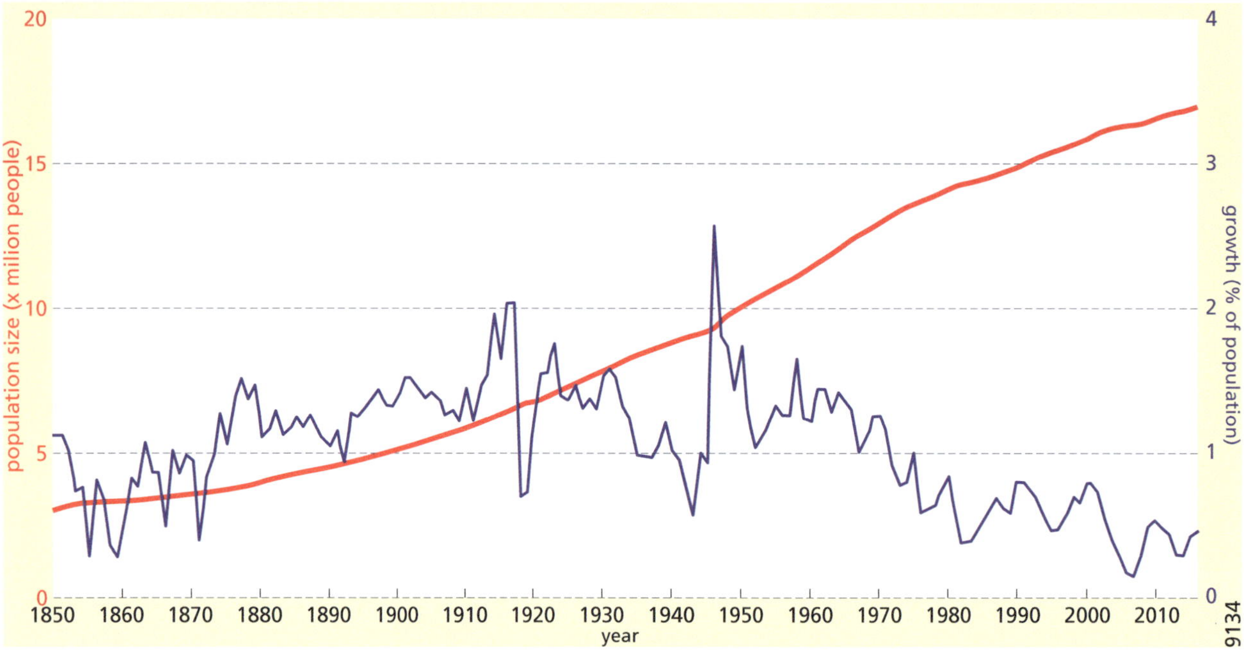

Fig. 11.1 Population growth 1850–2015. (Sources: Ekamper et al. 2003, p. 13; CBS Statline)

Reduced (mainly infant) mortality can be explained by better food provision resulting from innovations in agricultural production methods, better personal hygiene, sanitation, waste disposal, clean drinking water provision and improved health care (by the availability of vaccination programs). Hofstee (1978) pointed to the relevance of modernization of drinking water provision in the second half of the nineteenth century, in particular in the western part of the Netherlands. These western provinces used to retrieve drinking water from surface water from the rivers, instead of from groundwater wells which more commonly occurred in the northern and eastern parts of the country. The delta rivers were heavily polluted by industrial and household waste disposal and caused contagious diseases such as cholera, typhus and dysentery. Introduction of water mains was an important impetus for Dutch population growth in the late nineteenth century.

Mortality rates have been fluctuating during the nineteenth century. From 1880 onwards, death rates steadily decreased, except for 1918 with high mortality due to WWI and the Spanish influenza epidemic, and for 1943 and 1944 due to the last years of WWII (Van den Bie and Smits 2001; Fig. 11.2). Birth rates in the Netherlands steadily grew until the 1960s, but fertility (the number of births per woman) decreased from 1900 until the end of WWII (Fig. 11.3). Concurrently with other European nations the years after that war Dutch birth rates rapidly rose (and spiked in 1946 and 1947, the 'babyboom').

The First Demographic Transition model explains modifications in mortality by falling birth rates as parents cut birth numbers as they do no longer need many children to secure their old age. Infant mortality drops and more children survive until adulthood. The Second Demographic Transition model (Lesthaege and Van de Kaa 1986) explains why fertility rates have been decreasing from 1965 onwards in European and other developed countries and why these are now below reproduction level (i.e. below 2.1 children per woman) in all EU countries (Fig. 11.4). Industrialised countries experienced a transition in family values and bourgeois family types were substituted by more individualistic family types. Marriage and parenthood were no longer inevitable stages in a prescribed life cycle but rather belonged among many life course options. Individual women and men may choose to marry or not, to dissolve a marriage or not, to have children inside or beyond marriage or no children at all (Latten and Mulder 2013). Individualisation, secularisation and women emancipation were the main drivers for such changes (Lesthaege and Van de Kaa 1986; Liefbroer 2010). More women are highly educated, participate in the labour market and postpone decisions regarding marriage, partner relations and family formation (Wobma and Van Huis 2012). The mean age of women in the Netherlands giving birth to their first child increased from 24 years for children born in 1970 to 30 years in 2013, the highest in the world (De Graaf and Van Galen 2014). The percentage of childless women grew from 11.7% in the cohort of women born in 1935 to 17.6% women cohort born in 1970 (CBS Statline).

A well-developed welfare state that cares for older, disabled and unemployed people enables individual people to make personal life course choices. Dutch legal systems

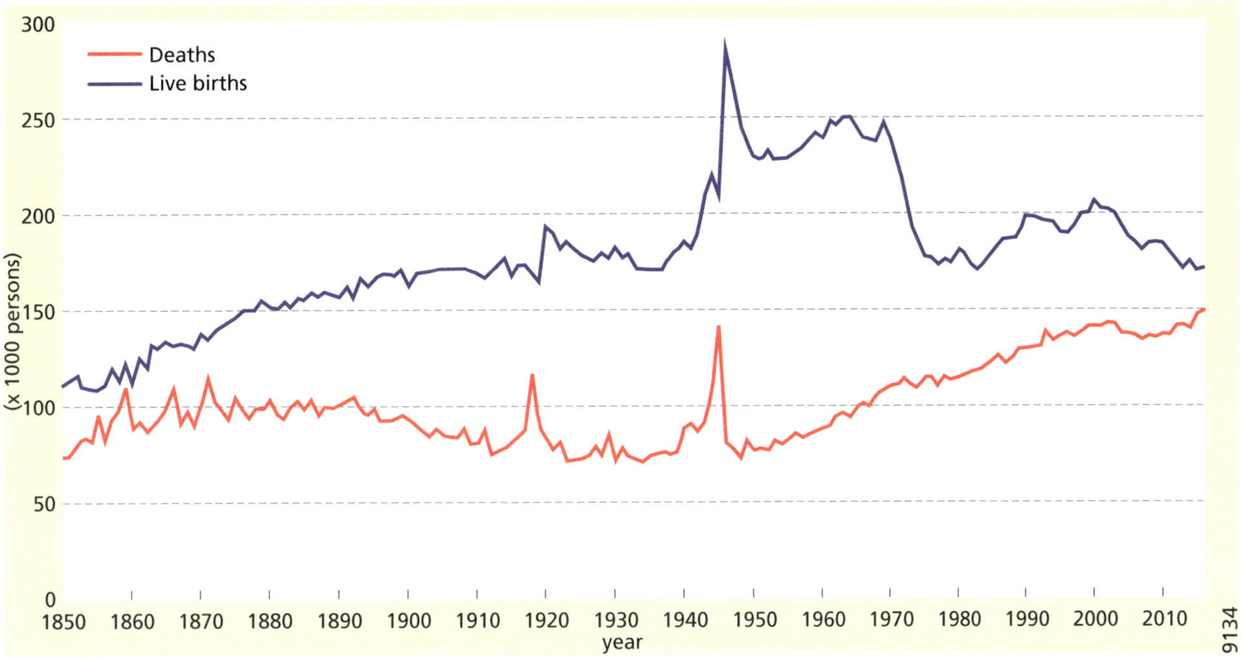

Fig. 11.2 Birth and death rates 1850–2015. (Sources: Ekamper et al., p. 17; CBS Statline)

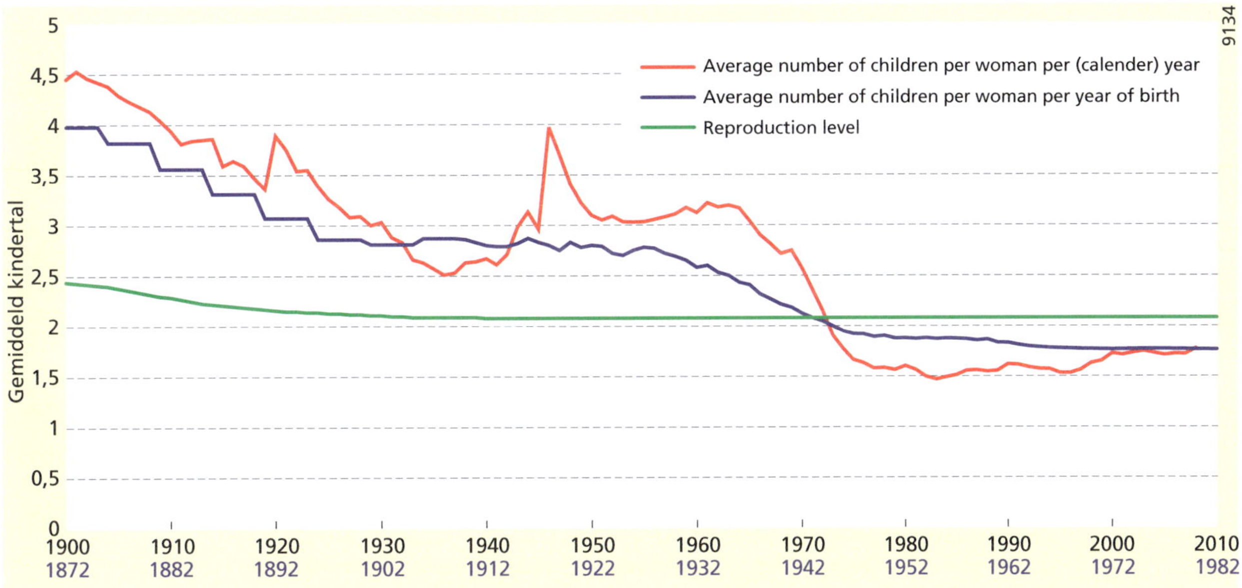

Fig. 11.3 Mean number of children per woman 1900–2010. (Source: Beets 2010, p. 3)

allow and support a variety of partner and family relations. The Dutch health care system makes cheap and reliable contraception available for everyone. Contraceptive pills and male and female sterilization are the usual family planning methods in the Netherlands where teenager mothers are an exception and among least in Europe together with Denmark and Switzerland. In 2015, no more than 0.3% of the 15–19 years old women gave birth to a child and their numbers declined since 2001 (CBS 2016a).

Until the end of the twentieth century, the Netherlands has been ageing less than most of the European nations as fertility rates were relatively high and its population comparatively young (Fig. 11.3). In 2015, fertility rates, life expectancy at birth and age structure were more in compliance with the demographic structure of other European nations (Figs. 11.4 and 11.5),

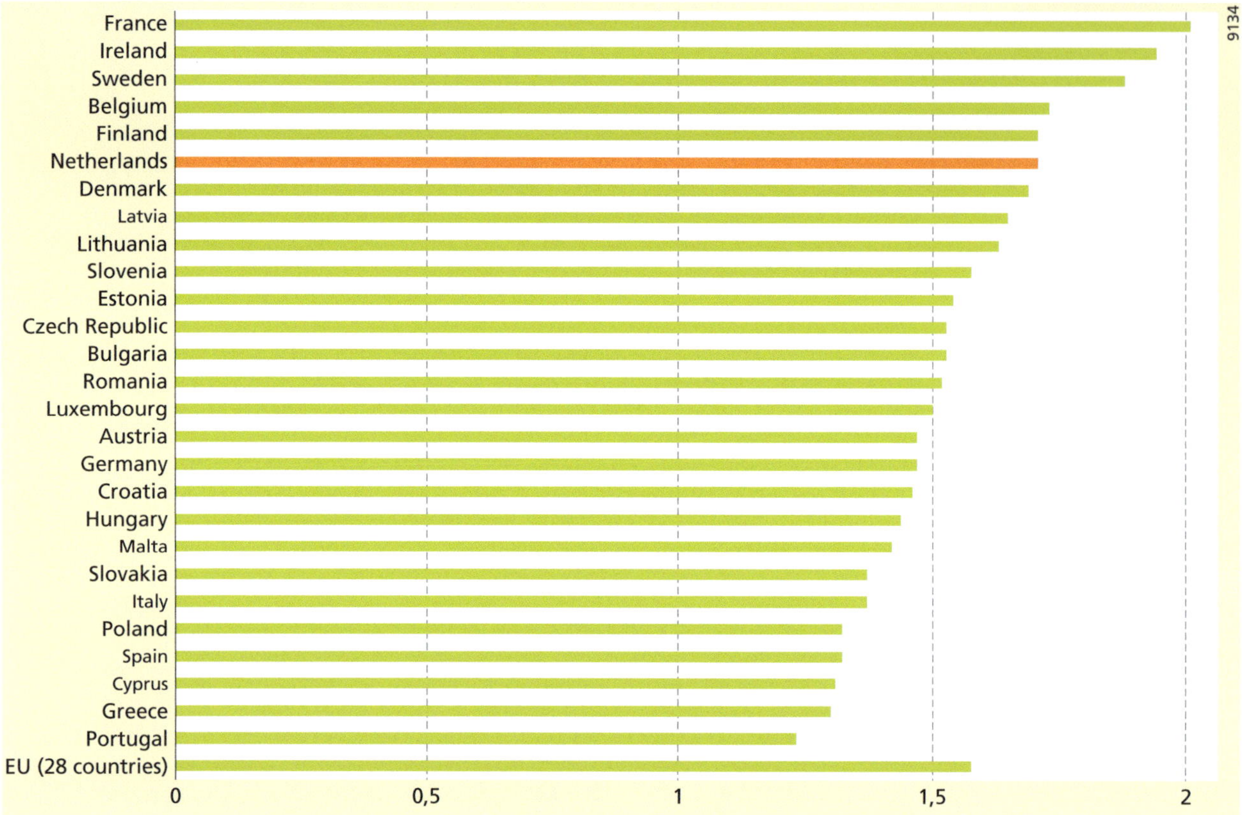

Fig. 11.4 Total fertility rate (number of children per woman), EU countries 2014. (Source: Eurostat)

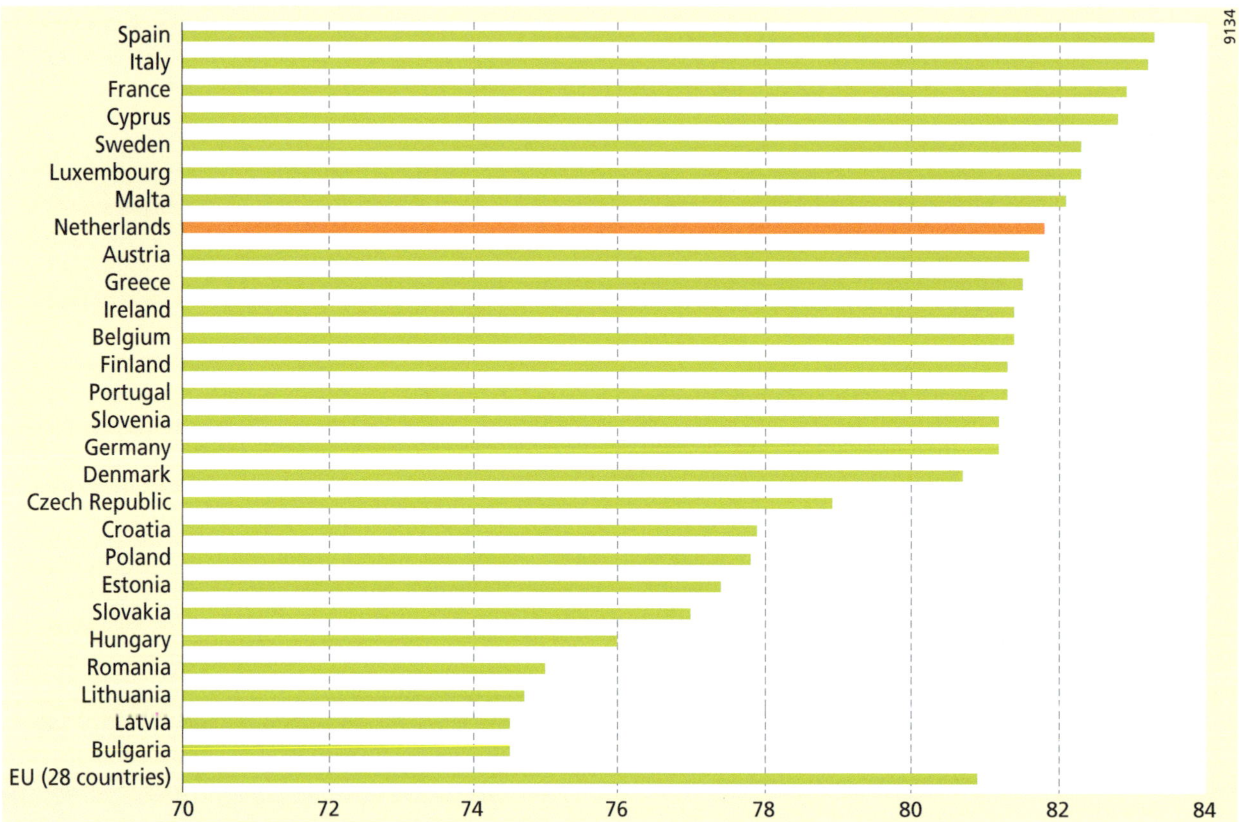

Fig. 11.5 Life expectancy at birth, men and women, EU countries 2014. (Source: Eurostat)

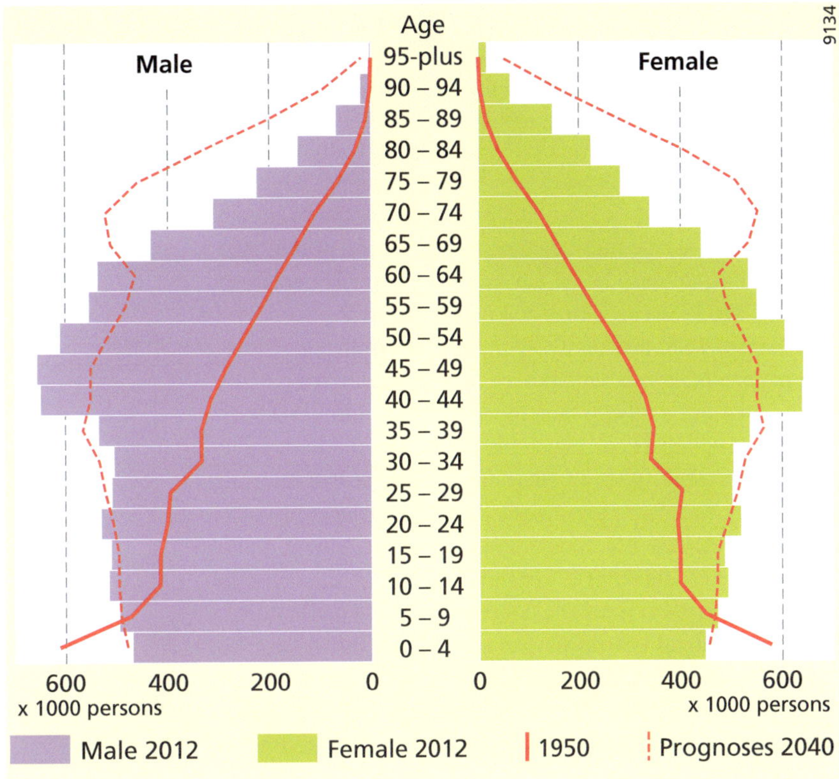

Fig. 11.6 Population pyramids 1950, 2012 and prognoses 2040. (Sources: Ekamper et al. 2003, pp. 132/3; CBS 2016b)

although fertility in the Netherlands is still higher than the EU average. 17.8% of the Dutch population is 65 years or older (EU average is 18.9%) while the share of the older population is expected to grow to about 25% by 2030 (De Jong 2015). Populations are aging as fertility rates fell between 1965 and 2000 and because life expectancy at birth rises and the post-WWII baby boom generation gets older (Fig. 11.6).

11.2 A Growing Diversity in Household Types

In the beginning of the twentieth century the Dutch household structure reflected the standard life cycle at that time (De Graaf 2011; Liefbroer 2010). People were born in a two-parent family. Young people left home in their early 20s to marry and start a family. After the youngest child left home their parents stayed behind as a couple until one of them died. Family households with two parents and two or more children were the dominant household type. Two-person households with no children were either young married couples or older couples in the empty-nest stage of their life cycle. One-persons households were mainly older widows and widowers. In some rural areas older couples and widows/ers lived in three generation households, but three-generation households have never been a dominant household type in the Netherlands.

In the early 1930s young men left home at 24 years and young women at 23 years on average (De Jong-Gierveld et al. 1991). In the 1940s, this age rose due to poor economic conditions prior to, during and immediately after WWII and to the lack of available houses after WWII. By the end of the 1940s, no more than 20% of the young people left home to live alone, 80% left home at marriage (Mulder and Hooimeijer 2002).

The process of individualisation as core characteristic for the Second Demographic Transition model resulted in diversification of household types. The standard life cycle was substituted by a more individualised life course with a range of options for partnership and parenthood. The most important demographic change in the second half of the twentieth century was the increase of single-person households and the emergence of a new stage in life course: living alone between leaving the parental home and marriage, or cohabitation. The age at which young adults left home decreased from 26 years for men and 25 years for women in the late 1940s to 24 years for men and 22 years for women in the late 1970s. By then, more than 50% of the home-leavers formed a one-person household (Mulder and Hooimeijer 2002). Rising educational levels have been the crucial factor to understand this modification in demographic behaviour. Highly educated young adults left home younger and postponed marriage more often than low educated young people. That caused a growing number of young, single-person households. Another reason for this growth was desire for privacy and autonomy.

Introduction of students grants for all students in higher education in 1986 (with a higher grant for independently living students), the availability of affordable housing for young people (student apartment, starters homes) and the growing affluence of parents stimulated many young adults to live alone.

Demographic behaviour of lower educated young Dutch adults is more traditional. They generally leave home later, marry and have children earlier than the higher educated peers (De Graaf 2011). From the last decades of the twentieth century onwards the mean age for young people, young men in particular, to leave home has increased. This is explained partly by the 1980s economic stagnation and by the economic crisis since 2008, and partly by the improved housing situation in parental homes with private rooms, permissive parents and ever caring mothers: 'hotel mamma' (Stoeldraijer 2014).

Increasing divorce rates modified Dutch household structures as well. Women emancipation has been a relevant factor in these growing rates. Beside marriage, divorce became a relevant life course option as well. A new, more liberal law introduced in 1971 facilitated that option. Prior to 1970, marriage was normally ended only once a partner died. Only 12% of all marriages were dissolved by divorce in 1960. This percentage rose to 40% in 2015 (CBS Statline). In the Netherlands, divorce rates are higher among low-educated people than among their high-educated peers who usually marry later, after a period of cohabitation. Divorced women with no salary from paid work or maintenance allowance from the ex-partner are entitled to social benefit in the Netherlands. Growing divorce rates generated more one-person and one-parent households and are largely responsible for the increasing number of older singles (Van Duin et al. 2013). By far most Dutch one-parent households are ruled by women and no more than 10% by men (Kalmijn and de Graaf 2000).

Total number of Dutch households grew from 2.9 million in 1950 to 7.7 million in 2016. Dutch household composition is more diverse than 50 years ago. No more than 26% of all households are families with two parents and one or more children (CBS Statline). Single-person households increased from 245,000 in 1950 (10% of total) to 2.9 million in 2016 (38% of all households (CBS Statline). The recent increase of single-person households is mainly caused by more older singles due to general ageing of the Dutch population, by the current trend for older people to 'age-in-place' rather than moving to an old-age-institution, to a rising number of divorced older people and people who never married (Van Duin et al. 2013). New types of household arrangements, such as co-parenting or shared-residence families, step families and compiled families further contribute to diversification of household types (Bakker 2015).

Less people get married and more are single or opt for cohabitation partnerships beyond marriage. In 2010, more than 80% of the 20–24 years old and 66% of the 25–29 years old couples had an unmarried relationship. Over half of the children born in 2010 were born out of wedlock (Latten and Mulder 2013). Registered partnership as an alternative for marriage has become a serious option since introduction of the Law on Registration of Partnership, in 1998. In the Netherlands, marriages and registered partnerships formally apply to both heterosexual and gay and lesbian couples (Box 11.1).

Box 11.1: Gay Marriages

The Netherlands was the first country in the world that legalized marriages for same-sex couples, since 2001 (Latten and Mulder 2013). Belgium (2003), Spain (2005), Canada (2005 and South Africa (2006) followed the Dutch example. By 2017, the list of nations extended to 21, including newcomers Colombia (2016), Finland and Germany (2017). The number of formally registered and married gay and lesbian couples increased from 39,000 in 1995 to 55,000 in 2009. Almost 6000 of these couples have children (De Graaf 2011).

The Netherlands has a long history of gay and lesbian emancipation. In 1912, Jacob Anton Schorer founded the first association for gay emancipation, the 'Nederlandse Afdeling van het Wissenschaftlich – humanitäre Komitee' (Dutch branch of the Scientific – Humanitarian Committee, a branch of a German Association). Note that names of the first associations for gay/lesbian people never referred to homosexuality. Schorer was a law historian, a gay nobleman from Zeeland who lost his job because of his homosexual orientation (Van der Meer 2013). In 1946, the current Dutch organization for the integration of homosexuality, COC, was founded as 'Shakespeareclub' first to be renamed 3 years later as Cultuur and Ontspannings Centrum ('Centre for Culture and Leisure', again with no direct reference to homosexuality). Abolition of penalties to homosexuality and social support and activities for gays and lesbians were their main objectives in the first period. Later, antidiscrimination and AIDS/HIV became prime topics. The first objective (penelisation of homosexuality) has been abolished in 1971, and the Dutch Law on Equal Rights (1995) was a milestone in the emancipation process of the LHBT community in the Netherlands (COC 2017). In the 1970s and 1980s the Netherlands, Amsterdam in particular, was an important destination and safe haven for gays and lesbians from around the world. Amsterdam was famous for its gay scene with a rather exuberant gay night life.

The Dutch population in general, is rather positive concerning homosexuals, bisexuals and transgenders

(continued)

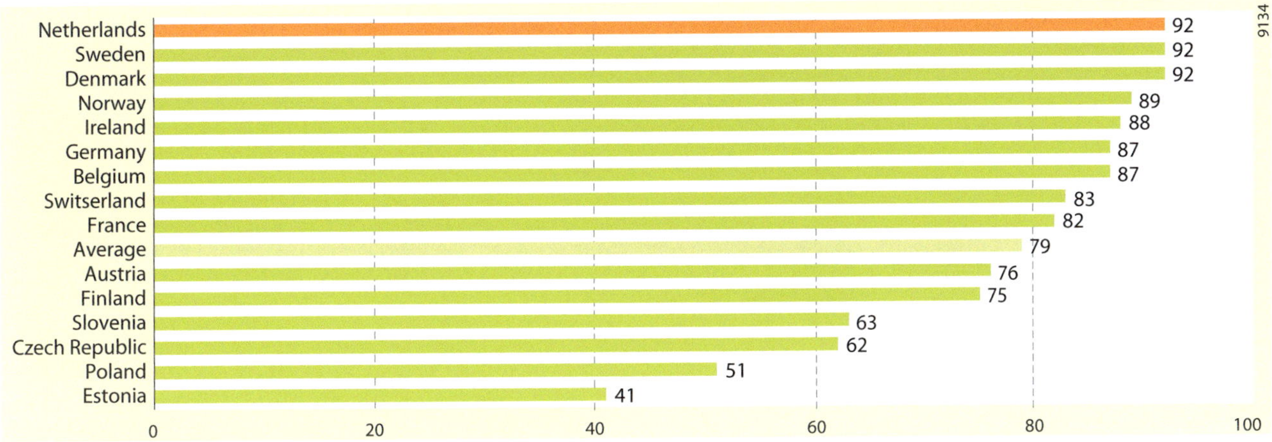

Fig. 11.7 Attitude towards homosexuality in Europe 2014 (weighed percentage of the population with a positive attitude). (Source: Kuyper 2016, p. 17)

Box 11.1 (continued)
and, together with the Danes and Swedes, most positive regarding these sexual orientations out of all EU countries (Fig. 11.7). Acceptation of homosexuality has grown. In 2014, 7% of the Dutch population (15% in 2006) was negative and 70% (53% in 2006) was positive about homo- and bisexuality. Women, young people, higher educated, non-religious and urban people are generally more positive than men, older people, lower educated, religious and rural people. Native Dutch, Surinamese and Antillians are in general more positive on homosexuality than immigrants from Morocco, Turkey, Poland and Somalia (Kuyper 2016).

11.3 Regional Differentiation: Growth, Stagnation and Decline

Regional differentiation in population growth, age and household structure may be explained partly by regional differences in fertility and mortality, and partly by processes of selective interregional migration. All regions in the Netherlands experienced the First and Second Demographic Transition as discussed before but not all at the same rate. Interregional migration often is the main factor in explaining regional and local differences in demographic structures. Long-distance migration is normally induced by employment or education factors. From 1900 onwards, mainly young adults migrated from the peripheral, rural regions in the north, east and south to the urbanized regions in the west of the country to find (better) jobs or for study at an Institution for higher education (Kooiman et al. 2013a). Short-distance migration is usually induced by residential reasons: to find a larger and better house, or in a greener and more quiet region, as migration processes are highly age-dependent. Most of the intranational migrants are young adults. Residential mobility was 27% among young, 20–25 in 2011, and no more than 4% among 65–85 years old adults in the Netherlands (De Groot et al. 2013). Residential mobility of 85 years and older adults has been slightly higher due to moves to a nursing home.

The Dutch population is still growing and is expected to further increase to at least 2030 (De Jong 2015). However, several Dutch regions currently face population decline and are expected to drop further. In particular south Limburg, east Groningen and Zeeuws-Vlaanderen are the regions where population started to decrease since around 2000 (Fig. 11.8). Population decline in other peripheral regions, as north Groningen, north Friesland and east Gelderland, began later. Regional differences in population decline may comprise demographic (population ageing, low birth rates, net migration loss), economic (deindustrialization) and political-institutional factors (national or regional policy to support old industrial sites or to stimulate new economic activities) (Bontje and Musterd 2012).

11.3.1 Population Decline in South Limburg, East Groningen and Zeeuws-Vlaanderen

South Limburg is a former mining area where the last coal mine closed in 1974 (Chap. 4). Coal mining attracted many young immigrants and their families in the first half of the twentieth century when coal mining boomed, generating high birth rates. Strong regional influence of the Roman-Catholic church urged people to have many children resulting in relatively high fertility in south Limburg. For the same reason, fertility and birth rates declined later here than elsewhere

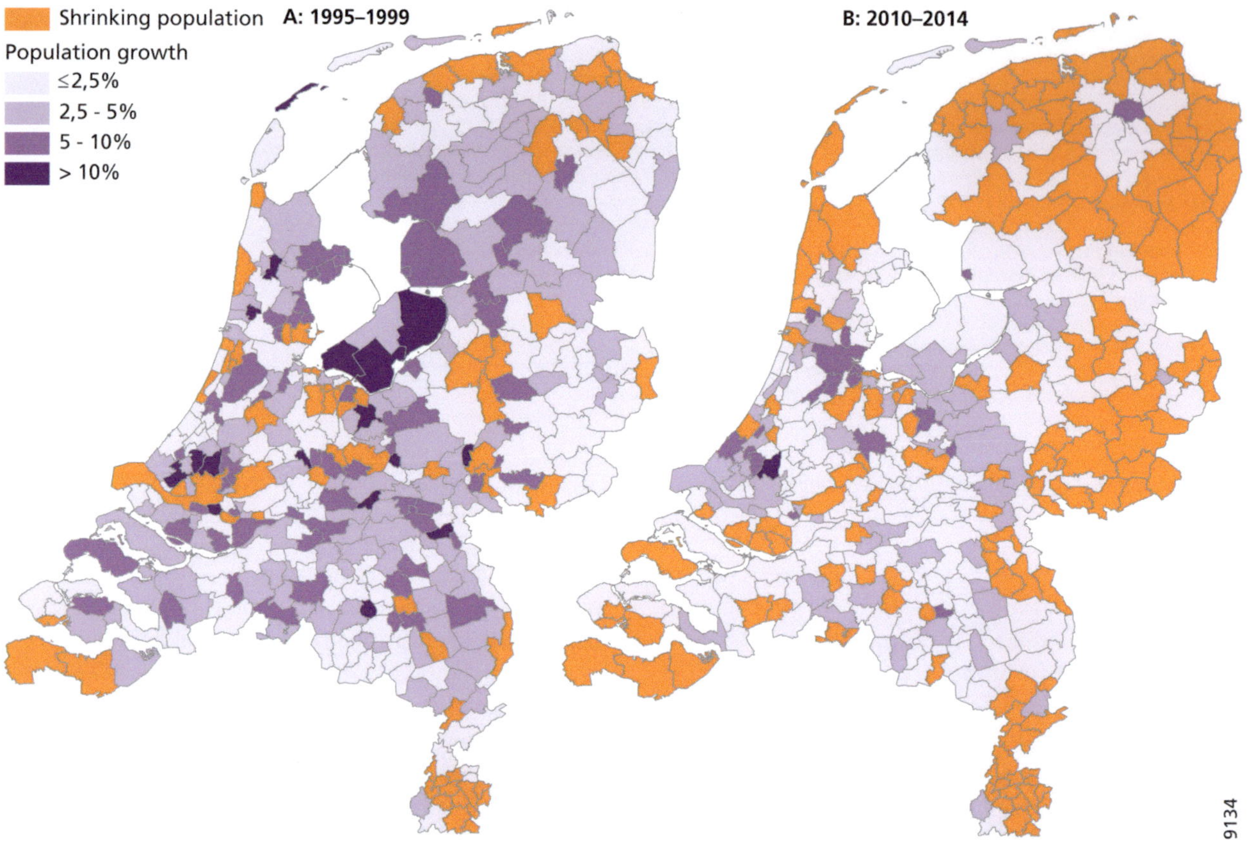

Fig. 11.8 Population growth, stagnation and decline by region 1995–1999 and 2010–2014. (Source: Kooiman 2015, p. 4)

in the nation. Upon closure of the south Limburg coal mines demographic structure rapidly transformed. In a relatively short time south Limburg changed from a high birth rates region to a region with the country's lowest birth rates. Selective emigration of young adults and young families, together with an ageing population has been the main cause of this transformation.

The national government prompted new industrial and non-industrial development and employment. To that end several national governmental agencies were relocated from Den Haag to south Limburg and two new institutions for higher education were installed in the region: Maastricht University (founded in 1976) and the Open University in Heerlen (1984). However, the new employment options offered did not always match with the educational level and skills of the regional labour force. That mismatch resulted in high unemployment rates, a negative migration balance and an ageing population. Population decline in south Limburg preceded diminishing regional employment. In recent years, suburbanisation from highly urbanised nearby German regions and labour migration from Eastern European countries generated net migration but that did not compensate for the negative natural growth in the region (Kooiman et al. 2013b).

Population decline in rural and peripheral east Groningen and Zeeuws-Vlaanderen did not begin before the onset of the twenty-first century (Hoekveld 2014). East Groningen is an agricultural area with major former peat extraction (Chap. 3) which was prosperous during the nineteenth century with modern farming, agriculture-based industries and shipyards. These industries flourished until the 1950s and agriculture industry-based employment dropped from 1960 onwards. Several new industries and non-industrial employment options were created here. However, shortly after a further process of de-industrialisation took place. By the end of the twentieth century job losses went hand-in-hand with emigration of younger people and falling birth rates. In contrast to south Limburg population decline was not caused here by low fertility but rather by population ageing. In east Groningen, no compensation for such falling populations took place by cross-border immigration from other regions as in south Limburg. The migration balance with the sparsely populated nearby German regions is negative.

Zeeuws-Vlaanderen is a peripheral area from the perspective of the Netherlands but nearby to the main Flanders cities of Antwerpen, Gent and Brugge. Zeeuws-Vlaanderen has been accessible either by ferry across the Westerschelde or via Antwerpen until the Westerschelde tunnel was opened in 2003. Zeeuws-Vlaanderen is an agricultural region with an industrialised zone alongside the Kanaal van Gent naar Terneuzen. The Dow Chemical plant in Terneuzen in 1964 has been an important impetus for industrialisation of the region which compensated the declining agricultural employment in the 1960s and 1970s (Thissen 1995). By the end of the twentieth century, however, industrial employment went down as well and increasing numbers of young people left the region. The migration balance is negative here since 2003 (Hoekveld 2014). Cross-border migration partly postponed population decline as more Belgians moved to Zeeuws-Vlaanderen as housing was relatively cheap, than Dutch people who migrated to Flanders to escape from high taxation in the Netherlands.

11.3.2 Two Types of Ageing Regions . . . .

Varying regional and local population age structures are the result of differences in fertility, birth rates and of selective migration. Figure 11.9 shows two different types of ageing regions and three different types of regions with young populations.

The first type of ageing regions are those with stagnating or diminishing populations due to selective migration. These are characterized by low incomes, low educational levels, declining employment, high unemployment rates and low housing prices. Young, single adults and young families leave the region while searching for better educational and employment opportunities elsewhere and older people stay behind. Birth rates are going down due to age selectivity of the migration process. South Limburg, east and north Groningen, north Friesland and Zeeuws-Vlaanderen are examples of this type of regions with an aged population.

A second type of regions with an aged population are characterised by high incomes, high educational levels, low unemployment rates and high housing prices. Populations in such areas are stagnating or are recently declining (Kooiman

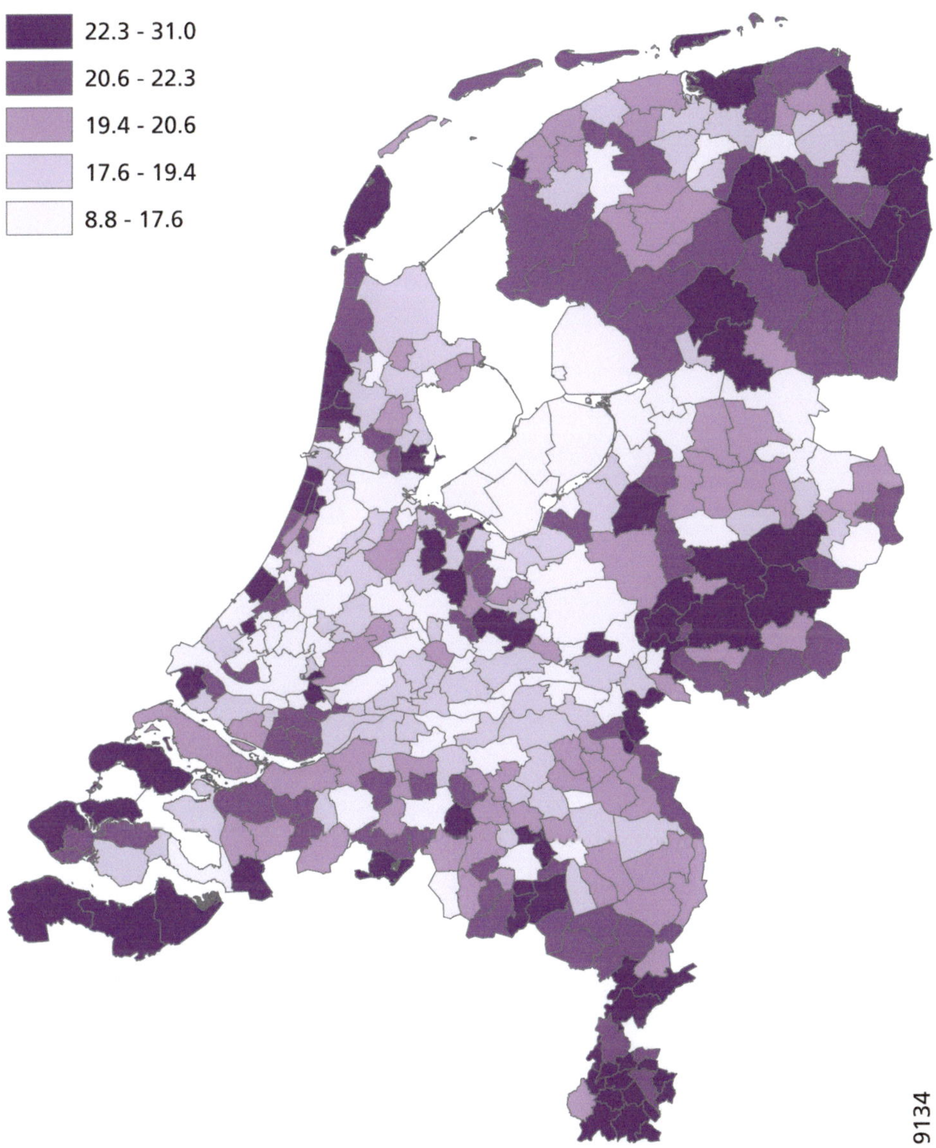

Fig. 11.9 Percentage of population of 65 years and older by municipality 2016. (Source: CBS Statline)

2015). These areas are attractive, wealthy suburbs not far from the main cities in the west and a few municipalities around cities elsewhere, as Haren near Groningen and Rozendaal near Arnhem. There, municipal policy is restrictive with respect to new residential development to preserve the local community's exclusive character (Ostendorf 1988). Housing prices are high and thus affordable to high income households only. People do not move easily and residential mobility is low, length of residence is high limiting housing opportunities for newcomers and younger households. The coastal areas of Noord- and Zuid-Holland, 't Gooi and De Utrechtse Heuvelrug are the best examples of this type of ageing regions in the Netherlands.

11.3.3 ...And Three Types of Young Regions

The first type of areas with a young population are regions or municipalities with high fertility and high birth rates. Religious factors may explain such age structures. The 'Bible Belt' is a zone of most rural municipalities, ranging from Walcheren (Zeeland) in the southwest, via the southern part of Zuid-Holland, the Rijn-Maas river belt, the high sands of the Veluwe, to the north-west of Overijssel (see Chap. 12). This zone is dominated by an orthodox-protestant population where people marry before they are 25 years old and have comparatively many children. Residential mobility is usually low, the number of newcomers is limited and population growth is mainly the result of high birth rates and high fertility. Urk is the youngest municipality in the Netherlands and may serve as a good example for this type of regions.

A second type of young population areas are municipalities with high immigration of young families where 30–45 years old adults and children under 15 years old dominate (Fig. 11.10). From 1965 onwards, the national government stimulated growth of new towns to compensate for massive housing shortages in large cities (see Chap. 8).

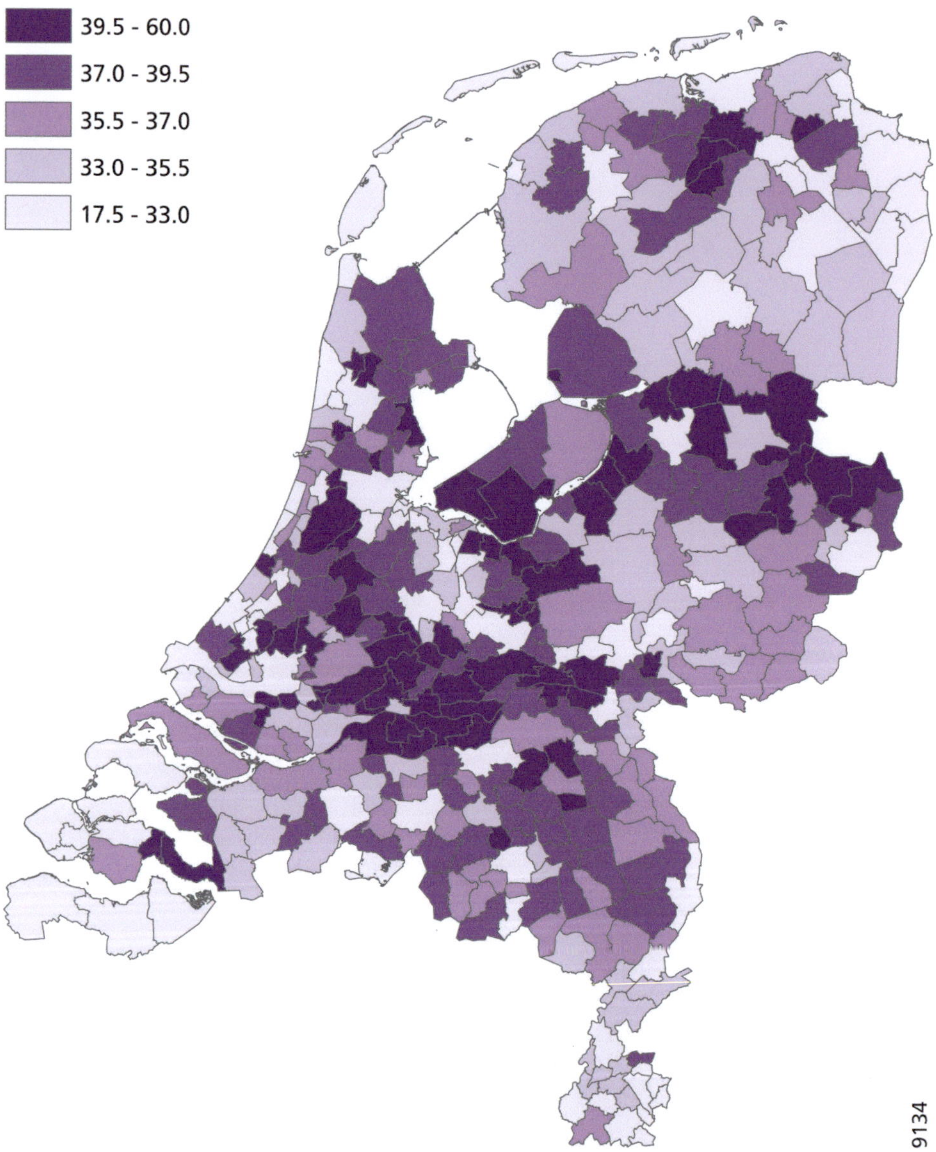

Fig. 11.10 Percentage of households with children by municipality 2016. (Source: CBS Statline)

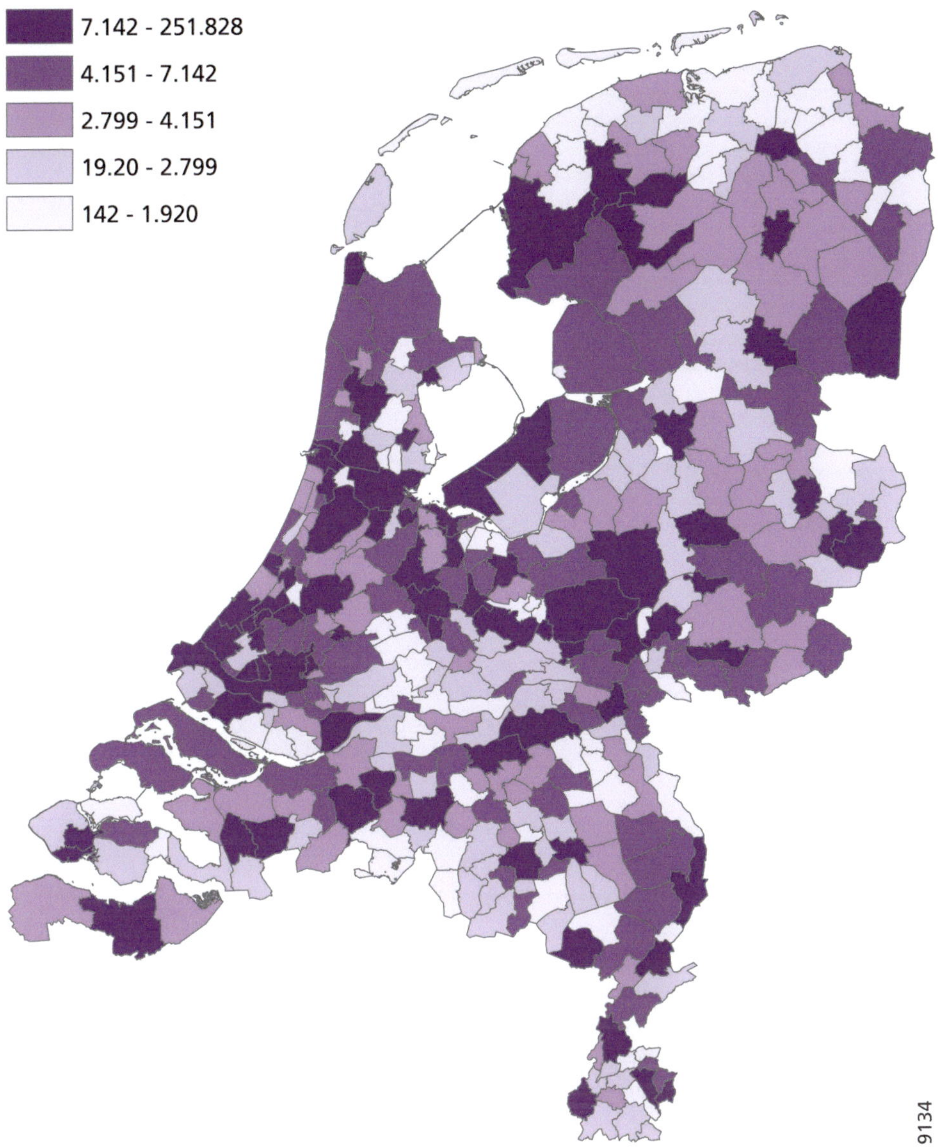

Fig. 11.11 Percentage single-person households by municipality 2016. (Source: CBS Statline)

Vast numbers of single family houses and apartments have been built in municipalities such as Almere, Purmerend, Spijkenisse, Zoetermeer and Nieuwegein to accommodate housing needs of young families for whom no affordable housing was available in the main cities.

(University) city centres are the last type of areas with dominating young populations. There, the 20–30 years old age group is overrepresented (Fig. 11.11 – see also Chap. 8) but children under 15 years old and young parents are underrepresented. Young adults from nearby municipalities or elsewhere migrate to these cities to start academic studies or to start their careers in the labour market. Amsterdam, Utrecht, Leiden, Delft, Groningen and Nijmegen are examples of young population areas.

11.4 International Migration

International migration has been a main factor in twentieth century's Dutch population growth. Apart from its first decade and the 1940–1960 period net migration balance has been positive. International migration has become a global process. In 1900, 2% of the Dutch population was not born in the Netherlands, most of them (80%) born in Germany and Belgium. In 2014, 11% of the Dutch population was born outside the Netherlands, while 22% has one or two parents not born in the Netherlands. Currently, main countries of origin are not only Germany and Belgium but Turkey, Morocco, Indonesia, Suriname, former Dutch Antilles and Poland as well (Fig. 11.12).

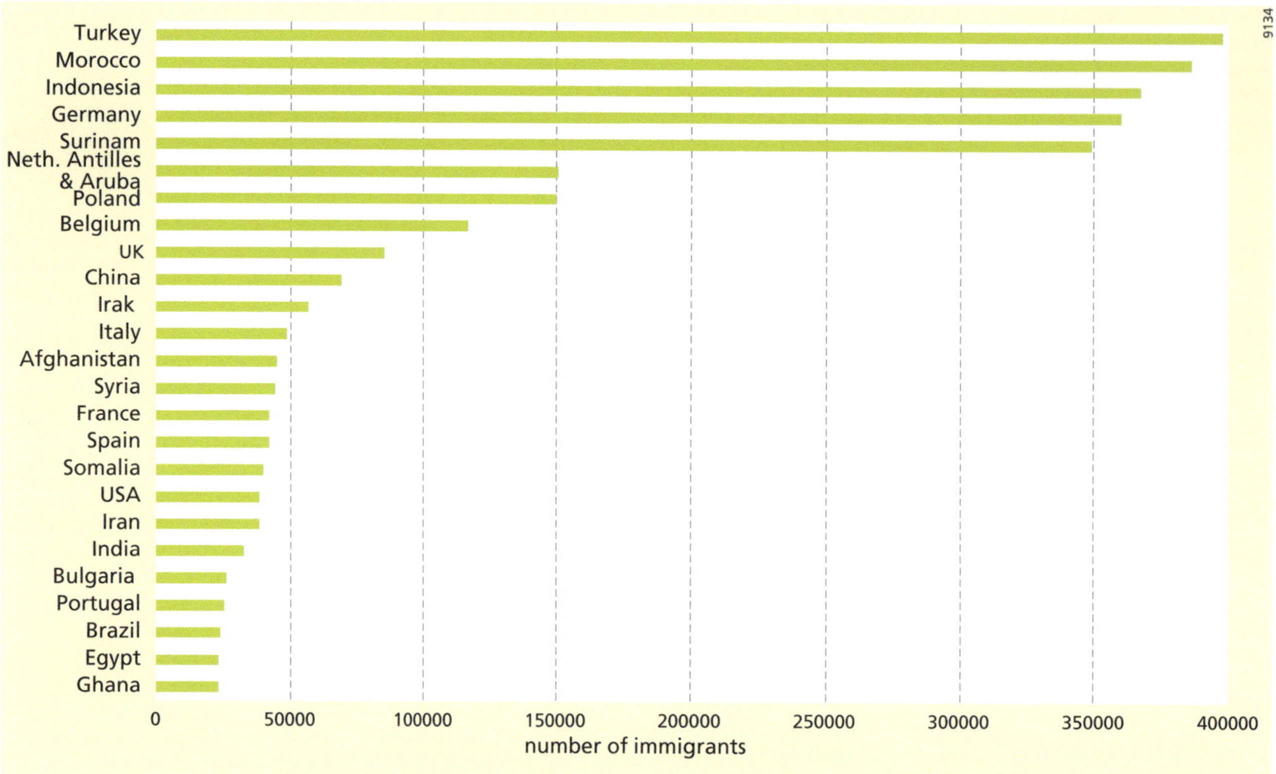

Fig. 11.12 Numbers of first and second generation immigrants by country of origin 2016 (25 most important countries). (Source: CBS Statline)

11.4.1 International Migration 1900–1960

During the nineteenth and the early twentieth century emigration exceeded immigration (Nicolaas and Sprangers 2007). Rural areas in the peat and higher sand areas beyond the west were hit by the agricultural crisis of the 1870s. That crisis developed from competition by cheap cereals from the United States and successive mechanisation in the agricultural sector. By then, most of the emigrants left for the emerging German Ruhr industries as seasonal workers or as permanent migrants. There, coal exploitation began already in the eighteenth century but large-scale mining did happen before 1870. Hundreds of thousand migrants worked there as miners or in mining related industries, most of them from Poland but Italians and Dutch as well. In 1907, 125,000–150,000 Dutch people lived in the German Rheinland-Westphalia Province. Not all of them worked in German mines but most were employed in industry (steel, machines, chemical, textile) or in the construction sector (De Pater 2002).

In the seventeenth century Dutch migrants first settled in North-America. The American census of 1790 registered about 100,000 citizens from Dutch origin, mainly in and around New York city (Krabbendam 2013). In the early nineteenth century, many Jews migrated from the Netherlands to the United States of America for economic reasons and lived in New York, Philadelphia and Boston. Other Dutch migrants, most peasants and farm workers from the rural provinces of Groningen, Friesland and Zeeland, moved to the USA for economic and religious reasons, later in the nineteenth century. Some reverends of orthodox protestant church communities, who were separated from the main reformed church in the Netherlands, left for the USA together with their followers. For example, reverend Van Raalte and a hundred followers went to Michigan and founded the city of Holland in 1846. They all joined the Dutch Reformed Church, founded in the seventeenth century by the first Dutch settlers. After a few years van Raalte cum suis separated from that Church and started the more orthodox The Christian Reformed Church in North America (GKM 2017). Their presence is reflected in the names of several Michigan towns and villages: Groningen, Vriesland, Overissel and Zeeland. Hope College in Holland, Rutgers University in New Brunswick, New Jersey, Vanderbilt University in Nashville (TN) and Dordt College in Sioux Center, Iowa have Dutch roots too.

The First World War (WWI) generated many war refugees in Europe. About one million Belgians fled to the neutral Netherlands soon after the beginning of WWI, in 1914 and in particular upon the Antwerpen bombardments. After a few months and certainly by the end of WWI in 1918, most of them returned home. In 1920, after WWI, no more than

30,000 Belgians still lived in the Netherlands (Ekamper et al. 2003).

The political and economic chaos following WWI brought many Germans to the Netherlands, mainly young women who found jobs as domestic servants and in health care. While 32,000 Germans lived in the Netherlands in 1900, that number more than tripled to 102,000 in 1930, many more female than male migrants. Around 1920, Polish, Slovenian and Italian men migrated to the Netherlands to work in the Limburg coal mines (Schoorl 2011) (Box 11.2).

German political refugees arrived in the Netherlands in the 1920s and 1930s and their numbers considerably increased once the German Nazi regime came into power, in 1933. Most of these refugees were Jews, but communists and socialists sought shelter in the Netherlands too. In the few years before German occupation of the Netherlands in 1940–1945, many people, Jews in particular, were in transit and left the country again to find a safe haven in the United Kingdom, United States of America, Canada and other countries. During the Second World War 110,000 Jews have been deported to the extermination camps in Germany and Poland and more than 90% of them were murdered. No more than 5500 Jews returned to the Netherlands when WWII was over (Nicolaas and Sprangers 2007).

After WWII, the Netherlands entered a period of post-war reconstruction. The first few years were chaotic and characterized by scarcity and low wages. Housing shortage was enormous and food and other essential citizen commodities were rationed. The political and economic conditions prompted many Dutch to consider emigration. In the 1950s, some 350,000 people migrated to Canada, Australia, the United States of America, South-Africa and New-Zealand. Although most of them intended to settle permanently some 20% did not manage and returned to the Netherlands eventually (Nicolaas and Sprangers 2007).

Concurrently, many immigrants from the Dutch East Indies arrived in the Netherlands. Between Japanese capitulation in 1945 and formal Dutch recognition of Indonesia's independency in 1949, an independence war occurred. In that period, some 125,000 people migrated from Indonesia to the Netherlands. About 76,000 of them eventually returned to Indonesia some years later. Upon transfer of Dutch sovereignty to Indonesia, expropriation of Dutch property in 1958, and transfer of Dutch New Guinea to Indonesia in 1962, even more people migrated from Indonesia (Nicolaas and Sprangers 2007).

The Dutch government called these immigrants 'repatriates'. They were considered Dutch people returning to the Netherlands. Some of them were born in the Netherlands indeed and returned, but most of them were born in the Dutch East Indies and arrived in the Netherlands for the first time in their lives. They entered while the Dutch suffered from severe post-war shortage of housing. Some of

Box 11.2: Chinese Immigrants

Chinese are among the first immigrant groups in the Netherlands. Today about 70,000 first and second generation immigrants from China and Hong Kong are living in the Netherlands and about 30,000 Chinese first and second generation immigrants from other countries. In 1911, Chinese seamen, hired by Dutch shipping companies as strike-breakers and firemen, came to live in the Netherlands. They were housed in boarding-houses near Rotterdam's and Amsterdam's harbour areas. Canteens in these boarding-houses became the first Chinese restaurants and their neighbourhoods developed into 'Chinatowns' with Chinese restaurants, shops, temples and schools.

During the 1930s economic depression many Chinese lost their jobs. Most Chinese seaman were deported to their home countries or transferred to ships of foreign companies. The few Chinese who remained in the Netherlands tried to make a living as street vendors, selling Chinese peanut-cookies (Wubben 1986). After WWII Chinese immigrants did not come from China and Hong Kong only, but also from the independent Republic of Indonesia as many of them were alleged to have collaborated with the Dutch colonisers. Later, Chinese from Suriname came to the Netherlands, mainly in relation to Surinam's independence.

In the 1970s and 1980s labour migrants arrived from the coastal regions of mainland China (Guandong and Zhejiang) and from Hong Kong who worked in one of the thousands of Chinese restaurants in the Netherlands. Men migrated first, and women followed some years later in a process of family reunification. Chinese restaurants were family enterprises where husband and wife worked together with relatives. In the same period refugees from Chinese minority groups, e.g. Tibetans and Uighurs, fled to the Netherlands as well. Today's Chinese immigrants, from 2000 onwards, usually are high educated and come to the Netherlands as students or as knowledge migrants.

The older, first generation Chinese migrants are a low-educated, hard working group with a strong orientation on their peer *group* of Chinese migrants, rather than on their *country* of origin. They have few contacts with native Dutch people and generally a low proficiency of the Dutch language. Nevertheless, they highly appreciate education, and the second generation of Chinese is usually high educated and well integrated in Dutch society. Today's highly educated Chinese immigrants usually do not speak Dutch and have few contacts with Dutch society (Gijsberts et al. 2011; Harmsen 2011).

them were accommodated by relatives, for others (some 135,000) the Dutch government paid board and accommodation fees to Dutch individuals willing to share their private homes with these immigrants. As they all spoke Dutch, most of them found jobs and have generally been successful in making a new living in Dutch society (Van Amersfoort 1974) (Box 11.3).

Box 11.3: Moluccan Immigrants

A separate group of immigrants from the Dutch East-Indies are those (some 13,000) of the Moluccan archipelago who were forced to migrate to the Netherlands in 1951. Almost all (90%) of them originated from the island of Ambon. Men served as regular soldiers in the Dutch East Indian army before and during the independence war. After sovereignty transfer the Dutch East-Indian army was abolished and the soldiers could choose for either being transferred to the Indonesian army, or for demobilisation. These Moluccan soldiers would accept entering the Indonesian army only on the condition of autonomy for the Moluccan province. The Indonesian government refused this option and the Moluccans proclaimed their independent state: the Republic Maluku Selatan.

The international community did not recognize this new republic and the Moluccan archipelago was taken over by the Indonesian government. Some Moluccan soldiers switched armies, others remained in the Dutch army to be demobilised. But demobilisation could happen only in the Netherlands. Thus, the soldiers and their families were transported to the Netherlands. Upon arrival they were housed in barracks, geographically isolated in peripheral areas of the Netherlands. From the point of view of the Dutch government this housing situation was temporary: upon demobilisation these migrants should find civil jobs and a house. However, these Moluccan soldiers refused demobilisation and many Moluccan families lived in these barracks until the early 1970s. Others were placed in neighbourhoods where new houses were built for them (Van Amersfoort 1974). Even today, Moluccan neighbourhoods still exist in about 45 municipalities, where the fourth generation of Moluccan immigrants is growing up (Duursma and Heck 2017).

11.4.2 International Migration 1960–2015

From 1960 onwards the Netherlands turned from an emigration into an immigration nation. Three major immigration processes have been responsible for this: (1) inflow of low-skilled migrants and their families from Mediterranean countries and from Middle and Eastern Europe; (2) migration from the former Dutch colonies, and (3) inflow of refugees in several groups. In addition, a growing number of highly educated workers ('expats') entered in the Netherlands as related to globalisation of the economy and the growing number of international companies in this country (Chap. 6). Immigrants generally are concentrated in the large cities, industrial towns and along the Belgium and German borders (Fig. 11.13).

11.4.2.1 Low-Skilled Labour Migrants

By the end of the 1950s economic growth and low labour market participation rates of women generated high demands for labour in the Netherlands. International recruitment for low-skilled male workers was done by several industrial companies, first in southern European nations as Spain, Italy and Yugoslavia, and later in other circum-Mediterranean countries as Turkey and Morocco. In the period 1964–1973 about 225,000 migrants from these countries entered to work in the Netherlands (Nicolaas and Sprangers 2007). More than 60% of the (former) Yugoslavian, Italian and Spanish migrants returned to their home countries within 10 years (Schoorl 2011).

However, migration from Turkey and Morocco proved to be permanent because of the poor economic conditions in their home countries. Moreover, a process of family reunification began after a few years (Van Amersfoort 1974). In the 1960s, most Turkish and Moroccan migrants were male, but female in the 1970s. Later, family formation, i.e. immigration of brides from their home countries, began dominating Turkish and Moroccan immigration (Sterckx et al. 2014). Due to these immigration processes and their high fertilities, Turks and Moroccans are the current main immigration groups (first and second generation immigrants) in the Netherlands.

Initially, migrants employing companies provided or arranged for housing. Later, most Turkish and Moroccan labour migrants moved to low-standard private rental housing in older urban neighbourhoods. Once they got access to social housing most of them were living in post-war urban neighbourhoods.

With the eastward extension of the European Union and associated free cross-border labour facilities among EU countries, many labour migrants from Poland, Bulgaria and Romania came to the Netherlands to work in agriculture, industry or construction. Since Poland became a member state of the European Union in 2004, the Polish have become the 7th largest immigration group in the Netherlands (Nicolaas 2010).

Dutch legislation distinguishes seasonal workers and immigrants based on length of residence, less or more than 4 months. Many low-skilled workers from Middle and Eastern Europe are temporary, seasonal workers in horticulture or in construction. Many stay in campsites, holiday camps, in rebuilt greenhouses, guesthouses or in shared rooms.

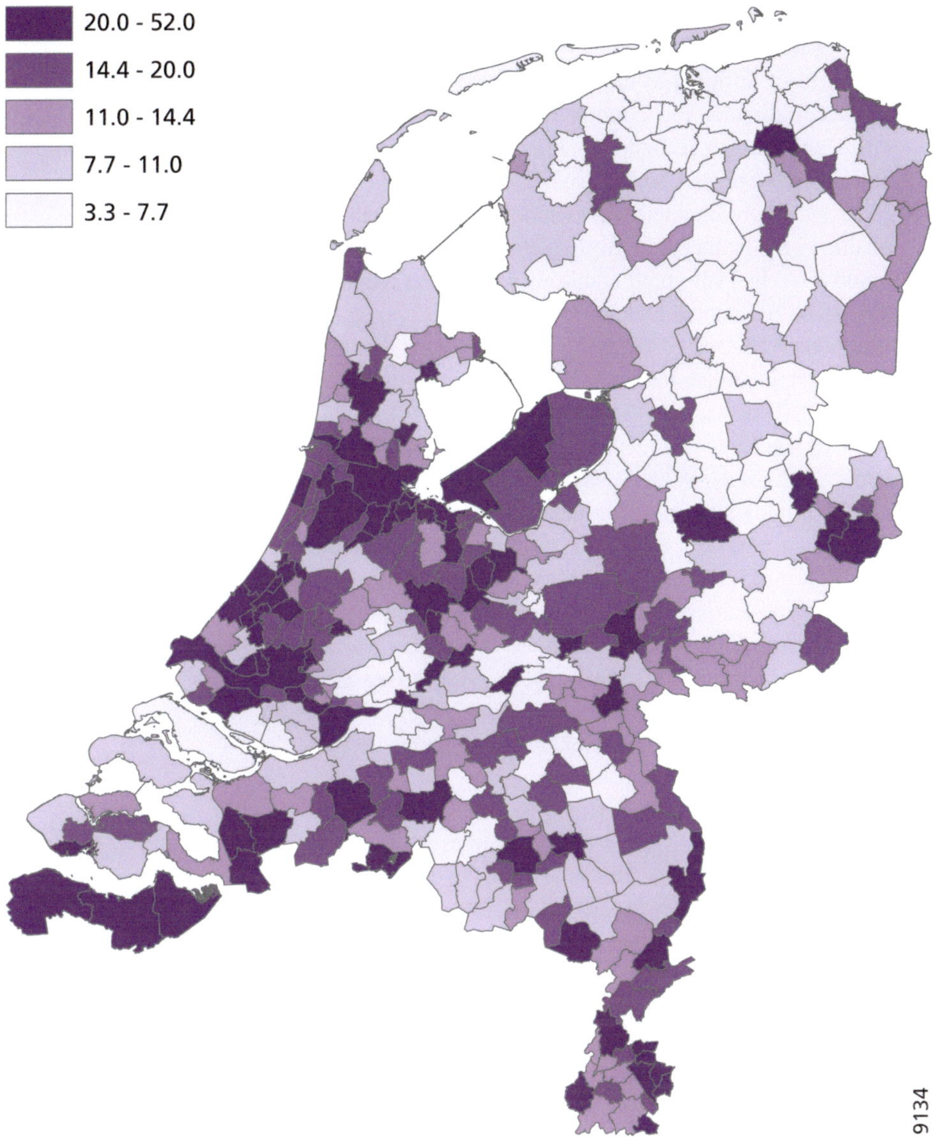

Fig. 11.13 Percentage first and second generation immigrants by municipality 2016. (Source: CBS Statline)

However, more and more Middle- and Eastern European workers stay for a longer period in the Netherlands and are housed in appropriate housing in the private or social rental sector (Harms 2014).

Although by far most immigrants from Middle and Eastern European countries are low-skilled workers, some are employed in high-skilled jobs and should be considered as 'expats' (Gijsberts and Lubbers 2015).

11.4.2.2 Expats

About one sixth of all 600,000 foreign workers in the Netherlands are highly educated specialists who usually stay temporarily in the Netherlands: expats. Most expats work in financial and commercial services, health care and universities and are employed more in international firms than higher educated, Dutch employees do (Groot et al. 2013). Most expats stay for less than 5 years in the Netherlands and return to their home countries or migrate elsewhere. Most expats are male and about half of them is born in the European Union, predominantly in the UK, Germany, Belgium, France, Italy and Poland. Non-EU countries with high numbers of expats are Turkey, Morocco, India, USA and Russia.

Expats mainly occur in the west, more specifically in Amsterdam, Den Haag, Rotterdam, Utrecht and in the Schiphol area, where headquarters of several multinational companies are situated (Chap. 7), and predominantly in the high-income neighbourhoods, e.g. in Amsterdam-Zuid and Amsterdam-Buitenveldert and in Den Haag-Benoordenhout. Other concentrations of expats are in university cities as Eindhoven, Enschede, Groningen and Maastricht.

11.4.2.3 Migration from Surinam and the Dutch Antilles

Migration from Surinam to the Netherlands and vice versa always occurred since that country became a Dutch colony after the end of the second Dutch-English war in 1667. Migration from Surinam to the Netherlands was limited to Dutch people returning home and to middle-class Surinamese who came to the Netherlands for higher education or to find well-paid jobs. In 1956, both countries agreed that Surinam people would be granted Dutch nationality. That agreement and the perceived lack of perspectives in Surinam, spurred many of them, including low-income groups, to migrate (Van Amersfoort 1974). Most of them arrived shortly before and after Surinam became an independent nation in 1975. In that particular year some 40,000 Surinam people (about 10% of their entire population) arrived in the Netherlands. Migration from Surinam peaked again 5 years afterwards when the agreement to opt for Dutch nationality was about to expire. Many Surinamese immigrants lived in the Bijlmermeer, a newly built Amsterdam neighbourhood with many vacant apartments by the time of Surinam independence (see Box 8.4 in Chap. 8). With their Dutch nationalities these migrants had access to the social rental sector and to the Dutch system of housing allowances.

The former Dutch Antilles is a group of six Caribbean islands (Curaçao, Aruba, Bonaire, St. Maarten, Saba and St Eustatius), colonised by the Dutch in 1634. In 1954, Dutch Antilles was granted autonomy within the Kingdom of the Netherlands. From 2010 onwards, Aruba, Curaçao and St Maarten are autonomous countries within the Kingdom of the Netherlands while the smaller islands of Bonaire, Saba and St Eustatius received the status of Dutch municipalities. Migration from the former Dutch Antilles has always been a circular migration where people migrate from the Antilles to the Netherlands and vice versa. Today, about half of the population of the former Dutch Antilles lives in the Netherlands (Oostindie 2016).

11.4.2.4 Refugees

Since WWII various groups of refugees arrived in the Netherlands and people from many nations suffering from (civil) war or repressive regimes currently live in the Netherlands. Until the mid-1980s rather few refugees arrived: some 3000 who left Hungary after the 1956 Soviet invasion, and some one thousand from Czechoslovakia who fled after a comparable invasion in 1968 (Nicolaas and Sprangers 2007).

From the 1980s onwards the number of refugees increased and has been fluctuating in the period 1985–2015 depending on the political situation in the world's war zones and the asylum policy in the Netherlands and its neighboring countries. In the 1980s and 1990s most refugees came from Iran, Iraq, Somalia, Afghanistan and Yugoslavia (Dourleijn and Dagevos 2011). Top years in that respect in the Netherlands were 1994 (53,000 refugees, most of them from Yugoslavia) and 2015 (57,000 refugees, most from Syria and Eritrea) (Box 11.4).

Box 11.4: Transit Migration: Somalian Refugees

In the late 1980s and the 1990s many Somalians fled from a violent civil war and found refuge in the Netherlands. In the first decade of the twenty-first century a new wave of Somalian refugees arrived in the Netherlands for comparable reasons. Today, some 39,000 Somalian immigrants are living in the Netherlands. Their numbers have been strongly fluctuating, more than in other groups of immigrants, and caused by changes in both immigration and emigration. Since 1999, thousands of Somalian immigrants left the Netherlands and migrated to the UK. Recently, many young, second generation Somalians left the UK and returned to the Netherlands.

Compared to other groups of immigrants and refugees, Somalians are low-educated, have low incomes, have low labour market participation and are poorly integrated in Dutch society. They are highly dependent on social benefits, don't speak Dutch well enough and are generally not satisfied with their situation. Many Somalians in the Netherlands believe that job opportunities and social support and respect for their culture and religion are better in the UK where larger Somalian communities occur than in the Netherlands. It seems, however, that some of the young, second generation Somalians prefer to return to the Netherlands, as they do speak Dutch and have friends in the Netherlands where they grew up, and rather wish to escape from the closed Somalian communities in the highly segregated UK cities (Klaver et al. 2010).

Individual assessments of asylum seekers are the core of the asylum system in the Netherlands. However, the Dutch asylum system includes a general procedure for people from 'war zones'. In 2001, the Dutch government introduced new and more restrictive asylum legislation. Subsequently, their numbers dropped, from 44,000 in 2000 to less than 10,000 in 2007. This decline has been particularly significant for former Soviet Union, Afghanistan and former Yugoslavia refugees. A similar reduction occurred when the general procedure for asylum seekers from Iraq (in 2008) and from Somalia (in 2009) were abolished. Numbers of refugees, mainly from Syria and Eritrea, rapidly and significantly increased again in the period 2013–2015 (Vluchtelingen in getallen 2015).

Most asylum seekers are housed in special refugee centers, dispersed over the country, but predominantly in peripheral regions with vacant accommodation in former health care institutions, monasteries or holiday centers. The procedure between asylum request and admission (or non-admission) takes at least 6 months to several years. Asylum is granted to more than half (about 60%) of the asylum seekers. As soon as asylum is granted he or she is entitled to housing in the social rental sector. Municipalities are required to make housing available for refugees in proportion to their number of inhabitants. That is why refugees are more dispersed over the Netherlands than other groups of immigrants. Refugees who are granted asylum are free to move to any other municipality but they generally move to places where relatives and compatriots live. After some time former refugees eventually tend to live in more segregated regions and in the big urban centres rather than before (Zorlu and Mulder 2008).

Refugees who are denied admission must leave the country. Some of them do this voluntarily but many keep on staying in the Netherlands as undocumented immigrants. Standing Dutch policy is to send them back to their countries of origin but some of these nations (e.g. Nigeria) refuse their entrance.

References

Bakker W (2015) Post-separation families; residential arrangements and everyday life of separated parents and their children. Dissertation, University of Amsterdam, Amsterdam

Beets N (2010) Kleinere gezinnen. DEMOS Bulletin over Bevolking en Samenleving 26(9):2–5

Bontje M, Musterd S (2012) Understanding shrinkage in European regions. Built Environ 38(2):153–161

CBS (2016a) Steeds minder tienermoeders. Centraal Bureau voor de Statistiek, Den Haag. https://wwwcbsnl/nl-nl/nieuws/2016/14/steeds-minder-tienermoeders. Accessed 2 June 2017

CBS (2016b) Bevolkingspiramide. Centraal Bureau voor de Statistiek, Den Haag. https://wwwcbsnl/nl-nl/visualisaties/bevolkingspiramide. Accessed 2 June 2017

CBS (2017a) Bevolking in 2016 relatief sterk gegroeid. Centraal Bureau voor de Statistiek, Den Haag. https://wwwcbsnl/nl-nl/nieuws/2017/01/bevolking-in-2016-relatief-sterk-gegroeid. Accessed 2 June 2017

CBS (2017b) Geen natuurlijke bevolkingsgroei in eerste kwartaal. Centraal Bureau voor de Statistiek, Den Haag. https://wwwcbsnl/nl-nl/nieuws/2017/18/geen-natuurlijke-bevolkingsgroei-in-eerste-kwartaal. Accessed 2 June 2017

COC (2017) https://www.coc.nl/over-ons. Accessed 2 June 2017

De Graaf A (2011) Gezinnen in cijfers. In: Bucx F (ed) Gezinsrapport 2011; een portret van het gezinsleven in Nederland. Sociaal en Cultureel Planbureau, Den Haag

De Graaf A, Van Galen R (2014) Ouderlijk gezin beïnvloedt timing eerste moederschap. Centraal Bureau voor de Statistiek, Den Haag. https://www.cbs.nl/nl-nl/achtergrond/2014/29/ouderlijk-gezin-beinvloedt-timing-eerste-moederschap. Accessed 2 June 2017

De Groot C, Van Dam F, Daalhuizen F (2013) Vergrijzing en Woningmarkt. Planbureau voor de Leefomgeving, Den Haag

De Jong A (2015) Toekomstverkenning welvaart en leefomgeving. Cahier Demografie. Planbureau voor de Leefomgeving/Centraal Planbureau, Den Haag

De Jong-Gierveld J, Liefbroer AC, Beekink E (1991) Leaving home among young adults. Eur Stat Rev 7(1):55–71

De Pater B (2002) Nederlanders naar en in het Ruhrgebied 1860–1914. In: Hoogers K, Hoogeveen J, Visser R (eds) 100 jaar Nederlandse Kerk in Duitsland. NKiD, Duisburg/Ruhrort, pp 17–30

Dourleijn E, Dagevos J (2011) Vluchtelingengroepen in Nederland. Over de integratie van Afghaanse, Iraakse, Iraanse en Somalische migranten. Sociaal en Cultureel Planbureau, Den Haag

Duursma M, Heck W (2017) Nog 45 gemeenten hebben aparte Molukse wijk. NRC/Handelsblad 19 May 2017. https://wwwnrcnl/nieuws/2017/05/19/nog-45-gemeenten-hebben-aparte-molukse-wijk-9344587-a1559706. Accessed 2 June 2017

Ekamper P, Van der Erf R, Van der Gaag N, Henkens K, Van Imhoff E, Van Poppel F (2003) Bevolkingsatlas van Nederland. Demografische ontwikkelingen van 1850 tot heden. NIDI Netherlands Interdisciplinary Demographic Institute. Uitgeverij Elmar, Rijswijk

Gijsberts M, Lubbers M (2015) Roemeense migranten; de leefsituatie in Nederland kort na de migratie. Sociaal en Cultureel Planbureau, Den Haag

Gijsberts M, Huijnk W, Vogels R (eds) (2011) Chinese Nederlanders; van horeca tot hogeschool. Sociaal en Cultureel Planbureau, Den Haag

GKM (2017) Dominee A. C van Raalte 1811–1876 Gereformeerde Kerken van Nederland. http://gereformeerdekerkeninfo/ds-a-c-van-raalte-1811-1876/. Accessed 2 June 2017

Groot S, Van Gestel G, Raspe O (2013) Foreign knowledge workers in the Netherlands. In: Internationalisation monitor. Centraal Bureau voor de Statistiek, Den Haag, pp 185–204

Harms E (2014) MOE-landers krijgen zicht op fatsoenlijke woonruimte. Tijdschrift voor de Volkshuisvesting 2014(2):6–10

Harmsen C (2011) Ruim 51 duizend Chinezen van de eerste generatie in Nederland. Centraal Bureau voor de Statistiek, Den Haag. https://wwwcbsnl/nl-nl/achtergrond/2011/17/ruim-51-duizend-chinezen-van-de-eerste-generatie-in-nederland. Accessed 2 June 2017

Hoekveld J (2014) Urban decline within the region; understanding the intra-regional differentiation in urban population development in the declining regions Saarland and Southern-Limburg. Dissertation, University of Amsterdam

Hofstee EW (1978) De demografische ontwikkeling van Nederland in de eerste helft van de negentiende eeuw; een historisch-demografische en sociologische studie. NIDI Netherlands Interdisciplinary Demographic Institute. Van Loghum Slaterus, Deventer

Kalmijn M, De Graaf PM (2000) Gescheiden vaders en hun kinderen. Bevolking en Gezin 29(2):59–84

Klaver J, Poel J, Stouten J (2010) Somaliërs in Nederland. Een verkenning van hun maatschappelijke positie en aanknopingspunten voor het beleid. Eindrapport. Regioplan Beleidsonderzoek, Amsterdam

Kooiman N (2015) Ruim een op de drie gemeenten is sinds 2010 gekrompen. Bevolkingstrends 2015. Centraal Bureau voor de Statistiek, Den Haag

Kooiman N, Latten J, Annema A (2013a) Bevolking van vertrekregio's blijft sociaal-economisch achter. Centraal Bureau voor de Statistiek, Den Haag

Kooiman N, Latten J, Nicolaas H (2013b) Bevolkingskrimp Limburg geremd door de toegenomen migratie. Centraal Bureau voor de Statistiek, Den Haag

Krabbendam H (2013) Nederlanders trekken naar het beloofde land. http://wwwisgeschiedenisnl/archiefstukken/nederlanders-trekken-naar-het-beloofde-land/. Accessed 2 June 2017

Kuyper L (2016) Lhbt-monitor 2016. Opvattingen over en ervaringen van lesbische, homoseksuele, biseksuele en transgender personen. Sociaal en Cultureel Planbureau, Den Haag

Latten J, Mulder CH (2013) Partner relationships in the Netherlands: new manifestations of the second demographic transition. Genus 69 (3):103–121

Lesthaeghe R, Van de Kaa DJ (1986) Twee demografische transities? In: Van de Kaa DJ, Lesthaeghe R (eds) Bevolking: groei of krimp? Mens en Maatschappij Boekaflevering 61. Van Loghum Slaterus, Deventer, pp 9–24

Liefbroer AC (2010) Veranderingen in partnerrelaties. DEMOS Bulletin over Bevolking en Samenleving 26(9):5–7

Mulder CH, Hooimeijer P (2002) Leaving home in the Netherlands; timing and first housing. J Housing Built Environ 17(3):237–268

Nicolaas H (2010) Ruim de helft van de Poolse immigranten vertrekt weer uit Nederland. Centraal Bureau voor de Statistiek, Den Haag. https://wwwcbsnl/nl-nl/nieuws/2010/51/ruim-de-helft-van-de-poolse-immigranten-vertrekt-weer-uit-nederland. Accessed 2 June 2017

Nicolaas H, Sprangers A (2007) Buitenlandse migratie in Nederland 1795–2006: de invloed op de bevolkingssamenstelling. Centraal Bureau voor de Statistiek, Den Haag

Oostindie G (2016) Geschiedenis en betekenis van de Antilliaanse migratie. DEMOS Bulletin over Bevolking en Samenleving 32 (2):5–7

Ostendorf W (1988) Het sociaal profiel van de gemeente; woonmilieudifferentiatie en de vorming van het stadsgewest Amsterdam: het ruimtelijk beleid van een achttal gemeenten na de Tweede Wereldoorlog. Dissertation, University of Amsterdam

Schoorl J (2011) Arbeidsmigratie door de jaren heen. DEMOS Bulletin over Bevolking en Samenleving 27(10):1–4

Sterckx L, Dagevos J, Huijnk W, Van Lisdonk J (2014) Huwelijksmigratie in Nederland; achtergronden en leefsituatie van huwelijksmigranten. Sociaal en Cultureel Planbureau, Den Haag

Stoeldraijer L (2014) Jongeren blijven langer thuis wonen. Bevolkingstrends 2014. Centraal Bureau voor de Statistiek, Den Haag

Thissen F (1995) Bewoners en nederzettingen in Zeeland: op weg naar een nieuwe verscheidenheid. Dissertation, University of Amsterdam. Nederlandse Geografische Studies 191. Koninklijk Nederlands Aardrijkskundig Genootschap/Instituut voor Sociale Geografie Universiteit van Amsterdam, Utrecht/Amsterdam

Van Amersfoort H (1974) Immigratie en minderheidsvorming. Een analyse van de Nederlandse situatie 1945–1973. Dissertation, University of Amsterdam

Van den Bie RJ, Smits JP (2001) Tweehonderd jaar statistiek in tijdreeksen 1801–1999. Centraal Bureau voor de Statistiek/Rijksuniversiteit Groningen, Den Haag/Groningen

Van den Brekel JC (1986) Bevolkingsbeleid op maat? In: Van de Kaa DJ, Lesthaeghe R (eds) Bevolking: groei of krimp? Mens en Maatschappij Boekaflevering 61. Van Loghum Slaterus, Deventer, pp 89–103

Van der Meer TAM (2013) Schorer, jhr Jacob Anton (1866–1957) In Biografisch Woordenboek van Nederland. http://resourceshuygensknawnl/bwn1880-2000/lemmata/bwn6/schorer. Accessed 2 June 2017

Van Duin C, Stoeldraijer L, Garssen J (2013) Huishoudensprognose 2013–2060: sterke toename oudere alleenstaanden. Bevolkingstrends 2013. Centraal Bureau voor de Statistiek, Den Haag

Vluchtelingen in getallen 2015. Vluchtelingenwerk Nederland, Amsterdam

Wobma E, Van Huis M (2012) Vruchtbaarheid van mannen en vrouwen naar opleidingsniveau. Bevolkingstrends 2012. Centraal Bureau voor de Statistiek, Den Haag

Wubben WJJ (1986) 'Chinezen en ander Aziatisch ongedierte'. Lotgevallen van Chinese immigranten in Nederland, 1911–1940. De Walburg, Zutphen103 2 286-288

Zorlu A, Mulder CH (2008) Initial and subsequent location choices of immigrants to the Netherlands. Reg Stud 42(2):245–264

12 Culture, Religion & Politics

Ben C. De Pater

Abstract

Since the end of the sixteenth century until the end of the nineteenth century, the Netherlands was a country of Protestants – the leading religion – and Roman-Catholics. Around 1900 two other 'pillars' came into existence: the Socialists and Liberals. In the 1960s processes of 'de-pillarisation' and secularization began. Old political parties languished, new political parties presented themselves successfully to the voters. In international perspective, Dutch culture is a Northern European culture: the cultural distance between the Netherlands and the Nordic countries is small, between the Netherlands and South- and East-Europe comparatively large. The Dutch cultural pallet got more colors with the arrival of immigrants since the 1960s. But one characteristic remained constant during all this changes: the relationship between culture and water – a 'delta culture'.

Keywords

Dutch culture in Europe · Pillars · Regional cultural differences · Frisian Culture · Painting · Philosophers · Political Parties · Drugs · Jews · Minorities

Once ice appears in the ditches after some days of frost in winter whispers are in the air (and in the media) about a forthcoming *Elfstedentocht* (11 cities tour), a very popular almost 200 km long skating tour in the Province of Friesland along lakes, rivers and canals connecting 11 historical cities in this province. If frost continues such whispers turn into a fever, certainly once the *ijsmeesters* (ice masters) meet. Public disappointment is overwhelming when they decide that the ice thickness cannot carry yet the thousands of professional and amateur speed skaters and the tens of thousands spectators who aim to witness this major event that overrules everything in the Netherlands the day that this tour will be held. Many Dutch take a day off and travel to Friesland or watch the tour on TV. Winners become national heroes. However, weather conditions are seldom favourable for holding such an event and since the first *Elfstedentocht* was held in 1909 this tour has been held only 14 times, last in 1997. The *Elfstedentocht* is a collective obsession, a passion, comparable with or exceeding the Super Bowl, the final of the American Football Competition.

The Netherlands is a nation of swimmers, rowers, sailors and notably of skaters – for many centuries, see for instance Hendrick Avercamps seventeenth-century paintings of Dutch winter landscapes, crowded with people skating on frozen lakes and rivers (Box 12.1). Today, many medals in these branches of sport are collected at international sports events including the Olympics. The Dutch live with water in leisure times and often professionally, e.g. as fisherman, dredgers, tug captains, life-boat crews, on river or canal transport barges, seamen, as dike builders, bridge watchmen, lock guards and ship builders. Such professions are considered typically Dutch. Water and 'water men' often figure as heroes in Dutch novels, for instance of K. Norel (e.g. *Vanglijn over*, 1960, about a boy on a salvage boat) or Jan Terlouw (*Oosterschelde; windkracht 10*, 1976, about the major flood of 1953 and the construction of the Oosterschelde barrage), or movies, for example *De Storm* (2009, also about the 1953-flood). Despite their many successes to counter the water, the Dutch remain alert as they are well aware of the fact that not all such battles were won and that water may and does strike back.

International reputation of the Dutch to fight the water has been fueled considerably by the heroic but untrue story of *Hans Brinker* (1865), by the American author Mapes Dodge, and by the impressive armies of Dutch dredging vessels and volumes of earth materials replaced by such vessels as depictured by Johan van Veen in his *Master of the Floods* (1948; Figs. 12.1 and 12.2; Meyer 2017). The struggle

B. C. De Pater (✉)
Department of Human Geography and Spatial Planning, Utrecht University, Utrecht, The Netherlands
e-mail: B.C.dePater@uu.nl

E. F. J. De Mulder et al., *The Netherlands and the Dutch*, World Regional Geography Book Series,
https://doi.org/10.1007/978-3-319-75073-6_12

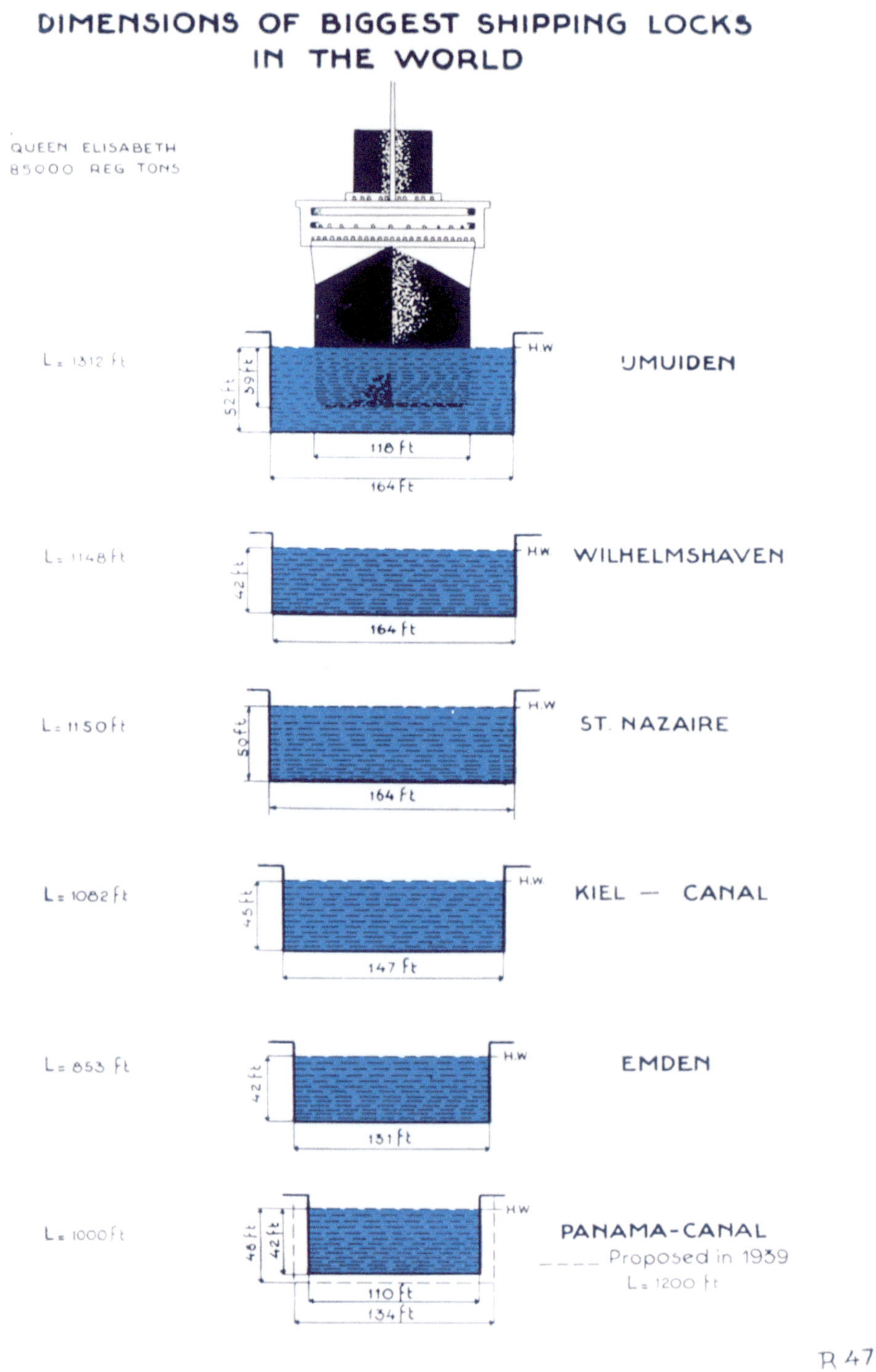

Fig. 12.1 Dimensions of biggest shipping locks in the world. National proud: the IJmuiden lock in Noordzeekanaal, connecting Amsterdam with the North Sea, is the largest. (Source: Van Veen 1948)

against the water as the symbol of the Dutch identity may be considered an 'invented tradition' of the nineteenth century (Knippenberg 1997) but this struggle is also a reality as exemplified for instance by the proclamation of seven (out of nine) UNESCO World heritage sites dealing with water works (see Chapter 2; Van Gorp and Renes 2003).

Water is an essential part of the cultural geography of the Netherlands. In this chapter we discuss the Dutch cultural geography from an international perspective; in what ways is the Dutch culture different from that in other (in particular surrounding) countries, and from a domestic perspective: the cultural diversity within the Netherlands. The chapter will focus particularly on everyday culture.

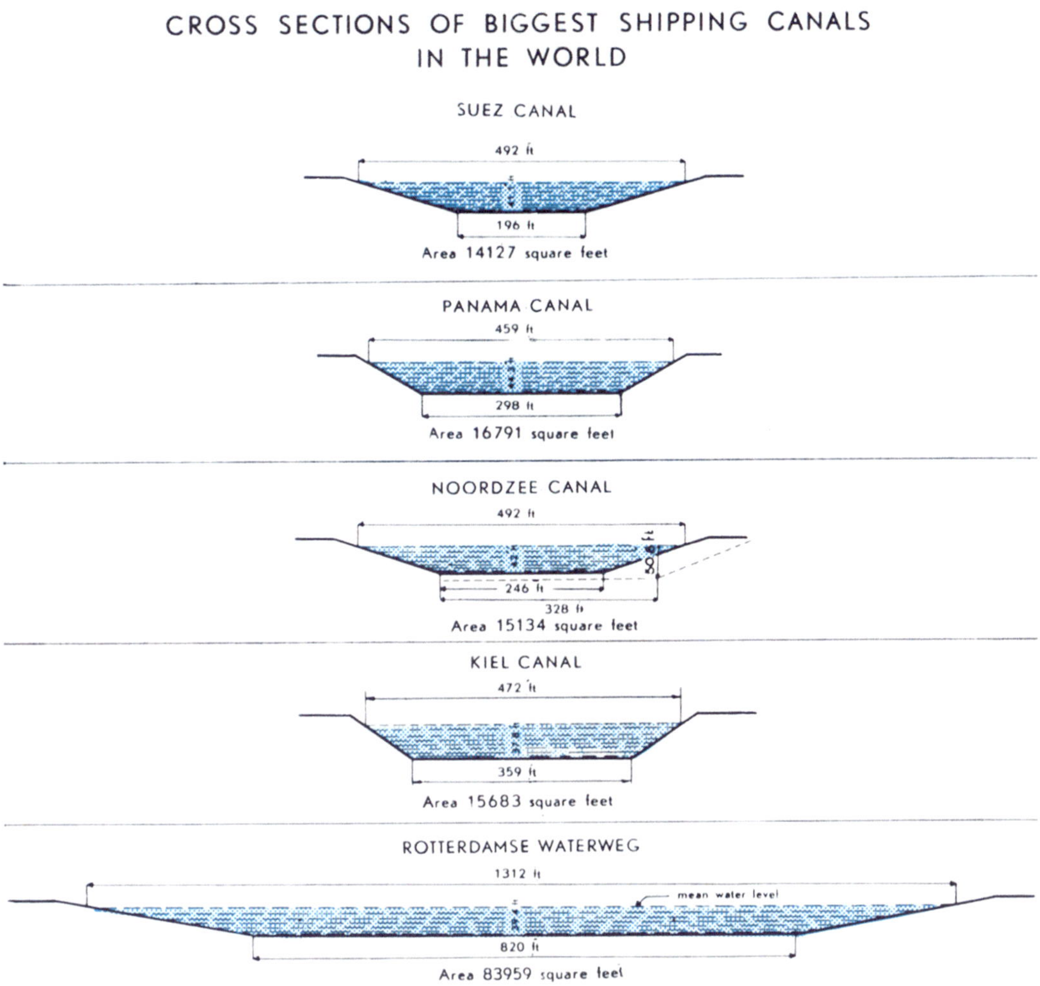

Fig. 12.2 Cross sections of biggest shipping canals in the world. National proud again: 'We constructed about 100 Suez canals of the size made by De Lesseps. All this *was done by hand*, whereas De Lesseps used 60 steam dredgers'. The Waterweg, connecting Rotterdam with the sea, is the biggest shipping canal. (Figure and quote from Van Veen 1948)

Box 12.1: The Dutch and the Art of Painting

The Fine Arts have always been an important issue in the Netherlands. This holds for active performing and implementing art expressions as in art conservation and exposition. Although the Dutch have been rather prominent in a wide variety of arts, Dutch painters are worldwide acknowledged. The earliest Dutch painters belonged to the Flemish School in the 15th and early 16th centuries when Flanders was a rather wealthy and urbanized part of the Netherlands. Painters as Jan van

Box 12.1 (continued)

Eyck (Brugge), Rogier van de Weyden (Brussels) and in particular Jeronimus Bosch (Den Bosch, also called 's Hertogenbosch, Brabant) were quite innovative and produced remarkable and fine examples of early renaissance painting in the Low Countries and beyond.

In the sixteenth and especially in the wealthy seventeenth centuries Dutch painting developed very well and a series of Dutch painters (eventually) received worldwide recognition. Peter Paul Rubens (Antwerp),

(continued)

Fig. 12.3 Winter landscape by Hendrick Avercamp, 1608. In Rijksmuseum, Amsterdam (Fragrent)

Box 12.1 (continued)
Rembrandt van Rijn (Amsterdam), Frans Hals (Haarlem), Jan Steen (Leiden and Haarlem), Johannes Vermeer (Delft) and Hendrick Avercamp (Kampen; Fig. 12.3) are among the best known. Without exception they introduced new styles and in particular Rembrandt displayed a fabulous technique to play with light and shadows. Paintings of landscapes and townscapes became popular, for instance the paintings of Jacob van Ruisdael of the city of Haarlem and its surroundings.

By the end of the nineteenth century, Vincent van Gogh and in the twentieth century Piet Mondriaan and Willem de Kooning belonged to the most prominent Dutch painters; well-known is also the graphic artist M.C. Escher. Some of these Dutch painters were well acknowledged (for instance, Frans Hals) during their live times, others only well after their death (Van Gogh).

The best visited of the Dutch museums are the Rijksmuseum, the Van Gogh Museum, the Stedelijk Museum (Modern Art) and the Hermitage in Amsterdam, the Mauritshuis and the Gemeente Museum in Den Haag, and the Kröller-Muller Museum in the National Park De Hoge Veluwe. All 1100 museums in the Netherlands together attracted some 38 million visitors in 2015. (EDM)

12.1 Dutch Culture, an International Perspective

The Dutch have a 'delta culture' as determined by its position at the mouths of two major European rivers. This delta culture has given a specific flavour to both economy and society. Sociologist Anton Zijderveld (1999) compared 'island cultures' (such as the British and Japanese) with 'delta cultures' as the Dutch. Island cultures are characterised by an internal orientation: 'their own history is often mythologized and their own traditions and institutions are more positively and stronger perceived than those on the continents'. According to Zijderveld, islanders are more conservative, chauvinistic and more often distrust (influences of) foreign cultures. Such island cultures have a tendency to maintain class distinction and internal inequality. Countries on a delta coast generally have quite different cultures. Delta cultures display a high degree of openness and an external orientation.

While social cohesion is fed by common traditions and institutions in island cultures, social cohesion in delta cultures is more often based on mutual needs rather than on traditions as their geographic position and openness to the outside world is not without risk in delta regions. Delta's may rather easily be seized by foreign powers. To jointly cope with such threats successfully, delta cultures should preferably avoid fierce internal conflicts and rather display some mutual tolerance – in daily life but also in intellectual circles (Box 12.2).

Box 12.2: Two Leading Dutch Philosophers

The Netherlands produced two internationally acknowledged philosophers and humanists,

Desiderius Erasmus was born out of a wedlock, in 1466, in Rotterdam. His father was a priest and he became a priest as well, in 1492, after his education at the Latin schools in Deventer where he learned Greek, and in 's-Hertogenbosch. As a priest he could dedicate much of his time to study as he got dispensation from most clerical duties. Supported by his bishop, Erasmus studied theology in Paris. He went to England where he met King Henry VIII's son and important humanists as Thomas More ('Utopia'). Back in Paris, Erasmus wrote his first book in 1500 about sayings, a bestseller in the very early times of the art of printing. Upon his return from a journey to Italy he wrote his famous book Praise of Fooly (*Moriae encomium, sive Stultitiae laus*), a satire on human ambitions, their extreme short-mindness and their mutual negative judgements. Back in Holland, Erasmus was appointed Councillor of Emperor Carl V. He lived in the southern Netherlands from 1516 to 1521. The last years of his life he stayed in Freiburg, Germany. He died in Basle, Switzerland, in 1536.

Erasmus was an early humanist and a strong advocate for personal freedom and peace. Erasmus fiercely opposed nationalism in particular among Christian people. Maarten Luther was a contemporary priest and although Erasmus sympathised with many of his anti-church ideas, the peace-loving and tolerant humanist Erasmus opposed Luther's rebellious actions. Erasmus argued that the Roman Catholic Church should remain united but rather changed. He remained a Roman Catholic throughout his long life while criticising the Church' bad manners. His numerous pleads for tolerance and mutual understanding could not prevent ongoing religious dogmatism, anti-Semitism and intolerant practises. Nevertheless, his ideas fructified in the later Dutch Republic of the Seven Provinces, where freedom of religion became institutionalised and where critics from many nations, both Christians and Jews, found a safe haven (Fig. 12.4).

More than one century later, *Baruch Spinoza* was born in Amsterdam in 1632 from Portuguese Jewish refugees. He was raised in the Portuguese Jewish community in Amsterdam. He developed highly controversial ideas regarding the authenticity of the Hebrew Bible and the nature of the Divine. The Jewish religious authorities banned (Jewish: *cherem*) him from Jewish society at age 23. However, the Dutch society as a

Box 12.2 (continued)

whole was more tolerant on religious issues and Spinoza continued his life outside Amsterdam where he became a grinder of fine lenses. But most of his time he spent as a philosopher and writer. His magnum opus, the posthumous 'Ethica' in which he opposed Descartes' mind–body dualism, has earned him recognition as one of Western philosophy's most important thinkers. 'Ethica' is Spinoza's last indisputable Latin masterpiece and one in which refined conceptions of medieval philosophy are finally turned against themselves and destroyed entirely. Hegel said: 'You are either a Spinozist or not a philosopher at all.' Baruch Spinoza may be considered one of the great rationalists of seventeenth-century philosophy. He was a very modest man and turned down rewards and honours throughout his life, including prestigious teaching positions. He died at the age of 44, in 1677 and is buried in the churchyard of the Christian Nieuwe Kerk in Den Haag.

Both philosophers were able to spread their rationalistic ideas and views in times when Holland developing as a prosperous and politically powerful regional body (Erasmus) or as an successful independent nation (Spinoza). Both philosophers were extremely critical about non-rational issues in their own religions, Spinoza also on basic conceptual issues of the Jewish religion. But public authorities in the Netherlands were quite tolerant and allowed freedom of religion explaining the large numbers of refugees who fled their home countries for religious reasons. (EDM)

In the twentieth century internal conflicts were normally neutralised by the mechanism of denominational segregation ('pillars'; see later in this Chapter). Much older than the 'pillarization' is the culture of mutual consultation and consensus between parties. This is known as the 'polder model', that, according to Zijderveld actually is a 'delta model'. During the Dutch 'Golden Age' (seventeenth century) many different groups of people were allowed to take part in discussions as equal partners (Frijhoff and Spies 1999). Meetings and consults, commissions and participations still characterize the Dutch professional culture. Belief that conflicts can be solved by consultation rather than by fight is wide spread in the Dutch population, also today. The Dutch identify themselves as a combination of merchants and priests rather than of generals and statesmen (Zijderveld 1999). Combating another enemy, the often aggressive water, also demands for a joint approach. The oldest Dutch autonomous and decentralised executive councils have been the water boards. Some of these date back even prior to the feudal ages (Box 2.3). Zijderveld consider

Fig. 12.4 Left: Portrait of Desiderius Erasmus (1466–1536), painted by his friend Hans the younger Holbein in 1523; Right: Portrait of Baruch Spinoza (1632–1677), painted around 1665

such boards to constitute the fundaments of Dutch democracy as they refrain from any kind of centralised power and promote participant's equality.

Although tolerance is considered part of the delta culture, Dutch tolerance should, however, not be overestimated. As in many nations these days there is a growing tendency towards intolerance viz-a-viz foreigners as excessive public tolerance towards foreign cultures may have partly eroded the Dutch identity. This, in turn, generated a growing concern among the Dutch people that their familiar identities may have been sacrificed. Such feelings fuel populism as reflected in democratic voting behavior in recent Dutch elections (Sect. 12.5).

12.1.1 Dutch Culture as a Northern European Culture

Except in essayistic and historical studies, Dutch identity has also been studied by means of systematic empirical research. A comprehensive collection of quantitative data of cultural differences between nations had been gathered by the Dutch researcher Geert Hofstede. As an IBM researcher Hofstede collected data on cultural differences in IBM staffs in over 40 countries worldwide. In *Culture's Consequences* (1980) he initially distinguished four cultural dimensions, two more added later. These cultural dimensions were (www.geert-hofstede.com):

1. High inequality versus low inequality cultures (power-distance: the (non)acceptance of an unequal distribution of power within society);
2. Uncertainty avoiding versus tolerant cultures (do cultures want to control the – unknown – future?);
3. Individualistic versus collectivistic cultures (priority to the individual/family or to the group?);
4. Masculine versus feminine cultures (competing with or caring for others?);
5. Long-term versus short-term orientation cultures (oriented on the own past or on the challenges of the present and future?),
6. indulgence versus restraint cultures (weak or strong control of personal desires and impulses).

Hofstede's analysis demonstrates that the Dutch in general reject inequality and can cope with uncertainty relatively well. Moreover, the Dutch prefer individualism over collectivism. These characteristics roughly comply with the characteristics of English speaking and Scandinavian countries. But the Dutch dislike masculinism as a cultural dimension and rather prefer feminism – helping the weak, instead of applauding the strong. This again complies with Scandinavian cultures but is quite different from the rather masculine oriented English speaking nations. By means of cluster analysis Hofstede ranks the Netherlands in the cluster of the Nordic Culture, along with Denmark, Finland, Norway and Sweden. Main characteristics of this culture are low power distance, low to medium uncertainty avoidance, medium to high individualism and low masculinism.

Neighbouring nations have a different culture and are part of other clusters. Germany, along with Austria and Switzerland, belong to the Germanic Culture, Great Britain along with other English-speaking nations to an Anglo

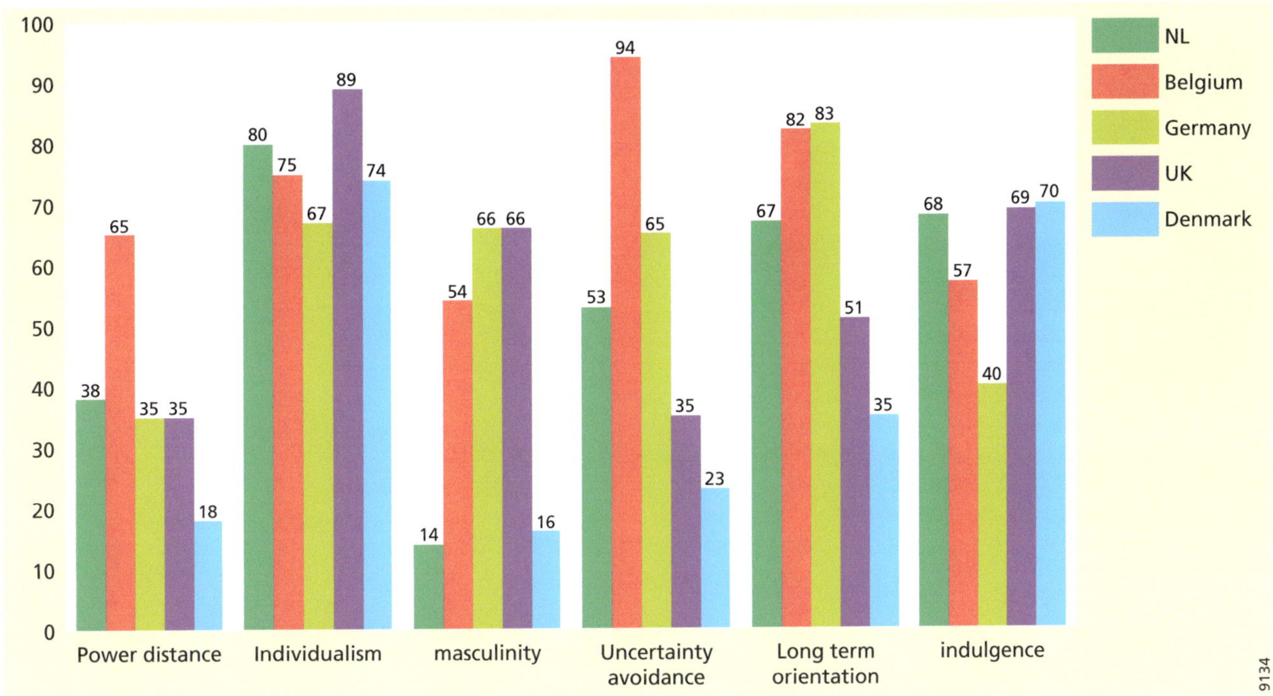

Fig. 12.5 Dutch culture in comparison of the cultures in neighbouring nations, according to Geert Hofstede. (Source: www.geert-hofstede.com)

Culture, and Belgium and France to the Latin culture (Fig. 12.5). In contrast to the Dutch, the Germans have, at least around 1970 (when Hofstede conducted this inquiry), problems to face uncertainty, are much stronger masculine and less individualistic. The Belgians and the French are much more hierarchically oriented, rather not live with uncertainty, and more masculine than the Dutch. The main difference between the British and the Dutch cultures is that, according to Hofstede, the Dutch care for weaker members of the society while the British rather admire its strong members.

Another approach to measure the 'cultural distance' of the Netherlands vis-a-vis other European nations was made by Kogut and Singh (1988) who examine the cultural relations with countries with respect to the United States of America and their foreign investment in the US. Researchers from the Dutch Social and Cultural Planning Agency (SCP) used the Kogut-Singh formula to calculate cultural distances between EU countries. Figure 12.6 shows the cultural distances between the Netherlands and 17 other nations; the scores of the countries on four of the Hofstede dimensions were used as input. The cultural distance between the Netherlands and the Nordic countries (measured by Hofstede's dimensions) is small, between the Netherlands and South- and East-European countries are comparatively large. Cultural differences between the Netherlands and its neighbouring countries is highest with respect to Belgium.

Cultural distances among inhabitants of nations may also be measured on basis of mutual trust (Fig. 12.7). SCP studied this parameter using primary data from *Eurobarometer*, a EUROSTAT (Statistical Office of the European Union) database used to determine opinions of EU citizens. An 1996 inquiry revealed that the Dutch trust Danes, Finns, Luxembourgers and Swedes most, but also the Belgians. Spaniards, Greeks and Italians were trusted least.

12.2 National: Regional Differences in Dutch Culture

Regrettably, Hofstede did not investigate cultural differences between regions within countries. Therefore, no evidence exists weather or not state borders coincide with cultural borders. The eastern and southern borders of the Netherlands appear rather diffuse and incidental, in contrast to the land-sea boundaries in the west and north. According to Dutch geographer Hans Knippenberg (1997), Dutch territorial borders resulted from military actions rather than from cultural or ethnic reasons. The Netherlands gained its present borders in 1839, when King William I (reluctantly) approved separation of Belgium from the Netherlands. By then, eastern Dutch and western German dialects were merged. Since, this Dutch-German state border developed into a language border as school children from both nations learned to speak the standard (=non dialect) languages. The Dutch standard has been set for the language spoken in the city of Haarlem, in the west and spread over the country over time (Box 12.3).

Fig. 12.6 The culture distance between the Netherlands and 16 other European countries and the United States of America. (Source: Dekker et al. 2007)

Fig. 12.7 Confidence of the Dutch in other peoples. (Source: Dekker et al. 2007, originally L. Guiso et al.)

Less obvious is the cultural impact of the southern border of the Netherlands. The main divide between the Nordic/Germanic and the Latin/Roman cultures is situated further south and coincides with the boundary between the northern, Dutch speaking Flanders and French speaking Wallonia in the south of Belgium. Another cultural border between people in the north and in the south may be found in the hearth of the Dutch delta, where both rivers Rijn and Maas flow westward. In Dutch, this boundary is typified as 'Moerdijk', southeast of the city of Rotterdam where a bridge crosses the *Hollands Diep*. There, Holland ends and the province of Noord-Brabant begins. The provinces of Noord-Brabant, Limburg and Flanders may be considered as a cultural transition zone between the typical Nordic/Germanic and Latin/Roman cultures.

The Dutch River belt ('Moerdijk') divides the Netherlands also in another aspect. This zone separates realms where most people traditionally adhere the Roman Catholic religion in the south from the predominantly Protestant realms in the north (mainly mighty Holland). This divide is the traditionally most stereotyped and vividly perceived regional cultural border zone in the Netherlands and exceeds religious differences which today have been diluted due to massive secularisation since the 1970s. Apart from traditional religious contrasts both regions may vary in lifestyles, in family life and leisure pursuits.

Protestants have seen themselves as the true representatives of the Dutch nation. The Netherlands would essentially be a Protestant nation, born at the end of the sixteenth century in the Dutch Revolt against the Roman Catholic Spaniards (King Philip II). The rebellion leader against Spain was the Protestant House of Orange. For a long time, Roman Catholics (and Socialists) have been greatly excluded from formal positions in the Dutch nation. A Roman Catholic leader of the House of Orange would still be unimaginable. However, the King's wife is Roman Catholic. But King Willem-Alexander's and Queen Máxima's three daughters are (liberal) protestants (*Nederlands Hervormd*).

It lasted until WWII that Roman Catholics and socialists were recognised as real patriots thanks to their resistance against the common enemy, the Germans. By then, also the House of Orange received more recognition as the national binder of the compartmentalised Dutch society.

Box 12.3: Language: Expansion of Standard Dutch and the Frisian Movement

Around 1600, the cradle of Standard Dutch language was in the region of Amsterdam-Haarlem. The language that was spoken in this core area of the Dutch Golden Age spread over surrounding areas during the next centuries, at the expense of regional languages and local dialects that there were (and often still are) used. The further away (geographically) from Amsterdam-Haarlem, the less pronounced was Standard Dutch's influence in the local language and the more the originally spoken language managed to survive.

Geographic diffusion of Standard Dutch over the entire Dutch territory has been studied by G.G. Kloeke in his thesis *De Hollandsche Expansie* (1927). He explained its victory by the choice made by individuals in peripheral regions. They (or their children) aimed at rising on the social ladder. A boundary condition for this was the ability to speak a language that may serve to communicate beyond their home region. But choosing for Standard Dutch was also a bit enforced by the national Government. In French times (around 1800) it was decided to have one single (written) language: 'one nation, one people, one language' was the adage of the new Kingdom of the Netherlands. Early in the nineteenth century, teachers were ordered to teach school children in Standard Dutch only and dialects were banned from school. In

Box 12.3 (continued)

the twentieth century, the media (newspapers, radio, television), national associations, conscription, the new means of transport, et cetera, all ensured a further advance of Standard Dutch (De Pater 1993).

The sole language in the Netherlands which is not a variant of Dutch, is Frisian. Frisian is spoken in the province of Friesland, particularly in its rural areas. In the nineteenth century, Frisian was not allowed in schools or in other public institutions (Court, City Hall). That generated significant resistance, and emergence of the *Fryske Beweging* ('Frisian movement'). In 1962, the *Fryske Nasjonale Partij* (Frisian National Party, FNP) was founded. Its ambition was to collect more political autonomy for Friesland. In provincial elections, FNP gained about 5% of the votes in the twentieth century (Penrose 1990), since the onset of the twenty-first century, about 10%. FNP has been active in municipal elections as well.

Since the 1960s, suppression of the Frisian language halted. Since a few decades, Frisian is taught at school, names of towns and cities now put their Frisian names first follow by its equivalent in Dutch (e.g. 'Snits' for Sneek). Legally, Frisian has been proclaimed as the second official language in the province of Friesland (Fryslân in Frisian). But, fewer and fewer people speak Frisian as their first language, Standard Dutch is gaining ground. One may assume that roughly half of the people (some 300,000) in Friesland speak Frisian as their mother tongue, mostly in rural areas.

Dutch writer and journalist Sietse van der Hoek (2014) observed another cultural divide in the Netherlands. He also describes two distinctly different regions in this small country, but rather along another geographical delineation, roughly along the topographical zero line (0 m – MSL, Fig. 2.1). Northwest of this line is the low-laying part of the Netherlands which would be flooded by the sea if all coastal defense systems would fail. The higher and dryer landscapes are in the southeast (Chap. 3). People in this last region are considered to be more conservative and populistic, more inward-oriented, while people in the northwest considered to be more mobile, tolerant, progressive, individualistic, egalitarian and democratic, rather living in a more open society.

Another cultural distinction among Dutch people is between citizens (in particular in Randstad Holland; Chap. 8) and dwellers of rural municipalities. Castenmiller and Knol (1989) classified Dutch municipalities into four categories: 1. municipalities in Randstad Holland region with >50,000; 2. Randstad Holland municipalities with

<50,000 people; 3. beyond Randstad Holland municipalities with >50,000; and 4. municipalities beyond Randstad Holland with <50.000 inhabitants. Largest differences were measured between people of the first and fourth categories while those in categories 2 and 3 were intermediate. But, upon correction for age, marital status, education and religion, differences between the four categories proved to be rather moderate overall. Inhabitants of large municipalities in the Randstad region were relatively less nationalistic, less authoritarian, and in favor of more political and social freedom and emancipation. This was confirmed by Brons (2006) who described Randstad Holland dwellers as more individualistic than people beyond, too. There, individual interest normally overrules collective interest, marriage and childbearing are being postponed, and liberal parties are more popular in elections. Statistically, Randstad Holland individualism appears to be more related to overrepresentation of higher education and higher incomes then with Randstad Holland as a region by itself.

Despite such relatively small statistical differences between (large) urban and rural areas, perceived differences in culture and lifestyle are rather large. Dutch historian Geert Mak (1994) describes major social and cultural changes in a small agricultural village, Jorwerd (Province of Friesland). Despite all modernisation processes that have hit the countryside over the past century, by the end of the twentieth century a deep cultural gap still exists in the Netherlands. He observes 'a fault line between city dwellers and farmers, between consumers and survivors' and that the differences between Jorwerd and Amsterdam 'were considerably larger than between Jorwerd and any village in England or Germany. Farmers have been part of a small community, adhered to their autonomy as independent and self-employed persons together with their family members working on the farm and thus frequently being in each other's geographical proximity while the farm's continuity was much more important than profit maximisation. According to Mak (1994) all cultural factors are still quite different in a village than in a city.

Major differences between people exist even at the lowest scale levels (between city districts and between villages). Old districts in Amsterdam (e.g. *de Jordaan*) and other cities used to have their own distinct folk culture. Today, original *Jordaan* people moved to the new, nearby city of Almere and their old houses were bought by Young Urban Professionals while introducing an entirely new culture in this district (the process of gentrification). Cultural differences occurred even at the level of individual streets, as has been e.g. the case in Den Haag's *Schilderswijk* (Klein Kranenburg 2013). However, original *Schilderswijk* working class people moved outside Den Haag and today this district is predominantly inhabited by migrants with their associated cultures (Box 8.4). Schilderswijk and other lower class neighbourhoods were built around 1900 in the low lying, peat and clay soils of Den Haag. The rich high-class neighbourhoods (with numerous embassy's, international organisations, and houses of expats) can be found on the high and dry sand ridges. The difference between fen and sand, poor and rich, corresponds with a difference between cultures (Fig. 12.8).

Villages close to each other may have quite different cultures. Often this is based on differences in religion which may lead to rivalry and mutual stereotyping, in the Netherlands primarily the Roman Catholics versus the Protestants. In turn, the Protestants were split in two groups: the orthodox *Gereformeerden*, and the more liberal *Nederlands Hervormden*. Significant tensions existed (and still exist) between the Roman Catholic village of Volendam and the nearby, small Protestant town of Edam. But even between villages with people of the same religious denomination significant cultural differences may exist, e.g. between the strictly orthodox Protestant village of stock farmers Westbroek (north of Utrecht) and the nearby, also but less strict Protestant village of Tienhoven where people are more sociable, extrovert, free and modernized. Such rather striking differences may be explained by different socio-economic histories of both villages. In Tienhoven, surrounded by water-covered relicts of peat digging activities, poor peat diggers, fishermen and laborers lived who daily commuted to industries in the nearby city of Utrecht. In Westbroek peat had been thinner among others making peat digging less attractive. In contrast to Tienhoven, stock farmers in Westbroek were never replaced by peat diggers (Eysberg and De Pater 1993).

12.3 Four Pillars: Catholic, Protestant, Socialist, Liberal

As diffuse regional cultural differences may be, so obvious are religious-cultural differences. Until the 1970's, the Netherlands has been a compartmentalized society. Each compartment (pillar; in Dutch: *zuil*) had been characterised by its own culture. Statistically, however, not always clear distinctions between pillar-driven and region-driven cultural influences can be made. For example, many more Roman Catholic people live in the south of the Netherlands than north of the River belt and it may therefore be questioned if their ways of life and values are driven by religion or by region.

The Netherlands is a society of classes and standings like in any other nation. Rich and poor, low- and high-skilled, from a simple origin or from circles of the *haute bourgeoisie* or nobility: together they form the Dutch society. Moreover, the Dutch society has been a building of pillars. These pillars originated from emancipation movements, first of groups of

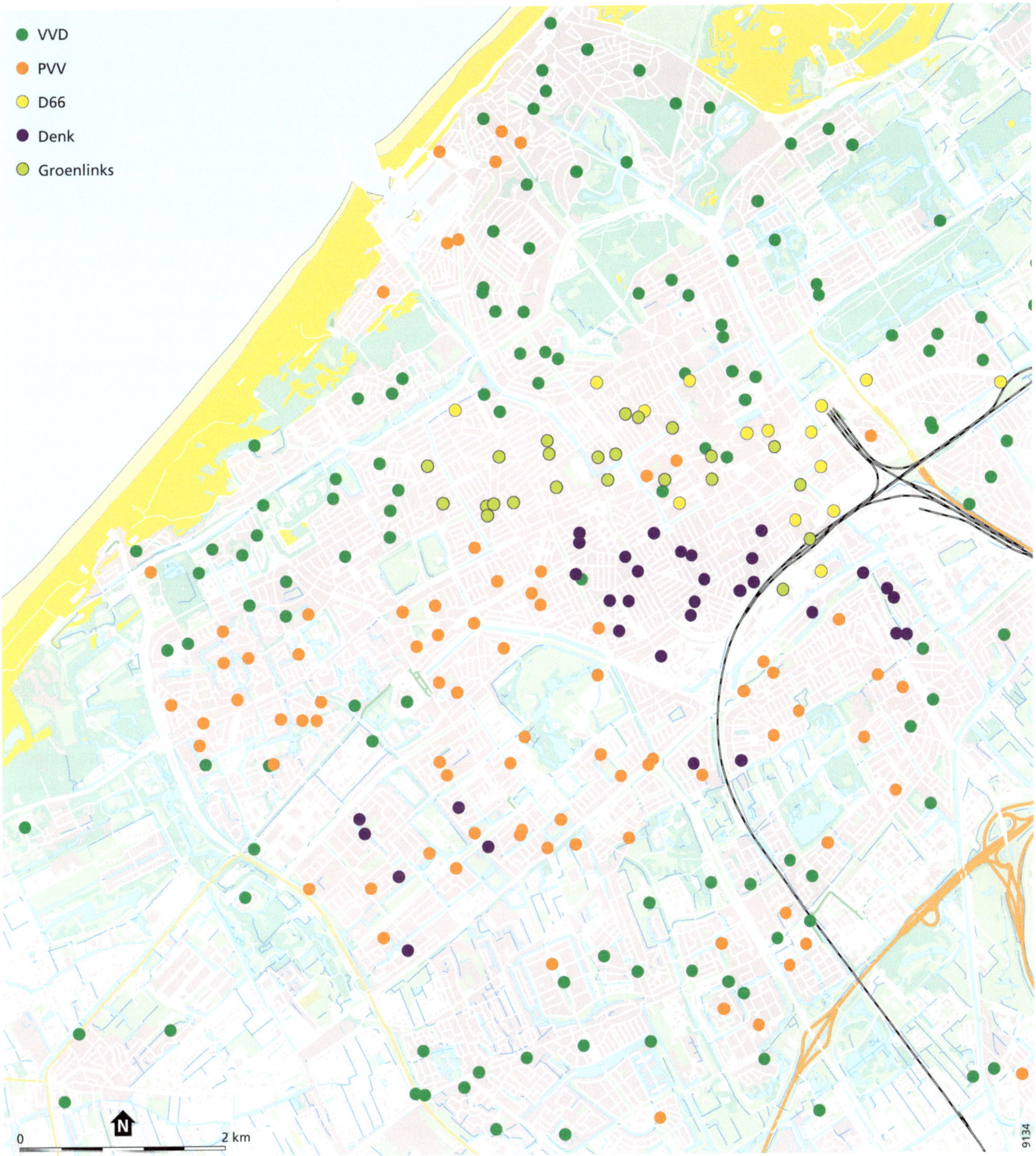

Fig. 12.8 Spatial differences in political preferences in Den Haag, in the 2017 national elections: largest party per polling-station. In the city as a whole, the liberal VVD is most popular, as in most municipalities (Fig. 12.14). But within Den Haag strong contrasts exists between the relatively rich districts on sands (VVD), Groen-Links (a 'green' party) in gentrified urban districts, DENK (the 'Turkish party for Erdogan-supporters') in poor immigrant areas (e.g. Schilderswijk) and populist party PVV in relatively poor Dutch neighbourhoods. (Source: Stichting Politieke Academie/Joost Smits & Frank van Dalen; nrc.nl/mijnstembureau)

other religious denominations, later of the working classes (see for an opposite development the case of the Jews, Box 12.4). The dominant three different religious groups that occurred at the end of the nineteenth century were the Roman Catholics, the orthodox Protestants (*Gereformeerd*) and the more liberal Protestants (*Nederlands Hervormd*). Often, the orthodox and liberal Protestant are taken together into one pillar: Protestant although there are significant internal disparities within Protestantism: from quite liberal to severely orthodox. Prior to the nineteenth century the liberal Protestant group dominated Dutch society and *Nederlands Hervormd* had been the State Religion. Initially, the orthodox Protestants had severe difficulties in getting influential positions in the liberal Protestant landscape and formed the first pillar. Roman Catholics were repressed initially and had to fight for their emancipation during the nineteenth and early twentieth centuries. Internally, Roman Catholics were much less diversified than the Protestants.

Box 12.4: The Jews in the Netherlands: History in a Nutshell

In the Dutch 'Golden Age' the Republic became a safe haven for Jews. The first Jews came from Spain and Portugal by the end of the sixteenth century. Since the 1630's, many Jews left Central and Eastern Europe. They settled mainly in Amsterdam where both groups built their synagogues, some 100 m apart from each other: the Portuguese and the German Synagogues. Both seventeenth-century buildings still exist at the Jonas Daniël Meijerplein (the enlarged German Synagogue – in Dutch: '*Grote Synagoge*'- hosting the Jewish Historical Museum today).

During Napoleon's French occupation (around 1800) Jews were given the legal status of citizens and assimilation began. A century later, many Jews primarily were no longer members of a Jewish nation, but rather Dutch who were members of the Israeli Church. Orthodox-Protestants and Roman Catholics emancipated in the nineteenth century as well, but rather by constructing their own respective pillars. In contrast, Jews escaped from their old pillar (Knippenberg 1992). Incidentally, the Jewish integration movement was not always welcomed; in the Netherlands mild or virulent anti-Semitism existed, e.g. in some children's books published by Christian publishing houses (Sanders 2017).

Around 1900, about 100,000 Jews lived the Netherlands, 2% of the total Dutch population. About 80% of them lived in Amsterdam, in particular districts near their synagogues where more than 75% was Jewish (Knippenberg 1992). The 'ghetto' was poor and

Box 12.4 (continued)

densely populated, theirs houses in bad conditions and traditional Jewish culture dominant. Many Jews worked in 'typically Jewish' professions, as the diamond industry. Successful, integrated and mostly secularised Jews often moved to new homes in Amsterdam-Zuid where Anne Frank lived as well before her family went for hiding.

In the 1930s, ten thousands of Jewish refugees from Germany arrived in the Netherlands, most of them migrated to English-speaking nations; some 15,000 of them were still in the Netherlands at the onset of WWII, in May 1940. Total number of Jews in the Netherlands in the first years of WW II was about 140,000, 110,000 of them were deported by the Germans, 104,000 of them were killed. Percentages of killed Dutch Jews in WWII were higher than in many other western European nations. Some of those surviving WWII settled in Israel in the 1950s. The 1960 Dutch census revealed 14,504 Jews living in the Netherlands (Knippenberg 1992). Today, there are some ten thousands.

These pillars became institutionalised since the last decades of the nineteenth century. Such pillars existed of the respective churches (if applicable), of a wide range of other societal organisations, as political parties, trade unions, newspapers, magazines, schools, hospitals, housing associations, leisure clubs, broadcasting societies (after WWI), sanatoriums, holiday homes, training centres, et cetera. That gave rise to the vertical societal pillar structure in the Netherlands. People of different classes and standings grouped together under a single, often religious banner. 'Ordinary' people stayed within their own pillar: Roman Catholic people only purchased their goods in shops of Roman Catholic owners, they only read Roman Catholic newspapers and magazines, they only voted for a Roman Catholic party during elections, they were members of a Roman Catholic Radio association and Trade union, they played football in a Roman Catholic football club or were a member of a Roman Catholic women's organisation. Contacts between members of different pillars hardly existed. Ninety percent of the marriages were between members of the same pillar (Lijphart 1968).

Apart from the two religion-based pillars (Catholics and Protestants), two more but non-religious pillars were added around 1900: the Socialist and the Liberal pillars. Both pillars, especially the Socialist, had their own systems of institutions. The Liberal pillar was less institutionalised and self-contained. In contrast to the Socialist pillar, the Liberal pillar attracted mostly people with higher incomes and standings. Decades after pillar decline, since the 1970s immigrants from Turkey and Morocco developed their own institutions (Box 12.5).

Pillars ruled themselves; 'sovereignty in own circle' was the device. Problems were resolved as much as possible internally by the pillar's authoritarian elite. This elite comprised a group of gentlemen (no ladies) who often combined a variety of related positions. Issues of national extent were discussed (and often agreed) collectively by the four pillars' elites, via their political parties, in the parliament and government. These elites in fact ruled the nation, aiming for pragmatic cooperation, accepting compromises and the agreement to disagree: 'the pragmatic acceptance of the ideological differences as basic realities which cannot and should not be changed. Fundamental convictions of other pillars should be tolerated at least, if not respected', as the political scientist Arend Lijphart (1968; this section is based on his book) concluded. It is 'the spirit of accommodation', also described by Zijderveld (1999; see Sect. 12.1), who explained this by referring to the geographical situation: the so-called delta-culture. Once taken, decisions were accepted by the rank and file in each pillar easily, for there was a high confidence that the 'gentlemen' were the best informed people.

Social modernisation arrived in the Netherlands in the capacity of institutionalisation of the four pillars, of which by far most people were members. It was not very unique; comparable processes of pillarisation took place in some other countries, for instance in Belgium, Austria and Germany (Hellemans 1993). In addition to these form of modernisation, the Netherlands underwent a general, Western-European, type of modernisation as well. With increasing prosperity, a materialistic mass culture gradually developed around 1900, first and foremost in (big) cities. This culture manifested itself in cinemas, theatres, shops, swimming pools, museums, seaside resorts, stadiums, restaurants, but also in 'mass dwellings' in the new neighborhoods in the outskirts of cities (Van der Woud 2015).

Box 12.5: Integration in Dutch Culture?

Since about 1600, numerous immigrants settled in Netherlands (see Chap. 11). This generated some cultural diversity, often temporary as most of them eventually integrated in the Dutch society. That happened e.g. with the Flemish people who came to the Netherlands after the conquest of Antwerp by the Spaniards (1585) and with the Huguenots (French Protestants) who escaped from France upon the revocation of the Edict of Nantes in 1685. Other groups, such as the Jews from Eastern Europe, Spain and

Box 12.5 (continued)

Portugal retained more their own culture, especially in Amsterdam.

More recently, Surinamese and Antilleans from former Dutch colonies in the Caribbean arrived in the Netherlands. Their languages are Dutch and most of them feel quite at home in the Netherlands. Surinamese Hindustanis (in Den Haag) and Surinamese Creoles (in Amsterdam) are seen by the authorities as 'model minorities'.

Turkish and Moroccan people immigrated since the late 1960s. Dutch industry was in demand of cheap labor and these people were welcomed as 'guest workers'. Unexpectedly, they did not return to their home countries over time but were given the opportunity to bring their families to the Netherlands as well ('family reunification'). Most of the Turkish and Moroccan people are Islamic. In 1960, 1400 Muslims lived in the Netherlands, now about 1 million – from orthodox to liberal.

During the past four decades, Turkish and Moroccan Muslims built their own pillars, often with support by the Dutch government. These comprised private schools, associations, and religious institutions. Dutch politicians in general are quite concerned about their poor integration in the Dutch society. Recent Islam-driven attacks in Europe significantly reinforced such concerns.

The Social and Cultural Planning Agency (SCP) investigated the level of integration of Turkish, Moroccan, Surinamese and Antillean people in the Netherlands (Huijnk et al. 2015). They were asked, in 2011, in interviews at home, about their social contacts, norms and values, and their emotional identification. Table 12.1 shows that Surinamese and Antilleans are much better integrated than Turks and Moroccans. However, variation within each group was large. It is mainly the elderly, first generation immigrants, less educated and those with strong religious orientation who stay far from the mainstream Dutch culture. They remain among themselves, a comparable situation as once happened in times of the traditional pillarized Dutch society. Socially segregated people feel less healthy and less wealthy, they believe that Dutch women are too free, show little interest in politics and have little confidence in others. Additional qualitative research makes clear that the differences are hard to bridge. Native Dutch people feel threatened by young, not infrequently criminal, Moroccans and Turks, while conversely many Moroccans and Turks feel excluded and discriminated. They live in 'parallel worlds', as Dutch people did in their pillars until the late 1960s.

Table 12.1 Socio-cultural categories in relation to ethnic origin, in 2011

Ethnic group	Social segregated (%)	Mix of segregation and integration (%)	Social integrated (%)
Turkish	24	64	12
Moroccan	22	65	13
Surinamese	8	49	44
Dutch Antilles	10	54	37

Source: Huijnk et al. (2015); (the 7 categories in the original report are taken together to three here)

12.3.1 Spatial Patterns of the Pillars

Each pillar had its own geographical pattern of members and followers (Fig. 12.9). Traditional contagious Roman Catholic populations were (and are) in the south, in the provinces of Noord-Brabant and Limburg. This pattern may be explained as these areas belonged to the Spanish Habsburg empire in the early seventeenth century. The border between Protestants and Roman Catholics coincides quite well with the frontline during the Twelve Year's Truce (1609–1621), as situated between the liberated part of the Netherlands which converted to Protestantism and the Roman Catholic parts occupied by the Spanish Habsburgs. Beyond both provinces were (and are) more or less extensive enclaves of Roman Catholics in rural areas of west Netherlands and in the east, in the provinces of Overijssel (especially Twente) and Gelderland (Achterhoek).

In the Protestant part of the Netherlands, orthodox and liberal Protestant regions are distinguished. No satisfactory general explanation for this spatial pattern exist (Knippenberg et al. 1989; Knippenberg 1992). Orthodox Protestants occur in a belt running from the southwest islands and holms of Zeeland and Zuid Holland via the peat areas of Holland and Utrecht, Gelderse Vallei, NW Veluwe, and NW Overijssel to the NE parts of Friesland and Groningen. The orthodox Protestant regions are now known as 'Bible belt'. Liberal Protestants, secularized people (members of the liberal pillar) and socialists mainly occur in the large cities in the west (now known as Randstad Holland) and in the north.

In particular SE Friesland (home region of Ferdinand Domela Nieuwenhuis, the founder of the socialist movement in the Netherlands), Drenthe and Groningen were (and still are) in the north an extensive regional stronghold of the socialists. In these relatively poor rural areas non-denominationalism has been larger than in the urban areas in the west. The socialists were relatively strong in big cities and industrial areas as well, as Zaanstreek. The Liberal pillar (well to do secular people and part of the liberal protestants) found its strongholds in urban areas too.

12.4 After the 1960s: De-pillarisation and Secularisation

Denominational segregation was at its peak between the onset of the twentieth century, when the Netherlands rapidly modernised (and thus built pillars), and the 1960s, when the pillars started to decline. Causes of secularisation and de-pillarisation include the increased prosperity and geographical mobility (thanks to the purchase of private cars), the introduction and spread of television (confrontation with other cultures), access to higher levels of education, and diminishing traditional authority – the elite of the own pillar and its local representatives such as a priest, pastor or vicar was no longer necessarily obeyed. Fueled by a new time spirit people became more critical, more independent and more individualistic, no longer saw themselves as part of a collective (or: pillar) but rather as an autonomous personality. For example, Roman Catholic parents did no longer obey pressure from the church to have large families. This was facilitated by the introduction of more effective contraceptives giving rise to the Second Demographic Revolution (Chap. 11).

By today and thanks to improved sexual education, the availability of birth control and contraceptives, the number of abortions in the Netherlands is relatively low. Abortion is no longer a hot issue as it was in the 1970s and is now legalised in principle. Women are much more emancipated thanks to feminist movements. Dutch people also tolerate other types than traditional nuclear families. All but outspoken orthodox groups today accept that young couples live together unmarried. Today, 44% of all babies are born out of wedlock, a multiple of what was considered normal until about 1990, and certainly before. The Netherlands was the first country in the world where gay marriages – a commitment between two men or two women – was legalised in 2001. Annually, approximately 1400 gay marriages are contracted, a number that has remained rather constant over the years.

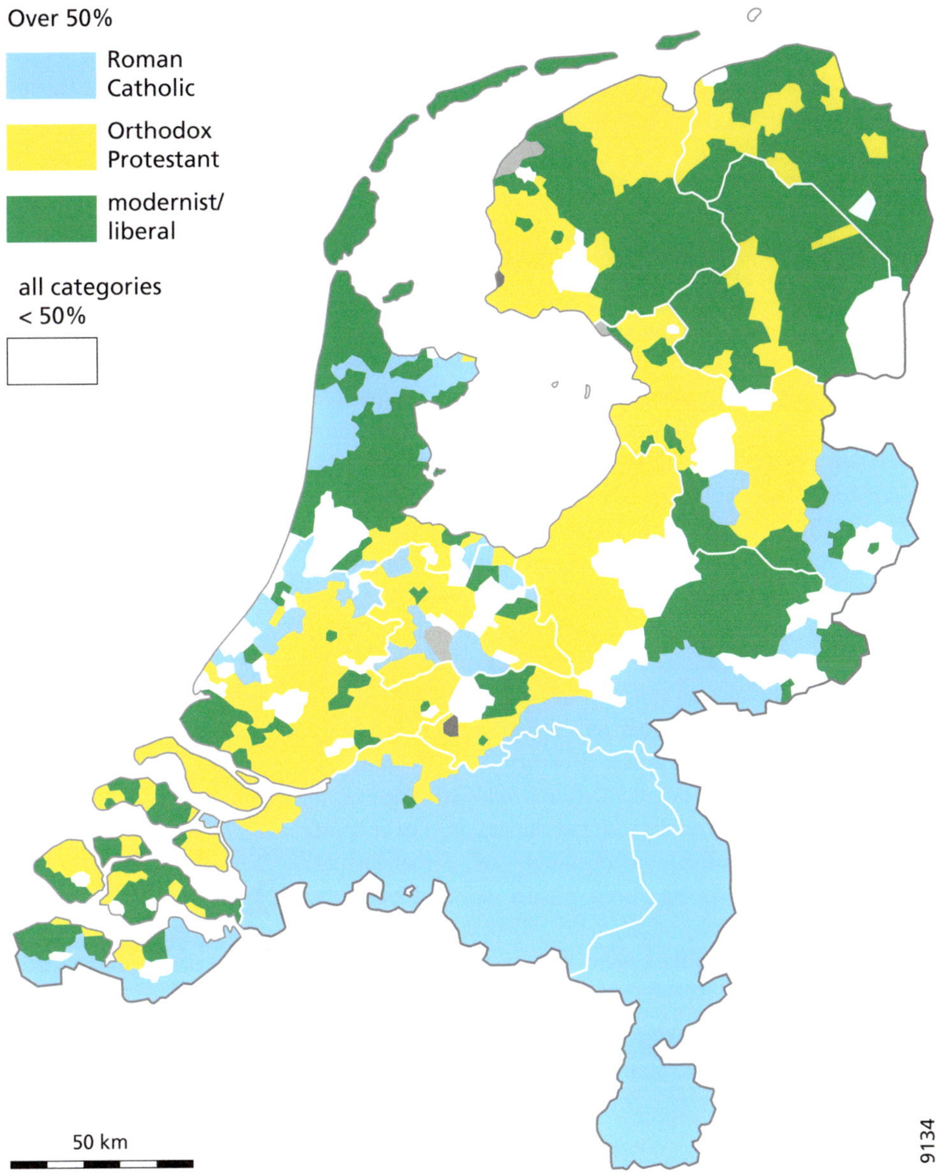

Fig. 12.9 Spatial patterns of religions in the Netherlands, about 1920. (Source: Van Amersfoort and Van der Wusten 1989)

Public debate in the Netherlands today is focused more on other ethical issues as euthanasia. Euthanasia is forbidden by law but is not prosecuted if at explicit request of the patient and done by a medical doctor under strict medical conditions. Currently, stretching these conditions are a matter of debate. Most of the people who die by euthanasia are cancer patients without chances of recovery. Annually, some 5500 cases of euthanasia are reported, which is almost 4% of all deaths. Decisions are made very carefully. The lion's share is carried out by family doctors, after consultations with the patient, the family and external experts.

Another item in public debate concerns legalisation of soft drugs like marihuana. Box 12.6 is dedicated to this issue.

In conjunction with crumbling of the pillars secularisation increased. Table 12.2 shows that in particular liberal Protestants began to abandon their church memberships as early as 1971. Starting from the 1970s, Roman Catholics followed in secularisation and also part of the orthodox

Table 12.2 Population according to religious groups in 1910, 1971 and 2014

	1910 (%)	1971 (%)	2014 (%)
Roman Catholic	35	40	24
Nederlands Hervormd (liberal protestants)	44	24	7
Gereformeerd (orthodox protestants)	9	9	3
Protestant Church in the Netherlands[a]			6
Other[b]	7	3	11
No religious affiliation	5	24	49

Sources: CBS/Atlas of the Netherlands (1963)
[a]Merge between part of the *Nederlands Hervormd* and part of the *Gereformeerd* parishes in 2004;
[b]Other: in 1910: Evangelical Lutheran, Mennonite, Remonstrant and other churches, and Jews; in 1971 and 2014 also Muslims

Protestants. Only the most determined orthodox Protestants maintained their own 'mini-pillar' beyond the main and crumbling Protestant pillar, and strengthened that even by establishing new institutions. Thanks to their high birth rates strictly orthodox Protestants certainly do not diminish in numbers – the 'bible belt' is a persistent phenomenon in modern times (Knippenberg and De Pater 1988).

In reality secularisation is even more prominent than suggested by Table 12.2. Often church members do no longer attend church services on a regular basis. According to statistics of the CBS, only one out of six Roman Catholics and liberal Protestants (*Nederlands Hervormd*) visit churches more frequently than every month (CBS Statline). The extent to which religion determines political choices, leisure or sport cubs, school choice, membership of broadcasting associations, et cetera has been greatly reduced between 1966 (perhaps the final year when the pillars were still intact) and 1987 (when the pillars were already mostly declined) (Fig. 12.10). After 1987, percentages dropped even further.

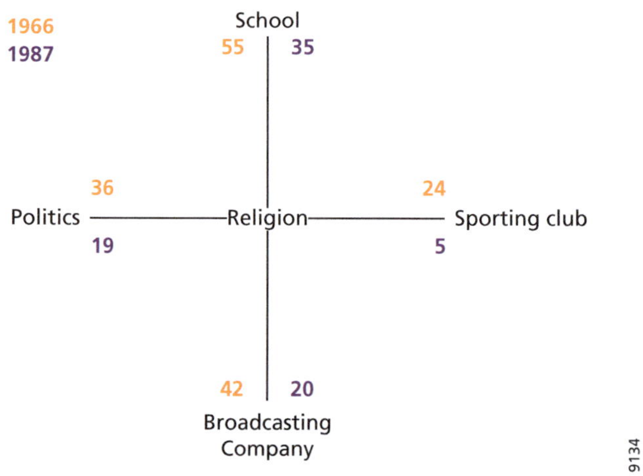

Fig. 12.10 Intertwining of religion and other domains of society, in 1966 and in 1987. For instance, 55% of 1966 respondents stated they were guided in their choice of a school for their children by the school's denomination. In 1987 it was only 35%. (Source: Eysberg and De Pater 1993)

Box 12.6: The Dutch and Narcotics

Dutch history with respect to narcotics goes back to the seventeenth century when they imported opium from Bengal and narcotics were incorporated in regular overseas trade practices. Opium trade was formally regulated since. In the first decades of the twentieth century a Dutch factory produced large quantities of cocaine from coca plantations in former Dutch colonies. Nevertheless, a restrictive international Opium Treaty was signed in 1919 by the Dutch government as well and narcotics were entirely prohibited from then. Perception concerning narcotics began to change in the early 1970s when use of narcotics exploded mainly among youngsters. Dutch authorities were no longer able to enforce the Opium Law and political pressure urged for a more liberal stance regarding law enforcing on the one hand and for better medical care to drugs addicts on the other.

Box 12.6 (continued)

Consequently, the Opium Law was considerably modified in 1976. A distinction was made between less dangerous 'soft drugs' as marihuana and hashish, and the strongly addictive 'hard drugs' as e.g. heroin, cocaine and opium. Free Methadon was provided to heroine addicted people but trafficking and trade was severely prosecuted. Although still illegal, possession and use of small quantities (< 5 g) of 'soft drugs' (marihuana, hashish) by adults (> 18 years old) was no longer prosecuted. Soft drugs were considered as 'recreational drugs' comparable to alcohol. Under strict conditions, Dutch adults may buy (and use) small amounts of hashish and marihuana in 'coffee shops'. The question of how such shops were supplied with their resources remained unanswered.

This liberal policy on narcotics seems rather pragmatic as hazardous drugs remain illegal and laws are implemented rather strictly, while wide-spread use

(continued)

Box 12.6 (continued)
would be impossible to combat effectively anyway. Dutch narcotics policy has been subject to fierce debates between governments in neighbouring countries. Nevertheless, more and more nations tend to follow more liberal policies and practices in this respect.

However, both production and trade of hard and soft drugs and their related criminality has become a serious issue in the Netherlands. Weed farms where marihuana is grown are wide spread and often found in private houses, many of which in the south of the country. Moreover, Dutch marihuana ('Nederwiet') has developed towards a hard drug as Nederwiet contains 3 times more of the effective THC than foreign marihuana (DRUGSINFO 2016). The Netherlands became one of the world largest illegal producers of the hard drug ecstasy (MDMA 2016) presently used by about a quarter of a million people in the Netherlands and increased between 2009 and 2015 (CBS 2015). In the Netherlands soft drugs and ecstasy are mostly used by urban, high educated people. More than 30% of the higher educated Dutch people at least once experienced cannabis, about double the numbers of people with lower educations. In contrast, daily use of cannabis more often occurs among lower educated people (MDMA 2016). As compared to other EU countries, the Dutch (15–64 years) score average on most soft and hard drugs, except for ecstasy where the Dutch belong to the most frequent users (DRUGSEU 2016). The 2017 Dutch Government aims at experiments to grow State-controlled Cannabis to safeguard product quality while reducing risks for additional toxic weed ingredients. Another ambition of such experiments is to compete illegal Cannabis production and related criminality. (EDM)

12.5 Political Shifts

The four pillars ruled their own political parties. In 1917, when universal suffrage for men over 25 years old was introduced, Roman Catholics and Socialists both had a single party to adhere to. (Liberal) Protestants could choose between two parties while orthodox Protestants had one party to vote on. Liberals could even choose among four parties.

In 1918, for the first time elections were through proportional representation which replaced the first-past-the-post system, in which each district delegated one person for the House of Representatives (*Tweede Kamer*). Figure 12.11 shows the largest party/pillar for each municipality in 1918.

After WWII, new parties emerged but links with the four pillars remained. The Labour Party (PvdA) and the Dutch Communist Party (CPN) together adhered to the socialist pillar; the People's Party for Freedom and Democracy (VVD) was an exponent of the Liberal pillar, the Catholic People's Party (KVP) of the Roman Catholic pillar. Protestant parties (ARP and CHU) remained unchanged as before WWII. Figure 12.12 shows where supporters of these parties lived. Characteristic for the pillarized society is the regional stability and minor shifts in number of votes over time. 'Floating voters' were a rarity by that time, major shifts in numbers of seats likewise. Roughly, in the decades of pillarization, the Roman Catholic party collected around 30% of the votes, the Socialist party as well, Protestant parties approximately 20% and the Liberal party 10% (Lijphart 1968). The remaining 10% went to splinter and protest parties.

Secularisation terminated both the temporal and spatial form of political stability. Unless very orthodox, people who used to attend services in church no longer inherently voted in favour of Christian parties. Roman Catholic KVP and Protestant political parties ARP and CHU jointly collected a majority of the votes (76 out of a total of 150 seats in the House of Representatives) in 1963. From then these three Christian parties declined. Their merger into a new Christian party in 1977 (CDA), did not halt this decrease. Further decline of the (liberal) Christian parties continued and resulted in 13 seats only in the 2012 elections. This party finds its voters predominantly in rural areas, CDA considers and supports traditional families as the cornerstones of society. In contrast to CDA, orthodox Christian parties as SGP and CU (their supporters mainly occur in the Bible belt) remained with 8 seats (in total) in 2012 and in 2017 again quite stable until today (Fig. 12.13).

Socialist party PvdA made an impressive advance in the 1970s, in particular in the formerly rather homogeneous Roman Catholic south. Since the 1980s however, its popularity declined. Its downturn was partly due to the emergence of another Socialist Party (SP) prioritising traditional socialist

Fig. 12.11 The largest party/pillar per municipality in 1918 general elections in the Netherlands. Only adult males were allowed to vote; women were not allowed to vote until 1922. (Source: De Jong et al. 2011 (map 28, generalised) and Databank Verkiezingsuitslagen)

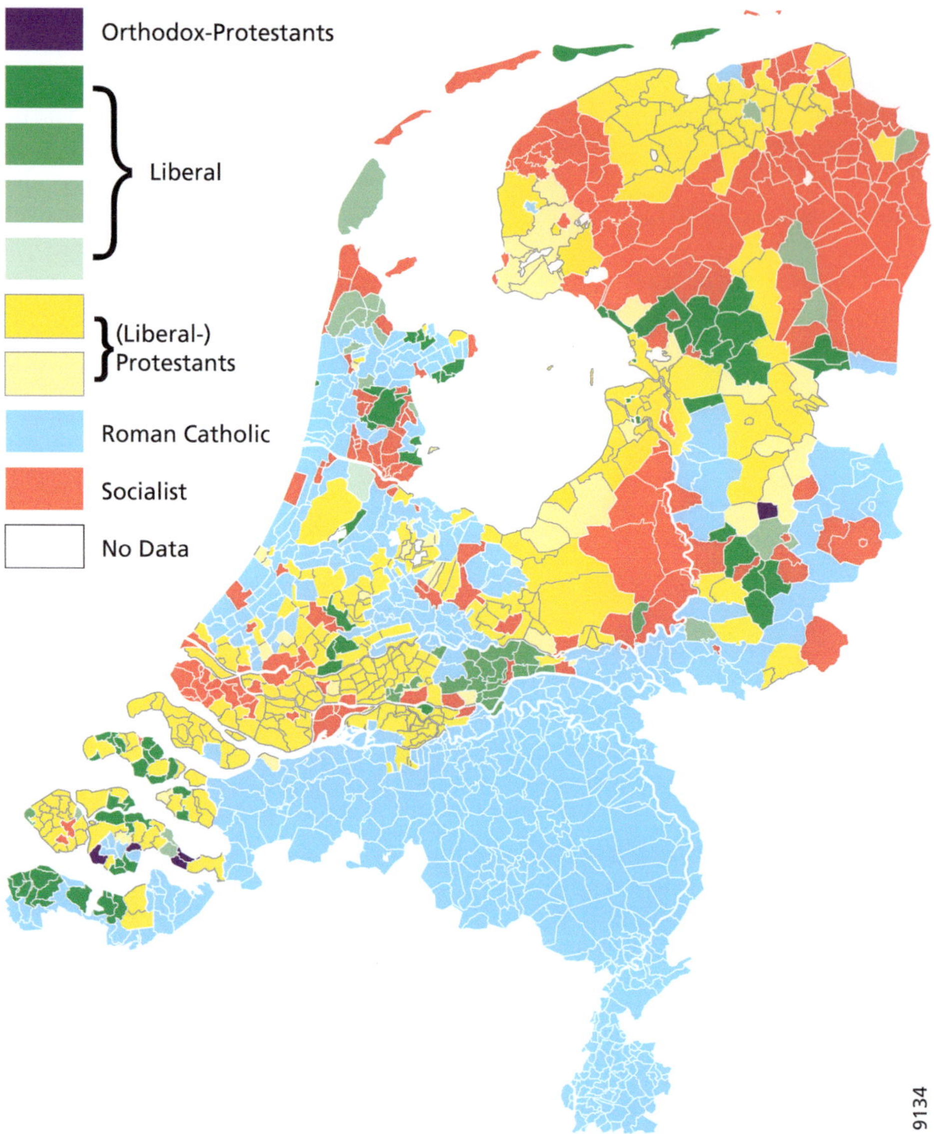

values of solidarity, equality and a strong caring government. Another new, but more liberal left-wing party (*Groen Links*) drained the PvdA voting public as well. Voting populations change together with changing societies. Most of Dutch heavy industries disappeared together with their large numbers of traditional blue-collar workers. Many low income people from Dutch origin now vote either for the left-wing Socialist Party (SP) or for the nationalistic, economically conservative, anti-(Muslim) immigrant and anti-European Union Freedom Party (PVV).

In the last decades both center-right-wing liberals (VVD) and center-left-wing liberals (D66) won many supporters. Especially VVD attracted voters not only from its traditional strongholds in metropolitan areas but far beyond as well (Fig. 12.14). In the times of pillarisation, liberals received 10% or a little more of the votes. During last general Dutch elections, in March 2017, VVD proved the largest Dutch political party, with 21% of all votes, giving them 33 out of 150 seats in Parliament. D66 received 12% (19 seats); the 'green parties' Groen Links 9% and Partij voor de Dieren ('Party for the Animals') 3% (14 respectively 5 seats). The socialists (PvdA and SP) collected 15% (together 23 seats), the Christian-Democrats (CDA) 12% (19 seats). The populistic PVV, led by Geert Wilders, became the second largest Dutch political party: 13% of the votes (20 seats) in March 2017. After 7 months of negotiations a new Dutch Government was installed in October 2017. This Government consists of a coalition of four political parties: two liberal

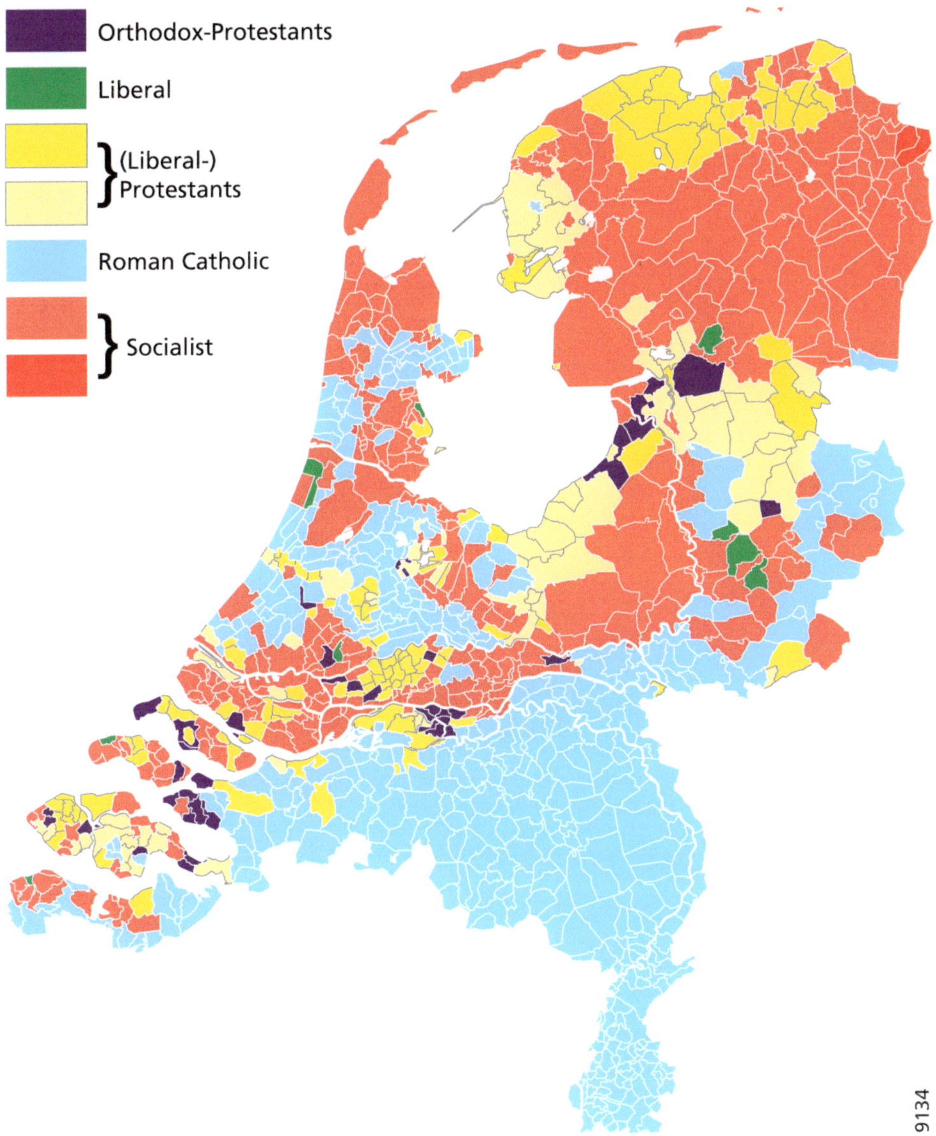

Fig. 12.12 The largest political party per municipality in 1952 general election in the Netherlands, as compared to the 1918 elections. The map shows significant stability. (Source: Databank Verkiezingsuitslagen)

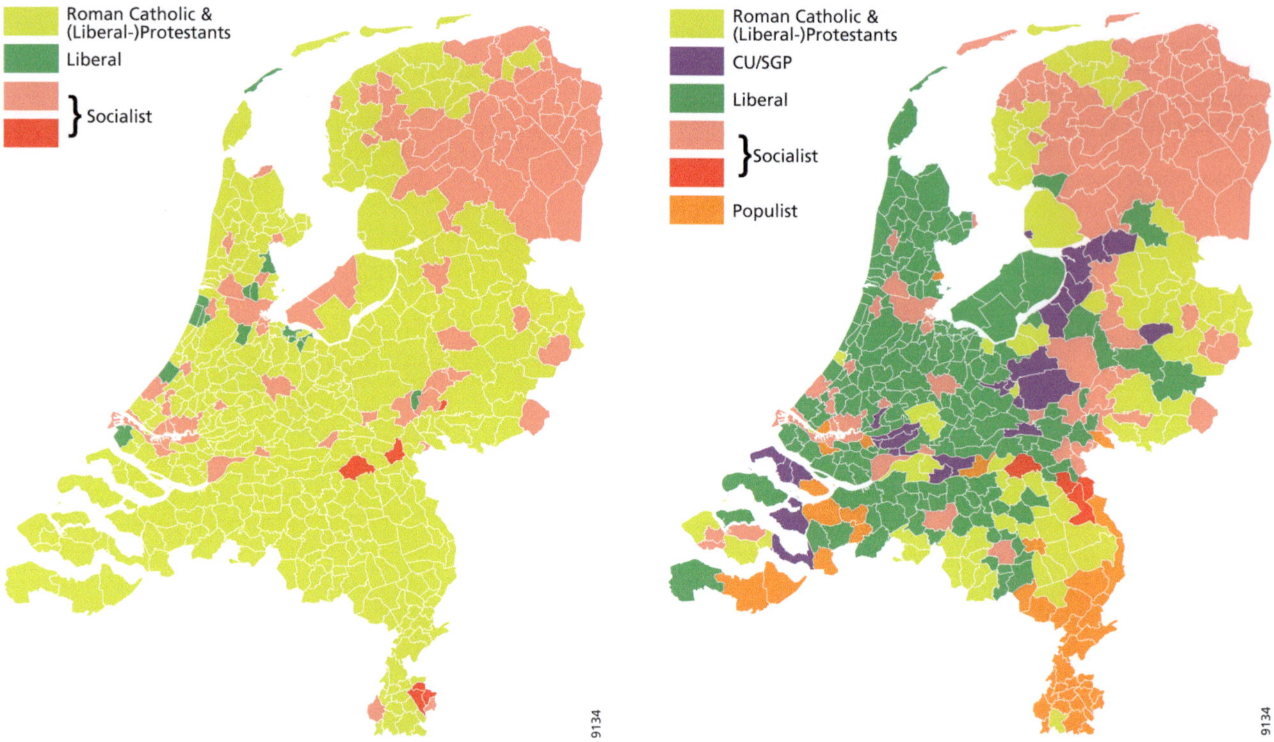

Fig. 12.13 The largest political parties per municipality at general elections in 2006 and in 2010 in the Netherlands. The gain of the liberal VVD party is impressive. (Source: Databank Verkiezingsuitslagen)

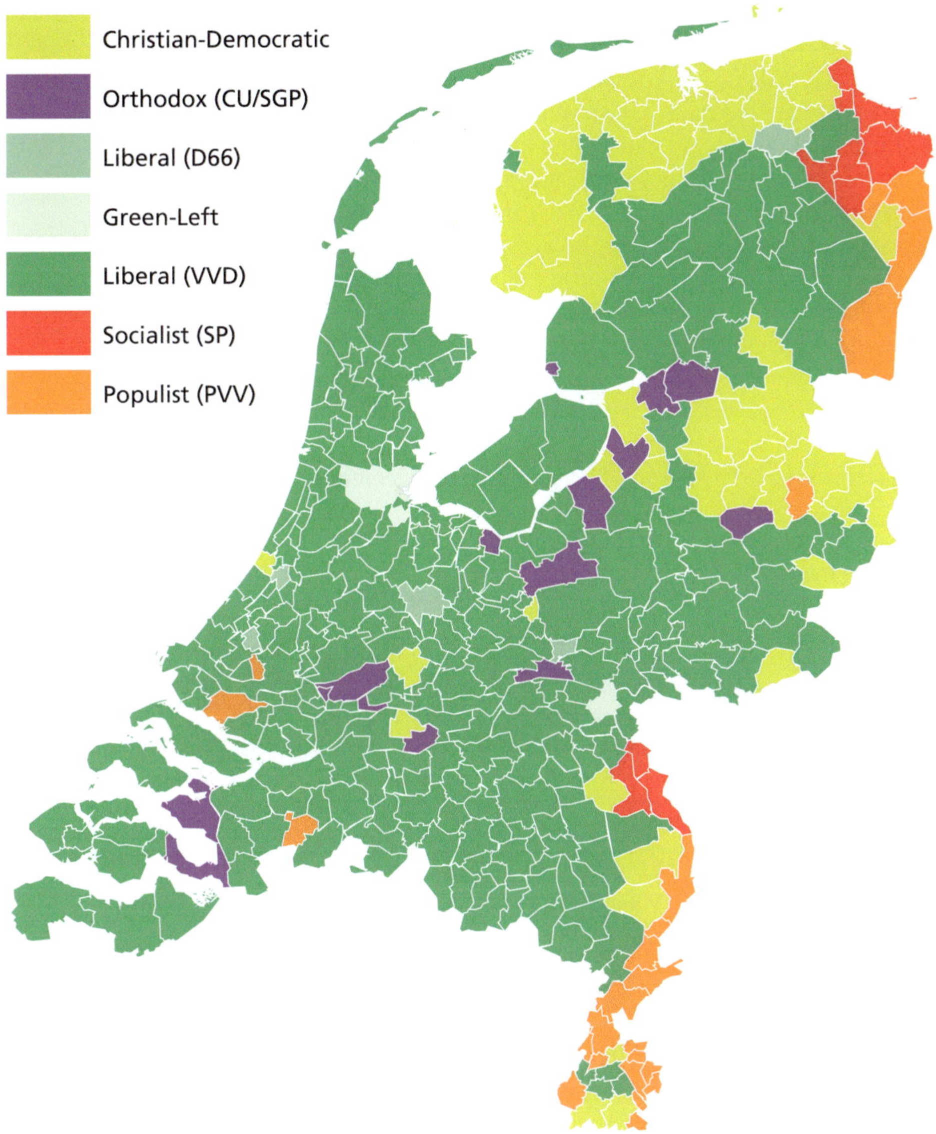

Fig. 12.14 The largest political party per municipality in 2017 general elections in the Netherlands

parties (VVD and D66) and two Christian parties (CDA and CU), which together hold a minimal majority in Parliament: 76 out of 150 seats. Mark Rutte (VVD) serves as a Prime Minster for the third time in succession.

In general, Dutch people support parties that promise economic stability, employment, and a rather restrictive immigrant policy. Today, many Dutch are concerned about economic and political instability in an ever closer European Union. But 'floating voters' dominated all recent elections – quite differently as compared with pillarisation times when their share was less than 10%. Remarkable has been the disconnection between religion and Christian parties on the one hand and between working class people and traditional political left on the other. Political landslides and fragmentation seems to have become the rule rather than the exception.

References

Atlas of the Netherlands (1963) Compiled by the foundation for the scientific atlas of the Netherlands. Staatsdrukkerij- en uitgeverijbedrijf, Den Haag

Brons LJ (2006) Indirect measurement of regional culture in the Netherlands. Tijdschr Econ Soc Geogr 97(5):547–566

Castenmiller P, Knol F (1989) Convergentie of divergentie. Sociale en culturele ontwikkelingen in stedelijke en landelijke gebieden. Cahier 72. Sociaal en Cultureel Planbureau, Rijswijk

CBS (2015) https://www.cbs.nl/nl-nl/nieuws/2015/29/kwart-miljoen-nederlanders-gebruikt-ecstasy. Accessed 28 Nov 2016

De Jong R, Van der Kolk H, Voerman G (2011) Verkiezingen op de kaart 1848–2010. Tweede Kamerverkiezingen vanuit geografisch perspectief. Matrijs, Utrecht

De Pater B (1993) De opmars van de standaardtaal in Nederland als ruimtelijk diffusieproces. Historisch Geografisch Tijdschrift 11 (1):1–8

Dekker P et al. (2007) Diverse Europe. Public opinion on the European Union & Cultural diversity, economy and policy. European Outlook 4. Centraal Planbureau/Sociaal en Cultureel Planbureau, Den Haag

DRUGSEU (2016) https://www.volksgezondheidenzorg.info/onderwerp/drugsgebruik/regionaal-internationaal/internationaal-0#node-gebruik-harddrugs-nederland-en-de-europese-unie. Accessed 28 Nov 2016

DRUGSINFO (2016) https://drugsinfoteam.nl/drugsinfo/cannabis/cannabis-feiten. Accessed 28 Nov 2016

Eysberg C, De Pater BC (1993) Regional geography and local cultures: theoretical approaches and the case of Tienhoven-Westbroek in the Dutch context. In: Hauer J, Hoekveld GA (eds) Moving regions, Netherlands geographical studies, vol 161. Koninklijk Nederlands Aardrijkskundig Genootschap/Universiteit Utrecht, Utrecht, pp 99–119

Frijhoff W, Spies M (1999) 1650. Bevochten eendracht. Sdu, Den Haag

Hellemans S (1993) Zuilen en verzuiling in Europa. In: Becker U (ed) Nederlandse politiek in historisch en vergelijkend perspectief. Het Spinhuis, Amsterdam, pp 121–150

Hofstede G (1980) Culture's consequences: international differences in work-related values. Sage, Beverly Hills

Huijnk W, Dagevos J, Gijsberts M, Andriessen I (eds) (2015) Werelden van verschil. Over de sociaal-culturele afstand en positie van migrantengroepen in Nederland. Sociaal en Cultureel Planbureau, Den Haag

Klein Kranenburg D (2013) 'Samen voor ons eigen'. De geschiedenis van een Nederlandse volksbuurt: de Haagse Schilderswijk. Verloren, Hilversum

Kloeke GG (1927) De Hollandse expansie in de zestiende en zeventiende eeuw en haar weerspiegeling in de hedendaagsche Nederlandsche dialecten. Noord- en Zuid-Nederlandsche Dialectbibliotheek, Den Haag

Knippenberg H (1992) De religieuze kaart van Nederland. Omvang en geografische spreiding van de godsdienstige gezindten vanaf de Reformatie tot heden. Van Gorcum, Assen

Knippenberg H (1997) Dutch nation-building: a struggle against the water? GeoJournal 43(1):27–40

Knippenberg H, De Pater BC (1988) De eenwording van Nederland. Schaalvergroting en integratie sinds 1800. SUN, Nijmegen

Knippenberg H, Stoppelenburg CM, Van der Wusten HH (1989) De protestantenband in Nederland: de geografische spreiding der orthodox-protestanten in 1920 en 1985/6. Geografisch Tijdschrift 23(1):12–22

Kogut B, Singh H (1988) The effect of national culture on the choice of entry mode. J Int Bus Stud 19(3):411–432

Lijphart A (1968) The politics of accommodation. Pluralism and democracy in the Netherlands. University of California Press, Berkeley/Los Angeles

Mak G (1994) Hoe God verdween uit Jorwerd. Contact, Amsterdam

MDMA (2016) https://en.wikipedia.org/wiki/MDMA. Accessed 28 Nov 2016

Meyer H (2017) The state of the delta. Hydraulic engineering, urban development and nation building in the Netherlands. Vantilt, Nijmegen

Penrose J (1990) Frisian nationalism: a response to cultural and political hegemony. Environ Plann D Soc Space 8(4):427–448

Sanders EW (2017) Levi's eerste kerstfeest. Jeugdverhalen over jodenbekering (1729–2015). Vantilt, Nijmegen

Van Amersfoort JMM, Van der Wusten HH (1989) De cultuurgeografie van Nederland na 1945. In: De Pater BC, Hoekveld GA, Van Ginkel JA (eds) Nederland in delen, een regionale geografie, vol 1, De Haan, Houten, pp 83–104

Van der Hoek S (2014) Langs de nullijn. De twee kanten van Nederland. De Arbeiderspers, Utrecht

Van der Woud A (2015) De Nieuwe Mens. De culturele revolutie in Nederland rond 1900. Prometheus – Bert Bakker, Amsterdam

Van Gorp B, Renes H (2003) The Dutch landscape: a way of seeing. In: Van Gorp B, Hoff M, Renes H (eds) Dutch windows. Cultural geographical essays on the Netherlands. Faculteit Ruimtelijke Wetenschappen, Universiteit Utrecht, pp 55–79

Van Veen J (1948) Dredge, drain, reclaim. The art of a nation. Martinus Nijhoff, Den Haag

Zijderveld AC (1999) De economie van een deltacultuur. ESB 84, 2 September 1999, nr. 4217, dossier Economie en Infrastructuur

Social Maps 13

Joos C. Droogleever Fortuijn

Abstract

The Netherlands is one of the wealthiest and most egalitarian societies in the world. The Dutch rank high in terms of income, income equality and human development. However, even in Dutch society (relatively small) inequalities occur among social groups and regions, cities and neighbourhoods. Spatial differences in income, poverty, employment, education, housing and health both reflect demographic variability and differences in geographical economic structure. The population in the western part of the country, Randstad Holland, has relatively high incomes, high labour market participation rates and high educational levels, while income, labour market participation and educational levels are relatively low in peripheral regions in the north (east Groningen) and the south (south Limburg and Zeeuws Vlaanderen).

Keywords

Human development · Income · Income equality · Employment · Education · Housing · Health care

13.1 A Wealthy and Egalitarian Society

A Google search on 'the richest countries of the world' consistently puts the Netherlands in the world's top-twelve. The Allianz Global Wealth Report 2016 ranks the Netherlands sixth in as to financial assets per capita. In terms of richest nations by Gross Domestic Product per capita purchasing power the International Monetary Fund ranks the Netherlands 12th. Only the oil states of Qatar, Kuwait, Brunei, United Arab Emirates and Saudi Arabia, and the non-oil states Luxembourg, Singapore, Norway, Switzerland, United States of America and Ireland rank higher.

The Netherlands has a long tradition of global economic orientation with dominant sectors as international trade, finances and business services (Chaps. 6 and 7). Together with well-developed infrastructures, strong rules of law, low corruption, highly educated workforce and rich natural resources (primarily natural gas), these factors explain the Netherlands' high ranks in wealth.

The Dutch rank even higher if a broader concept of wealth would be considered as measured by the UN Human Development Index, which includes life expectancy and education in addition to income. The Dutch are relatively high educated and their life expectancy at birth is more than 80 years for both men and women. Together with Norway, Australia, Switzerland, Germany, Denmark, and Singapore, the Netherlands belongs to the leading nations in the UN Human Development Index Report 2016 (Fig. 13.1). The Dutch rank high on 'soft' indicators as well-being too: 7th as to overall life satisfaction, 12th as to living standard and 12th again as to safety perception, and 7th as to satisfaction with health care quality. And the Netherlands ranks 7th, after Denmark, Switzerland, Iceland, Norway, Finland and Canada, in the UN World Happiness Report (Helliwell et al. 2016). That index includes indicators for income, social support, healthy life expectancy, freedom to make choices in life, donations to charity, perceptions of corruption in government and business, and positive affect (previous day feelings of happiness, laughter and enjoyment). According to UNICEF (2013), Dutch children are the happiest in the world as to material well-being, health and safety, education, behaviours and risks, and housing and environment.

Poverty in material sense, i.e. lack of housing, food and clothing, is almost non-existent in the Netherlands. The European Union defines poverty not only in terms of basic material needs, but in terms of social exclusion as well. Consistently, the Netherlands, Czech Republic and Sweden lead the ranks of non-poor EU households, with 83% of them

J. C. Droogleever Fortuijn (✉)
Department of Geography, Planning and International Development Studies, University of Amsterdam, Amsterdam, The Netherlands
e-mail: J.C.DroogleeverFortuijn@uva.nl

E. F. J. De Mulder et al., *The Netherlands and the Dutch*, World Regional Geography Book Series,
https://doi.org/10.1007/978-3-319-75073-6_13

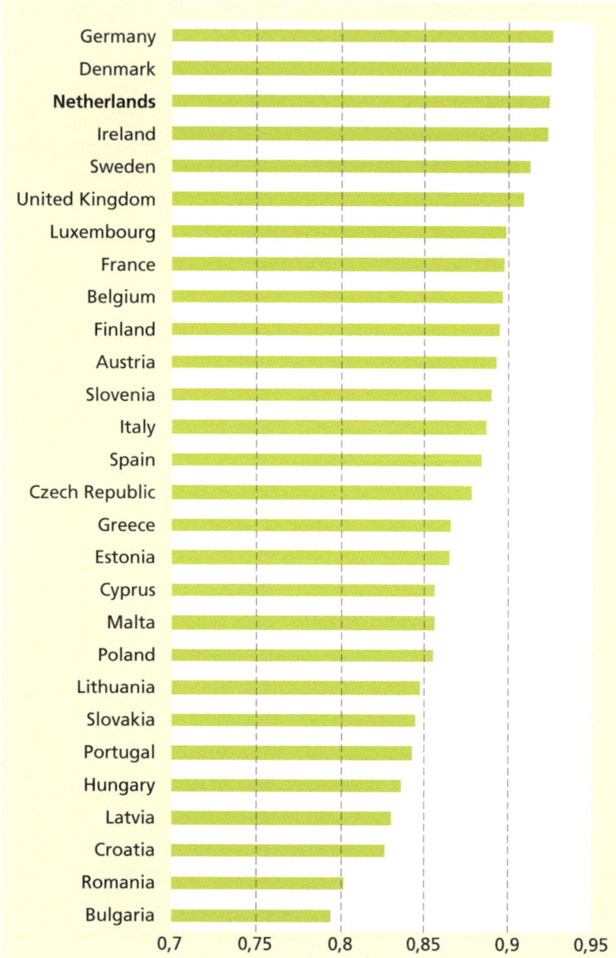

Fig. 13.1 Human development index EU countries 2015. (Source: UNDP 2016)

that are neither at risk of poverty, nor severely materially deprived, nor living in a household with low work intensity (Eurostat 2017a). Almost all households in the Netherlands live in houses or apartments with heating, bath or shower, electricity, and sewage, and have enough food and clothing. In 2013, 5.4% of Dutch households had incomes less than the social minimum level of 969 € per month, a level that is considered necessary for decent housing, heating, food, clothing and insurance, while 1.5% of the Dutch live in long-term poverty (SCP 2016).

Since long, the Netherlands maintain a social-democratic or Nordic welfare state model (Esping-Anderson 1990), including universal provision of social security based on individual citizenship. Income-dependent taxation, old age state-pension, social security in case of unemployment, illness and disability, allowances for children, health costs and housing costs, and free primary and secondary education belong to the instruments to reduce financial and social inequality. However, the Dutch moved toward a more Anglo-Saxon welfare state model with a more modest social security system in recent years. Together with the financial crisis that began in 2008, this resulted in a slight increase of poverty and inequality. But the Netherlands still rank high on all equality lists, with a 3rd position on inequality adjusted income index, 3rd on gender equality (UNDP 2016) and 3rd on happiness equality (UNDP 2015).

Notwithstanding the egalitarian character of Dutch society, significant inequalities can be identified, socially as well as spatially.

13.2 Income and Poverty

French economist Piketty (2013) argues that inequality of wealth is increasing as the rate of capital return is higher than the rate of economic growth. This happens not only in liberal welfare states as USA and UK but in European countries with a more corporatist welfare state model as well. The high eighteenth and nineteenth centuries inequality of wealth decreased after World War I, but began growing again during the last decades of the twentieth century and continued to increase in the twenty-first century. The Netherlands is a highly egalitarian state in terms of income but its distribution of wealth (financial and housing assets) is unequal and growing. With 0.3, the Dutch Gini-coefficient for income is lower than most EU nations, and one of the lowest in the world, while its Gini coefficient for financial assets (0.8) is higher than average among EU countries, and similar to the USA (Van Bavel 2014).

The 'welfare state paradox' explains why low-income households in the Netherlands have no or negative assets (more debts than savings) (Kremer et al. 2014). Low-income households do not save money as they do not need personal savings to address financial emergencies as unemployment, long-lasting illness, or disability. Such emergencies are covered by the Dutch welfare state which provides a security net to its inhabitants. The lowest earnings decile of the Dutch population has negative assets and the second, third and fourth decile has zero assets, while the richest ten percent has more than 60% of all assets. In contrast to its high income taxation, property taxation in the Netherlands is below EU average and increasing capital mobility provides options to the very rich to avoid taxation by moving their capital to tax heavens (Van Bavel 2014).

Taxation levels in the Netherlands are related to income levels. In 2017, 35.5% income tax is paid over incomes of less than 19,982 € and 52% over incomes of more than 67,071 €. A system of collective and mandatory (for employed people) social insurances protect employees against risks of unemployment, illness and disability. However, that does not apply for the rising number of self employed people. Old-age pensions partly consist of state-pensions and partly of collective pensions for employees. Private, individual insurances and savings might supplement income for retired people. All over 65 years old Dutch citizens receive state-pension (in 2017, full state pension was 1153 € per month for singles and 1592 € for couples).

Immigrants usually receive lower state-pensions as most of them haven't been Dutch citizens since they were 15 years old. Only those who have lived in the Netherlands uninterruptedly from age 15 to 65 receive full state pensions. This upper age limit dates back from 1947 when this state pension system was introduced. However, with increasing life expectancies this upper limit is being leveled-up to 67 years in 2022, with a further increase dependent on future rising life expectancies.

Apart from collecting taxes, the Dutch government pays several types of allowances. All households with children under 18 years old receive family allowances. Low-income households are entitled to receiving additional allowances for health costs and house rentals. Social benefit (in 2017: 983 € per month for singles and 1404 € for couples) constitutes a social security net for 21 years old and more with no income.

Single parent households, families with more than three children, singles below 65 years old, non-western immigrants and people on benefit are categories at risk of poverty. In contrast to most EU and other nations, older Dutch people do not belong to such a risk category as to the relative generous Dutch state-pensions. However, several categories of working people have quite low incomes, e.g. if they are self-employed or if they work part-time. Often, Dutch women have lower incomes than men. Single mothers live on benefit more often than single fathers. More women work part-time than men, their careers are interrupted more often than men and more often women have low-salary jobs than men (De Beer 2014). Non-western immigrants, in particular immigrants of Moroccan background, are at risk of poverty resulting from lower employment rates and high prevalence of invalidity pensions. Non-western immigrants are more often employed in heavy labour or physically demanding jobs with high risk of accidents and sick leave. As large families and one-parent families are categories at risk of poverty, children are more often at risk than adults. High incomes are often found in households with two partners in their fifties at the end of their careers.

13.2.1 Spatial Dimensions of Prosperity

Spatial distribution of income may partly reflect regional demography and partly local and regional economic structures (Fig. 13.2). In Randstad Holland incomes are higher than elsewhere in the Netherlands. Low household incomes particularly occur in the northern provinces, Zeeland and in regions along the German border, reflecting weak economic structures of these regions and their lower educational levels of their ageing populations. Older, suburban municipalities with a high share of end-of-career couples and families as Bloemendaal, Blaricum, Laren, Rozendaal and Wassenaar, host people with the highest Dutch household incomes. Nuclei of the lowest incomes occur in urban municipalities with many students and non-western immigrants. This explains why university cities as Maastricht, Delft, Nijmegen, Enschede, Wageningen and Groningen belong to the top-ten lowest-income municipalities (CBS 2016a).

Spatial inequality within cities is usually higher than between cities. Amsterdam hosts most (14.4% in 2013) poor households, with incomes less than 1063 € per month for singles and 2000 € for families with two children. However, the districts of Crooswijk (30.8%) and Tussendijke (29.7%) in Rotterdam lead the list of poorest districts in the Netherlands (SCP 2016) (Box 13.1).

Box 13.1: Urban Segregation: 'High' and 'Low' Den Haag

Den Haag is the most segregated big city of the Netherlands and segregation is highly correlated with the physical-geographic segmentation between the higher sands and the lower peat and clays. This geographical denomination of 'high' and 'low' refers to social-economic segregation between high and low-income groups as well as to the physical structure and topography of high sand dune and beach sands area on the one hand and the low peat and marine clays on the other (Fig. 3.14).

In early Medieval Times, the count of Holland had his hunting lodge in the dune area of Den Haag that developed later into his court. Den Haag became the seat of the federal government ('Staten-Generaal') at independence of the Dutch Republic in 1588 and later the seat of the royal court and national government of Kingdom of the Netherlands since 1815.

During the nineteenth century, the sandy parts of Den Haag became residence for a considerable number of wealthy people: nobility, government officials, diplomats, rentiers and returned entrepreneurs from the (former) Dutch colonies, close to the royal court and the centre of political power of the Dutch national government. Noble land owners of the sandy soils, including the king, sold land for urban expansion and stipulated the building of villas.

Concurrently, Den Haag faced a growth of industries, mainly of luxurious consumption goods, and a growth of low-income people employed in domestic services or as industrial workers. Such industries were often based in the low peat areas, at some distance from court and government, where canals were dug for transport of industrial materials and products. Private land owners of the peat and clay areas sold land for building mass housing for working class families.

Present income segregation in Den Haag reflects the past and the ongoing segmentation between high-income households on the high sands and the low-income people on the lower peat and clay areas in the city (Schmal 1995).

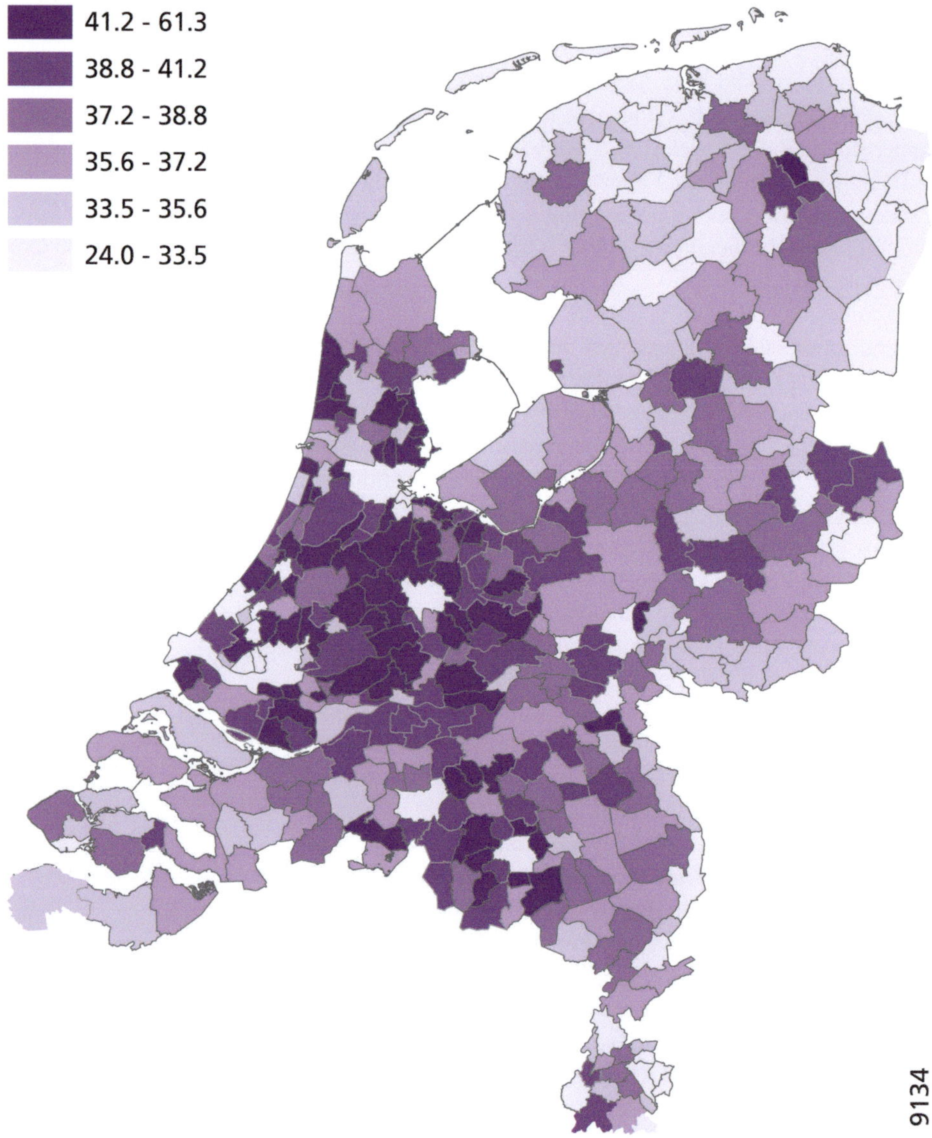

Fig. 13.2 Average annual household income (*1000 €) by municipality 2014. (Source: CBS Statline)

Furthermore, income segregation within cities is connected with twentieth century's urban expansion cycles and spatial distribution of social housing. Around 1900, rich households used to live in the city centers or in their nearby high-income districts. In the late 19th and early twentieth century, many such households left the city and moved to suburban municipalities (Chap. 8). However, some urban neighbourhoods remained nuclei of high income households. From the outset, Concertgebouwbuurt in Amsterdam, Hilligersberg in Rotterdam, Benoordenhout in Den Haag and Wilhelminapark in Utrecht are residential areas for the richest urban households.

Initially, new urban expansions rank high but often deteriorate once cities growth further. Newly built neighbourhoods usually charge higher housing prices. Skilled workers with stable jobs belonged to the first residents of the new districts built by the end of the nineteenth century, with private rental apartments for the growing numbers of factory workers. By then, the poorest households lived in slum dwellings in the city's interior. The Netherlands was a 'kingdom full of slums' (Van der Woud 2012). According to Van der Woud about 2 out of 5 million Dutch people were poor and hundreds of thousands people were living in comparable conditions as in current South-Asian or African metropolises.

Between WWI and WWII (1920–1940), new urban expansions were developed with social rental housing built by housing associations. These houses were primarily built for skilled workers but social rental housing of the 1950s and 1960s were predominantly built for lower and middle income households. Mass suburbanization processes of the 1960s and 1970s, with new residential areas beyond cities caused a selective migration process with mainly middle-income households leaving cities while the lowest income groups stayed behind (Chap. 8). Departure of middle income groups to the suburbs created space in the older inner cities for the lowest income groups among which Mediterranean immigrants, students and other low-income households as one-parent families on benefit (Ostendorf and Droogleever Fortuijn 2006).

By the end of the 1980s a gentrification or upgrading process transformed the nineteenth century and pre-WWII twentieth century districts. In this process new urban households of highly educated singles, heterosexual and homosexual couples, and later of middle-class families with symmetrical task divisions between men and women, revalued urban life. Old houses and apartments were reconstructed and revitalized these districts. Large-scale urban renewal of the post-WWII housing stock by municipalities and housing associations in the 1990s and early twenty-first century improved and diversified such neighbourhoods.

Currently, lowest income groups are concentrated in apartment blocks, built in the 1960s and 1970s and which often are under reconstruction. Today, higher income groups live in the city centers, in the nineteenth century and pre-WWII twentieth century gentrified neighbourhoods and in the youngest urban expansions. Segregation levels are increasing, this holds in particular for the highest income groups. A welfare state system shift toward a more liberal system resulted in less housing options for lowest income groups and in more segregation (Musterd and Van Gent 2016).

13.3 Employment

Labour market participation in the Netherlands is high for both men and women as compared with other EU countries. The Netherlands ranks 7th in the EU with respect to participation rates for 20–64 years old men (81.9% in 2015) and 3rd concerning similar aged women (70.8%; Eurostat 2017b). This reflects a rather recent development as labour market participation rates of women used to be quite low as compared with other European nations until the end of the 1980s.

Former low female participation rates may be explained by the 'welfare state paradox' to a certain degree. Labour unions successfully negotiated *family* wages for male workers, i.e. sufficient for family costs of living. The social security system formed a safety net in case of unemployment and disability for male breadwinners. Therefore, there was no urgent need for married women with children to be employed. Cultural factors played an important role as well. In a pillarized Dutch society religious values regarding marriage and motherhood formed the basis for the welfare state model with a priority for traditional male breadwinner – housewife families (Pott-Buter 1993). The former welfare state model in the Netherlands has been a combination of the more individualistic, social-democratic Nordic model and the more traditional corporatist model of Germany and Austria. From the 1970s onwards the Dutch welfare state system developed more into an individual and less family-based model. As a consequence, labour participation rates of women, of mothers in particular, increased considerably during the 1980s and 1990s.

Health care and social services, trade and business services are the most important sectors in the Netherlands in terms of employed persons (Table 13.1). The Dutch labour market is highly gendered. Health and social services, education, hotels and restaurants and domestic and other services are sectors with a high level of female employment, while construction, mining, water and waste disposal, industry, information and communication, transport and energy have relatively low levels of female employment.

The Netherlands is a part-time working country par excellence: part-time employment rate of women is extremely high as compared with other European countries and even Dutch men are more often working part-time than in other countries (Fig. 13.3). Deschacht and Tijdens (2014) explain the high part-time rate partly by the service-oriented economic structure (traditionally, the service sector offers more part-time jobs than the manufacture sector), and partly by the specific mixture of a universal provision of social security in an institutional context with conservative values regarding the role of women in families.

13.3.1 Spatial Dimensions of Employment

As in other European nations labour market participation rates are lower for lower-educated, older people and non-western immigrants. Low participation rates are found in regions with weak economic structures (north and east Groningen, east Drenthe, south Limburg and Zeeuws-Vlaanderen), in big cities with many students and non-western immigrants (Den Haag, Rotterdam, Arnhem

Table 13.1 Employment (number of jobs) in the Netherlands as by gender and economic sector 2015

Sector	Total (×1000)	Men (×1000)	Women (×1000)	% women
Agriculture, forestry, fishery	101	68	33	32.7
Mining	11	9	2	18.2
Industry	737	574	163	22.1
Energy	27	20	7	25.9
Water, waste	33	27	6	18.2
Construction	298	264	34	11.4
Trade	1300	690	610	46.9
Transport, storage	364	273	91	25.0
Hotels, restaurants, cafes	374	180	194	51.9
Information, communication	238	174	64	26.9
Financial services	283	167	116	41.0
Real estate	65	34	31	47.7
Specialised business services	472	281	191	40.5
Other business services	921	528	393	42.7
Civil service	502	306	196	39.0
Education	510	185	325	63.7
Health care, social services	1293	207	1086	84.0
Culture, sports, recreation	128	61	67	52.3
Domestic and other services	128	45	83	64.8
Total	7785	4093	3692	47.4

Source: CBS Statline

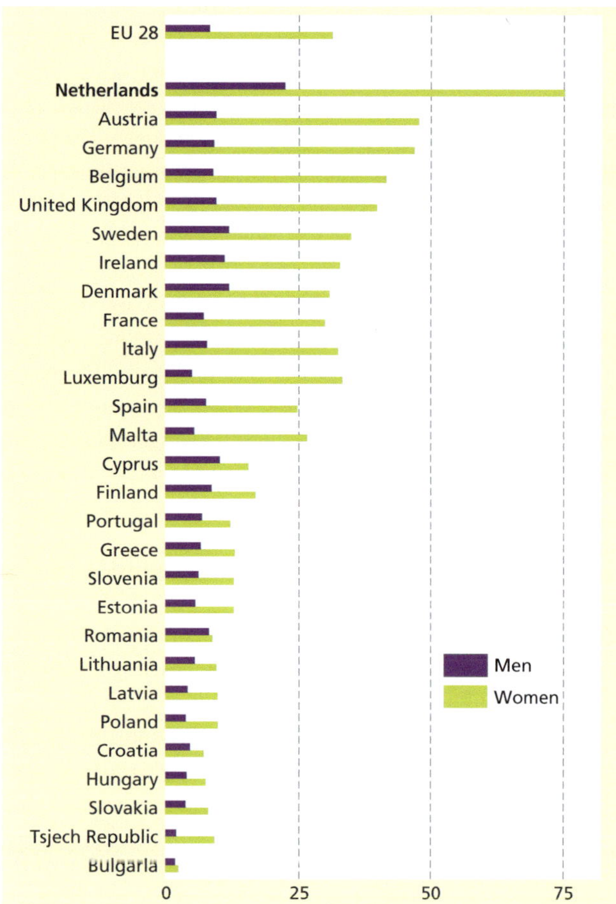

Fig. 13.3 Part-time labour participation rates for men and women EU countries 2015 (percentages of all employed men and women). (Source: Eurostat 2017c)

and Enschede) and some older suburban municipalities in Randstad Holland with an ageing population (Fig. 13.4).

Women labour market participation is higher in Randstad Holland than in peripheral regions and higher in urbanized than in rural municipalities. Noback and Van Dijk (2007) argue that differences in educational level, religiosity, household size (number of children) and employment opportunities are the main factors explaining such regional differences. De Meester et al. (2007) and Boterman and Karsten (2014) demonstrate that symmetrical arrangements by two full-time or two part-time working partners is rather typical for higher educated families in urban environments, in particular in the big cities interiors in Randstad Holland. Today, many highly educated parents of young children, especially working in non-profit sectors, are both working four days per week with a 'daddy-day', a 'mum-day' and three days of external child care, including a 'grandparents-day'.

In the last decade of the twentieth and the first decade of the twenty first centuries Dutch unemployment rates were lower than average in OECD-countries. But unemployment increased since the 2008 economic crisis that hit the Dutch export-oriented economy relatively hard and is like OECD' average now (Fig. 13.5). Youth and middle-aged adults unemployment rates is below OECD's average, but unemployment rates for the group of 55–65 years old is higher in the Netherlands than OECD-average, both for men and women (OECD 2016a).

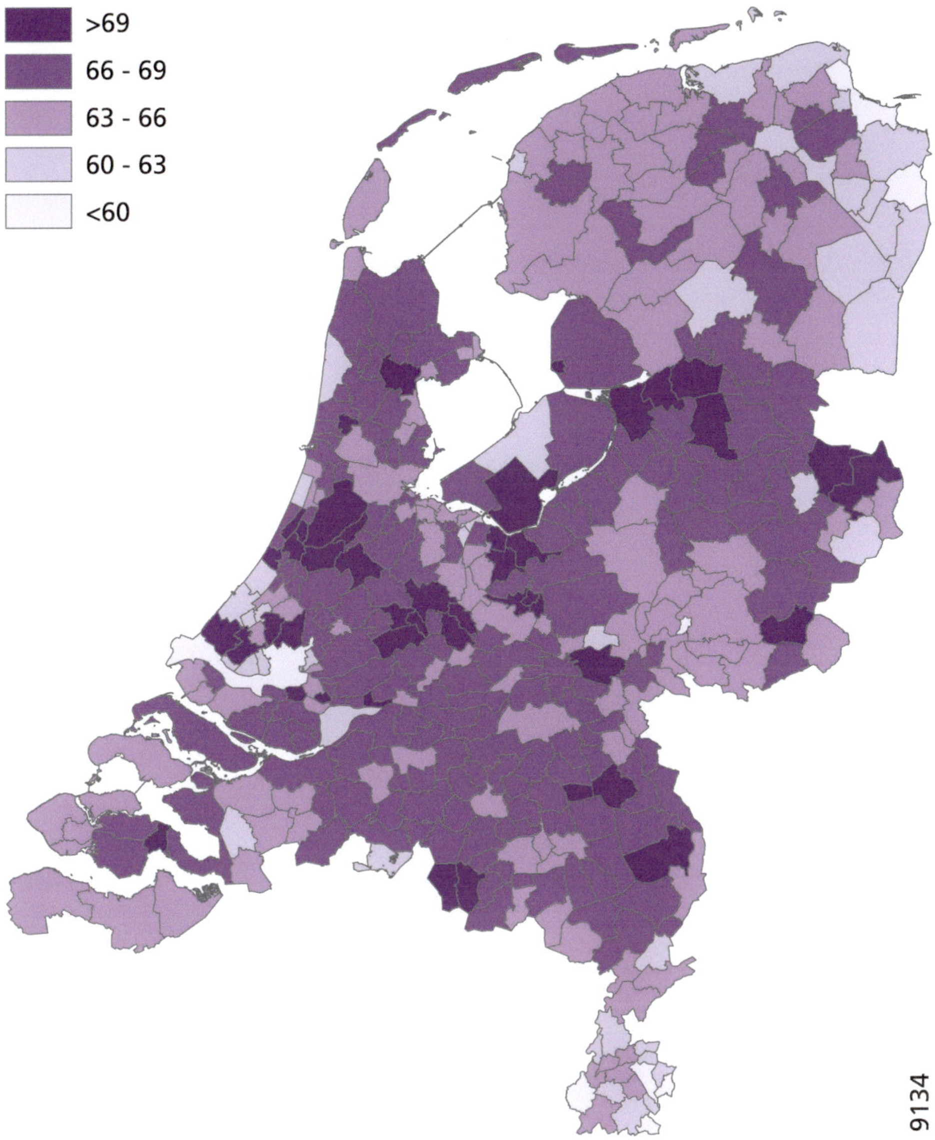

Fig. 13.4 Net labour participation rate by municipality 2015. (Source: CLO 2016a)

Geographically, high unemployment rates occur in the northern provinces and in Flevoland, in south Limburg and in some big cities, places with many low-educated people and non-western immigrants (Dirven et al. 2015). Rotterdam, Den Haag, Almere, Lelystad and Arnhem rank highest in terms of unemployment rates (Fig. 13.6).

13.4 Education

Free primary and secondary education is available for all Dutch children. The Law on Compulsory Education was introduced in 1900 and only concerned 6–11 years old. Much earlier, in the beginning of the nineteenth century, most Dutch children went to school already and literacy rates (75% for men and 60% for women) was higher than in other European nations. That may be explained by the dominant protestant ideology (with an emphasis on reading the bible), a trade-oriented economy, a commercialized agricultural sector and a relatively wealthy and urbanized society (Knippenberg 1986). Compulsory education was extended to the age of 15 in 1969 and currently is from 5 to 16 years old with a further extension for 16–18 years old without secondary school diplomas. Almost all children (98%) enter primary school at the age of four.

About one-third of the primary and secondary schools are public schools, as governed by municipalities or by the state; two-thirds are private schools, predominantly religion-based (Onderwijs in cijfers 2015). Most such private schools are Roman-catholic or reformed schools; some Jewish, Islamic and Hindu schools occur in larger cities. Eight percent out of all primary private schools in the Netherlands are

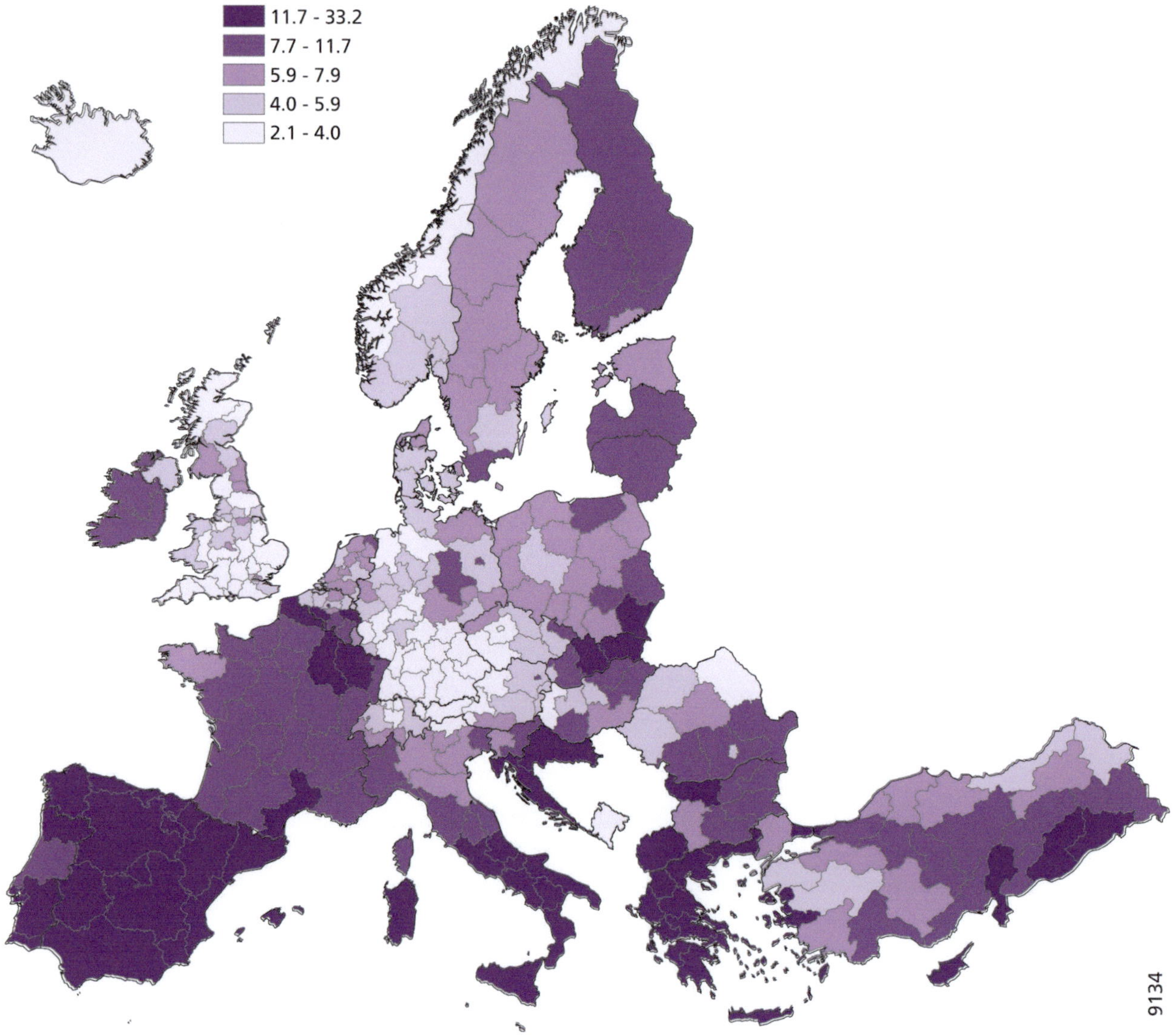

Fig. 13.5 Unemployment by region EU countries in 2015 (percentage unemployed of the total labour force). (Source: CBS 2016a)

non-religious schools, initiated by groups of parents, often schools with a specific didactic signature: Montessori, Dalton, Jenaplan and others.

Not only municipal and state-governed schools, but almost all private religious and non-religious schools as well are fully paid by the state. This arrangement resulted from pillarization in Dutch society (Chap. 12) and the so-called 'School Pacification' of 1917. In the beginning of the nineteenth century, under Napoleonic law, state and church responsibilities were strictly divided and only public schools were financed by the state. Revision of the Dutch Constitution in 1848 introduced freedom for parents to initiate new schools, but with no state financial support. Successive emancipation processes of orthodox reformed and Roman-catholic populations resulted in the establishment of many reformed and Roman-catholic schools, while their associated political parties struggled for state-support for non-public schools too. The 1917 constitutional revision established equal financial status for public and private schools, both types are paid by the state. Private schools with high school entrance fees are exceptional in the Netherlands: no more than 2% of all primary school children visit private schools without state funding, for example international schools.

Unlike some nations with free higher education as Germany, Austria and the Scandinavian countries, school fees for universities in the Netherlands have increased considerably over the past decades from about 100 € annually in 1970 to almost 2000 € annually in 2017. Previously, all students received state grants but this has been substituted in 2015 by a system of cheap loans and an additional grant for low-income parent students.

The Dutch school system with free and compulsory education for all 5–18 years old results in overall high

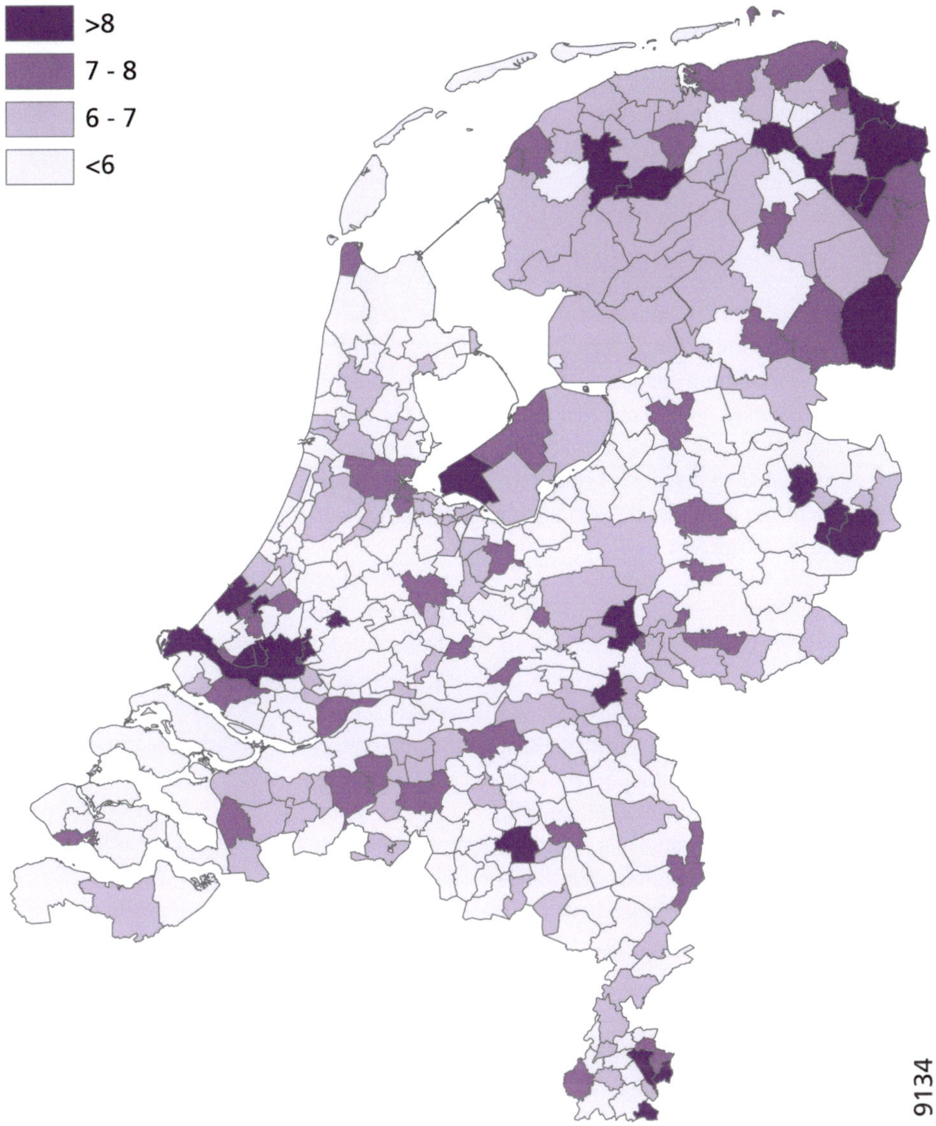

Fig. 13.6 Unemployment by municipality in 2015 (percentage unemployed of the total labour force). (Source: CBS 2016a)

educational levels and relatively high scores in international comparative assessments. The share of the Dutch who completed higher education is about equal to OECD's average (35% of the 25–64 years old) and slightly higher for the 25–34 years old groups. Fifty six percent of Dutch bachelor graduates, 57% of master graduates and 47% of doctoral graduates are female, comparable with OECD's averages (OECD 2016b).

The Netherlands ranks 9th in the PISA (Programme for International Student Assessment of the OECD) for mathematics and sciences in 2015 and 3rd within the EU (after Finland and Estonia). The OECD considers the Dutch education system as one of the best in the world with relatively few poor performers, good vocational education, room for experiments and innovation, and a balance between centralization and decentralization in education (OECD 2016c).

The Netherlands has an early tracking system where 12 years old follow a track either in general or in vocational education. A review of international comparative literature on educational inequality by Van de Werfhorst and Mijs (2010) shows that early track school systems increase inequality as compared to comprehensive school systems. While the Netherlands ranks high on educational level indicators (expected years of schooling), the nation does not belong to the world's best in terms of educational equality (23rd in the Human Development Index Report – UNDP 2016). Levels of education highly depend on their parent's education levels. Immigrants, in particular first generation immigrants from Morocco and Turkey, have lower education levels than native Dutch (CBS 2016b).

13.4.1 Spatial Dimensions of Education

Schools for primary and secondary education and institutions for higher education are distributed geographically according to population spreads which is uneven but rather centred in the west. This generates more nearby educational opportunities for people in Randstad Holland than for those beyond.

Small villages face closing primary schools due to declining numbers of 4–12 years old and to recently modified law-imposed minimum levels for schools to operate. Dutch law requires minimum numbers for creating or closing schools which may vary between 31 children for small villages and 194 for big cities. Until recently such minimum levels were related to a school's religious identity and were lower for schools with equal denomination at distances of more than five kilometers.

Schools for secondary education are located in smaller towns and in cities. Because of the highly urbanized character and dense system of public transport in the Netherlands, distance is seldom a decisive factor in choosing secondary schools (Box 13.2). Average distance to the nearest school for secondary education is 2.4 km and 85% of the population has a secondary school within 5 km. Distances are larger in the northern provinces and Zeeland (Fig. 13.7). In some cases, secondary school students commute long distances to schools of a particular religious denomination, didactic concept, or to special schools for students with mental or learning disabilities. School busses provide transport for these students.

Box 13.2: School Segregation in Amsterdam

Two or three generations ago distance to school and religion were the main decision factors in school choices. Although more than half of all Dutch primary schools are religion-based (Roman-catholic, protestant, Jewish, Islamic, Hindu), religion plays a minor role in school choice today. Roman-catholic and protestant schools have a mixed population of protestant, Roman-catholic, non-religious, Islamic and other denominations. Jewish, Islamic, Hindu and orthodox reformed schools are an exception and have rather homogeneous populations of children with the same religion.

Today, schools in big cities and more and more in medium-sized towns are segregated along ethnicity lines, in 'white' and 'black' schools. School segregation correlates with residential segregation but

Box 13.2 (continued)

segregation levels in school are higher. Neighbourhoods with many 'white' native-Dutch people normally host schools with an almost exclusively 'white' population, while schools in districts with many immigrants are usually 'black' schools. However, even in mixed neighbourhoods segregation between 'white' and 'black' schools occurs. This may be explained partly by demographic structure and partly by parent's preferences. Such mixed neighbourhoods usually consist of a mixture of immigrant families (often with many children) and native-Dutch singles and couples without children. The rather numerous immigrant children thus dominate school populations there and often are less 'white' than the residential population. Moreover, both groups may prefer different schools. Immigrant parents (as well as lower educated native-Dutch parents) usually prefer nearby schools or schools with special programs for non-Dutch speakers. Highly educated native-Dutch parents often prefer schools with children of like-minded parents, high reputation schools, or schools with specific didactic concepts, as Montessori, Dalton or Jenaplan (Karsten et al. 2006).

Parents in the Netherlands have the right of free school choice. But, parents in the big cities, Amsterdam in particular, are faced with limited numbers of places or with long waiting lists in popular 'white' schools. Such schools do select often on the distance criterium. In anticipation, parents may move to places nearby a school of their preference (Boterman 2013).

Until 1880, the Netherlands hosted four universities. Three universities were in Randstad Holland (Leiden, founded in 1575, Utrecht, founded in 1636 and Amsterdam, founded as Atheneum Illustre in 1632 and as university in 1877) and one in the north (Groningen, founded in 1614). Dutch pillarization resulted in creation of a reformed university in Amsterdam (Free University, now VU University, founded in 1880) and a Roman-catholic university in 1921 beyond Randstad Holland, Radboud University in Nijmegen. Further expansion of higher education resulted from upgrading existing specialised institutes. The current technical university in Delft was created in 1842 as a school for education of engineers and was upgraded to an institute for higher education in 1905. The current Agricultural University in Wageningen was founded in 1876 as National Agricultural

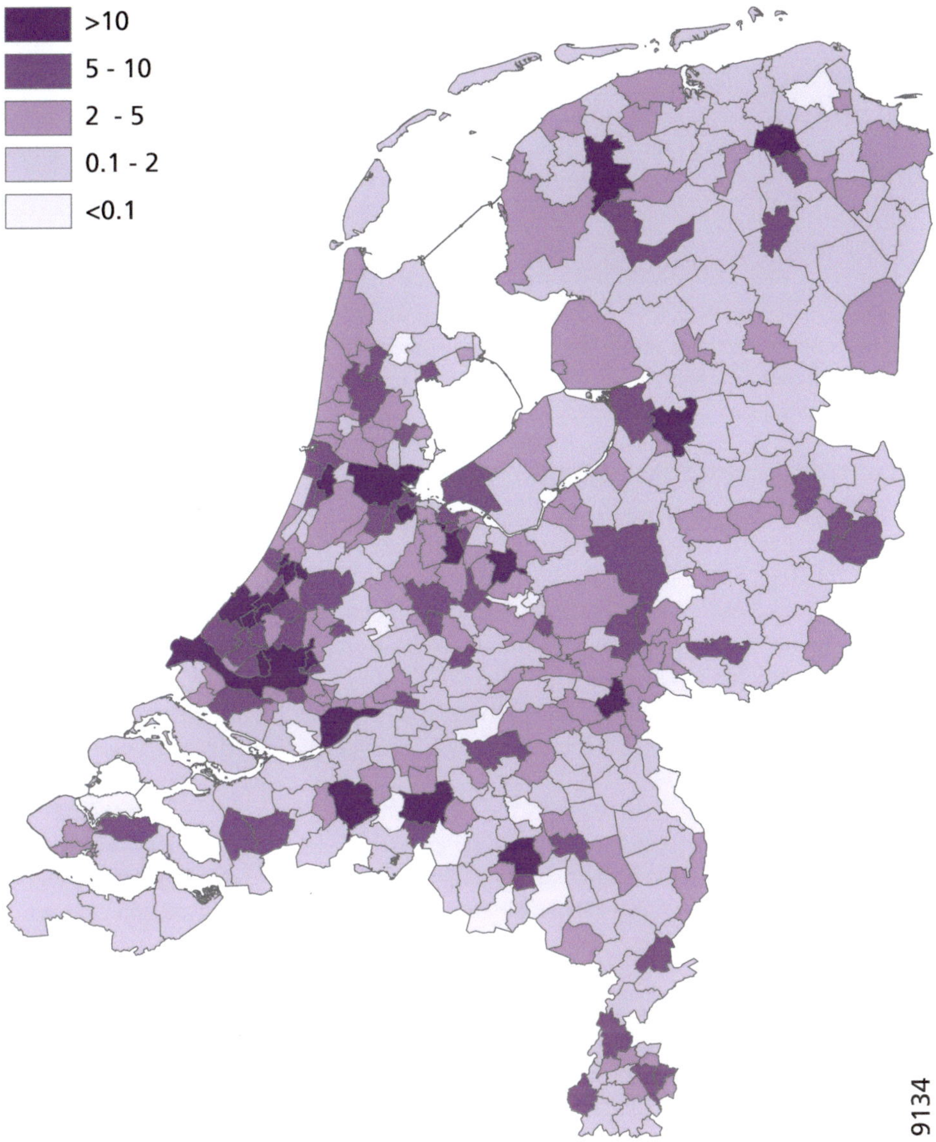

Fig. 13.7 Distance to secondary education 2015 (in percentage of the population with a school for secondary education within a distance of 5 kms). (Source: CLO 2016b)

School and became an institute for higher education in 1904. In the harbour city Rotterdam, a school for higher economic (trade) education settled in 1913 and a Roman-catholic school for higher economic education in textile city Tilburg in 1927. After WWII new universities were created beyond Randstad Holland as part of regional policy to stimulate peripheral region economies: technical universities in Eindhoven (1956) and Enschede (1964), a university in Maastricht (originally a medical university, later a general university) in 1976 and the Open University in Heerlen (1984).

Broad universities of applied sciences are merger institutes of former specialised institutes for higher education (former technical schools, economic schools, agricultural schools, normal schools, teacher colleges and schools for social work) locate in one or more cities in all parts of the country (Boxes 13.3 and 13.4).

Regional differences in Dutch educational attainment show high levels of completed tertiary education in the big cities and in some nearby high-income suburban municipalities (Fig. 13.8). Education levels in Randstad Holland are higher than beyond and spatial differences in educational level have increased since the beginning of this century (Latten et al. 2017). University cities hold more people with completed tertiary educations; Utrecht, Leiden, Amsterdam, Wageningen and Nijmegen belong to the Dutch top-ten cities with a highly educated population. University cities of Rotterdam, Enschede and Tilburg are exceptions: as their educational levels mainly reflect their manufacture-based economic structures.

Box 13.3: Research and Education in the Netherlands

Despite of its low number of inhabitants (0.0022% of the world population) the Netherlands produced a considerable number of internationally renowned scientists. Out of a total of nineteen Nobel laureates eight were for Physics, three for Chemistry and two Dutchmen received a Nobel Price for Peace. Moreover, many Dutch scientists have been awarded other important acknowledgements for their scientific achievements. Except for one, all 14 Dutch universities occur in the main international academic rankings. According to the Academic Ranking of World Universities top 500 (ARWU 2016), the universities of Utrecht, Groningen and Leiden score best at positions 65, 72 and 93 respectively in 2016. Such relatively high positions may be considered remarkable as Dutch universities do not select just the best students as many high-level foreign universities do, and Dutch university budgets are relatively low.

In contrast to many high-ranking foreign universities all Dutch universities are public institutions. They are all charged with the same set of societal tasks and are all financed by the Dutch Ministry of Education according to the same set of conditions. Dutch universities teach relatively large numbers of students towards bachelor, master or doctor degrees. Dutch universities operate both in scientific education and research.

Applied research in the Netherlands is done by some large private institutions as TNO, DELTARES and ALTERRA. Often these institutions conduct research projects on request of (and paid by) one or more Dutch ministries. In addition, they may receive some general public funding for basic research. Such institutions may also maintain large databases to keep collective memories of relevant data. (EDM).

People in Dutch rural areas are generally less educated. Many people with no more than basic education occur in east Groningen, east Drenthe, Zeeuws-Vlaanderen and in south Limburg, regions with weak economic structures, high unemployment and high emigration. Moreover, low educated people but with a strong work ethos and a tradition of early labour market entrance frequently occur in the 'Bible Belt'. In 1996, the Department of Geography, Planning and International Development Studies of the University of Amsterdam organised a field study with second year bachelor students Geography in several villages in Walcheren

Box 13.4: Geography Education

Geography is part of the educational programs in primary, secondary and higher education. All children learn geography at primary school. They learn about their own neighbourhood, city or village and country as well as about the world. They learn how to use maps and atlases and digital geography sources.

Geography is compulsory at initial levels in general secondary schools and optional in their last years. Secondary school students learn geography skills at a more advanced level (how to formulate questions, how to use maps, atlases, digital sources, ICT, how to do field work and how to analyse and present data), they learn about urban issues, floods, water management, physical structures, landscapes, Dutch spatial planning, and about globalization, time-space compression, global issues regarding social and economic inequality, ethnic and cultural diversity, political conflicts, development, climate change and sustainability worldwide (SLO 2011).

Geography in higher education is taught at teachers colleges of universities of applied sciences (Amsterdam, Nijmegen, Rotterdam, Utrecht, Tilburg and Zwolle) and in bachelor, master and PhD programmes at universities and as part of interdisciplinary programmes, e.g. environmental studies, marine sciences, international development studies and urban studies.

Physical and human geography are integrated in primary and secondary schools and at teachers colleges, but separated at universities. University of Amsterdam, Groningen University, Radboud University Nijmegen and Utrecht University have bachelor programmes in human geography and planning, while VU University in Amsterdam, Utrecht University, Wageningen University and Technical University Delft have programmes in earth sciences. A few bachelor programmes (Earth and Economics at VU University and Future Planet Studies at University of Amsterdam) are integrative programmes. Master and PhD programmes are usually either specialised (e.g. cultural geography, economic geography, GIS and hydrology) or interdisciplinary.

Geography in the Netherlands has a good international reputation. In 2017, Utrecht University globally leads the ranks of universities of the Center for World University Rankings for physical geography. The University of Amsterdam ranks 14th and Utrecht University 23rd in the QS World University Rankings 2016 in geography and area studies.

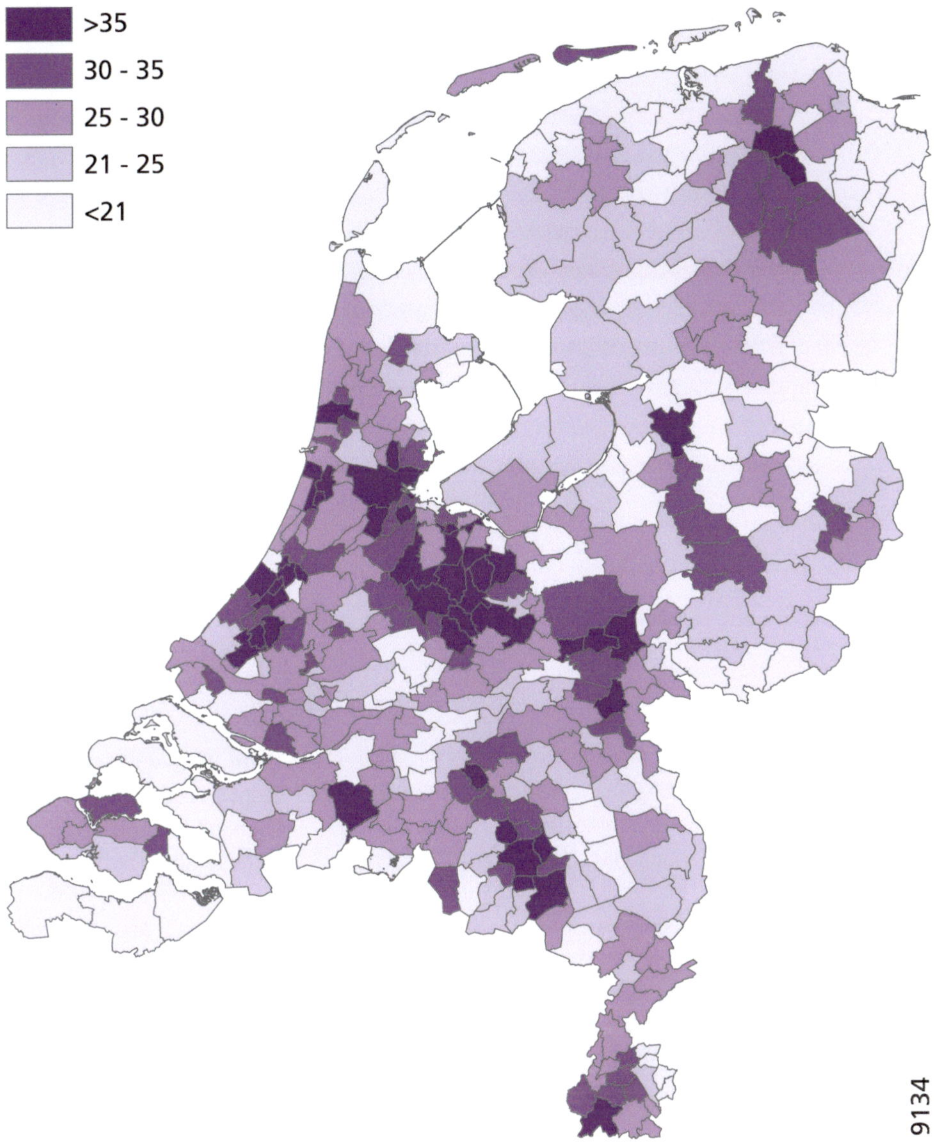

Fig. 13.8 Percentage of 15–74 years old with completed higher education by municipality 2015. (Source: CLO 2016c)

(province of Zeeland), an orthodox reformed 'Bible Belt' region. Twenty-years old Amsterdam students were quite surprised (and even shocked) to interview their married peers in age being a parent, a self-employed entrepreneur and owning a single-family house.

13.5 Housing

Internationally, the Dutch housing market is quite exceptional with about one-third of the total housing stock existing of social rented housing (Fig. 13.9). Unlike other countries, social housing in the Netherlands is not limited to low-income households but may contain middle-income households as well.

The Housing Act (1901) constituted the basis for Dutch social housing although special houses for low-income families had been built by some employers and charity organisations already in the nineteenth century. This new Law had been a response by the Dutch government to a 1896 report on "The Problem of Social Housing": an overview of social distress in housing conditions of the poor (Van der Cammen and De Klerk 2012). This Housing Act included a system of housing subsidies for private, non-profit housing associations, a system of housing regulations by municipalities, an obligation for municipalities to expropriate dilapidated houses and to plan construction of new houses (Chap. 10). Social housing in the Netherlands is not provided by the state, although big cities used to provide some council housing to the very poor, but rather a mixture of decentralized, municipal planning and regulation in combination with state-funded housing provision by private housing associations (Box 13.5).

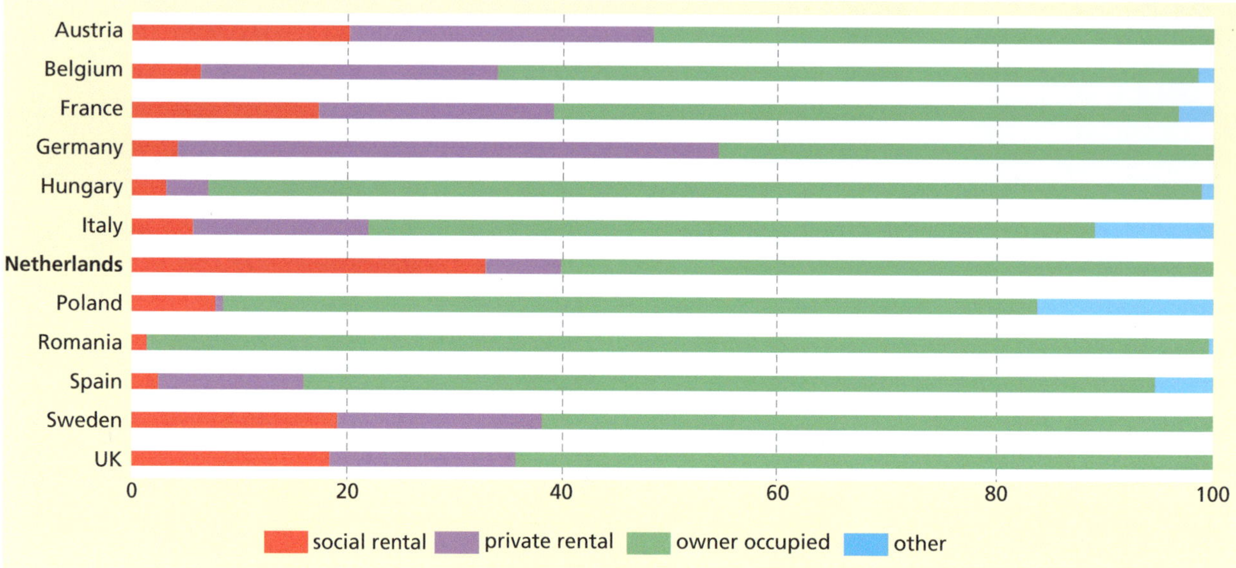

Fig. 13.9 Social rental, private rental and owner-occupied housing EU countries 2012/13 (percentage of the total housing stock). (Source: Pittini et al. 2015)

Box 13.5: Garden Villages: Agnetapark in Delft

In reaction to the miserable housing and living conditions of the working class by the end of the nineteenth century, a few manufacturers built garden villages with houses in green surroundings for their workers. Their intentions were to improve their workers' life conditions while making them more productive.

A first example of a Dutch garden village had been Agnetapark, initiated in 1982 by the director of the Dutch Yeast and Spirit factory in Delft, J.C. van Marken. Agneta was his wife's name. That garden village included 78 houses built in blocks of three or four single-family houses, all with private gardens where workers could grow their own vegetables. The houses were positioned in a romantic landscape design, in fashion in the late nineteenth century with green spaces covered with trees and winding lanes, ponds and watercourses. Agnetapark hosted several facilities, as schools, cooperative grocery shops, a community center and various recreational facilities. Agnetapark hosted the director's family villa too.

An extension of 156 houses was built in 1924–1926, but in a more (modern) rectangular design. As the time spirit by then required the director to live elsewhere, the director's villa was used as an Industrial and Domestic Science school until the villa was demolished in 1981. Around 1990, houses and green spaces in Agnetapark were renovated (Rijksmonumenten 2015) (Fig. 13.10).

In line with pillarization of Dutch society Reformed, Roman-catholic and socialist housing associations were founded. It took some time before the Housing Act resulted in large-scale building of new social rented houses. By 1910 no more than 2% of the total housing stock was built in the social sector and that sector stayed under 10% until the end of WWII. Social housing was not affordable for the poorest households, but was mainly built for skilled workers and office workers.

The 1960–1990 period was the heyday of Dutch social rented housing in large-scale urban expansions and urban renewals of 19th and early 20th centuries districts. In 1990, 41% of total Dutch housing stock was social rented housing (Table 13.2). Social housing had become regular housing for low- and middle-income households. Social housing is accessible for households with an income below a certain level, but people have the right to stay in their house even when their income increases. Therefore, many people prefer staying in their cheap but decent houses instead of moving to a house in the expensive private rental sector or to buy a house. Especially in big Randstad Holland cities private rental houses and houses for sale are quite scarce and therefore expensive.

From 1990 onwards, housing policy changed. A new Housing Act was introduced in 1995 and responsibility for housing was transferred to municipalities and housing associations; property subsidies for new social housing were no longer available. Priority was given to owner-occupation. While new urban expansions of the 1960s contained around 80% social housing, that percentage

Fig. 13.10 Agnetapark Delft (Source: http://mapio.net/pic/p-87318577/)

Table 13.2 Housing tenure in the Netherlands 1947–2016 (percentage of the total housing stock)

	Social rent	Private rent	Owner-occupation
1947	12	60	28
1960	23	47	30
1970	31	34	35
1980	36	22	42
1990	41	14	45
2000	36	12	52
2016	30	13	57

Sources: Van Kempen and Priemus (2002) and CBS Statline

dropped to 20 in 1990s and 2000s urban expansions and 80% of the houses were for sale. The Dutch social rented sector is diminishing and more and more restricted to the poorest households. A system of rent allowances for low-income households as introduced in 1975, makes social rented houses affordable for the lowest-income households.

13.5.1 Spatial Dimensions of Housing

Social rented housing (with a monthly rent of less than 711 € in 2017) is unevenly distributed over the country. Cities, in particular the big Randstad Holland cities, host more social rental houses and fewer owner-occupied houses than elsewhere (Fig. 13.11). While 57% of the 2016 Dutch housing stock is privately owned nation-wide, Amsterdam has no more than 30% and Rotterdam 35% of its houses occupied by owners. In 2016, 43% of the total housing stock in Amsterdam and 45% in Rotterdam is social housing. The large (in numbers) and broad (for low and middle-income households) social rental sector prevented a development of high income segregation in Dutch cities. According to Musterd (2014), relative decrease of the social housing sector since 1989 caused more segregation in terms of income, educational level and ethnicity, although Amsterdam's large social housing sector appears to be more diversified than the – smaller – social housing sector in the Netherlands as a whole.

The share of house owners steadily increased since WWII and most Dutch households live in their own house now. Housing prices increased significantly between 1947 and 1978. After a housing and economic crisis, Dutch housing prices went up again until the next economic crisis in 2008. By today, the housing market is recovering rapidly but regionally unequally. Housing prices are currently rising in

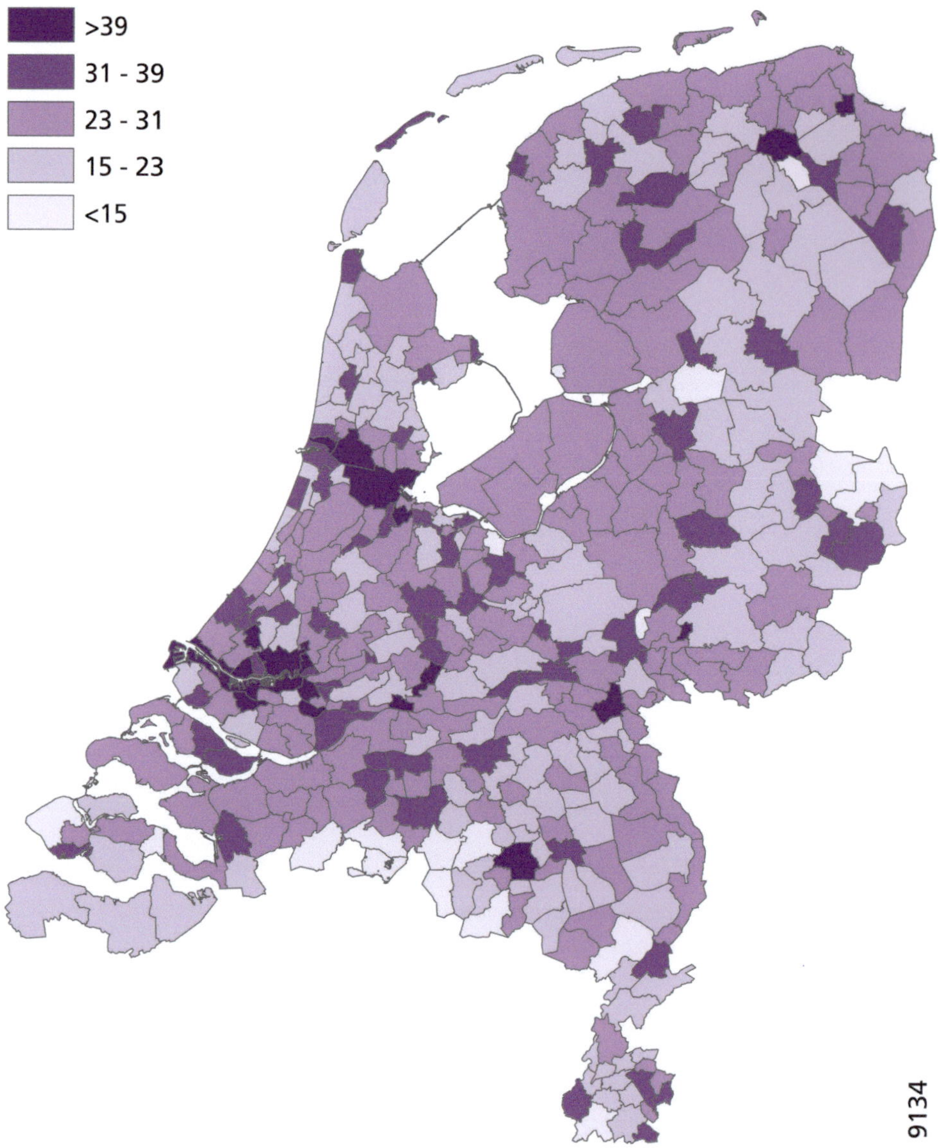

Fig. 13.11 Social housing by municipality 2012 (percentage of the total housing stock). (Source CBS Statline)

Randstad Holland (mainly in Amsterdam and Utrecht) but much less in the north and in Zeeland (Fig. 13.12). The municipality of Amsterdam initiated a policy shift to accommodate cheaper housing opportunities for middle income households by building 40% social housing, 40% medium-priced (rental and owner-occupied) housing and 20% expensive housing in the next decade (Woonagenda 2017).

13.6 Health

Internationally, the Netherlands belongs to the leading nations in self-perceived health and satisfaction with health care. According to the UN Human Development Index Report 2016, the Netherlands ranks 12th worldwide in terms of satisfaction with health care quality and leads with Ireland, Sweden and Cyprus, in self-perceived health in the EU (Volksgezondheidenzorg 2016). This has not always been the case. The Dutch health situation has been quite poor in the early nineteenth century, as in most other nations. Life expectancy at birth was low and perinatal mortality high due to poverty, poor food quality, lack of hygiene, bad water quality and sanitation, prevalence of diseases as cholera, typhus and dysentery, and poor health care. Dutch health situation improved considerably during the second half of the nineteenth century with the introduction of clean drinking water by water mains, sewage systems and with introduction of vaccination (Boxes 13.6 and 13.7).

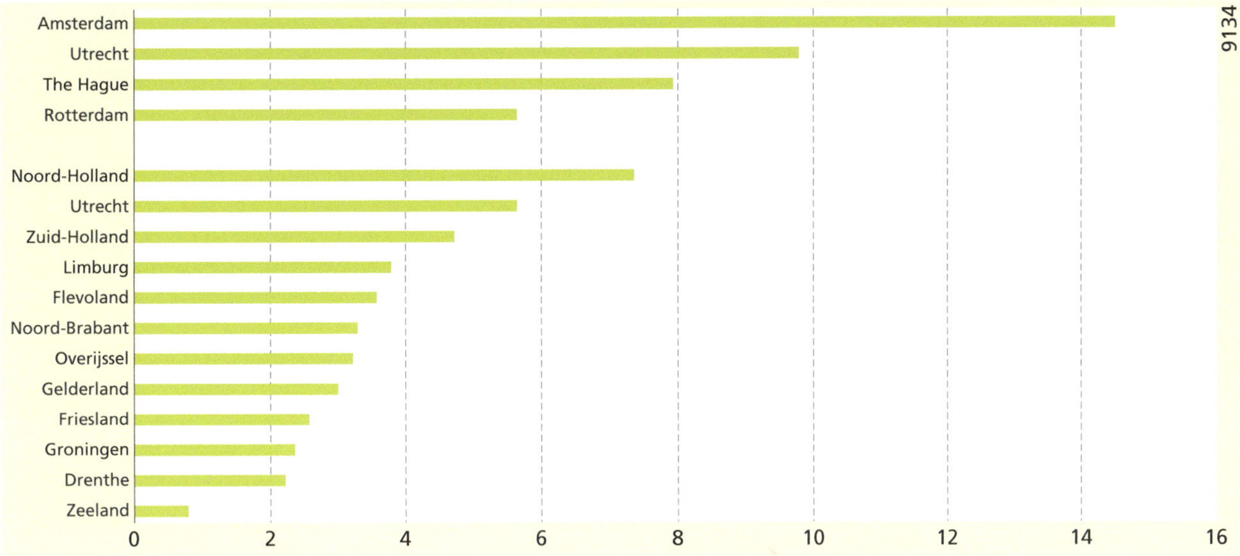

Fig. 13.12 Regional differences in housing price increase 2015–2016 (increase as percentage of housing prices 2015). (Source: CBS 2016c)

Box 13.6: Tall People

The excellent Dutch health situation is reflected in the length of Dutch men and women. Conscripts born in 1800 were 1.63 m tall and contemporary women were 6 cm shorter. Average length increased until 1840 when hunger from blight, high grain prices and epidemics hampered further increase. From the end of the nineteenth century onwards the Dutch became taller and taller: 1.69 for (male) conscripts born in 1900 and 1.75 for conscripts born in 1950. Today, the Dutch belong to the tallest in the world: they rank 1st world-wide for men born in the late twentieth century (1 m83) and 2nd (after Latvia) for women (1 m68). Nutrition (a diet of many calories and dairy products), in particular during pregnancy, early childhood and adolescence, limited exposure to infectious diseases and genetic factors are the main determinants of differences in length between populations (Bentham et al. 2016).

Box 13.7: Vaccination in the Netherlands

Vaccination began in Napoleonic times (early nineteenth century) in the Netherlands when King Lodewijk Napoleon urged the Dutch to vaccinate against small pox. In the late nineteenth century more vaccines became available, e.g. against rabies, cholera, typhus and plague. Vaccines against diphtheria, pertussis (whooping cough) and tetanus were introduced here in the 1920s and 1930s. The poliomyelitis epidemic of 1956 (with 2206 patients and 70 casualties) prompted introduction (in 1957) of a national vaccination

Box 13.7 (continued)

programme for all babies and young children with vaccinations against diphtheria, pertussis, tetanus and polio. The national program gradually extended with vaccines against other diseases and today includes vaccinations against haemophilus influenza, hepatitis B, mumps, measles, rubella, meningococci and cervical cancer. According to Van Wijhe et al. (2016), vaccination saved about 9000 children's lives in the first 35 years upon introduction of the national programme in the Netherlands. About 95% of the Dutch is vaccinated, but this number is decreasing in the last few years. Lower vaccination levels (< 95%) presently occur in certain urban districts with high educated people among which relatively many vaccination-skeptics, and in the 'Bible Belt' with many orthodox protestants refusing vaccination for religious reasons. As a consequence, local epidemics of polio emerged, mainly in the 'Bible Belt'. In 1961, there were 83 polio patients, 13 of them in Kerkwijk Waardenburg, a small village in the Bible Belt. The 1971 polio epidemic caused 39 patients in Staphorst, five of whom died. The 1978 epidemic with 110 polio patients (1 died) developed from Elspeet and Uddel while spreading over the nation, mainly in the Bible Belt. Fifty-nine percent of the patients were members of a small orthodox reformed community (Gereformeerde Gemeenten in Nederland). Even nine Canadians were infected after having contact with community members in the Netherlands (Bijkerk et al. 1979). The latest polio outbreak occurred in 1992/3 in Streefkerk with 71 patients, two of them died.

Cholera epidemics of 1848 boosted introduction of clean drinking water provision in Amsterdam, with a water mains system bringing sand-filtered water from the coastal dune area to the city, preventing a next cholera epidemics (1866) to affect the Amsterdam population as severely as elsewhere in the country. Originally, clean drinking water provision was a private initiative, but from 1896 onwards the Amsterdam municipality took responsibility for this system. Later, private drinking water provision companies began to operate in other Dutch cities. In the 1920s all Dutch cities had drinking water mains and in the 1960s all Dutch buildings were connected, with just a few exceptions. Provision by private companies had been replaced by public organisations nationwide.

Health insurance originated from Mediaeval times when part of the urban population had access to health insurance provided by the guilds. Poor people had no access to health insurance and depended on charity by the municipality or by the church. Moreover, several cities had almshouses, initiated by religious and non-religious charity organisations. Health insurance became available for low-income households in the nineteenth century. Medical doctors and pharmacists established the first health insurance organisation for workers in Amsterdam in 1846, while the first workers' initiative, the Algemene Onderlinge Ziekenfonds Voor en Door Werklieden (General Mutual Insurance For and By Workers), was founded in 1874 (KHZ 2017).

A compulsory health insurance system for low-income employees, including civil servants, was imposed in 1941 during German occupation but continued after WWII. High income households and self-employed people could opt for private insurance (ZIC 2016). Distinction between public and private health care was abolished in 2006 with the introduction of a public-private health care system. Today, health insurance is compulsory for all Dutch citizens and the national government compensates low-income households with health care allowances. Health care insurance companies are private commercial firms which determine insurance fees in mutual competition and negotiate best prices for medical treatment including surgical operations and medicines with medical doctors, pharmacies and hospitals.

Most Dutch hospitals are private organisations, many of them originally developed as Roman-catholic or reformed institutions, in line with their corresponding pillars. Originally, these hospitals provided health care exclusively to people of their pillars. But, with de-pillarization of Dutch society, hospitals became accessible to anyone. Several Roman-catholic and reformed hospitals became general hospitals or merged with hospitals from other denominations, e.g. the Roman-catholic hospital in Groningen that merged with the reformed Diakonessenhuis into the general Martiniziekenhuis in 1991.

Health care in the Netherlands is organised at two levels. Primary health care by family doctors, midwives, district nurses and local child consultation clinics is free of charge and often nearby. Patients have access to a second level of medical specialists and hospitals but after consulting a family doctor only. The Netherlands has a tradition of home-birth and the number of home-births is higher than in all other EU countries; but their numbers are decreasing. In 1990, thirty-eight percent of all child births were home-births, in 2011 no more than 25% (CBS 2011).

13.6.1 Spatial Dimensions of Health and Health Care

In an urbanised nation as the Netherlands, hospitals are well accessible for almost all people (Fig. 13.13). All except 0.2% of the Dutch can reach a hospital with emergency care within 45 min (Kommer et al. 2015). In general, health care in the Netherlands is well organised and accessible to everybody. However, some regional differences occur. Self-perceived health is in particular high in regions with rich suburbs: Zuid-Kennemerland, Gooi & Vechtstreek and Hollands-Midden, but also in Friesland. Self-perceived health is relatively low in low-income areas, low-educated, ageing households and by unemployed persons, often in south Limburg, Rijnmond, Flevoland, west Brabant and Zeeland (Fig. 13.14).

In addition, regions with many low-educated people and an ageing population have high average levels of obesity: south Zuid-Holland, Twente, Drenthe and south Limburg. Smoking is decreasing in the Netherlands since a smoking ban in public spaces was enforced by Dutch parliament in 2004. Prevalence of smoking is relatively high in low-income urban regions as Amsterdam, Rotterdam, Den Haag and Utrecht, and in peripheral regions Groningen, Friesland, Drenthe, Twente and south Limburg. Alcohol consumption is high in the north of Noord-Holland, in the middle of Noord-Brabant, Amsterdam and Utrecht, and relatively low in Flevoland and in the 'Bible Belt' (CBS 2013). Regional differences in age, education, income and religion may explain such geographical differences in these life-style related health indicators.

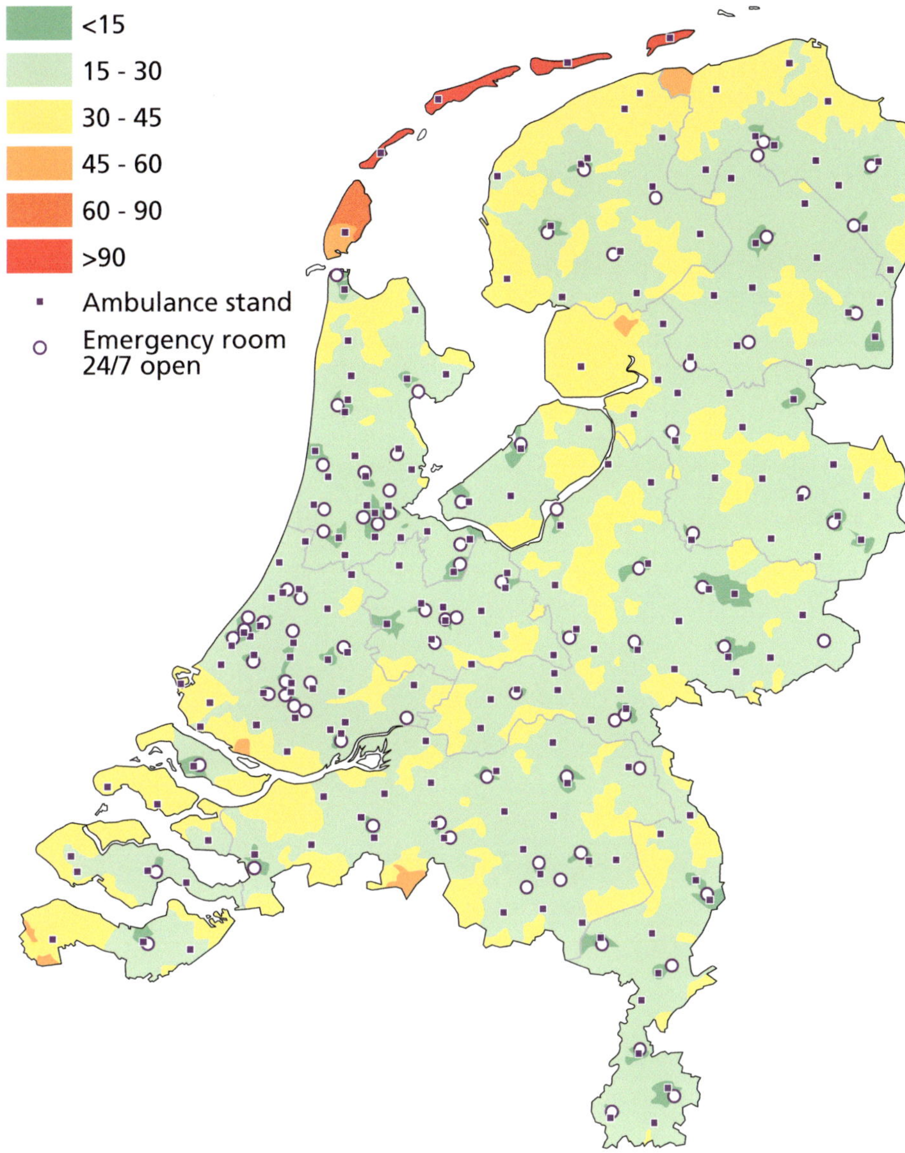

Fig. 13.13 Accessibility of emergency rooms 2014 (in minutes travel time). (Source: Kommer et al. 2015)

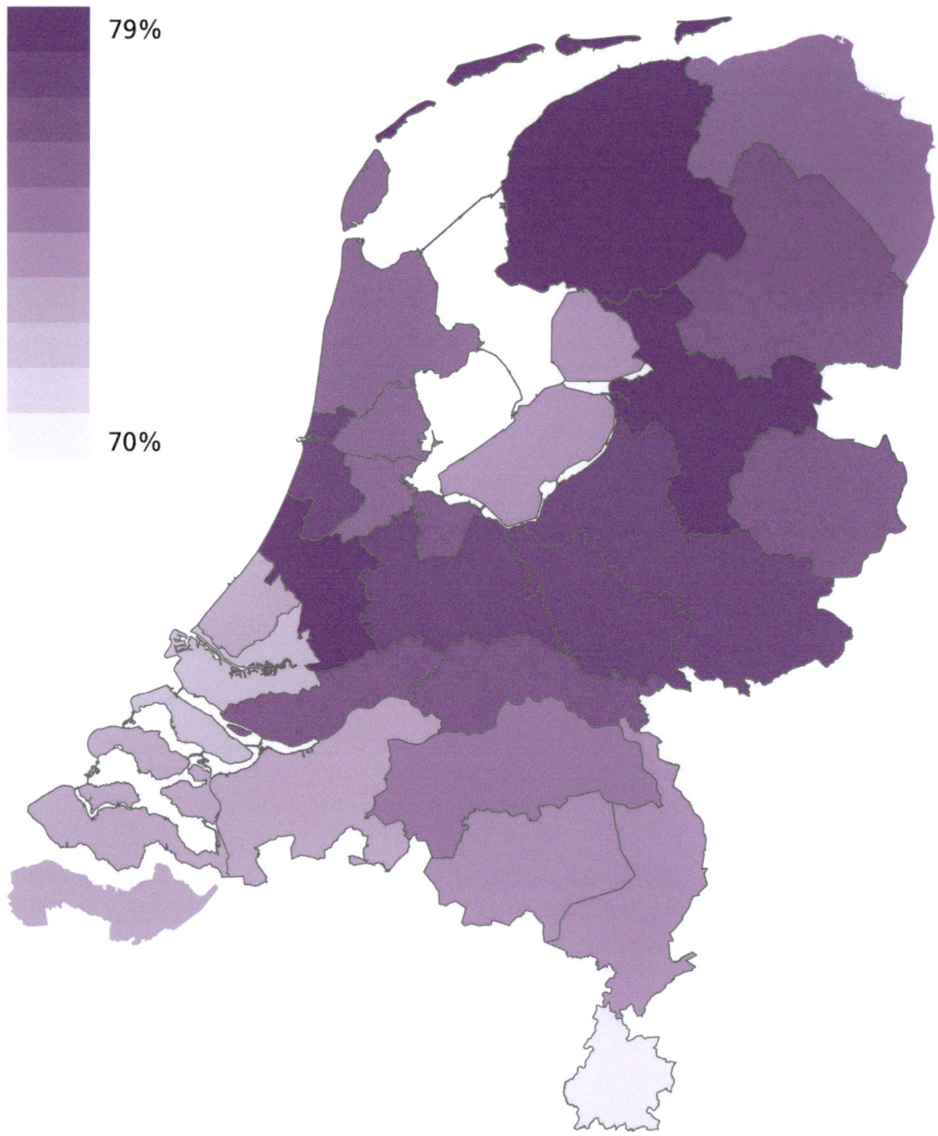

Fig. 13.14 Self-perceived health by region 2016 (percentage 19 years and older with good self-perceived health). (Source: RIVM 2017)

References

ARWU (2016) http://www.vsnu.nl/nl_NL/f_c_rankings.html. Accessed 25 Nov 2016

Bentham J, Di Cesare M, Stevens GA, Zhou B, Bixby H, Cowan M et al. (2016) A century of trends in adult human height. eLife 5. https://doi.org/10.7554/eLife.13410.001. Accessed 6 June 2017

Bijkerk H, Draaisma FJ, Van der Gugten AC, Van Os M (1979) De poliomyelitis-epidemie in 1978. Ned Tijdschr Geneeskd 123 (39):1700–1714

Boterman W (2013) Dealing with diversity: middle-class family households and the issue of 'black' and 'white' schools in Amsterdam. Urban Stud 50(6):1130–1147

Boterman WR, Karsten L (2014) On the spatial dimension of the gender division of paid work in two-parent families: the case of Amsterdam, the Netherlands. Tijdschr Econ Soc Geogr 105(1):107–116

CBS (2011) Driekwart van bevallingen in ziekenhuis. Centraal Bureau voor de Statistiek, Den Haag

CBS (2013) Grote regionale verschillen in overgewicht, roken en alcoholgebruik. Centraal Bureau voor de Statistiek, Den Haag

CBS (2016a) Factsheet Nederland en de provincies. Centraal Bureau voor de Statistiek, Den Haag

CBS (2016b) Welvaart in Nederland 2016. Inkomen, bestedingen en vermogen van huishoudens en personen. Centraal Bureau voor de Statistiek, Den Haag

CBS (2016c) Groot regionaal verschil in prijsstijging koopwoningen. Centraal Bureau voor de Statistiek, Den Haag

CLO (2016a) Netto arbeidsparticipatie 2015. Compendium voor de Leefomgeving. Rijksoverheid, Den Haag

CLO (2016b) Afstand tot voortgezet onderwijs. Compendium voor de Leefomgeving. Rijksoverheid, Den Haag

CLO (2016c) Hoogopgeleiden. Compendium voor de Leefomgeving. Rijksoverheid, Den Haag

De Beer P (2014) Groeiende beloningsverschillen in Nederland. In: Kremer M, Bovens M, Schrijvers E, Went R (eds) Hoe ongelijk is Nederland? Een verkenning van de ontwikkeling en gevolgen van economische ongelijkheid. Wetenschappelijke Raad voor het Regeringsbeleid/Amsterdam University Press, Den Haag/ Amsterdam, pp 59–77

De Meester E, Mulder C, Droogleever Fortuijn J (2007) Time spent in paid work by women and men in urban and less urban contexts in the Netherlands. Tijdschr Econ Soc Geogr 98(5):585–602

Deschacht N, Tijdens K (2014) Why Dutch women work part-time: a oaxaca-decomposition of differences in European female part-time work rates. Working papers 146. Amsterdams Instituut voor Arbeids Studies (AIAS), University of Amsterdam

Dirven HJ, Michiels J, Ter Steege D (2015) Regionale verschillen in arbeidsparticipatie, werkloosheid en vacatures. Sociaal-Economische Trends 2015-10. Centraal Bureau voor de Statistiek, Den Haag

Esping-Anderson G (1990) The three worlds of welfare capitalism. Policy Press, Cambridge

Eurostat (2017a) Eurostat Intersection of Europe 2020 Poverty target Indicators. https://www.google.nl/search?q=Eurostat+Intersection+of+Europe+2020+Poverty+Target+Indicators&sourceid=ie7&rls=com.microsoft:nl-NL:IE-Address&ie=&oe=&gfe_rd=cr&ei=iag2WbWyA5DSXprIiuAH&gws_rd=ssl. Accessed 6 June 2017

Eurostat (2017b) http://ec.europa.eu/eurostat/statistics-explained/index.php/File:Employment_rates_for_selected_population_groups,_2004%E2%80%9314_(%25)_YB16.png. Accessed 6 June 2017

Eurostat (2017c) http://ec.europa.eu/eurostat/statistics-explained/index.php/File:Persons_employed_part-time,_age_group_20%E2%80%9364,_2015_(%C2%B9)_(%25_of_total_employment)_YB16_III.png. Accessed 6 June 2017

Helliwell J, Layard R, Sachs J (eds) (2016) World happiness report 2016. United Nations Sustainable Development Solutions Network, New York

Karsten S, Felix C, Ledoux G, Meijnen W, Roeleveld J, Van Schooten E (2006) Choosing segregation or integration? The extent and effects of ethnic segregation in Dutch cities. Educ Urban Soc 38 (2):228–247

KHZ (2017) http://www.kenniscentrumhistoriezorgverzekeraars.nl/bronarchief/canon/1850.html. Accessed 6 June 2017

Knippenberg H (1986) Deelname aan het lager onderwijs in Nederland gedurende de negentiende eeuw. Nederlandse Geografische Studies 9. Amsterdam, Koninklijk Nederlands Aardrijkskundig Genootschap/Instituut voor Sociale Geografie Universiteit van Amsterdam, Amsterdam

Kommer GJ, Gijsen R, Lemmens LC, Kooistra M, Deuning C (eds) (2015) Beschikbaarheid, specialisatie en bereikbaarheid van spoedeisende hulp; analyse gevoelige ziekenhuizen. Rijksinstituut voor Volksgezondheid en Milieu, Bilthoven

Kremer M, Bovens M, Schrijvers E, Went R (eds) (2014) Hoe ongelijk is Nederland? Een verkenning van de ontwikkeling en gevolgen van economische ongelijkheid. Wetenschappelijke Raad voor het Regeringsbeleid/Amsterdam University Press, Den Haag/Amsterdam

Latten J, Kooiman N, Bontje M (2017) Toenemende ruimtelijke verschillen in opleiding. DEMOS Bulletin Over Bevolking en samenleving 33(6):1–4

Musterd S (2014) Public housing for whom? Experiences in an era of mature neo-liberalism: the Netherlands and Amsterdam. Hous Stud 20(4):467–484

Musterd S, Van Gent WPC (2016) Changing welfare context and income segregation and its metropolitan area, 2004-2022. In: Tammaru T, Marcińczak S, Van Ham M, Musterd S (eds) Socio-economic segregation in European capital cities. East meets West. Routledge, London, pp 55–79

Noback I, Van Dijk J (2007) Regionale verschillen in arbeidsparticipatie van vrouwen: stad versus platteland. In: Van Dijk J, Schutjens V (eds) De Economische Kracht van de Stad. Van Gorcum, Assen, pp 59–69

OECD (2016a) Employment outlook 2016. OECD, Paris

OECD (2016b) Education at a glance. OECD, Paris

OECD (2016c) Reviews of national policies for education. Netherlands 2016. Foundations for the future. OECD, Paris

Onderwijs in cijfers (2015) Aantal en omvang van instellingen in het primair onderwijs. https://www.onderwijsincijfers.nl/kengetallen/primair-onderwijs/instellingenpo/aantal-instellingen. Accessed 6 June 2017

Ostendorf W, Droogleever Fortuijn J (2006) Amsterdam: gender and poverty. In: Musterd S, Murie A, Kesteloot C (eds) Neighbourhoods of poverty; urban social exclusion and integration in Europe. Palgrave Macmillan, Houndmills, pp 52–69

Piketty T (2013) Capital in the twenty-first century. Belknap Press, Cambridge, MA

Pittini A, Ghekière L, Dijol J, Kiss I (2015) The state of housing in the EU 2015. Housing Europe, Brussels

Pott-Buter H (1993) Facts and fairytales; a seven country comparison 1850–1990. Amsterdam University Press, Amsterdam

Rijksmonumenten (2015) Agnetapark. http://rijksmonumenten.nl/monument/46887/agnetapark-arbeiderskolonie-met-in-kleine-blokjes-gegroepeerde-woningen-elk-voorzien-van-eigen-door-heggen-begrensde-tuintjes-omvat-slingerend-stratenpatroon-landschappelijk-aangelegd-park-arbeiderswoningen-drukkerij-gemeenschapshuis/delft/. Accessed 6 June 2017

RIVM (2017) Goed ervaren gezondheid 2016. Rijksinstituut voor Volksgezondheid en Milieu, Bilthoven. https://www.volksgezondheidenzorg.info/onderwerp/ervaren-gezondheid/regionaal-internationaal/regionaal. Accessed 6 June 2017

SCP (2016) Armoede in kaart 2016. Sociaal en Cultureel Planbureau, Den Haag

Schmal H (1995) Den Haag of 's-Gravenhage? De 19[de]-eeuwse gordel, een zone gemodelleerd door zand en veen. Dissertation, University of Amsterdam. Matrijs, Utrecht

SLO (2011) Vakdossier aardrijkskunde. Nationaal Expertisecentrum Leerplanontwikkeling, Enschede

UNDP (2015) Human development report 2016. Human development for everyone. United Nations Development Programme, New York

UNDP (2016) Human development report 2016. Human development for everyone, United Nations Development Programme, New York

UNICEF (2013) Child well-being in rich countries; a comparative overview. Innocenti report Card 11. UNICEF Office of Research, Florence

Van Bavel B (2014) Vermogensongelijkheid in Nederland; de vergeten dimensie. In: Kremer M, Bovens M, Schrijvers E, Went R (eds) Hoe ongelijk is Nederland? Een verkenning van de ontwikkeling en gevolgen van economische ongelijkheid. WRR/Amsterdam University Press, Den Haag/Amsterdam, pp 79–99

Van der Cammen H, De Klerk L (2012) The selfmade land. Culture and evolution of urban and regional planning in the Netherlands. Spectrum, Houten/Antwerpen

Van der Woud A (2012) Koninkrijk vol sloppen. Achterbuurten en vuil in de negentiende eeuw. Bert Bakker, Amsterdam

Van de Werfhorst HG, Mijs JJB (2010) Achievement inequality and the institutional structure of educational systems: a comparative perspective. Annu Rev Sociol 36:407–428

Van Kempen R, Priemus H (2002) Revolution in social housing in the Netherlands: possible effects of new housing policies. Urban Stud 39 (2):237–253

Van Wijhe M, McDonald SA, De Melker HE, Postma MJ, Wallinga J (2016) Effect of vaccination programmes on mortality burden among children and young adults in the Netherlands during the 20th century: a historical analysis. Lancet Infect Dis 16(5):592–598

Volksgezondheidenzorg (2016) Ervaren gezondheid in de Europese Unie. https://www.volksgezondheidenzorg.info/onderwerp/ervaren-gezondheid/regionaal-internationaal/internationaal. Accessed 6 June 2017

Woonagenda 2025 (2017) Woonagenda 2025. Voldoende, betaalbare en goede woningen. Gemeente Amsterdam

ZIC (2016) Ontstaan zorgverzekering in Nederland. Zorgverzekering Informatie Centrum. http://www.zorgverzekering.org/algemene-informatie/ontstaan. Accessed 6 June 2017

The Future

14

Joos C. Droogleever Fortuijn, Eduardo F.J. De Mulder, and Ben C. De Pater

Abstract

This book depicts a country with a prosperous population and a globalised economy, located along the Noordzee and situated on a European river delta bestowed with rich natural resources. The previous chapters provide an overview of differences within the Netherlands, between regions and between social groups, in terms of economy, demography, culture and social conditions; however, in general social and spatial differences are rather moderate. Nevertheless, Dutch society is confronted with some challenges for its future.

Keywords

Challenges for the future · Scenarios · Future geography of the Netherlands

14.1 Challenges

A prospective study on wealth and the environment, a joint initiative of PBL Netherlands Environment Assessment Agency (Planbureau voor de Leefomgeving) and CPB Netherlands Bureau for Economic Policy Analysis (Centraal Planbureau) identifies six major challenges for future spatial development of the Netherlands in the 2015–2050 period (Manders and Kool 2015).

J. C. Droogleever Fortuijn (✉)
Department of Geography, Planning and International Development Studies, University of Amsterdam, Amsterdam, The Netherlands
e-mail: J.C.DroogleeverFortuijn@uva.nl

E. F. J. De Mulder
Delft Technical University (emeritus), Haarlem, Noord-Holland, The Netherlands

B. C. De Pater
Department of Human Geography and Spatial Planning, Utrecht University, Utrecht, The Netherlands
e-mail: B.C.dePater@uu.nl

14.1.1 Demographic Challenges

After a long period of rapid population growth, the increase of the population is currently slowing down and a decline is foreseen for 2030 or 2050, almost solely controlled by changes in international migration. In turn, international migration depends on economic growth and political preparedness to accommodate immigrants or refugees in the Netherlands.

The Netherlands has an ageing population. The share of the older population (65+) is expected to increase from 16% in 2015 to about 25% in 2050. Population ageing has consequences for economic productivity, dependency ratio (the ratio between workforce and non-employed young and retired people) and arrangements for (health) care. A major challenge is creating attractive and affordable housing and residential environments for an ageing and ethnic diverse population.

14.1.2 Economic Challenges

The Netherlands is a rather wealthy nation with a globalised, technologically highly developed and diversified economy. WWII was followed by a relatively long period of high economic growth (with some interruptions in the 1970s and 1980s) until the economic crisis that begun in 2008. The Dutch economy recovered again recently. However, rapid and major economic growth in the Netherlands is rather unlikely in the near future due to the character of the nation's demography. Economic growth is anticipated for the urbanised western part of the country, in the cities in Randstad Holland in particular. One challenge is the increasing inequality in wealth and economic growth between Randstad Holland and Eindhoven on the one hand, and peripheral regions on the other hand. Another important challenge is to strengthen the competitive position of

E. F. J. De Mulder et al., *The Netherlands and the Dutch*, World Regional Geography Book Series,
https://doi.org/10.1007/978-3-319-75073-6_14

Randstad Holland vis-à-vis other European metropolitan regions.

14.1.3 Urbanisation and Regional Development

Cities in Randstad Holland are growing and are expected to grow in the near future because of uneven regional economic growth and a relatively young population in the Randstad Holland cities. A few regions in the periphery of the country, east Groningen, south Limburg and Zeeuws Vlaanderen, currently face declining populations while populations in several more regions in the north, east and south are anticipated to stagnate or to decline in the near future. Main challenges are the lack of space, housing deficits and high house prices in Randstad Holland cities, Amsterdam and Utrecht in particular. As more activities by more people require more space in and near the main Dutch cities other ways to accommodate such needs, including underground space is anticipated to further develop in the future. In contrast, shrinking regions are fighting vacancies, falling house prices, weak economic structures and relatively high unemployment ratios. The main challenge in these regions is to accommodate shrinkage without a loss of quality of life.

14.1.4 Mobility and Accessibility

Economic growth and population growth go hand in hand with growing mobility, increasing air pollution by motorised traffic and transport and more space claims for infrastructure in a small country. In the Netherlands, all types of mobility are increasing: private car use, public transport, cycling (in particular cycling on electric bicycles), air movements and transport of goods. The Dutch government has and still is investing in infrastructural improvements; however, traffic jams are a daily nuisance, particularly in Randstad Holland. Due to its geographic position nearby north wing Randstad Holland cities, further growth of Schiphol airport is problematic. Combatting congestion, accommodating increased cycling, and allocating space for further growth of air traffic and transport are the main mobility challenges.

14.1.5 Climate and Energy

As everywhere else, fossil energy is still dominant in the Netherlands. To comply with the 2015 Paris Agreement on Climate Change, the Netherlands is challenged to turn from fossil fuel based to renewable energy. This energy transition will not only impact energy provision for residential use, commercial property and mobility, but also economic development, as for example the Rotterdam harbour highly depends on imports and (re)exports of coal and hydrocarbons and on oil refining industries.

The long-term, tectonically controlled Noordzee basin subsidence, isostatic rebound, and both natural, Holocene and human-induced sea level rise and risk for marine and river floods, together with ongoing human-induced land subsidence will generate considerable challenges for future water management in the Netherlands.

14.1.6 Agricultural Land Use

Although less than 2% of the Dutch labour force is employed in agriculture, about three quarters of the land is used for agriculture. Agricultural land use in the Netherlands is highly intensive. Intensive livestock production and industrial farming for cows, pigs and poultry, and horticulture in greenhouses are the dominant types of agricultural production and important sources of Dutch export. Manure surplus contributes to emission of methane, and CO_2 production and livestock production is an important source of particulates emissions. Intensive livestock production negatively impacts biodiversity in the Dutch landscape. Transition to more sustainable ways of agricultural production, combatting landscape integrity disturbing factors as deep ploughing and other invasive types of agriculture, preservation of a diversified agricultural landscape and extension of ecological corridors and agri-environmental schemes belong to the main challenges for the future.

14.2 Scenarios for the Future Geography of the Netherlands

The earlier-mentioned joint PBL and CPB study presents two different scenarios concerning the years 2030 and 2050 for the Netherlands with respect to the six above-mentioned challenges: one scenario with both a low population increase and a low economic growth ('Scenario Low') and one scenario with both a high population and a high economic growth ('Scenario High'). In 'Scenario High', a continuing growth of the population (mainly from immigration) and a continuing ageing of the Dutch population is foreseen. Economic growth will go hand in hand with growing business services, more congestion, more urbanisation in Randstad Holland but population decline in some peripheral areas, cheaper renewable energy, further progress of the energy transition and a more sustainable agricultural land use. 'Scenario Low' anticipates a declining population after 2030, less immigration and a further ageing of the population, limited economic growth, predominantly in the governmental and (health) care sectors, population decline in most regions beyond Randstad Holland, less mobility growth and thus less congestion, high dependency on fossil fuels due to expensive renewable energy and more traditional types of agricultural production.

Both scenarios summarize potential future developments as designed for policy making. Although the future of a small country as the Netherlands highly depends on external factors, political decision making might make a difference. According to an earlier PBL scenario exploration (Hilbers and Sneller 2010) two dimensions constitute the basis for the political spectrum regarding decision making about the future geography of the nation: 1. a market versus collective values dimension, and 2. a globalisation versus regionalisation dimension. These two dimensions result in four scenarios: 'Global Economy', 'Strong Europe', 'Regional Communities' and 'Transatlantic Market'.

The 'Global Economy' scenario prioritises an open globalised economy. In this scenario European economic integration is successful, high economic growth goes hand in hand with high incomes, low unemployment, high economic inequality, low levels of social security and low sustainability.

In the 'Strong Europe' scenario collective values such as equity, social security and sustainability are highly valued. This scenario combines a strong European integration with moderate economic growth, moderate unemployment levels and low income inequality.

The 'Regional Communities' scenario values more regional and national interests rather than globalisation and is more sustainable than the other scenarios. Low economic growth, high unemployment, a robust social security system and low inequality are main characteristics of this scenario.

Finally, the 'Transatlantic Market' scenario is market oriented with a focus on regional and national interests with less globalisation and less European integration. Moderate economic growth, low unemployment, low levels of social security and high income inequality characterizes the 'Transatlantic Market' scenario.

In 2017, the Dutch government published its first report on the National Vision on the Environment (Ministerie van Infrastructuur en Milieu 2017). This report summarizes the main strategic hurdles for the future:

1. Towards a sustainable and competitive economy
2. Towards a climate-proof society
3. Towards accessible living and working spaces
4. Towards valuable landscapes

A new Dutch coalition government to be effective upon the 2107 general elections is supposed to make strategic choices to overcome such hurdles. Such choices will reflect the political preferences of the new government on the 'green' versus 'growth' dimension, the 'globalisation' versus 'regionalisation' dimension and the 'free market' versus 'social and spatial equity' dimension.

Such scenarios may give inclinations on how the Netherlands may look like in the relatively near future. As discussed at the onset of this book several links between physical and human geographical factors may be uncovered. Many such links have been quite significant in the past. However, over the centuries human beings, and perhaps the Dutch in particular, have become less and less dependent on their soils in terms of suitability for construction. They learned to build almost any structure on any type of surface (even in the subsurface or on water). They also learned to grow almost any crop on any type of soil (even on artificial substratums). Dutch students seem to be optimistic about their future (Box 14.1). As extensively described in this book the Dutch have learned to cope with aggressive waters that threaten their lands and homes and, instead, to turn such potential water threats in all their diversity into assets to the benefit of the people. Finally, international trade has made the country less dependent of the natural resources in the national subsurface.

Box 14.1: Scenarios for the Future Geography of the Netherlands: A Competition for Secondary School Students

In 2015, 12–15 years old secondary school students participated in a competition initiated by the Year of Spatial Environment and the Royal Dutch Geographical Society (KNAG) by designing 'Atlases of the Future' for the Netherlands in 2040. The students were invited to design a map for each of four scenarios: (1) Growth & Concentration, (2) Growth & Spread, (3) Green & Concentration, and (4) Green & Spread. In addition to the designing task, these students were invited to jointly select their most probable and most preferred scenario. More than 2500 students submitted a total of 92 atlases (see Fig. 14.1 for examples from price winning atlases).

An analysis of the atlases demonstrates that the young students are creative and optimistic about the future (Adriaens et al. 2015). As the most probable future they foresee a country with a strong international position and a growing economy, high priority to sustainability, a successful transition to renewable energy (windmills, solar panels), recycling of natural resources, regional specialisation, urban growth of 'smart' cities with people living in apartment blocks with green roofs, regions with shrinking populations providing space to nature and recreation, local food production, good railway systems and motorways for

(continued)

> **Box 14.1** (continued)
> electric cars, more cycling and home working to reduce congestion. These students believe that sufficient money will be available to protect the country against flooding and they are confident that Dutch society continues to be a global pioneer in water management.

More than all other countries in the world, the Netherlands is a planned, man-made country, where people seem to be detached from their subsurface and in control of nature and natural hazards. However, even the Netherlands is never fully in control and not totally detached from the powers of Planet Earth as humanity cannot effectively handle some typical Earth-related threats as (deep) earthquakes, volcanic eruptions, storms and flooding.

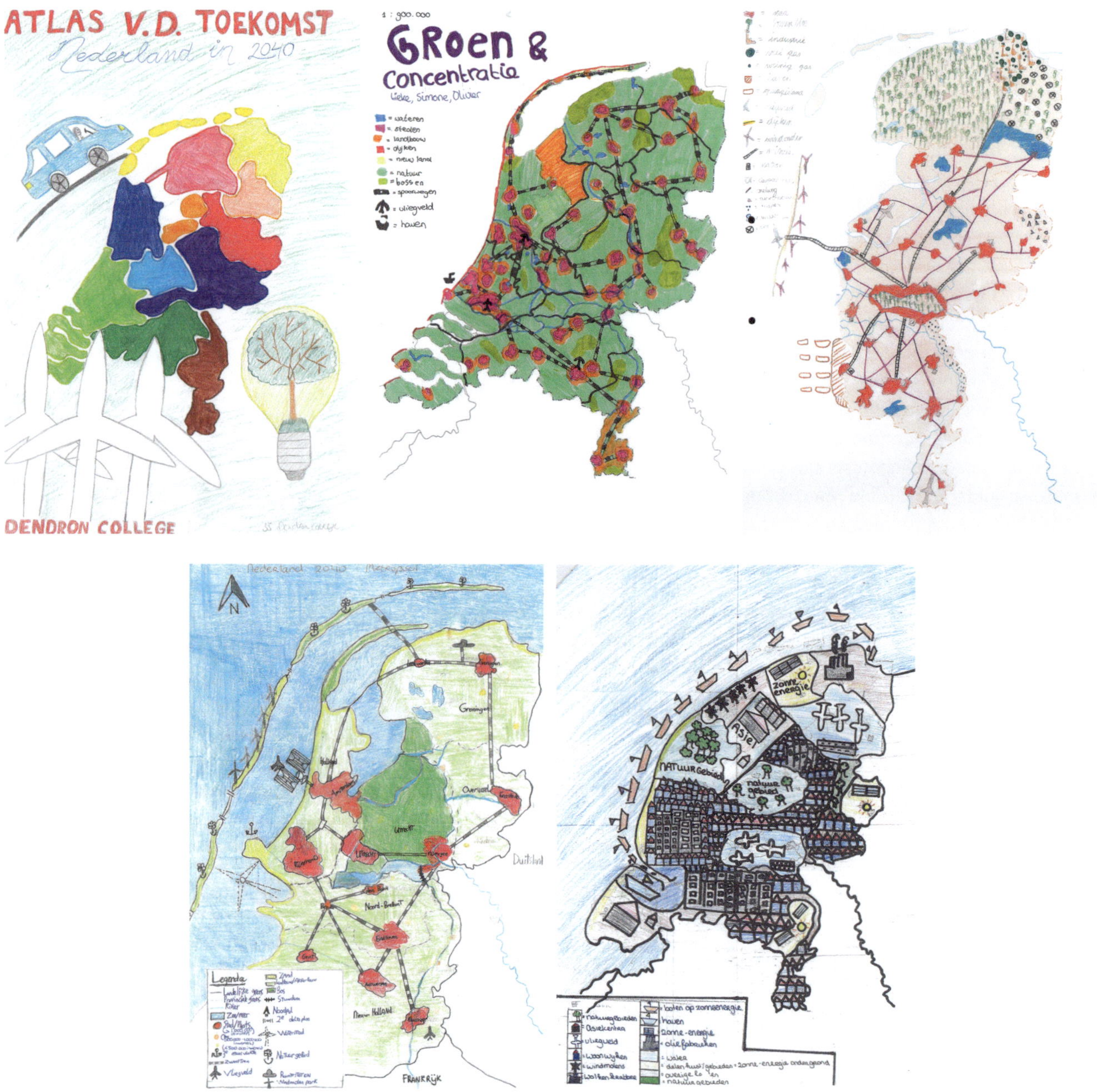

Fig. 14.1 A map of the future geography of the Netherlands, designed by secondary school students. (Source: Atlas van de toekomst 2015)

References

Adriaens R, Beek A, Béneker T, Palings H, Knuivers R (2015) Atlas van de toekomst, een analyse. Koninklijk Nederlands Aardrijkskundig Genootschap/Jaar van de Ruimte, Utrecht. http://geografienl/sites/default/files/paragraph/attachment/file/Brochure%20-%20Atlas%20van%20de%20Toekomst_1pdf. Accessed 9 June 2017

Atlas van de toekomst (2015) Koninklijk Nederlands Aardrijkskundig Genootschap, Utrecht. http://geografienl/onderwijs/lesmateriaal/atlas-van-de-toekomst. Accessed 9 June 2017

Hilbers H, Sneller D (2010) Bestendigheid van de WLO-scenario's. Planbureau voor de Leefomgeving, Den Haag

Manders T, Kool C (2015) Nederland in 2030 en 2050. Centraal Planbureau/Planbureau voor de Leefomgeving, Den Haag

Ministerie van Infrastructuur en Milieu (2017) De opgaven voor de National Omgevingsvisie. Ministerie van Infrastructuur en Milieu, Den Haag

MIX
Papier aus verantwortungsvollen Quellen
Paper from responsible sources
FSC® C105338

If you have any concerns about our products,
you can contact us on
ProductSafety@springernature.com

In case Publisher is established outside the EU,
the EU authorized representative is:
Springer Nature Customer Service Center GmbH
Europaplatz 3, 69115 Heidelberg, Germany

Printed by Libri Plureos GmbH
in Hamburg, Germany